Resource Management and Performance Analysis of Wireless Communication Networks

Shunfu Jin • Wuyi Yue

Resource Management and Performance Analysis of Wireless Communication Networks

Shunfu Jin
School of Information Science and
Engineering
Yanshan University
Qinhuangdao
Hebei, China

Wuyi Yue
Department of Intelligence and
Informatics
Konan University
Kobe, Japan

ISBN 978-981-15-7758-1   ISBN 978-981-15-7756-7   (eBook)
https://doi.org/10.1007/978-981-15-7756-7

This Springer imprint is published by the registered company Springer Nature Singapore Pte Ltd.
The registered company address is: 152 Beach Road, #21-01/04 Gateway East, Singapore 189721,
Singapore

This book provides analytical methods, approaches, algorithms, and simulation to evaluate numerically the quality of service and optimize the resource allocation in various wireless communication networks including broadband wireless access networks, cognitive radio networks, and cloud computing systems. The methods, techniques, and algorithms provided in this book include traffic analysis, applications of queueing theory and Markov chain theory, game theory, intelligent optimization, and operations research. In order to understand these methods, algorithms and techniques, basic concepts of computer communication networks and knowledge of queueing theory are needed. A familiarity with stochastic processes would also be useful.

Preface

Resource management techniques, primarily power, energy consumption minimization, and network design have become increasingly important in wireless communication networks (WCNs), since the explosion of demand for mobile devices. Specifically, the typical optimization problem in WCNs is to efficiently reduce the necessary energy consumption while maintaining the quality of service (QoS) of network users.

In the operation of WCNs, ensuring green wireless communication, managing resources, maintaining efficient energy production and conservation, while at the same time guaranteeing the best possible QoS are all key factors being implemented. Other important issues in resource management and energy conservation of WCNs are how to share the limited wireless resources and how to apply sleep mode technology.

Queueing theory and Markov chains are commonly used as powerful methods for analyzing performance and evaluating communication networks. There are several books on the topic of WCNs using queueing theory and Markov chains. However, although queueing theory and optimization techniques play vital roles in the deployment and operation of almost every type of network, none of the existing books covers the topics of resource management and energy conservation in WCNs. Resource management and energy conservation are the keys to producing successful WCNs in the future.

This book provides the fundamental concepts and principles underlying the study of queueing systems as they apply to resource management and energy conservation in modern WCNs. This book gives analytical methods, approaches, algorithms, and simulation to evaluate numerically the QoS and optimize the resource allocation in various WCNs including broadband wireless access (BWA) networks, cognitive radio networks (CRNs), and cloud computing systems.

This book explains the constructed stochastic models that are at the core of evaluating system performance and presents intelligent searching algorithms to optimize the strategy under consideration. This book also provides sufficient analytical methods, approaches, and numerical simulation for students, analysts, managers, and industry people who are interested in using queueing theory to model congestion problems.

This book will be the first to provide an overview of the numerical analysis that can be gleaned by applying queueing theory, traffic theory, and other analytical methods. It will provide readers with information on recent advances in the resource management of various WCNs, such as BWA networks, CRNs, and cloud computing systems.

The authors of this book have been engaged in researching the performance evaluation of communication networks for nearly twenty years. This book has grown out of the authors' collaborative research on the performance evaluation for the resource management and energy conservation in WCNs, including BWA networks, CRNs, and cloud computing systems. The subject area discussed in the book is timely, gave the recent remarkable growth in wireless networking and the convergence of personal wireless communications, Internet technologies, and real-time multimedia. Each chapter of this book will give a detailed introduction of key topics in resource management of WCNs. The technical depth of the knowledge imparted in each chapter aims to satisfy experts in the field.

The methods, algorithms and techniques provided in this book include traffic analysis, applications of queueing theory and Markov chain theory, game theory, intelligent optimization, and operations research. In order to understand these methods, algorithms and techniques, basic concepts of computer communication networks and knowledge of queueing theory are needed. A familiarity with stochastic processes would also be useful.

We organize the book into three parts with a chapter on Introduction. In the Introduction, we briefly explain the background of our topic and give an overview of WCNs as they relate to the networks covered in this book. Then we introduce the queueing systems in general terms, as well as basic concepts and analysis methods that relate to the general theory of the stochastic processes used to capture the important properties and the stochastic behaviors of the network systems under research. We also show how to obtain the rate matrices, to solve the matrices and vectors numerically, and to optimize the system performances in queueing models and Markov chain models, effectively. These analytic methods, approximation methods, and techniques are used to analyze the system performance for resource management and energy conservation strategies on WCNs. Moreover, we define some of the important performance measures and common definitions to use in this book to present the performance analysis and optimization of the system models on WCN systems.

As an overview, we also outline the organization of the book. In particular, we explain the analysis methods for single- and multiple-vacation models, priority queueing systems, evaluation measures, and analytical methods and processes relating to performance optimization of the network systems.

In Parts I, II, and III, we present the performance evaluation and optimization for different WCNs, such as BWA networks, CRNs, and cloud computing systems, by applying queueing theory and the Markov chain models. Common performance measures like the energy conservation level, the energy-saving rate, the average response time, the system throughput, the spectrum utilization, and the switching rate are used in the evaluation. The optimization methods used are the steepest descent method, Newton's method, and the Intelligent optimization algorithm.

Specifically, Part I discusses the sleep mode in BWA networks, including Worldwide Interoperability for Microwave Access (WiMAX), and Long-Term Evolution (LTE). Part I includes 7 chapters beginning with Chap. 2, looking at how under the sleep mode operation, an MS operates two modes: the awake mode and the sleep mode. Revolving around this standard, we investigate some vacation queueing models with two types of busy periods: busy period in the listening state and busy period in the awake state, with a sleep-delay period, with a wake-up procedure, and with batch arrivals.

Part II discusses the dynamic spectrum allocation and energy-saving strategy in CRNs. There are 7 chapters in Part II, beginning with Chap. 9. In these chapters, we present an analytic framework to evaluate the system performance by constructing priority queueing models with possible service interruptions, using multiple channels, with several types of vacation mechanisms, and possible transmission interruptions.

Part III discusses the virtual machine (VM) allocation and sleep mode in cloud computing systems aiming to realize green cloud computing. Part III includes 6 chapters. Beginning with Chap. 16, we illustrate how from the perspective of multiple servers, we have an insight into queueing models with task migrations, wake-up thresholds, variable service rates, partial vacations, and second optional services.

The system model and the performance analysis offered in each chapter of Parts I–III are independent of others offered in other chapters, although depending on the class of queueing system involved, there is some common ground between the techniques employed. Each chapter contains its own system and offers important analysis methods and numerical results. The readers or students will find it helpful to refer to Chap. 1 initially, but after that the remaining chapters are stand-alone units which can be read in any order.

We would like to thank all the publishers to grant permission for our original publications.

The authors also would like to thank the editorial and publishing staff of Springer Nature, in particular, Celine Lanlan Chang, executive editor, computer science, and Beracah John Martyn and Nick Zhu, production editors, for their support and cooperation.

Finally, the authors would like to thank National Natural Science Foundation (No. 61872311 and No. 61973261) and Hebei Province Natural Science Foundation (No. F2017203141), China, for supporting this publication. The authors are also grateful for GRANT-IN-AID FOR SCIENCE RESEARCH, MEXT, Japan which led to this book.

Qinhuangdao, China
Kobe, Japan
May 2020

Shunfu Jin
Wuyi Yue

Contents

1 Introduction 1
  1.1 Overview of Wireless Communication Networks 1
      1.1.1 Broadband Wireless Access Networks 2
      1.1.2 Cognitive Radio Networks 3
      1.1.3 Cloud Computing 4
  1.2 Resource Management 6
      1.2.1 Static Spectrum Allocation 6
      1.2.2 Dynamic Spectrum Allocation 7
      1.2.3 Virtualization 7
      1.2.4 Virtual Machine Migration 8
  1.3 Queueing Models and Performance Analyses 9
      1.3.1 Basic and Vacation Queueing Models 9
      1.3.2 Queueing Model with Multiple-Class Customers 13
      1.3.3 Matrix-Geometric Solution Method 15
      1.3.4 Jacobi Iterative Method 19
      1.3.5 Gauss-Seidel Method 20
      1.3.6 Performance Optimization 21
  1.4 Organization of This Book 23

Part I Resource Management and Performance Analysis on Broadband Wireless Access Networks

2 Sleep Mode for Power Saving Class Type I 35
  2.1 Introduction 35
  2.2 Working Principle and System Model 37
      2.2.1 Working Principle 37
      2.2.2 System Model 39
  2.3 Performance Analysis 40

2.4 Performance Measures 43
    2.4.1 System Energy 43
    2.4.2 Average Response Time of Data Packets 45
    2.4.3 System Cost 45
2.5 Numerical Results 45
2.6 Conclusion 49

3 Sleep Mode for Power Saving Class Type II 51
3.1 Introduction 51
3.2 Working Principle and System Model 54
    3.2.1 Working Principle 54
    3.2.2 System Model 55
3.3 Analysis of Busy Cycle 57
    3.3.1 Busy Period in Listening State 57
    3.3.2 Busy Period in Awake State 58
    3.3.3 Time Length of Busy Cycle 59
3.4 Analysis of Waiting Time 60
    3.4.1 Waiting Time in Listening State 60
    3.4.2 Waiting Time in Awake State 61
    3.4.3 System Waiting Time 62
3.5 Performance Measures and Numerical Results 63
    3.5.1 Performance Measures 63
    3.5.2 Numerical Results 64
3.6 Optimal Sleep Window Length 66
3.7 Conclusion 68

4 Sleep Mode for Power Saving Class Type III 69
4.1 Introduction 69
4.2 Working Principle and System Model 71
    4.2.1 Working Principle 71
    4.2.2 System Model 72
4.3 Performance Analysis 74
    4.3.1 Number of Data Packets and Batches 74
    4.3.2 Queue Length and Waiting Time 74
    4.3.3 Busy Cycle 78
4.4 Performance Measures 80
    4.4.1 Handover Rate 80
    4.4.2 Energy Saving Rate 80
    4.4.3 System Utilization 81
    4.4.4 Average Response Time 81
4.5 Numerical Results 81
4.6 Conclusion 85

5 Bernoulli Arrival-Based Sleep Mode in WiMAX 2 87
5.1 Introduction 87
5.2 Working Principle of Sleep Mode in IEEE 802.16m 89

5.3   System Model and Performance Analysis   90
           5.3.1   System Model..   90
           5.3.2   Performance Analysis   91
  5.4   Numerical Results..   95
  5.5   Conclusion..   96

6   Markovian Arrival-Based Sleep Mode in WiMAX 2   97
  6.1   Introduction..   97
  6.2   System Model and Performance Analysis 100
           6.2.1   System Model... 100
           6.2.2   Number of Data Packets................................... 103
           6.2.3   Busy Cycle.. 107
  6.3   Performance Measures and Optimization 107
           6.3.1   Performance Measures 108
           6.3.2   Performance Optimization 110
  6.4   Numerical Results.. 111
  6.5   Conclusion.. 116

**7   Two-Stage Vacation Queue-Based Active DRX Mechanism
     in an LTE System** 119
  7.1   Introduction.. 119
  7.2   Enhanced Energy Saving Strategy................................ 121
  7.3   System Model and Performance Analysis 122
           7.3.1   System Model... 123
           7.3.2   Busy Period 125
           7.3.3   Queue Length and Waiting Time 127
           7.3.4   Busy Cycle.. 128
           7.3.5   Performance Measures 129
  7.4   Numerical Results and Performance Optimization 130
           7.4.1   Numerical Results 130
           7.4.2   Performance Optimization 136
  7.5   Conclusion.. 139

**8   Multiple-Vacation Queue-Based Active DRX Mechanism
     in an LTE System** 141
  8.1   Introduction.. 141
  8.2   Enhanced Active DRX Mechanism 143
  8.3   System Model and Performance Analysis 146
           8.3.1   System Model... 146
           8.3.2   Transition Probability Sub-Matrices for Case I 148
           8.3.3   Transition Probability Sub-Matrices for Case II 151
           8.3.4   Performance Measures 154
  8.4   Numerical Results and Performance Optimization 155
           8.4.1   Numerical Results 155
           8.4.2   Performance Optimization 159
  8.5   Conclusion.. 162

Part II   Resource Management and Performance Analysis on Cognitive Radio Networks

9   Channel Aggregation Strategy with Perfect-Sensing Results 165
   9.1   Introduction 165
   9.2   Channel Aggregation Strategy and System Model 167
      9.2.1   Channel Aggregation Strategy 167
      9.2.2   System Model 168
   9.3   Performance Analysis and Numerical Results 170
      9.3.1   Steady-State Distribution 170
      9.3.2   Performance Measures and Analysis of System Cost 174
      9.3.3   Numerical Results 176
   9.4   Analysis of Admission Fee 181
      9.4.1   Nash Equilibrium Behavior 181
      9.4.2   Socially Optimal Behavior 183
      9.4.3   Pricing Policy 184
   9.5   Conclusion 185

10   Spectrum Reservation Strategy with Retrial Feedback and Perfect-Sensing Results 187
   10.1   Introduction 187
   10.2   Spectrum Reservation Strategy and System Model 189
      10.2.1   Spectrum Reservation Strategy 189
      10.2.2   System Model 191
   10.3   Performance Analysis and Numerical Results 192
      10.3.1   Performance Analysis 192
      10.3.2   Performance Measures 198
      10.3.3   Numerical Results 199
   10.4   Performance Optimization 204
      10.4.1   Analysis of System Cost 205
      10.4.2   Optimization of System Parameters 205
   10.5   Conclusion 208

11   Opportunistic Spectrum Access Mechanism with Imperfect Sensing Results 209
   11.1   Introduction 209
   11.2   Opportunistic Spectrum Access Mechanism and System Model 211
      11.2.1   Activity of PU Packets 211
      11.2.2   Activity of SU Packets 212
      11.2.3   System Model 214
   11.3   Performance Analysis 215
      11.3.1   Mistake Detections and False Alarms 215
      11.3.2   Transition Probability Matrix 215

11.4 Performance Measures and Numerical Results 220
     11.4.1 Performance Measures 220
     11.4.2 Numerical Results 221
11.5 Analysis of Admission Fee 225
     11.5.1 Behaviors of Nash Equilibrium and Social
             Optimization 225
     11.5.2 Pricing Policy 227
11.6 Conclusion... 228

**12 Mini-Slotted Spectrum Allocation Strategy with Imperfect
Sensing Results**... 229
12.1 Introduction.. 229
12.2 Mini-Slotted Spectrum Allocation Strategy and System Model ... 231
     12.2.1 Mini-Slotted Spectrum Allocation Strategy 231
     12.2.2 System Model.. 233
12.3 Performance Analysis.. 235
     12.3.1 Transition Probability Matrix 235
     12.3.2 Steady-State Distribution................................ 238
12.4 Performance Measures and Numerical Results 239
     12.4.1 Performance Measures 239
     12.4.2 Numerical Results 240
12.5 Performance Optimization 243
12.6 Conclusion... 244

13 Channel Reservation Strategy with Imperfect Sensing Results 247
13.1 Introduction.. 247
13.2 Channel Reservation Strategy and System Model 249
     13.2.1 Channel Reservation Strategy 249
     13.2.2 System Model.. 252
13.3 Performance Analysis and TLBO-SOR Algorithm................. 252
     13.3.1 Performance Analysis 253
     13.3.2 TLBO-SOR Algorithm 257
13.4 Performance Measures and Numerical Results 261
     13.4.1 Performance Measures 262
     13.4.2 Numerical Results 263
13.5 Conclusion... 269

**14 Energy Saving Strategy in CRNs Based on a Priority Queue
with Single Vacation**.. 271
14.1 Introduction.. 271
14.2 Energy Saving Strategy and System Model........................ 273
     14.2.1 Energy Saving Strategy 273
     14.2.2 System Model.. 275

14.3   Performance Analysis and Numerical Results 276
       14.3.1   Performance Analysis 276
       14.3.2   Performance Measures 281
       14.3.3   Numerical Results 281
14.4   Analysis of Admission Fee 283
       14.4.1   Behaviors of Nash Equilibrium and Social
                Optimization 283
       14.4.2   Pricing Policy 287
14.5   Conclusion 288

**15 Energy Saving Strategy in CRNs Based on a Priority Queue
     with Multiple Vacations** 291
15.1   Introduction 291
15.2   Energy Saving Strategy and System Model 294
       15.2.1   Energy Saving Strategy 294
       15.2.2   System Model 295
15.3   Performance Analysis 296
15.4   Performance Measures and Numerical Results 299
       15.4.1   Performance Measures 299
       15.4.2   Numerical Results 301
15.5   Performance Optimization 308
       15.5.1   Analysis of System Cost 308
       15.5.2   Optimization of System Parameters 310
15.6   Conclusion 312

**Part III   Resource Management and Performance Analysis
            on Cloud Computing**

16 Speed Switch and Multiple-Sleep Mode 315
16.1   Introduction 315
16.2   Virtual Machine Scheduling Strategy and System Model 317
       16.2.1   Virtual Machine Scheduling Strategy 317
       16.2.2   System Model 320
16.3   Performance Analysis 321
       16.3.1   Transition Rate Matrix 321
       16.3.2   Steady-State Distribution 324
16.4   Performance Measures and Numerical Results 326
       16.4.1   Performance Measures 326
       16.4.2   Numerical Results 327
16.5   Performance Optimization 332
16.6   Conclusion 336

17 Virtual Machine Allocation Strategy 337
17.1   Introduction 337
17.2   Virtual Machine Allocation Strategy and System Model 340
       17.2.1   Virtual Machine Allocation Strategy 340
       17.2.2   System Model 342

17.3   Performance Analysis 344
    17.3.1   Transition Rate Matrix 344
    17.3.2   Steady-State Distribution 348
17.4   Performance Measures and Numerical Results 350
    17.4.1   Performance Measures 350
    17.4.2   Numerical Results 351
17.5   Performance Optimization 354
17.6   Conclusion 356

18   Clustered Virtual Machine Allocation Strategy 359
18.1   Introduction 359
18.2   Clustered Virtual Machine Allocation Strategy
    and System Model 361
    18.2.1   Clustered Virtual Machine Allocation Strategy 361
    18.2.2   System Model 363
18.3   Performance Analysis 365
    18.3.1   Transition Rate Matrix 365
    18.3.2   Steady-State Distribution 369
18.4   Performance Measures and Numerical Results 371
    18.4.1   Performance Measures 372
    18.4.2   Numerical Results 372
18.5   Performance Optimization 376
18.6   Conclusion 379

19   Pricing Policy for Registration Service 381
19.1   Introduction 381
19.2   Cloud Architecture and System Model 384
    19.2.1   Cloud Architecture 384
    19.2.2   System Model 385
19.3   Performance Analysis 386
    19.3.1   Transition Rate Matrix 386
    19.3.2   Steady-State Distribution 391
19.4   Performance Measures and Numerical Results 393
    19.4.1   Performance Measures 393
    19.4.2   Numerical Results 394
19.5   Analysis of Registration Fee 397
    19.5.1   Behaviors of Nash Equilibrium and Social
        Optimization 397
    19.5.2   Pricing Policy 401
19.6   Conclusion 403

20   Energy-Efficient Task Scheduling Strategy 405
20.1   Introduction 405
20.2   Energy-Efficient Task Scheduling Strategy and System Model 408
    20.2.1   Energy-Efficient Task Scheduling Strategy 408
    20.2.2   System Model 409

20.3 Performance Analysis 411
    20.3.1 Transition Rate Matrix 411
    20.3.2 Steady-State Distribution.................................. 414
20.4 Performance Measures and Numerical Results 416
    20.4.1 Performance Measures 416
    20.4.2 Numerical Results 417
20.5 Performance Optimization 419
20.6 Conclusion... 422

21 Energy-Efficient Virtual Machine Allocation Strategy 423
21.1 Introduction.. 423
21.2 Energy-Efficient Virtual Machine Allocation Strategy and
    System Model 425
    21.2.1 Energy-Efficient Virtual Machine Allocation Strategy ... 425
    21.2.2 System Model.. 427
21.3 Performance Analysis 428
    21.3.1 Transition Rate Matrix 428
    21.3.2 Steady-State Distribution.................................. 431
21.4 Performance Measures and Numerical Results 433
    21.4.1 Performance Measures 433
    21.4.2 Numerical Results 434
21.5 Analysis of Admission Fee 437
    21.5.1 Behaviors of Nash Equilibrium and Social
          Optimization 437
    21.5.2 Pricing Policy 440
21.6 Conclusion... 442

References... 445

Index... 461

Abbreviations

3GPP	3rd Generation Partnership Project
5G	5th Generation
BS	Base station
BWA	Broadband wireless access
CDC	Cloud data center
CRN	Cognitive radio network
CTMC	Continuous-time Markov chain
CDMA	Code-division multiple access
D-MAP	Discrete-time Markovian arrival process
DPM	Dynamic power management
DRX	Discontinuous reception
DTMC	Discrete-time Markov chain
DVFS	Dynamic voltage and frequency scaling
EAS	Early arrival system
eNodeB	Evolved node B
FCFS	First-come first-served
FDM	Frequency division multiplexing
IaaS	Infrastructure as a service
i.i.d.	Independent and identically distributed
LAS	Late arrival system
LTE	Long-Term Evolution
LTE-A	LTE-Advanced
MAC	Medium access control
M-ADRX	Multi-threshold Automated configuration DRX
MS	Mobile station
NRT-VR	Non-real-time variable rate
PaaS	Platform as a service
PGF	Probability generating function
PH	PHase type
PM	Physical Machine
PSO	Particle swarm optimization

PU	Primary user
QBD	Quasi-birth–death
QoE	Quality of experience
QoS	Quality of service
RT-VR	Real-time variable rate
SaaS	Software as a service
S-ADRX	Single-threshold Automated configuration DRX
SLA	Service level agreement
SNR	Signal-to-noise ratio
SOR	Successive over relaxation
SU	Secondary user
TLBO	Teaching-learning-based optimization
UE	User equipment
UGS	Unsolicited grant service
UMC	Underlying Markov chain
VIP	Very important person
VM	Virtual machine
WCN	Wireless communication network
WiMAX	Worldwide Interoperability for Microwave Access

List of Figures

Fig. 2.1   Working principle of power saving class type I
in IEEE 802.16e 37
Fig. 2.2   Energy saving rate of system versus initial-sleep window 46
Fig. 2.3   Energy saving rate of system versus close-down time 46
Fig. 2.4   Average response time of data packets versus initial-sleep
window 47
Fig. 2.5   Average response time of data packets versus
close-down time 47
Fig. 2.6   System cost function versus close-down time 49

Fig. 3.1   Working principle of power saving class type II
in IEEE 802.16e 55
Fig. 3.2   State transition of system model 56
Fig. 3.3   Handover rate versus time length of sleep window 64
Fig. 3.4   Energy saving rate of system versus time length of sleep
window 65
Fig. 3.5   Average response time of data packets versus time length
of sleep window 65
Fig. 3.6   System cost function versus time length of sleep window 67

Fig. 4.1   State transition of system model 72
Fig. 4.2   Handover rate versus time length of sleep-delay timer 82
Fig. 4.3   Energy saving rate of system versus time length of
sleep-delay timer 83
Fig. 4.4   System utilization versus time length of sleep-delay timer 83

Fig. 5.1   Embedded Markov points chosen in system model 91
Fig. 5.2   Average response time of data packets versus arrival rate of
data packets 95

Fig. 5.3   Energy saving rate of system versus arrival rate of data
            packets 96

Fig. 6.1   State transition of system model 101
Fig. 6.2   Various time epochs in LAS with delayed access 102
Fig. 6.3   Average response time of data packets versus time length
            of sleep cycle 111
Fig. 6.4   Energy saving rate of system versus time length of sleep cycle ... 112
Fig. 6.5   Average response time of data packets versus correlation
            parameter 113
Fig. 6.6   Energy saving rate of system versus correlation parameter 113
Fig. 6.7   Standard deviation versus arrival rate of data packets 114
Fig. 6.8   System cost function versus time length of sleep cycle 116

Fig. 7.1   Time sequence of proposed enhanced energy saving strategy 122
Fig. 7.2   State transition of system model 124
Fig. 7.3   Handover rate versus threshold of short DRX stages 131
Fig. 7.4   Energy saving rate of system versus threshold of short
            DRX stages 133
Fig. 7.5   Average response time versus threshold of short DRX stages 135
Fig. 7.6   System cost function versus threshold of short DRX stages 137
Fig. 7.7   System cost function versus time length of sleep-delay timer 138

Fig. 8.1   Time sequence of proposed enhanced Active DRX
            mechanism 145
Fig. 8.2   Energy saving rate of system versus time length of
            sleep-delay timer 156
Fig. 8.3   Blocking rate of data packets versus time length of
            sleep-delay timer 157
Fig. 8.4   Average response time versus time length of sleep-delay
            timer 158
Fig. 8.5   System profit function versus sleep interval 160
Fig. 8.6   System profit function versus time length of sleep-delay timer ... 160
Fig. 8.7   System profit function versus number of logical channels 161

Fig. 9.1   Proposed channel aggregation strategy 168
Fig. 9.2   Blocking rate of SU packets versus channel aggregation
            intensity 177
Fig. 9.3   Average latency of SU packets versus channel aggregation
            intensity 178
Fig. 9.4   Channel utilization versus channel aggregation intensity 179
Fig. 9.5   System cost function versus channel aggregation intensity 180
Fig. 9.6   Individual benefit function versus arrival rate of SU packets 183
Fig. 9.7   Social benefit function versus arrival rate of SU packets 184

Fig. 10.1   Proposed adaptive spectrum reservation strategy 189
Fig. 10.2   Interruption rate of SU packets versus adaptive control factor 199
Fig. 10.3   Interruption rate of SU packets versus admission threshold 200
Fig. 10.4   Average latency of SU packets versus adaptive control factor 201
Fig. 10.5   Average latency of SU packets versus admission threshold 202
Fig. 10.6   Blocking rate of PU packets versus adaptive control factor 203
Fig. 10.7   Throughput versus adaptive control factor 203
Fig. 10.8   Normalized throughput increment versus admission threshold ... 204
Fig. 10.9   System cost function versus adaptive control factor 206
Fig. 10.10   System cost function versus admission threshold................. 206

Fig. 11.1   Transmission process of PU packets 212
Fig. 11.2   Transmission process of SU packets 213
Fig. 11.3   Throughput of SU packets versus energy threshold 221
Fig. 11.4   Blocking rate of SU packets versus energy threshold 223
Fig. 11.5   Average latency of SU packets versus energy threshold.......... 224
Fig. 11.6   Individual benefit function versus arrival rate of SU packets 226
Fig. 11.7   Social benefit function versus arrival rate of SU packets.......... 226

Fig. 12.1   Working principle of proposed mini-slotted spectrum
             allocation strategy 232
Fig. 12.2   Interruption rate of PU packets versus slot size................... 240
Fig. 12.3   Throughput of SU packets versus slot size........................ 241
Fig. 12.4   Switching rate of SU packets versus slot size..................... 242
Fig. 12.5   Average latency of SU packets versus slot size 242
Fig. 12.6   System profit function versus slot size 244

Fig. 13.1   Transmission process of user packets in system 250
Fig. 13.2   Throughput of SU packets versus energy detection threshold 263
Fig. 13.3   Average latency of SU packets versus energy detection
             threshold 264
Fig. 13.4   Switching rate of SU packets versus energy detection
             threshold 265
Fig. 13.5   Channel utilization versus energy detection threshold............. 265
Fig. 13.6   Throughput of SU packets versus number of reserved
             channels... 266
Fig. 13.7   Average latency of SU packets versus number of reserved
             channels... 266
Fig. 13.8   Switching rate of SU packets versus number of reserved
             channels... 267
Fig. 13.9   Channel utilization versus number of reserved channels 267

Fig. 14.1   State transition of proposed energy saving strategy 274
Fig. 14.2   EAS with possible arrival and departure instants 275
Fig. 14.3   Average latency of SU packets versus sleep parameter.......... 282

Fig. 14.4   Energy saving degree versus sleep parameter...................... 283
Fig. 14.5   Individual benefit function versus arrival rate of SU packets 284
Fig. 14.6   Social benefit function versus arrival rate of SU packets.......... 285

Fig. 15.1   State transition on one port in BS 295
Fig. 15.2   Normalized throughput of SU packets versus sleep parameter ... 301
Fig. 15.3   Average latency of SU packets versus sleep parameter........... 302
Fig. 15.4   Energy saving rate of system versus sleep parameter............. 302
Fig. 15.5   Channel utilization versus sleep parameter 303
Fig. 15.6   Normalized throughput of SU packets versus service rate
             of one channel 305
Fig. 15.7   Average latency of SU packets versus service rate of one
             channel.. 305
Fig. 15.8   Energy saving rate of system versus service rate of one
             channel.. 306
Fig. 15.9   Channel utilization versus service rate of one channel 307
Fig. 15.10   System cost function versus sleep parameter 309
Fig. 15.11   System cost function versus service rate of one channel 309

Fig. 16.1   Transition among three CDC cases in proposed strategy.......... 319
Fig. 16.2   State transition of system model................................... 322
Fig. 16.3   Energy saving level of system versus arrival rate of tasks........ 329
Fig. 16.4   Average latency of tasks versus arrival rate of tasks.............. 331
Fig. 16.5   System profit function versus sleep parameter.................... 333

Fig. 17.1   State transition in CDC with proposed VM allocation strategy ... 341
Fig. 17.2   Average latency of tasks versus sleep parameter.................. 352
Fig. 17.3   Energy saving rate of system versus sleep parameter............. 353

Fig. 18.1   Working flow of VM with proposed strategy 362
Fig. 18.2   Average latency of tasks versus sleep parameter.................. 374
Fig. 18.3   Energy saving rate of system versus sleep parameter............. 375

Fig. 19.1   Sleep mode-based cloud architecture.............................. 384
Fig. 19.2   Average latency versus arrival rate of anonymous users 395
Fig. 19.3   Energy saving rate of system versus arrival rate of
             anonymous users 396
Fig. 19.4   Individual benefit function versus arrival rate of anonymous
             users.. 399
Fig. 19.5   Social benefit function versus arrival rate of anonymous users ... 400

Fig. 20.1   Transition among three states in proposed strategy............... 410
Fig. 20.2   Average latency of tasks versus sleep parameter.................. 418
Fig. 20.3   Energy saving rate of system versus sleep parameter............. 419

Fig. 21.1   State transition of proposed VM allocation strategy 426
Fig. 21.2   Average latency of tasks versus arrival rate of tasks 435
Fig. 21.3   Energy saving rate of system versus arrival rate of tasks 435
Fig. 21.4   Individual benefit function versus arrival rate of tasks 439
Fig. 21.5   Social benefit function versus arrival rate of tasks 439

List of Tables

Table 1.1   System models in Part I... 26
Table 1.2   System models in Part II.. 29
Table 1.3   System models in Part III....................................... 31

Table 3.1   Optimum time length of sleep window 68

Table 4.1   Average response time of data packets......................... 84

Table 6.1   Parameter settings in numerical results 111
Table 6.2   Optimum time length of sleep cycle 116

Table 7.1   Optimum threshold of short DRX stages 137
Table 7.2   Optimum time length of sleep-delay timer 138

Table 8.1   Optimum parameters in proposed enhanced Active DRX
            mechanism 161

Table 9.1   Optimum channel aggregation intensity 181
Table 9.2   Nash equilibrium and socially optimal arrival rates of SU
            packets... 184
Table 9.3   Numerical results for admission fee............................ 185

Table 10.1 Optimum combination of parameters in proposed strategy 208

Table 11.1 Parameter settings in numerical results 221
Table 11.2 Numerical results for admission fee............................ 227

Table 12.1 Optimum slot size in proposed strategy......................... 244

Table 14.1   Socially optimal arrival rate of SU packets   287
Table 14.2   Numerical results for spectrum admission fee.....................   288

Table 15.1   Optimum combination of parameters in proposed strategy   312

Table 16.1   Relation between system level and system case   321
Table 16.2   Parameter settings in numerical results   328
Table 16.3   Optimum sleep parameter in proposed strategy   336

Table 17.1   Parameter settings in numerical results   351
Table 17.2   Optimum combination of parameters in proposed strategy   356

Table 18.1   Optimum combination of parameters in proposed strategy   379

Table 19.1   Numerical results for registration fee   403

Table 20.1   Parameter settings in numerical results   418
Table 20.2   Optimum combination of parameters in proposed strategy   422

Table 21.1   Numerical results for admission fee................................   442

Chapter 1
Introduction

This book provides analytical methods, approaches, algorithms, and simulation techniques to evaluate numerically the quality of service and optimize the resource allocation in various Wireless Communication Networks (WCNs), including Broadband Wireless Access (BWA) networks, Cognitive Radio Networks (CRNs) and cloud computing systems.

In this chapter, we briefly explain the background of our topic and give an overview of WCNs, including BWA networks, CRNs and cloud computing systems, as they relate to the networks covered in this book. We also introduce the queueing systems in general terms, as well as basic concepts and analysis methods that relate to the general theory of the stochastic processes used to capture the important properties of the problems under research. We show how to efficiently obtain the transition rate matrix, thereby complementing the matrix analytical and system optimization approach in queueing models and Markov chain models that are used for WCN system analysis, resource management, and proposals for energy conservation strategies. We also define some of the important performance measures and common definitions to use in this book to present the performance analysis and optimization of the system models on WCN networks.

1.1 Overview of Wireless Communication Networks

With the development and progress of the times, intelligent age for mankind is dawning. In the era of the development of new technologies, cloud computing, artificial intelligence, big data and block chain applications are continually emerging. An expanding future for WCNs is suggested by the explosive growth in wireless systems coupled with the proliferation of mobile terminals, laptop and palmtop computers. However, many technical challenges remain in designing robust wireless networks that deliver the performance necessary to support emerging applications.

© Springer Nature Singapore Pte Ltd. 2021
S. Jin, W. Yue, *Resource Management and Performance Analysis of Wireless Communication Networks*, https://doi.org/10.1007/978-981-15-7756-7_1

This section presents information mainly on recent advances in various WCNs, such as Broadband Wireless Access networks, Cognitive Radio Networks and cloud computing systems relating to the networks covered in this book.

1.1.1   Broadband Wireless Access Networks

BWA networks include mobile Worldwide Interoperability for Microwave Access (WiMAX) and Long Term Evolution (LTE). Under the sleep mode operation, a Mobile Station (MS) operates two modes: the awake mode and the sleep mode. Revolving around this standard, we investigate some vacation queueing models with two types of busy periods: busy period in the listening state and busy period in the awake state, with a sleep-delay period, with a wake-up procedure, and with batch arrivals.

IEEE 802.16e [IEEE06a] is one of the latest standards for mobile BWA network system. The sleep mode proposed in IEEE 802.16e [IEEE06a] is intended to minimize the MS power usage and to decrease usage of serving Base Station (BS) air interface resources. There are three types of power saving classes (say type I, II and III) based on sleep mode operation as follows:

(1) *Power Saving Class Type I*: Power saving class type I is recommended for connections of Best Effort (BE), Non-Real-Time Variable Rate (NRT-VR) type. For definition and/or activation of one or several power saving classes of type I, the MS shall send MOB_SLP-REQ or a Bandwidth Request (BR) and an uplink sleep control header; the BS shall respond with a MOB_SLP-RSP message or a downlink sleep control extended sub-header.

(2) *Power Saving Class Type II*: Power saving class type II is recommended for connections of Unsolicited Grant Service (UGS), Real-Time Variable Rate (RT-VR) traffic. This Power Saving Class becomes active at the frame specified as "start frame number for first sleep window". All sleep windows are the same size as the initial window.

(3) *Power Saving Class Type III*: Power saving class type III is recommended for multicast connections as well as for management operations, for example, periodic ranging, Dynamic Service Addition/Change/Deletion (DSx) operations, MOB_ NBR-ADV, etc. Power saving class type III is defined/activated by MOB_SLP-REQ/MOB_SLP-RSP or BR and an uplink sleep control header/downlink sleep control extended sub-header transaction.

With the development of communication industry, how to conserve the energy consumption and to extend the lifetime of the battery in an MS are now key questions to solve for WiMAX [Xue11]. IEEE 802.16m [IEEE11] is an evolution of mobile WiMAX and is currently being processed for standardization as an amendment of IEEE 802.16e. The aim of IEEE 802.16m is to reduce energy consumption and to improve the system performance. Unlike IEEE 802.16e, in the sleep mode of IEEE 802.16m, the BS will negotiate with the MS by using Traffic

Indication. In this way, messages for the sleep request and sleep response, which are used in IEEE 802.16e, are omitted, and the state transition overhead between the listening window and the sleep window is therefore minimized.

The LTE project for Universal Mobile Telecommunications Systems (UMTs) has been initiated by the Third Generation Partnership Project (3GPP) [3GPP, Wiga09]. The purpose of LTE is to accommodate more users in every cell, accelerate the data transmission rate, and reduce the energy consumption and the cost of the network. Many telecom operators have deployed LTE networks and concentrated their research into LTE productions [Abet10]. Compared with 3G technology, LTE has an ability to operate at a higher transmission rate [Abet10]. However, the improvement in the transmission rate leads to excessive energy consumption at the mobile terminal. In order to reduce the energy consumption and to achieve more efficient and greener communication, a Discontinuous Reception (DRX) mechanism was introduced into the LTE technology [Koc13]. This mechanism influences the downlink transmission at the User Equipment (UE).

In this book, we focus on the sleep mode in BWA networks, including WiMAX and LTE system. Under the sleep mode operation, an MS operates two modes: the awake mode and the sleep mode. Revolving around this standard, we investigate various vacation queueing models: those with two types of busy periods: busy period in the listening state and busy period in the awake state, with a sleep-delay period, with a wake-up procedure, and with batch arrivals.

1.1.2 Cognitive Radio Networks

Currently, spectrum allocation for wireless services indicates that most frequencies below 6 GHz have already been occupied.

Considering future wireless trends, the integration of emerging 5th Generation (5G) technologies will require a special task force, especially for large-scale networks. Cognitive radio aims at using spectrum holes by dynamic spectrum access to enhance spectrum efficiency. This technique can greatly improve the spectrum efficiency in WCNs. Technology of CRNs has emerged as an effective method to enhance the utilization of the radio spectrum where the Primary Users (PUs) have priority to use the spectrum, and the Secondary Users (SUs) try to exploit that part of the spectrum unoccupied by the PUs. In CRNs, by sensing the network condition, and collecting the environment information with space-time, the spectrum hole can be utilized reasonably.

In CRNs, SUs search for unoccupied channels and build a list of candidate channels. However, for transmission, every communicating SU pair needs to agree on which channels to use. Therefore, careful coordination between SUs as well as between SUs and PUs to choose and access the channels is required. Consequently, spectrum sensing is an important requirement for the design and implementation of CRNs. The capability of spectrum sensing is particularly relevant in the cases of out-of-band sensing and in-band sensing. For out-of-band sensing, SUs try to find available spectrum holes over a wide frequency range as transmission opportunity. For in-band sensing, SUs monitor transmissions in spectrum bands to detect the presence of primary networks and avoid interferences.

Based on the different network structures [Sult16], the management technology of channel allocation in CRNs can be classified into a centralized channel allocation strategy and a distributed channel allocation strategy as follows:

(1) *Centralized Channel Allocation Strategy*: With centralized schemes, the spectrum access is controlled by a fusion center. By coordinating and controlling SUs' access, this fusion center helps prevent inter-user collisions and reduces energy consumption that would result from those collisions. Typically, a database is created at the fusion center. Such a database could be established via the help of SUs. Any SU that wants to access a channel must consult with the fusion center. Obviously, additional overheads will be introduced due to the mandatory communication between SUs and the fusion center. In addition, PUs' activity is always changing, and thus, the database needs to be updated periodically.

(2) *Distributed Channel Allocation Strategy*: Distributed schemes do not rely on a centralized BS. However, SUs must cooperate with each other to coexist and access the available bands. In particular, each cooperating SU must perform local spectrum sensing and share the results with other SUs. SUs must coordinate with each other for a fair sharing of the available spectrum resources. The SUs will compete for the unoccupied spectrum, and employ some technologies, such as Carrier Sense Multiple Access/Collision Avoidance (CSMA/CA) technology, to avoid inter-user collisions.

This book focuses on the spectrum allocation and energy saving strategy in CRNs. We present an analytic framework to evaluate the system performance by constructing priority queueing models with possible service interruptions, using multiple channels, with several types of vacation mechanisms, and possible transmission interruptions.

1.1.3 Cloud Computing

Network-based cloud computing is rapidly expanding as an alternative to conventional office-based computing. The service mode in cloud computing systems involves the provision of large pools of high performance computing resources and high-capacity storage devices that are shared among end users.

Cloud computing as a new type of business computing service has been widely used by enterprises and individual users. Also, cloud computing technology can deal with a wide range of services across the Internet [Madn16]. According to National Institute of Standards and Technology (NIST), service modes in cloud computing systems are classified as Software as a Service (SaaS), Platform as a Service (PaaS) and Infrastructure as a Service (IaaS) [Huan14, Lin14]. IaaS providers supply storage space, computing and network resources with which users can execute Operating System (OS), applications and any software. PaaS providers supply the software programming languages and system development tools to users so they can deploy their own applications. SaaS providers supply applications to users through a client interface, such as a Web browser. SaaS providers may own a small local data center, and can also acquire resources from the public IaaS cloud [Davi15, Li15c].

Architectures in cloud computing systems can be either public or private as follows:

(1) *Public Cloud*: The public cloud is defined as computing services offered by third-party providers over the public Internet, making them available to anyone who wants to use or purchase them. They may be free or sold on-demand, allowing customers to pay only per usage cost for the CPU cycles, storage, or bandwidth they consume. Public clouds can save companies the expensive costs of having to purchase, manage, and maintain on-premises hardware and application infrastructure. Public clouds can also be deployed faster than on-premises infrastructures and with an almost infinitely scalable platform. Every employee of a company can use the same application from any office or branch using their device of choice as long as they can access the Internet.

(2) *Private Cloud*: The private cloud is defined as computing services offered either over the Internet or a private internal network to only select users instead of the general public. Also called an internal or corporate cloud, private cloud computing gives businesses many of the benefits of a public cloud, including self-service, scalability, and elasticity. In addition, private clouds deliver a higher level of security and privacy through both company firewalls and internal hosting to ensure operations and sensitive data are not accessible to third-party providers.

As a direct result of the rapid growth in the number of cloud users, some cloud providers have already built large numbers of data centers to satisfy the resources demand. However, the data centers consume a lot of electricity, resulting in excessive increases in carbon emissions and a reduction in benefits for the cloud providers. Green cloud computing solutions that cannot only minimize the operational costs but also reduce the environmental impact have become a necessary.

Green cloud computing is also called green information technology (GREEN IT). Green cloud computing is used not only for efficient processing and utilization of computing infrastructure, but also for minimizing energy consumption. This is essential for the future growth in cloud computing to be sustainable. Otherwise, cloud computing with the increasingly prevalent front-end client devices interacting with back-end data centers causes a huge escalation of energy usage. Study on green cloud computing includes research and practice of designing, manufacturing, using, and disposing of computing resources with minimal environmental impact.

This book, in part III discusses the Virtual Machine (VM) allocation and sleep mode in cloud computing systems to trade off the energy consumption against the performance of the system for achieving more efficient and greener cloud computing.

1.2   Resource Management

WCNs are fundamental to many applications, such as the Internet of Things (IoT), big data, and cloud computing. The Quality of Service (QoS) offered by a WCN is often measured by how well it satisfies the end-to-end requirements of tasks executed in the WCNs. Resource management in WCNs plays a critical role in achieving the desired QoS.

In this section, we introduce some basic concepts of resource management in terms of static spectrum allocation, dynamic spectrum allocation, virtualization and VM migration.

1.2.1   Static Spectrum Allocation

Allocated spectrum is a scarce and precious resource in WCNs. Traditional WCNs feature static spectrum allocation policies, according to which licensees are approved to the PUs exclusively for the use of spectrum bands on a long-term basis over huge geographical regions. Basic methods of static spectrum allocation are given as follows:

(1) *Time Division Multiplexing (TDM)*: For a TDM link, time is divided into frames of fixed duration, and each frame is divided into a fixed number of time slots. When the network establishes a connection across a link, the network dedicates one time slot in every frame to this connection. These slots are allocated for the sole use of that connection, with one time slot available for use to transmit the connection's data.

(2) *Frequency Division Multiplexing (FDM)*: With FDM, frequency spectrum of a link is divided up among the connections established across the link. Specifically, the link allocates a frequency band to each connection for the duration of the connection. In telephone networks, this frequency band typically has a width of 4 kHz. FM radio stations also use FDM to share the frequency spectrum between 88 MHz and 108 MHz, with each station being allocated a specific frequency band.

(3) *Code Division Multiple Access (CDMA)*: CDMA assigns a different code to each node. Each node then uses its unique code to encode the data bits it sends. If the codes are chosen carefully, CDMA networks have the beneficial property of different nodes being able to transmit simultaneously and yet have their respective receivers correctly receive a sender's encoded data bits. CDMA has been used in military systems for some time and now has widespread civilian use, particularly in phone technology.

Technologies to improve spectrum efficiency include adaptive coding and modulation, multiple-antenna technology, and multiple-carrier technology. At present, CDMA air interface technologies, such as high-speed downlink package access can achieve spectrum efficiency of 1 b/s/Hz. The application of Orthogonal Frequency Division Multiplexing (OFDM) and Multiple Input Multiple Output (MIMO) technologies can achieve spectrum efficiency of 3-4 b/s/Hz.

To some extent, these technologies have alleviated the contradiction in spectrum requirements, but the improvement of spectral efficiency is limited by the Shannon channel capacity. However, for the greater success of wireless applications, unlicensed band usage is required, which results in possible shortage of wireless spectrum. This spectrum crisis has motivated the development of dynamic spectrum allocation policies.

1.2.2 Dynamic Spectrum Allocation

Different from static spectrum allocation, dynamic spectrum access encompasses various approaches to spectrum management. Dynamic spectrum allocation strategies can be classified into three models as follows:

(1) *Dynamic Exclusive Use Model*: This model maintains the basic structure of the current spectrum regulation policy, namely, spectrum bands are licensed to services for exclusive use. The main idea of this model is to introduce flexibility to improve spectrum efficiency.

(2) *Open Sharing Model*: This model employs open sharing among peer users as the basis for managing the spectrum. Centralized spectrum sharing strategy and distributed centralized spectrum sharing strategy have been initially investigated under this spectrum management model.

(3) *Hierarchical Access Model*: This model adopts a hierarchical access structure with PUs and SUs. The basic idea is to open licensed spectrum to SUs while limiting the interference perceived by PUs. The underlaying approach imposes severe constraints on the transmission power of SUs so that they operate below the noise floor of PUs. Spectrum overlay was using the DARPA (Defense Advanced Research Projects Agency) next generation social science (NGS2) program under opportunistic spectrum access.

1.2.3 Virtualization

Virtualization is a resource management technology that abstracts and transforms various physical resources, such as server, network, memory and storage, and presents them to break the non-cutting barriers between entity structures so that users can use these resources in a better way than the original configuration. In general, virtualization is complemented on the following three levels:

(1) *Hardware Virtualization*: Depending on the host OS, hardware virtualization can be categorized into two main types. The first type is the native hypervisors

where the guest OS works on virtualized hardware. The second type is the hosted hypervisors where a standard OS is required and hosted hypervisors work as a regular application.

(2) *Containers*: The most characteristic aspect of containers, also known as para-virtualization, is that the host OS only separates processes and resources so that it is impossible to run another OS other than the host OS.

(3) *Programming Language Abstraction*: Large-scale distributed e-service systems generally use either .NET or JEE. Both .NET and JEE use a specific level of programming language abstraction, namely, .NET framework and Java Virtual Machine (JVM). In the case of JVM, the portability level is high; however, it is achieved at the cost of a negative impact on the system performance.

1.2.4 Virtual Machine Migration

Virtualized resources include computing power and data storage. Load balancing and power conservation can be resolved by VM migration. VM migration is the process of transferring a VM from an overloaded or under-loaded Physical Machine (PM) to another PM to balance the load or to conserve the consumption of resources.

Based on the migration process, VM migration is classified into the following categories:

(1) *Non-Live Migration*: Non-live migration, also called cool migration, is the migration of a powered off VM from one PM to another PM. The drawbacks of this method are the loss of VM status and the interruption of service to the user.

(2) *Live Migration*: Live migration, also called hot migration, is the process of transferring a running VM from one PM to another without disconnecting the system. Storage, network connectivity, and memory of the VM are transferred from the source machine to the destination machine. In live migration, while the VM is running, data cannot be lost during migration. Local disks are not required to hold VM images, rather, network attached storage is needed to act as a hard drive for the VMs and is accessed by PMs. Total migrating time is less than in non-live migration and down time is seamless.

Based on the migration target, VM migration is classified into the following categories:

(1) *Migrating to Another PM*: When deploying or creating a VM, each VM "attaches" to a PM, but as the number of VMs increases, sometimes, the load of some PMs exceeds their performance, or when the unified planning or adjustment of the VM is needed, the VM may need to be migrated between different PMs.

(2) *Migrating to Other Storage on the Same PM*: When there are multiple storage sites on the same PM, some storage space may not meet the operating conditions

of the VM, or the performance of the physical disk or storage server to which the storage space belongs is limited. In this case, we can migrate the VM to another storage space on the same PM.

1.3 Queueing Models and Performance Analyses

Performance analyses based on mathematical models are needed to performance prediction of design of WCNs and performance evaluation of given WCNs.

In this section, we introduce some general terms for some queueing models and analysis methods relating to the general stochastic processes, as well as basic methods of performance optimization used in this book.

1.3.1 Basic and Vacation Queueing Models

It is evident that much more detail review of queueing models and analysis methods needs to be provided before we can analyze successfully queueing systems in the context of WCNs. Accordingly, in this subsection, we first address elementary queue. And then we introduce the stochastic decomposition property for both single-server vacation and multiple-server vacation queueing models.

1.3.1.1 Basic Queueing Model

Queueing theory has its root in the seminal work of A. K. Erlang, who worked at Copenhagen Telephone company and studied telephone traffic in the early 20th century. The basic queueing theory is in an effort to make resource management and performance analysis of WCNs.

Here, we first briefly summarize the arrival and service used in this book are given as follows:

(1) *Arrival Process*: The stochastic description of customer arrivals is where customers might have any abstract or physical meaning depending on the considered system. In general, we assume that the interarrival times are independent of each other and have a common distribution.
(2) *Service Process*: The stochastic description of customer service is just like customer arrivals, where customer service might depend on the considered system's properties. In basic queueing models, the service times are independent and identically distributed (i.i.d.) random variables.

1.3.1.2 Performance Measures and Common Definitions

When we analyze a queueing system, we need to obtain the values of certain system properties [Ross95]. Additionally, the optimal operation of queueing systems can be determined by analyzing several performance parameters [Whit78]. Some important performance measures and common definitions in queueing theory as they are used to evaluate the QoS of WCNs in this book are summarized as follows.

(1) *Number of Customers in the System*: Let N be the random variable indicating the number of customers in the system in the steady state. The probability p_n that the number of customers present in the system is n in the steady state is given by

$$p_n = \Pr\{N = n\}, \tag{1.1}$$

and the average number $E[N]$ in the system at the steady state is given by

$$E[N] = \sum_{n=0}^{\infty} n p_n. \tag{1.2}$$

In this book, we denote the average number of data packets, the average number of SU packets, and the average number of tasks in the system as $E[N_d]$, $E[N_s]$ and $E[N_t]$, respectively. These average values are used to analyze the response time of the data packets as well as other performance measures that are defined in Item (3) below.

(2) *Queue Length*: The queue length is defined as the number of the customers queueing in the system queue. In this book, to analyze the waiting time spent in the system buffer by the data packets, as defined in Item (4) below, we denote the average number of data packets, the average number of tasks, and the average number of anonymous users queueing in the system buffer, as $E[L_d]$, $E[L_t]$ and $E[L_a]$, respectively.

(3) *Sojourn Time*: The sojourn time, also called response time or system time, is the total time that a customer spends in the system, namely, the time from when the customer arrives at the system buffer until that customer's service completion. The sojourn time is therefore the waiting time *plus* the service time. In this book, the average response time of data packets, the average latency of SU packets, the average latency of tasks, and the average latency of anonymous users are denoted by $E[Y_d]$ in Part I, $E[Y_s]$ in Part II, $E[Y_t]$ and $E[Y_a]$ in Part III, for each applicable network system. By applying Little's law, the average sojourn time can be obtained using the average number $E[N]$ defined in Item (1) above.

(4) *Waiting Time*: The waiting time, also called the queueing time, is the time that a customer spends waiting in the system buffer, that being from the instant a customer arrives at the system buffer to the instant that customer begins receiving service from the system server. The average waiting time of the data

packets in the system buffer is denoted by $E[W]$ used in Parts I and II in this book. By applying Little's law, $E[W]$ can be obtained using the average number defined in Item (2) above.

(5) *Utilization*: In a queueing system with a single server, the utilization U is defined as the fraction of time that the server is busy. If the rate at which customers arrive at the system is λ, and the rate at which a customer is served is μ, then the utilization is calculated as λ/μ. In many queueing systems with a single server, the traffic load is defined as $\rho = \lambda/\mu$. Consequently, the traffic load is identified with the utilization. The system utilization in Part I and the channel utilization in Part II are denoted by U_s and U_c, respectively, in this book.

(6) *Throughput*: The throughput θ is the average number of data packets successfully transmitted per unit time on a network, device, port, or virtual circuit. In this book, the throughput θ is used to represent an SU packet's successful transmissions per unit time as per slot or per second in CRNs of Part II.

(7) *Blocking Rate*: The blocking rate, also called the loss rate, is the probability that an arriving customer will not be able to enter the system because the system is full or the number of customers in the queue buffer has reached a pre-determined threshold. In this book, the blocking rate of data packets in Part I, the blocking rate of SU packets and the blocking rate of PU packets in Part II are denoted by B_d, B_s and B_p, respectively.

(8) *Energy Efficiency*: The energy efficiency is considered a major factor when evaluating the system performance of a resource management strategy in WCNs. In this book, the energy efficiency is defined as the energy saving rate of the system presented in all Parts I, II, III, the energy saving degree of the system presented in Part II, and the energy saving level of the system presented in Part III, denoted as γ, γ_d and γ_l, respectively.

(9) *Handover Rate*: In this book, the handover rate denoted by ζ_h is defined as the number of handovers when the system changing from a sleep state to an awake state in a unit time as a slot or a second. ζ_h is a performance measure for evaluating the additional energy consumption caused by the sleep mode in Part I.

(10) *Switching Rate*: The switching rate denoted by ζ_s is defined as the average number of switches between different spectrums or the average number of switches where SU packets switch from the channels to the buffer in a unit time as a slot or a second in Part II of this book.

(11) *Interruption Rate*: The interruption rate is defined as the number of users that are interrupted by other users in a unit time as a slot or a second. In part II of this book, we define the interruption rate of SU packets, denoted by β_s, and we define the interruption rate of PU packets, denoted as β_p.

1.3.1.3 Vacation Queueing Model

In many practical applications of WCNs, servers may become unavailable for occasional periods of time while working on some other jobs, or just taking a break to reduce energy consumption. In order to describe these types of applications, a vacation queueing model is used to evaluate the system performance of the resource management in WCNs. Based on the number of servers, vacation queueing model is classified as the following categories:

(1) *Single-Server Vacation Queueing Model*: For a classical single-server queueing system that has reached the steady state, we denote the number of customers in the system, the queue length and the waiting time as N_0, L_0 and W_0, respectively, and denote the same performance measures as N_v, L_v and W_v, respectively, for the corresponding steady-state vacation system.

Let $X(z)$ and $X^*(s)$ be the probability generating function (PGF) and the Laplace-Stieltjes Transform (LST), respectively, of the stationary random variable X. Using these notations, the stochastic decomposition can be written as follows:

$$N_v = N_0 + N_d, \quad N_v(z) = N_0(z) \times N_d(z), \tag{1.3}$$

$$L_v = L_0 + L_d, \quad L_v(z) = L_0(z) \times L_d(z), \tag{1.4}$$

$$W_v = W_0 + W_d, \quad W_v^*(s) = W_0^*(s) \times W_d^*(s) \tag{1.5}$$

where N_d, L_d and W_d are the additional number of customers in the system, the additional queue length and the additional waiting time, respectively, introduced in single-server vacation queueing model.

(2) *Multiple-Server Vacation Queueing Model*: To expand the applications of vacation models, multiple-server queues with vacations were also studied after numerous achievements in single server vacation queueing model. However, it seems extremely difficult to establish the unconditional stochastic decomposition properties in multiple-server queueing models. When all servers in a multiple-server queueing model are busy, the conditional stochastic decomposition properties can be obtained [Tian99].

When we consider a classical multiple-server queueing model with c servers, we define $L_v^{(c)}$ to be the number of customers waiting in the buffer and $W_v^{(c)}$ to be the waiting time of a customer. Then when given that all the servers in the system are busy, we can have that

$$L_v^{(c)} = \{N_v - c | J = c\}, \tag{1.6}$$

$$W_v^{(c)} = \{W_v | N_v \geq c, J = c\} \tag{1.7}$$

where J is the number of busy servers, N_v is the number of customers in the model, and W_v is the waiting time of a customer.

Next, let $L_0^{(c)}$ be the same random variable as $L_v^{(c)}$ for the corresponding queueing model without vacations, and $W_0^{(c)}$ be the same random variable as $W_v^{(c)}$ for the corresponding queueing model without vacations. Then the conditional stochastic decomposition properties are given as follows:

$$L_v^{(c)} = L_0^{(c)} + L_d^{(c)}, \quad L_v^{(c)}(z) = L_0^{(c)}(z) \times L_d^{(c)}(z), \tag{1.8}$$

$$W_v^{(c)} = W_0^{(c)} + W_d^{(c)}, \quad W_v^{(c)*}(s) = W_0^{(c)*}(s) \times W_d^*(s) \tag{1.9}$$

where $L_d^{(c)}$ and $W_d^{(c)}$ are the additional queue length and additional waiting time due to multiple-server vacation, respectively.

1.3.2   Queueing Model with Multiple-Class Customers

In some applications, customers are grouped into different classes based on the necessary service time. For example, in an Internet email system, emails with attachments and those without attachments can be regarded as two different classes, since the mail with attachments requires more processing time than that without attachments. Considering that all the customers are served by the same servers, we can build queueing models with multiple-class customers.

For the queueing model with K ($K < \infty$) classes of customers, we assume that there are K sources generating customers, where one source generating each class customer. That is, each class of customer corresponds to one arrival process and one service time. The customer arrivals of kth ($1 < k < K$) class are assumed to follow a Poisson process with parameter λ_k ($\lambda_k > 0$), and the service time of the kth customer is assumed to follow an exponential distribution with parameter μ_k ($\mu_k > 0$). Moreover, we assume the random variables for different classes of customers are independent of each other.

Based on the Poisson' property, the arrivals, including all classes of customers, at the server follow a Poisson process with parameter $\lambda = \sum_{k=1}^{K} \lambda_k$. The arriving customer is the kth customer with probability λ_k/λ, the server is occupied by the kth customer with probability $\rho_k = \lambda_k/\mu_k$.

From the perspective of the system, the service time of a customer is no longer an exponential distribution, but a hyper-exponential distribution. The Probability Density Function (PDF) $h(x)$ of the service time x is given as follows:

$$h(x) = \begin{cases} \sum_{k=1}^{K} \alpha_k \mu_k e^{\mu_k x}, & x \geq 0 \\ 0, & x < 0. \end{cases} \tag{1.10}$$

We define the state space Ω of the queueing model with multiple-class customers as follows:

$$\Omega = \{(n_1, n_2, n_3, \ldots, n_k, \ldots, n_K) : n_k \geq 0, 1 \leq k \leq K\} \tag{1.11}$$

where n_k is the number of customers in class k and K is the number of customers' classes in the queueing models under consideration.

The sufficient and necessary condition for this queueing model to be positive recurrent is $\sum_{k=1}^{K} \rho_k = < 1$ [Rubi87]. We denote $p(\boldsymbol{n})$ as the probability of the model being at state $\boldsymbol{n} = (n_1, n_2, n_3, \ldots, n_K)$.

Consider the queueing model M/M/1 with multiple-class customers, we denote $q(n, k)$ as the probability that the number of customers in the model is n, and the customer being served is of class k. We have

$$q(n, k) = 1 - \sum_{k=1}^{K} \rho_k \sum_{i=1}^{K} g_{ki} \omega_i^{-n}. \tag{1.12}$$

$\{\omega_1, \omega_2, \omega_3, \ldots, \omega_K\}$ in Eq. (1.12) are K roots (arranged in order from small to large) of the following equation:

$$\sum_{k=1}^{K} \frac{\alpha_k}{\psi_k - x} = 1, \quad 1 \leq k \leq K \tag{1.13}$$

where

$$\psi_k = \frac{\lambda + \mu_k}{\lambda}.$$

g_{ki} in Eq. (1.12) satisfies the following equation:

$$g_{ki} = \frac{\alpha_k c_i}{\psi_k - \omega_i}, \quad 1 \leq k \leq K \tag{1.14}$$

where

$$c_i = \frac{\displaystyle\prod_{j=1}^{K} (\psi_j - \omega_i)}{\displaystyle\prod_{j=1, j \neq i}^{K} (\omega_j - \omega_i)}.$$

For $1 \leq i, j \leq K$ and $\mu_i \neq \mu_j$, we have

$$
\begin{cases}
p(\boldsymbol{n}) = \dfrac{(n_1 + n_2 + n_3 + \cdots + n_K)!}{n_1! n_2! n_3! \cdots n_K!} \alpha^{\boldsymbol{n}} \dfrac{1}{\displaystyle\sum_{k=1}^{K} n_k} \sum_{k=1}^{K} \dfrac{n_k}{\alpha_k} q \left(\sum_{k=1}^{K} n_k, k \right), \\
\qquad\qquad\qquad\qquad\qquad \boldsymbol{n} \neq 0 \\
p(0) = 1 - \displaystyle\sum_{k=1}^{K} \rho_k
\end{cases}
$$

$$(1.15)$$

where $\boldsymbol{\alpha} = (\alpha_1, \alpha_2, \alpha_3, \ldots, \alpha_K)$ and $\boldsymbol{\alpha^n} = \alpha_1^{n_1} \times \alpha_2^{n_2} \times \alpha_3^{n_3} \times \cdots \times \alpha_K^{n_K}$.

In the queueing model with multiple-class customers mentioned above, the service time of a customer with a different class is not the same, but the queueing discipline is First-Come First-Served (FCFS). However, in some practical applications, where different classes of customers have different service priorities, customers with a high priority are served before those with a low priority.

In a non-preemptive queue, customers are served continuously until its completion. The server will provide service to a low priority customer if no customer with a higher priority is present. Once a customer has been selected by the server, the service for this customer will be continued even if new customers with a higher priority arrive at the model during the service period.

In a preemptive queueing model, when a customer with a high priority arrives at the system, and finds a customer with a lower priority is being served, the service of the customer with a lower priority will be interrupted, and the preempted customer will be inserted into the queue. When a previously preempted customer returns to service, it is possible for the service to continue from the point at which the preemption occurred. Some previously completed work may have been lost. Moreover, in some cases it may be necessary to begin the service all over again.

1.3.3 Matrix-Geometric Solution Method

In [Neut81a, Neut81b], the author systematically developed the structural matrix analysis method, which made the random model analysis develop from the exponential distribution as the core to a new stage that widely used PHase type (PH) distribution. Classical method for solving the steady-state distribution of classical birth-death process is developed into a matrix-geometric solution method. The matrix-geometric solution method is widely used in modeling the resource management in advanced WCNs.

1.3.3.1   Birth-Death Process

If the transition rate matrix of a Continuous-Time Markov Chain (CTMC) $\{X(t),\ t \geq 0\}$ has the following tridiagonal form:

$$
Q = \begin{pmatrix}
-\lambda_0 & \lambda_0 \\
\mu_1 & -(\lambda_1 + \mu_1) & \lambda_1 \\
& \mu_2 & -(\lambda_2 + \mu_2) & \lambda_2 \\
& & \ddots & \ddots & \ddots \\
& & \mu_n & -(\lambda_n + \mu_n) & \lambda_n \\
& & & \ddots & \ddots & \ddots
\end{pmatrix},
\tag{1.16}
$$

we call $\{X(t),\ t \geq 0\}$ as a birth-death process with birth rate λ_n, $n \geq 0$ and death rate μ_n, $n \geq 1$.

In most applications, we are more concerned with what a process will look like long after it has run. If $\pi_j = \lim_{t \to \infty} \Pr\{X(t) = j\}$, $j = 0, 1, 2, \ldots$, always exists and $\pi_0 + \pi_1 + \pi_2 + \cdots = 1$, we say this birth-death process is positive recurrent. For this case, the steady-state distribution of this birth-death process is denoted as $\{\pi_j,\ j \geq 0\}$.

Letting $\mathbf{\Pi} = (\pi_0, \pi_1, \pi_2, \ldots)$, we construct the steady-state equation and the normalization equation as follows:

$$
\begin{cases}
\mathbf{\Pi} Q = 0 \\
\mathbf{\Pi} e = 1
\end{cases}
\tag{1.17}
$$

where e is a column vector with infinite elements and all elements of the vector are equal to 1.

The birth-death process $\{X(t), t \geq 0\}$ is positive recurrent if and only if

$$
\sum_{k=1}^{\infty} \frac{\lambda_0 \lambda_1 \lambda_2 \cdots \lambda_{k-1}}{\mu_1 \mu_2 \mu_3 \cdots \mu_k} < \infty,
\tag{1.18}
$$

and the steady-state distribution is given as follows:

$$
\pi_j = \begin{cases}
K, & j = 0 \\
\dfrac{\lambda_0 \lambda_1 \lambda_2 \cdots \lambda_{j-1}}{\mu_1 \mu_2 \mu_3 \cdots \mu_j} K, & j \geq 1
\end{cases}
\tag{1.19}
$$

where

$$K = \left(1 + \sum_{k=1}^{\infty} \frac{\lambda_0 \lambda_1 \lambda_2 \cdots \lambda_{k-1}}{\mu_1 \mu_2 \mu_3 \cdots \mu_k}\right)^{-1}.$$

1.3.3.2 Quasi Birth-Death Process

In order to describe the process evolution with multiple levels, multiple phases and variable parameters, the classical birth-death process is generalized from one-dimensional state space to two-dimensional state space, thus a Quasi Birth-Death (QBD) process is introduced [Rhee97].

For a two-dimensional Markov chain $\{(X(t), J(t)), \, t \geq 0\}$ with state space $\boldsymbol{\Omega} = \{(k, j) : k \geq 0, j = 1, 2, 3, \ldots, m\}$, the state set $\{(k, 1), (k, 2), (k, 3), \ldots, (k, m)\}$ is called level $k, k \geq 0$. Using the lexicographical sequence for the states, if the transition rate matrix $\boldsymbol{Q}$ of a Markov chain $\{(X(t), J(t)), \, t \geq 0\}$ can be written in a block-tridiagonal matrix form as follows:

$$\boldsymbol{Q} = \begin{pmatrix} A_0 & C_0 & & & & \\ B_1 & A_1 & C_1 & & & \\ & B_2 & A_2 & C_2 & & \\ & & \ddots & \ddots & \ddots & \\ & & & B_n & A_n & C_n \\ & & & & \ddots & \ddots & \ddots \end{pmatrix}, \tag{1.20}$$

we call $\{(X(t), J(t)), \, t \geq 0\}$ as a QBD process [Ozaw13].

All the sub-matrices of the transition rate matrix $\boldsymbol{Q}$ have order of $m \times m$, $A_k, k \geq 0$ have negative diagonal elements and nonnegative off-diagonal elements, $B_k, k \geq 1$ and $C_k, k \geq 0$ are all nonnegative matrices satisfying the following equation:

$$(A_0 + C_0)e = (A_k + B_k + C_k)e = \boldsymbol{0}, \quad k \geq 1 \tag{1.21}$$

where e is a column vector with m elements and all elements of the vector are equal to 1.

We note that the solving method for this type of QBD process is completely dependent on the special structure of the transition rate matrix $\boldsymbol{Q}$.

1.3.3.3 Matrix-Geometric Solution Method

In many applications, we have a special case of QBD process where the sub-matrices keep unchanged from a certain level, such as level c. transition rate matrix $\boldsymbol{Q}$ is then written as follows:

$$Q = \begin{pmatrix} A_0 & C_0 & & & & & \\ B_1 & A_1 & C_1 & & & & \\ & \ddots & \ddots & \ddots & & & \\ & & B_{c-1} & A_{c-1} & C_{c-1} & & \\ & & & B & A & C & \\ & & & & B & A & C \\ & & & & & \ddots & \ddots & \ddots \end{pmatrix}. \tag{1.22}$$

Assume that a QBD process with the transition rate matrix Q as in Eq. (1.22) is positive recurrent. We denote the steady-state distribution of the QBD process by

$$\pi_{k,j} = \lim_{t \to \infty} \Pr\{X(t) = k, J(t) = j\}, \quad (k, j) \in \Omega. \tag{1.23}$$

We define π_i as the steady-state probability vector of the system level being equal to i. π_i can be given as follows:

$$\pi_i = (\pi_{i,1}, \pi_{i,2}, \pi_{i,3}, \ldots, \pi_{i,m}), \quad i \geq 0 \tag{1.24}$$

where m is the number of phases for the level being at i.

The steady-state distribution Π of the CTMC is composed of π_i $(i \geq 0)$. Π is given as follows:

$$\Pi = (\pi_0, \pi_1, \pi_2, \ldots). \tag{1.25}$$

The irreducible QBD process is positive recurrent if and only if the following matrix equation:

$$R^2 B + RA + C = 0 \tag{1.26}$$

has the minimum nonnegative solution R, with spectral radius $\mathrm{Sp}(R) < 1$.

It is a difficult to derive the mathematical expression of the rate matrix R in closed form using a higher order matrix equation. In most of the literature, the iterative method has been employed to the numerical results of the rate matrix R. The main steps of the iteration process are listed as follows:

Step 1: Initialize a small constant ε (for example, $\varepsilon = 10^{-6}$) related to calculation accuracy and the rate matrix $R = 0$.

Step 2: Input A, B and C.

Step 3: Calculate R^*.
$$R^* = (R^2 \times B + C) \times (I - A)^{-1}$$
% I is an identity matrix.

Step 4:

> **if** $\{ \ ||R - R^*||_\infty > \varepsilon \}$
> > $R = R^*$
> > $R^* = (R^2 \times B + C) \times (I - A)^{-1}$
> > go to **Step 4**
> **else**
> > $R = R^*$
> **endif**

Step 5: Output R.

Using the rate matrix R obtained above, we construct a set of linear equations as follows:

$$\begin{cases} (\pi_0, \pi_1, \pi_2, \ldots, \pi_c)B[R] = 0 \\ \pi_0 e + \pi_1 e + \pi_2 e + \cdots + \pi_{c-1} e + \pi_c (I - R)^{-1} e = 1 \end{cases} \tag{1.27}$$

where e is a column vector with m elements and all elements of the vector are equal to 1. $B[R]$ is a matrix given by

$$B[R] = \begin{pmatrix} A_0 & C_0 & & & & \\ B_1 & A_1 & C_1 & & & \\ & \ddots & \ddots & \ddots & & \\ & & B_{c-1} & A_{c-1} & C_{c-1} \\ & & & B_c & RB + A \end{pmatrix}. \tag{1.28}$$

Applying the Gauss-Seidel method, we obtain $\pi_0, \pi_1, \pi_2, \ldots, \pi_c$. Furthermore, the steady-state distribution of the QBD can be expressed using a matrix-geometric form as follows:

$$\pi_k = \pi_c R^{k-c}, \quad k \geq c. \tag{1.29}$$

1.3.4   Jacobi Iterative Method

In numerical linear algebra, the Jacobi iterative method, also called the Jacobi method, is an algorithm for determining the solutions of a diagonally dominant system of linear equations [Slei00]. Each diagonal element is solved, and an approximate value is plugged in. The process is then iterated until it converges. This algorithm is a stripped-down version of the Jacobi transformation method of matrix diagonalization. The method is named after Carl Gustav Jacob Jacobi.

Let $Ax = b$ be a square system of n linear equations, where

$$A = \begin{pmatrix} a_{11} & a_{12} & \cdots & a_{1n} \\ a_{21} & a_{22} & \cdots & a_{2n} \\ \vdots & \vdots & \ddots & \vdots \\ a_{n1} & a_{n2} & \cdots & a_{nn} \end{pmatrix}, \quad x = \begin{pmatrix} x_1 \\ x_2 \\ \vdots \\ x_n \end{pmatrix}, \quad b = \begin{pmatrix} b_1 \\ b_2 \\ \vdots \\ b_n \end{pmatrix}.$$

A can be decomposed into a lower triangular matrix L, a diagonal matrix D and an upper triangular matrix U as follows:

$$A = L + D + U \tag{1.30}$$

where

$$L = \begin{pmatrix} 0 & & & & \\ a_{21} & 0 & & & \\ a_{31} & a_{32} & 0 & & \\ \vdots & \vdots & \vdots & \ddots & \\ a_{n1} & a_{n2} & a_{n3} & \cdots & 0 \end{pmatrix}, \quad D = \begin{pmatrix} a_{11} & & & \\ & a_{22} & & \\ & & \ddots & \\ & & & a_{nn} \end{pmatrix}, \quad U = \begin{pmatrix} 0 & a_{11} & a_{12} & \cdots & a_{1n} \\ & 0 & a_{23} & \cdots & a_{2n} \\ & & 0 & \cdots & a_{3n} \\ & & & \ddots & \vdots \\ & & & & 0 \end{pmatrix}.$$

The solution is then obtained iteratively via the following equation:

$$x^{(k+1)} = -D^{-1}(L+U)x^{(k)} + D^{-1}b \tag{1.31}$$

where $x^{(k)}$ is the kth approximation or iteration of x, $x^{(k+1)}$ is the next or $k+1$ iteration of x.

The element-based formula is given as follows:

$$x_i^{(k+1)} = \frac{1}{a_{ii}}\left(b_i - \sum_{j=1,\ j\neq i}^{n} a_{ij}x_j^{(k)} \right), \quad i = 1, 2, 3, \ldots, n. \tag{1.32}$$

The standard convergence condition for the Jacobi iterative is that the spectral radius of the iterative matrix is less than 1, given by

$$\text{Sp}(D^{-1}(D-A)) < 1. \tag{1.33}$$

1.3.5   Gauss-Seidel Method

In numerical linear algebra, the Gauss-Seidel method, also known as the Liebmann method or the method of successive displacement, is an iterative method used to solve a linear system of equations [Gree97, Usui94]. It is named after the German mathematicians Carl Friedrich Gauss and Philipp Ludwig von Seidel.

Similar to the Jacobi iteration method, the Gauss-Seidel method can be applied to any matrix with non-zero elements on the diagonals. Convergence is guaranteed if the matrix is either diagonally dominant, or symmetrical and positive definite. It was only mentioned in a private letter from Gauss to his student Gerling in 1823. A publication was not delivered before 1874 by Seidel.

The Gauss-Seidel method is an iterative technique for solving a square system of n linear equations with unknown $x = (x_1, x_2, x_3, \ldots, x_n)$ as follows:

$$a_{i1}x_1 + a_{i2}x_2 + a_{i3}x_3 + \cdots + a_{in}x_n = b_i, \quad i = 1, 2, 3, \ldots, n. \tag{1.34}$$

Gauss-Seidel iteration formula is given by

$$x_i^{(k+1)} = \frac{1}{a_{ii}} \left(b_i - \sum_{j=1}^{i-1} a_{ij}x_j^{(k+1)} - \sum_{j=i+1}^{n} a_{ij}x_j^{(k)} \right),$$

$$i = 1, 2, 3, \ldots, n, \ k = 0, 1, 2, \ldots \tag{1.35}$$

where $a_{ii} \neq 0 \ (i = 1, 2, 3, \ldots, n)$.

In many cases, it converges faster than the simple iterative method.

It is different to the simple iteration method in that the computation of $x_i^{(k+1)}$ uses the value of $x_1^{(k+1)}, x_2^{(k+1)}, x_3^{(k+1)}, \ldots, x_{i-1}^{(k+1)}$ just iterated. The Gauss-Seidel method must converge when the coefficient matrix A is strictly diagonally dominant or symmetrically positive definite.

1.3.6 Performance Optimization

In classical queueing theory, optimal design models may be classified according to arrival rate, service rate, inter-arrival time and service time distributions and queueing discipline. Performance optimization of WCNs considered in this book include the optimization of the system parameters by trading off different performance measures as well as Nash equilibrium and social optimization.

1.3.6.1 Optimization of System Parameters

We know that with an increase in the system investment, users will be more satisfied with the QoS. However, the system utilization will be reduced. This means that when designing or operating a WCN, there is a trade-off among different performance measures. In this book, we consider to put these performance measures together into an objective function to quantify their trade-offs, as a system cost function or a system profit function, to minimize the system cost or maximize the system profit.

For example, in a BWA network, a lower average response time of data packets can be obtained with a greater sleep parameter, whereas a higher energy saving rate of the system can be obtained with a smaller sleep parameter. In the actual application, both the response performance and the energy saving effect should be taken into consideration. For this, we establish a simple model with linear cost as follows:

$$F(\delta) = f_1 \times E[Y_d] + f_2 \times \frac{1}{\gamma} \tag{1.36}$$

where δ ($\delta > 0$) is the sleep parameter, f_1 is the factor of the average response time $E[Y_d]$ of data packets to the cost, and f_2 is the factor of the energy saving rate γ of the system to the system cost.

By minimizing the system cost function $F(\delta)$, the optimal sleep parameter δ^* is given as follows:

$$\delta^* = \underset{\delta \geq 0}{\operatorname{argmin}} \{F(\delta)\} \tag{1.37}$$

where "argmin" stands for the argument of the minimum.

1.3.6.2 Nash Equilibrium and Social Optimization

In some BWA networks, each data packet's optimal behavior is affected by acts taken by the network managers and by the other data packets. The result is an aggregate "equilibrium" pattern of behavior which may not be optimal from the point of view of society as a whole [Hass03, Jin17c].

Considering a non-cooperative game between data packets who want to receive a service benefit, we discuss the Nash equilibrium and social optimization in BWA networks of Part I. For this, we give the following hypothesis:

(1) Before getting service, a data packet has no information on the system state of the BWA networks.
(2) The reward for a submitted data packet is R_g.
(3) The time cost for a data packet staying in the BWA networks is C_g per unit time as per slot or per second.
(4) A decision to join the system is irrevocable, and reneging is not allowed.

We calculate the individual benefit function $G_{ind}(\lambda)$ as follows:

$$G_{ind}(\lambda) = R_g - C_g \times E[Y_d] \tag{1.38}$$

where λ is the arrival rate of data packets, $E[Y_d]$ is the average response time of data packets.

By setting the lowest arrival rate λ_{min} of data packets and the highest arrival rate λ_{max} of data packets, we discuss the Nash equilibrium behavior of data packets within the closed interval $[\lambda_{min}, \lambda_{max}]$ as follows:

(1) If $G_{ind}(\lambda_{min}) \leq 0$, the value of the individual benefit function $G_{ind}(\lambda)$ is negative. Obviously, not to queue is a domain strategy for a user.
(2) If $G_{ind}(\lambda_{max}) \geq 0$, the value of the individual benefit function $G_{ind}(\lambda)$ is positive. Therefore, to queue is a domain strategy for a user.
(3) For the case of $G_{ind}(\lambda_{min}) > 0$ and $G_{ind}(\lambda_{max}) < 0$, if $\lambda = \lambda_{max}$, a data packet cannot get a positive benefit by queueing. A unique arrival rate λ^e exists for data packets subject to $G_{ind}(\lambda^e) = 0$. We call that the arrival rate λ^e of data packets with zero benefit the Nash equilibrium arrival rate of data packets.

The social benefit is defined as the total benefit to all the data packets and the system. If no admission fees are imposed, the social benefit is the sum of the individual benefits of all data packets. The social benefit function $G_{soc}(\lambda)$ is given as follows:

$$G_{soc}(\lambda) = \lambda(R_g - C_g \times E[Y_d]). \tag{1.39}$$

The socially optimal arrival rate λ^* of data packets is then given as follows:

$$\lambda^* = \operatorname*{argmax}_{\lambda_{min} \leq \lambda \leq \lambda_{max}} \{G_{soc}(\lambda)\} \tag{1.40}$$

where "argmax" represents the argument of the maximum.

In general, Nash equilibrium and socially optimal arrival rates of data packets are not consistent. When the Nash equilibrium arrival rate of data packets is lower, incentives should be introduced to encourage more data packets to access the system. Otherwise, higher admission fees should be imposed to data packets, then the socially optimal arrival rate will define an equilibrium.

1.4 Organization of This Book

We organize the remainder of this book in three parts.

Part I discusses the sleep mode in BWA networks, including WiMAX and LTE. Under the sleep mode operation, a MS operates two modes: the awake mode and the sleep mode. Revolving around this standard, we investigate some vacation queueing models with two types of busy periods: busy period in the listening state and busy period in the awake state, with a sleep-delay period, with a wake-up procedure, and with batch arrivals.

There are seven chapters in Part I, beginning with Chap. 2.

Chapter 2 establishes a multiple-vacation Geom/G/1 queueing model with a close-down period to analyze the sleep mode for power saving class type I. The operation mechanism of the sleep mode for downlink traffic in type I is based on IEEE 802.16e standard. In this system model, after a MS receives a MOB-SLP-RSP message, the MS switches from an awake mode to a sleep mode, and sleeps during the initial-sleep window. If no traffic arrives, the MS will double the sleep window. If a sleep window size reaches its maximum, the sleep window size will be not doubled, but fixed. The setting of the close-down time reduces the response time, because a packet arriving during the close-down time can be transmitted directly without going through a sleep window.

In Chap. 3, considering the attractive feature that some data packets can be transmitted during the listening state, we present a queueing model with two types of busy periods: busy period in the listening state and busy period in the awake state, to capture the sleep mode for power saving class type II in IEEE 802.12e. In this system model, the time lengths of the sleep window and the listening window in power saving class type II are fixed, and a certain number of data packets can be transmitted during the listening state. The listening state in power saving class type II must be seen as a special period, as some parts where a transmission occurs are regarded as busy periods, and others parts where no transmission occurs are seen as vacation periods.

In Chap. 4, by taking into account the self-similar nature of massive multimedia traffic, we build a batch arrival multiple-vacation queue with a vacation-delay to capture the sleep mode in IEEE 802.12e. The operation mechanism of the sleep mode is based on the enhanced power saving class type III with sleep-delay in IEEE 802.16e standard. In the sleep mode with sleep-delay for power saving class type III, when there are no data packets to be sent to the MS in the buffer of the serving BS, a timer called the sleep-delay timer will be trigged and a sleep-delay period will begin. If there is a packet arrival in the serving BS within the time length of the sleep-delay timer, the system will return to the awake state immediately without going through the sleep state.

Chapter 5 presents an analytical approach to evaluate the Bernoulli arrival-based sleep mode in IEEE 802.16m by constructing a queueing model with heterogeneous multiple vacations and regarding the initial sleep window as one half of the subsequent sleep window. If there is no pending data packet, the MS will receive a negative Traffic Indication from the BS, and then the MS will enter a sleep window and power down immediately to save energy. If there are pending data packets in the buffer for the MS, the MS will receive a positive TRF-IND message from the BS, and the MS will be able to receive and send data packets as long as there are data packets in the buffer. For the "real-time traffic-only" and "real-time and BE-traffic mixed" scenarios, the final sleep cycle is equal to the initial sleep cycle, and the time length of the sleep cycle is fixed. A listening period without any data packet transmission and its subsequent sleep window is regarded as a vacation period. An extended listening period with data packet transmission is regarded as a busy period. A queueing model with heterogeneous multiple vacations is constructed, and by using a discrete-time embedded Markov chain, we derive the system model in the steady state and investigate the trade-off between the average response time of data packets and the energy saving rate of the system.

Chapter 6 establishes a Discrete-Time Markovian Arrival Process (D-MAP) based queueing model with multiple vacations to evaluate the sleep mode of IEEE 802.16m in a scenario where the real-time traffic includes a mixture of the real-time traffic and the BE traffic. Considering the correlation of the data packets shown in real-time traffic with multimedia applications, the arrival of data packets is assumed to follow a D-MAP. The steady-state distribution for the queueing model is derived by using an embedded Markov chain method. For the performance measures, the average response time of data packets, the energy saving rate of the system and the standard deviation of the number of data packets are given to evaluate the system performance with different correlation parameters.

In Chap. 7, by introducing a sleep-delay timer, we propose enhanced energy saving strategy based on the Active DRX mechanism in an LTE system to improve the sleep strategy for a better balance between response performance and energy efficiency. By regarding the sleep-delay period as a vacation-delay period, the short DRX stage as the short vacation period, the long DRX stage as the long vacation period, and the wake-up procedure as a set-up period, we establish a two-stage multiple-vacation queueing model with a vacation-delay period and a set-up period. By using an embedded Markov chain method, we present an exact analysis of the system model. We then derive performance measures of the system in terms of the handover rate, the energy saving rate of the system and the average response time of data packets to evaluate the sleep strategy of the Active DRX mechanism in LTE system. Based on the trade-off between different performance measures when setting the threshold of the short DRX stages and the time length of the sleep-delay timer, we construct a system cost function to optimize the sleep strategy under consideration.

Chapter 8 introduces a sleep-delay strategy to an LTE system and proposes an enhanced Active DRX mechanism influencing the control of the downlink transmission at the UE. The network system consists of a finite number of logical channels having the awake period, the sleep-delay period and the wake-up period for the data packets' transmissions. Chapter 8 models the system with the enhanced Active DRX mechanism as a synchronous multiple-vacation queueing system with a vacation-delay period and a set-up period. Accounting for the number of data packets in the system, the system period, and the sequence number of the current slot, we construct a three-dimensional Markov chain to evaluate the performance measures, such as the energy saving rate of the system, the average response time of data packets and the handover rate. By constructing a system profit function, we optimize the enhanced Active DRX mechanism in LTE system.

We summarize the system models established in each chapter of Part I in Table 1.1 with regard to what types of the queueing model are applied, what performance measures are evaluated, and what system parameters are optimized.

In Table 1.1, we use some abbreviated signs as follows: "Chap." for "Chapter", "P.S.C.T." for "Power Saving Class Type", "S.M." for "Sleep Mode", "Sleep Delay" for "S.D.", "M.V." for "Multiple-Vacation", "B.P.A.S." for "Busy Period in Awake

Table 1.1 System models in Part I

	Strategies	Model types	Performance measures				
			E.S.R.	A.R.T.	S.U.	H.R.	B.R.
Chap. 2	P.S.C.T.-I in IEEE 802.16e	Closed-down time and M.V.	○	○	×	×	×
Chap. 3	P.S.C.T.-II in IEEE 802.16e	Queue with B.P.A.S. and B.P.L.S.	○	○	×	○	×
Chap. 4	P.S.C.T.-III in IEEE 802.16e	Batch arrival queue with M.V. and V.D.	○	○	○	○	×
Chap. 5	S.M. in IEEE 802.16m	Heterogeneous M.V.	○	○	×	×	×
Chap. 6	S.M. in IEEE 802.16e	D-MAP heterogeneous M.V.	○	○	×	×	×
Chap. 7	S.M. in IEEE 802.16m	Two stages of vacations, V.D. and set-up	○	○	×	○	×
Chap. 8	S.M. in IEEE 802.16m	M.S. queue with M.V., V.D. and set-up	○	○	×	×	○

State", "B.P.L.S." for "Busy Period in Listening State", "V.D." for "Vacation-Delay", "M.S." for "Multiple-Server", "E.S.R." for "Energy Saving Rate", "A.R.T." for "Average Response Time", "S.U." for "System Utilization", "H.R." for "Handover Rate", and "B.R." for "Blocking Rate".

Part II discusses the dynamic spectrum allocation and energy saving strategy in CRNs. We present an analytic framework to evaluate the system performance by constructing priority queueing models with possible service interruptions, using multiple channels, with several types of vacation mechanisms, and possible transmission interruptions.

There are seven chapters in Part II, beginning with Chap. 9.

Chapter 9 proposes a channel aggregation strategy in which all the channels in a spectrum are aggregated as one channel for the transmission of a PU packet, while each SU packet occupies only one of the channels in the spectrum for its transmission. Considering the stochastic behavior of SU packets with the proposed strategy, we build a discrete-time preemptive retrial queueing model with multiple channels, a retrial buffer and synchronous transmission interruptions. Taking into account the number of PU packets and the number of SU packets in the system, we construct a two-dimensional Markov chain. We then evaluate the system performance for the channel aggregation strategy, and validate the model analysis with numerical results. From an economic perspective, we establish a system cost function to balance different performance measures and optimize the channel aggregation intensity. One approach that would oblige the SU packets to adopt the socially optimal arrival rate is to charge a fee to the SU packets for joining the system. In discussing the Nash equilibrium and socially optimal behaviors for SU packets, we present a pricing policy to regulate the arrival rate of SU packets and maximize the value of the social benefit function.

Chapter 10 proposes an adaptive control approach to determine the reservation ratio of the licensed spectrum for SUs and presents an adaptive spectrum reservation strategy to better adapt to systemic load changes in CRNs. In such a strategy, the licensed spectrum is separated into two logical channels, namely, the reserved channel and the shared channel, respectively. Combining the total number of SUs in the system and on the reserved channel, respectively, and the state of the shared channel, we construct a three-dimensional Discrete-Time Markov Chain (DTMC) model to record the stochastic behavior of PUs and SUs. By using a method similar to that of the matrix-geometric solution method, we obtain the steady-state distribution of the system model and derive the formulas for some required performance measures for two types of packets, the PU packets and the SU packets. We present numerical results to evaluate the influence of the adaptive control factor and the admission threshold on the system performance. By using a Teaching-Learning-Based Optimization (TLBO) based intelligent searching algorithm, we optimize the adaptive control factor and the admission threshold with a global minimum system cost.

Chapter 11 establishes a priority queueing model in which two types of packets, the PU packets and the SU packets, may interfere with each other. In this priority queueing model, we take into account the impatient behavior of the interrupted SU packets, the tolerance delay of an SU packet, the sensing errors of SUs and the preemptive priority of PU packets. By using the matrix-geometric solution method, we derive some important performance measures of the system and numerically evaluate the proposed mechanism. Considering the reward for an SU packet to be transmitted successfully and the cost of an SU packet staying in the system, we build a reward function and investigate the behaviors of SU packets for both the Nash equilibrium and the social optimization. Numerical results show that the equilibrium arrival rate is greater than the socially optimal arrival rate. Accordingly, we provide a pricing policy for SU packets to coordinate these two behaviors.

Chapter 12 presents a mini-slotted spectrum allocation strategy with the purpose of improving the normal throughput of SU packets and reducing the spectrum switching frequency in CRNs. Due to the mistake detections in practice, the PU packet and the SU packet will occupy the spectrum simultaneously, namely, a collision will occur on the spectrum. For this case, we establish a heterogeneous discrete-time queueing model with possible collisions to model the system operation. By using a matrix-geometric solution method, we derive performance measures of the system in terms of the disruption rate of PU packets, the throughput of SU packets, the switching rate of SU packets and the average latency of SU packets. In comparison to the conventional spectrum allocation strategy with a homogeneous structure, the throughput of SU packets and the switching rate of SU packets is improved when using the mini-slotted spectrum allocation strategy proposed in this chapter. However, the response performance of SU packets degrades to some extent. That is to say that there is a trade-off when setting the slot size in the mini-slotted spectrum allocation strategy. Based on this observation, we construct a net benefit function and optimize the slot size for the proposed strategy.

Chapter 13 establishes a two-dimensional CTMC model to record the stochastic behavior of two types of user packets, the PU packets and the SU packets, with a channel reservation strategy. In this channel reservation strategy, part licensed channels are reserved for SU packets for the purpose of properly controlling the interference between the PU packets and the SU packets. In order to make full use of the reserved licensed channels and enhance the QoS of SU packets, we introduce an admission threshold. If the number of SU packets aggregated in the buffer is greater than the admission threshold, all the licensed channels can be used opportunistically by SU packets, otherwise, only the reserved licensed channels can be used opportunistically by SU packets. In order to obtain numerical solutions for the QBD process, we present a new algorithm for solving the Markov chain model that effectively fuses the Teaching-Learning-Based Optimization (TLBO) algorithm and the Successive Over Relaxation (SOR) method, namely TLBO-SOR algorithm. Based on the energy detection method, we mathematically evaluate the system performance in terms of the throughput of SU packets, the average latency of SU packets, the switching rate of SU packets and the channel utilization in relation to the energy detection threshold and the number of reserved channels.

Chapter 14 proposes an energy saving strategy using a single-sleep mode in CRNs with the aim of alleviating the spectrum scarcity crisis and reducing the energy consumption. By establishing a preemptive priority queueing model with a single-vacation to capture the stochastic behavior of the proposed strategy, and by using the matrix-geometric solution method, we derive performance measures of the system in terms of the average latency of SU packets and the energy saving degree. Moreover, by using a searching algorithm based on gravitation, we investigate the Nash equilibrium and socially optimal behaviors of SU packets. We also present a pricing policy for SU packets, obliging them to adopt the socially optimal arrival rate.

Chapter 15 establishes a preemptive priority queueing model with multiple vacations to capture the stochastic behavior of user packets, the PU packets and the SU packets, and presents analyses to numerically evaluate the energy saving strategy using a multiple-sleep mode in CRNs. By using the matrix-geometric solution method, we obtain the steady-state distribution of the system model and derive performance measures of the system in terms of the throughput of SU packets, the average latency of SU packets, the energy saving rate of the system and the channel utilization. We construct a system cost function and develop an improved Jaya algorithm employing an insect-population model to optimize the energy saving strategy proposed in this chapter. We also show the optimal combination and global minimum of the system cost using numerical results.

We summarize the system models established in each chapter of Part II in Table 1.2 with regard to what types of queueing model are applied, what performance measures are evaluated, and what types of optimization are presented.

In Table 1.2, we use some abbreviated signs as follows: "Chap." for "Chapter", "S.M." for "Sleep Mode", "P.Q." for "Priority Queue", "S.V." for "Single Vacation", "M.V." for "Multiple-Vacation", "E.S.R." for "Energy Saving Rate", "A.L." for "Average Latency", "U." for "Utilization", "Thr." for "Throughout" and "S.W." for "Switching Rate".

Table 1.2 System models in Part II

	Strategies	Model types	Performance measures				
			E.S.R.	A.L.	U.	Thr.	S.W.
Chap. 9	Channel aggregation	P.Q.	×	○	○	×	×
Chap. 10	Spectrum reservation	P.Q.	×	○	○	×	×
Chap. 11	Opportunistic spectrum access	P.Q.	×	○	×	○	×
Chap. 12	Mini-slotted spectrum allocation	P.Q.	×	○	×	○	○
Chap. 13	Channel reservation	P.Q.	×	○	○	○	○
Chap. 14	S.M.	P.Q. with S.V.	○	○	×	×	×
Chap. 15	S.M.	P.Q. with M.V.	○	○	○	○	×

Part III discusses the VM allocation and sleep mode in cloud computing systems aiming to realize green cloud computing. From the perspective of multiple channels, we have an insight into queueing models with task migrations, wake-up thresholds, variable service rates, partial vacations, and second optional services.

There are six chapters in Part III, beginning with Chap. 16.

Chapter 16 proposes a VM scheduling strategy with a speed switch and a multiple-sleep mode to improve the energy efficiency of Cloud Data Center (CDC). Commensurate with our proposal, we develop a continuous-time queueing model with an adaptive service rate and a partial synchronous vacation. We construct a two-dimensional Markov chain based on the total number of requests in the system and the state of all the VMs. By using the matrix-geometric solution method, we mathematically estimate the energy saving level of the system and the response performance of the system for the VM scheduling strategy proposed in this chapter. Additionally, we establish a system profit function to trade off different performance measures and determine the optimal sleep parameter by developing an improved Firefly algorithm.

In Chap. 17, aiming to achieve greener, more efficient computing in CDC, we propose an energy-efficient VM allocation strategy with a partial asynchronous multiple-sleep mode and an adaptive task-migration scheme. The VMs hosted in a virtual cluster are divided into two modules, namely, Module I and Module II. The VMs hosted in a virtual cluster are divided into two modules, namely, Module I and Module II. The VMs process tasks independently of each other. When no tasks are processed at a VM, the VM will go into a sleep period as a vacation, or go into multiple sleep periods as multiple vacations. We model this system as a queueing model with partial asynchronous multiple vacations by using the proposed strategy to quantify the effects of the VMs in Module II and the sleep parameter. Moreover, we build a system cost function to investigate a trade-off between different performance measures. By introducing a chaotic mapping mechanism and a nonlinear decreasing inertia weight, we develop an improved Particle Swarm Optimization (PSO) algorithm, and jointly optimize the number of VMs in Module II and the sleep parameter with the minimum system cost.

Chapter 18 proposes a clustered VM allocation strategy based on a sleep mode with a wake-up threshold. Under the proposed strategy, all the VMs are dominated by a control server, where several sleep timers, a task counter, and a VM scheduler are deployed. To capture the stochastic behavior of tasks when using the proposed strategy, we establish a queue with an N-policy and asynchronous vacations for partial servers, and we derive performance measures of the system in terms of the average latency of tasks and the energy saving rate of the system. Furthermore, we present numerical results to evaluate the performance of the system using the clustered VM allocation strategy proposed in this chapter. Considering the trade-off between the average latency of tasks and the energy saving rate of the system, we establish a system cost function. By introducing a cube chaotic mapping mechanism for the grade initialization and an exponentially decreasing strategy for the teaching process, we develop an improved TLBO algorithm and optimize the proposed strategy with the minimum system cost function.

In Chap. 19, considering the high energy consumption and the establishment of a loyal client base in cloud computing systems, we propose a sleep mode-based cloud architecture with a free service and a registration service. Regarding the free service as the essential service, the registration service as the second, optional service, and the sleep state as the vacation, we establish an asynchronous multiple-vacation queueing model with a second optional service. We construct a three-dimensional Markov chain to derive the steady-state distribution of the queueing model, and estimate the energy saving rate of the system and the average latency of the anonymous users who select the registration service. From the perspective of economics, we construct an individual benefit function to investigate the Nash equilibrium behavior of anonymous users. Furthermore, by introducing an adaptive step adjusted by the number of iterations, we develop an improved Bat algorithm to obtain the socially optimal arrival rate of anonymous users.

Chapter 20 proposes a task scheduling strategy with a sleep-delay timer and a wake-up threshold aiming to satisfy the response performance of cloud users while reducing the energy consumption in a cloud computing system. In order to capture the stochastic behavior of tasks with the proposed strategy, we establish a synchronous vacation queueing system combining a vacation-delay and an N-policy. Taking into account the total number of tasks and the state of the PM, we construct a two-dimensional CTMC, and produce a transition rate matrix. Moreover, by using the matrix-geometric solution method we analyze the queueing model in the steady state, and then, we derive performance measures of the system in terms of the average latency of tasks and the energy saving rate of the system. Moreover, we develop a system cost function to trade off different performance measures and develop an improved Genetic algorithm to search for the optimal system's parameter combination.

Chapter 21 proposes an energy-saving VM allocation scheme with the constraint of response performance to aim a green cloud computing system. We establish a queueing model with multiple channels to capture the stochastic behavior of tasks in the CDC with the proposed scheme. Based on the reward for a processed task and the cost of a task waiting in the system buffer, we investigate the Nash equilibrium

Table 1.3 System models in Part III

	Strategies	Model types	Performance measures	
			E.S.R.	A.L.
Chap. 16	VM scheduling with with speed switch and S.M.	S.V.	○	○
Chap. 17	VM allocation with task migration and S.M.	A.V.	○	○
Chap. 18	Clustered VM allocation with S.M. and W.U.	A.V.	○	○
Chap. 19	Cloud architecture with registration service and S.M.	A.V.	○	○
Chap. 20	Task scheduling with S.M., S.D. and W.U.	S.V.	○	○
Chap. 21	VM allocation with S.M. and S.D.	S.V.	○	○

behavior. Considering also the saved income derived by a cloud service provider due to the energy conservation, we build a revenue function to investigate the socially optimal behavior of tasks. In order to maximize the value of the social benefit, we develop an improved Genetic algorithm to obtain the socially optimal arrival rate of tasks and impose an appropriate admission fee on tasks.

We summarize the system models established in each chapter of Part III in Table 1.3 with regard to what types of queueing model are applied, what performance measures are evaluated, and what types of optimization are presented.

In Table 1.3, we use some abbreviated signs as follows: "Chap." for "Chapter", "S.M." for "Sleep Mode", "W.U." for "Wake Up", "S.D." for "Sleep Delay", "S.V." for "Synchronous Vacation", 'A.V." for "Asynchronous Vacation", "E.S.R." for "Energy Saving Rate" and "A.L." for "Average Latency".

Part I
Resource Management and Performance Analysis on Broadband Wireless Access Networks

Part I discusses the sleep mode in Broadband Wireless Access (BWA) networks, including Worldwide Interoperability for Microwave Access (WiMAX), and Long Term Evolution (LTE). Under the sleep mode operation, a Mobile Station (MS) operates two modes: the awake mode and the sleep mode. Revolving around this standard, we investigate some vacation queueing models with two types of busy periods: busy period in the listening state and busy period in the awake state, with a sleep-delay period, with a wake-up procedure, and with batch arrivals.

There are seven chapters in Part I, beginning with Chap. 2.

In Chap. 2, we establish a multiple-vacation Geom/G/1 queueing model with a close-down period to analyze the sleep mode for power saving class type I. The operation mechanism of the sleep mode for downlink traffic in type I is based on IEEE 802.16e standard. In Chap. 3, considering the attractive feature that some data packets can be transmitted during the listening state, we present a queueing model with two types of busy periods: busy period in the listening state and busy period in the awake state, to capture the sleep mode for power saving class type II in IEEE 802.12e. In Chap. 4, by taking into account the self-similar nature of massive multimedia traffic, we build a batch arrival multiple-vacation queue with a vacation-delay to capture the sleep mode in IEEE 802.12e. The operation mechanism of the sleep mode is based on the enhanced power saving class type III with sleep-delay in IEEE 802.16e standard. In Chap. 5, we present an analytical approach to evaluate the Bernoulli arrival-based sleep mode in IEEE 802.16m by constructing a queueing model with heterogeneous multiple vacations and regarding the initial sleep window as one half of the sub-sequent sleep window. In Chap. 6, we establish a Discrete-Time Markovian Arrival Process (D-MAP) based queueing model with multiple vacations to evaluate the sleep mode of IEEE802.16m in a scenario where the real-time traffic includes a mixture of the real-time traffic and the Best Effort (BE) traffic. In Chap. 7, we propose an enhanced energy saving strategy based on the Active DRX mechanism in an LTE system for a better balance between response performance and energy efficiency by introducing a sleep-delay timer. We establish a two-stage multiple-vacation queueing model with a vacation-delay period and a set-up period to investigate the system performance. In Chap. 8, we introduce a sleep-delay strategy to an LTE system and propose an enhanced Active DRX mechanism influencing the control of the downlink transmission at the User Equipment (UE).

Chapter 2
Sleep Mode for Power Saving Class Type I

IEEE 802.16e is the latest standard of Broadband Wireless Access (BWA) systems designed to support mobility. In mobile networks, how to control energy consumption is one of the most important issues for the battery-powered Mobile Station (MS). The standard proposes an energy saving mechanism named a "sleep mode" for conserving the power of the MS. According to the operation mechanism of the sleep mode for downlink traffic in power saving class type I, in this chapter, we build a discrete-time Geom/G/1 queueing model with close-down time and multiple vacations. By employing an embedded Markov chain method and Little's law, we give the average queue length, the average sojourn time, and the average busy cycle of the queueing model. We derive performance measures of the system in terms of the energy saving rate of the system and the average response time of data packets, respectively. Then, we develop a system cost function to trade off these two performance measures to perform the system optimization numerically.

2.1 Introduction

IEEE 802.16e [IEEE06a] is the latest standard for mobile wireless broadband access systems in WCNs. In mobile networks, the energy consumption of the battery-powered MS is one of the most important factors for the application of the broadband wireless networks [Hwan07]. Sleep mode proposed in IEEE 802.16e is intended to minimize the MS's power usage and to decrease usage of serving Base Station (BS) air interface resources. There are three types of power saving classes (say types I, II and III) based on sleep mode operation. Among them, power saving class type I is recommended for connections of Best Effort (BE) service and NRT-VR service. In this chapter, we focus on power saving class type I.

Several authors have shown interest in the performance of the sleep mode operation, either in the case of IEEE 802.16e or other technologies. In [Xiao05],

© Springer Nature Singapore Pte Ltd. 2021
S. Jin, W. Yue, *Resource Management and Performance Analysis of Wireless Communication Networks*, https://doi.org/10.1007/978-981-15-7756-7_2

the author obtained the average energy consumption of the MSs only in case of downlink traffic, as well as an approximate expression for the average response time of data packets. In [Zhan06], the authors analyzed the energy consumption of the MSs by considering both downlink and uplink. In [Han06], the authors modeled the BS buffer as a continuous-time finite-capacity queue with a Poisson arrival process and deterministic service times, a semi-Markov chain analysis leads to expressions for the average delay of packets and the average energy consumption by the MSs.

However, all of the above performance analyses were based on continuous time, and the close-down period experienced by the MSs before starting the sleep windows was not considered. In [Turc07], the authors built a D-BMAP/G/1 queue with multiple vacations and derived the distribution of the number of data packets in the queue at various sets of time epochs, as well as deriving the average delay of data packets and the average number of consecutive vacations by assuming a D-BMAP arrival process. However, the close-down time of the sleep mode was not considered in [Han06, Turc07].

The generally accepted view is that discrete-time queueing systems can be more complex to analyze than equivalent continuous-time systems [Yue02]. However, in [Alfa10, Li07a, Li07b, Ma07, Ma09, Ma11, Taka93, Tian06], the authors indicated that it would be more accurate and efficient using discrete-time models than continuous-time counterparts when analyzing and designing digital transmitting systems. The classical discrete-time queueing analysis can be found in [Taka93, Tian06]. Analysis of discrete-time queueing models with server vacation or close-down time can be found in [Jin07, Jin12a, Tian06]. The queueing model built in this chapter is a discrete-time Geom/G/1 queue with close-down time and multiple vacations.

Taking into account the closed-down time of the MSs, in this chapter we build a discrete-time Geom/G/1 queue with close-down time and multiple vacations according to the operating mechanism of the sleep mode in IEEE 802.16e standard. By using an embedded Markov chain method, we give the average queue length, the average sojourn time, and the average busy cycle of the system model. We describe two key performance measures for the energy saving rate of the system and the average response time of data packets. We also perform the dependency relationships between the performance measures and the system parameters through some numerical results and develop a cost model to determine the optimum close-down length and the optimal system cost. The results obtained have potential applications in network control and the design of optimal systems.

The main contribution of this chapter is the fact that we take into account the close-down time of the sleep mode when building the queueing model, which is a very important factor in energy consumption. The model built in this chapter is also generic with respect to the sizes of the subsequent sleep windows. Instead of restricting the exponential increase strategy of IEEE 802.16e standard, we assume that the deterministic lengths of the first, second, third, ..., sleep window are free parameters. In this way, the model can capture any deterministic updating strategy of the sleep windows.

The chapter is organized as follows. In Sect. 2.2, we describe the working principle for the sleep mode of power saving class type I in IEEE 802.16e. Then, we present the system model built in this chapter. In Sect. 2.3, we present a performance analysis of the system model in the steady state. In Sect. 2.4, we obtain performance measures in terms of the energy saving rate of the system and the average response time of data packets. In this section, we also present a system cost function to trade off different performance measures. In Sect. 2.5, we present numerical results to evaluate the system performance. Our conclusions are drawn in Sect. 2.6.

2.2 Working Principle and System Model

In this section, we first explain the working principle for the sleep mode of power saving class type I in IEEE 802.16e. Then, we describe the system model built in this chapter.

2.2.1 Working Principle

To reserve the energy consumption, IEEE 802.16e standard proposes sleep mode operation. Under the sleep mode operation, a MS operates two modes: awake (normal) mode and sleep mode. The awake (normal) mode is the state in which the MS or the BS can transmit data packets. The sleep mode is the state in which the MS conducts pre-negotiated periods of absence from the serving BS air interface and any arriving traffic must be buffered there until the MS's sleep window ends.

IEEE 802.16e indicates that the MS is capable of waking up at any time when uplink traffic arrives, namely, the delay of uplink traffic is independent of the sleep mechanism. Thus, we focus on the downlink traffic only. The working principle of the sleep mode operation in power saving class type I is illustrated in Fig. 2.1. Since the listening window size is short, in this chapter, during any listening window, the MS is also considered to be in the sleep mode although it is physically awake.

In Fig. 2.1, the MS receives data packets from its serving BS until its buffer become empty, then the MS enters a constant time period called a close-down time

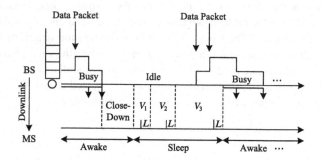

Fig. 2.1 Working principle of power saving class type I in IEEE 802.16e

D. If no data packet arrives during the close-down period, the MS sends a sleep request message to its serving BS ordering it to go to sleep, and obtains approval through a sleep response message from the BS.

The sleep mode in power saving class type I involves two operational windows: sleep window and listening window. After the MS receives an asleep response message, it switches from the awake mode to the sleep mode, and sleeps during the initial-sleep window. The size of the initial-sleep window is V_1, which is the minimum value among all the sleep window sizes. For this, we denote it by V_{min}.

At the following listening window with the length of T_L, the MS wakes up to listen to a traffic indication message from the BS, which indicates whether the BS has any buffered downlink traffic destined to it. In power saving class type I, if there are no such traffic, the MS will double the sleep window size, namely, $V_2 = 2V_1, V_3 = 2V_2, V_4 = 2V_3, \ldots, V_k = 2V_{k-1}, \ldots$, and sleep until the next listening window, where V_k is the size of the kth sleep window, $k = 1, 2, 3, \ldots$. If a sleep window size reaches up to the maximum value V_{max}, called the final-sleep window, then the sleep window size is not doubled, but fixed. We let $V_1 = V_{min}$, then both V_{min} and V_{max} can be considered as the system parameters.

These sleeping-and-listening procedures repeat with updated sleep window sizes until the MS is notified of the buffered messages destined to itself via a traffic indication message. If the BS has any buffered downlink traffic destined to the MS, the BS sends a positive TRF-INF message at the listening window and starts transmissions in an awake mode state. For the analysis of the system model, we ignore both the procedures from the awake mode to the sleep mode and from the sleep mode to the awake mode.

An inherent drawback of the sleep mode operation is the degradation of QoS, as the transmission of data packets arriving at the BS must be postponed until the current sleep window of the MS is finished. It is clear that the overall data packet delay will suffer. Hence, a trade-off needs to be made with respect the sleep window sizes. Short sleep windows result in too much unneeded activation of the MS radio interface, which is less energy efficient. On the other hand, while sleep windows that are too long result in excessive data packet delays. Therefore, it is important to be able to predict the influence of the sleep mode parameters on the system performance measures.

The setting of the close-down time reduces the response time, because a data packet arriving during the close-down time can be transmitted directly without going through a sleep window. The key point is, how long the time length of the close-down period should be set for. Obviously, if the time length is too short, the MS will switch between awake mode and sleep mode too frequently. On the other hand, if the time length is too long, power energy will be wasted due to the creation of excessive lengths of time where no data packets are transmitted.

2.2.2 *System Model*

In this discrete-time system model, the time axis is segmented into a sequence of equal intervals of unit duration, called slots. We assume that data packets arrive only just before the end of slot $t = k^-$ ($k = 1, 2, 3, \ldots$), and depart only just after the end of the slot $t = k^+$ ($k = 2, 3, 4, \ldots$). This is called a Late Arrival System (LAS) with delay access.

We assume the arrival process of data packets as a Bernoulli process. In a slot, a data packet arrival occurs with probability λ, and no arrival occurs with probability $\bar{\lambda} = 1 - \lambda$, where $0 < \lambda < 1$. We call λ the arrival rate. The data packet arrival in one slot is independent of the data packet arrivals in other slots. Thus, the probability distribution of inter-arrival T can be specified by

$$\Pr\{T = k\} = \lambda\bar{\lambda}^{k-1}, \quad k = 1, 2, 3, \ldots. \tag{2.1}$$

The transmission time of a data packet is assumed to be an independent and identically distributed random variable denoted as S (in slots). The probability distribution s_k, the probability generating function (PGF) $S(z)$ and the average value $E[S]$ of S are given as follows:

$$s_k = \Pr\{S = k\}, \quad k \geq 1, \tag{2.2}$$

$$S(z) = \sum_{k=1}^{\infty} z^k s_k, \tag{2.3}$$

$$E[S] = \sum_{k=1}^{\infty} k s_k. \tag{2.4}$$

A single channel is supposed and the data packets are transmitted according to a First-Come First-Served (FCFS) discipline. Based on the above definitions, let A_S be the number of data packets arriving during the transmission time of a data packet. The probability distribution and the PGF of A_S are obtained as follows:

$$\Pr\{A_S = k\} = \sum_{j=k}^{\infty} s_j \binom{j}{k} \lambda^k \bar{\lambda}^{j-k}, \quad k = 0, 1, 2, \ldots, \tag{2.5}$$

$$A_S(z) = \sum_{k=0}^{\infty} z^k \Pr\{A_S = k\} = S(\bar{\lambda} + \lambda z). \tag{2.6}$$

We define the probability that no arrival will occur within the close-down time D is $D(\bar{\lambda})$, then $D(\bar{\lambda})$ can be denoted as follows:

$$D(\bar{\lambda}) = \bar{\lambda}^D. \tag{2.7}$$

We define a sleep cycle C as the combination of a sleep window and a listening window. We assume that the size of the listening window is a fixed value of T_L. Let C_k be the size of the kth sleep cycle, then we have that

$$C_k = V_k + T_L, \quad k = 1, 2, 3, \ldots \tag{2.8}$$

where V_k is the size of the kth sleep-window defined in Sect. 2.2.1.

For convenience, we also define the time length τ_i of the first i sleep cycles as follows:

$$\tau_i = \begin{cases} \displaystyle\sum_{j=1}^{i} C_j, & i \geq 1 \\[2mm] 0, & i = 0. \end{cases}$$

In addition, it is assumed that the inter-arrival T, the transmission time S, the close-down time D and the size of the sleep window are mutually independent.

2.3 Performance Analysis

Consider an embedded Markov chain at the transmission completion instants. The number of data packets in the system at the mth transmission completion instant is denoted by Q_m ($m \geq 1$). The sufficient and necessary condition for this Markov chain to be positive recurrent is $\rho = \lambda E[S] < 1$, where $E[S]$ is the average of the transmission time S. Obviously, we have

$$Q_{m+1} = \begin{cases} Q_m - 1 + A_s, & Q_m \geq 1 \\ \eta, & Q_m = 0 \end{cases} \tag{2.9}$$

where η is the number of data packets left after the first departure in a busy period. To obtain the distribution of η, consider the following cases:

1. Let e_1 represent the event that there is at least an arrival during D. The data packet arriving during D is transmitted immediately. We have

$$\Pr\{e_1\} = 1 - D(\bar{\lambda}), \tag{2.10}$$

$$E[z^{\eta}|e_1] = S(\bar{\lambda} + \lambda z). \tag{2.11}$$

2. Let V denote a vacation period including multiple sleep cycles and let $e_{2,n}$ be the event that there is no arrival during D and the vacation period V consists of n sleep cycles. It follows that

$$\Pr\{e_{2,n}\} = D(\bar{\lambda})\bar{\lambda}^{\tau_{n-1}}(1 - \bar{\lambda}^{C_n}). \tag{2.12}$$

Under the condition $e_{2,n}$, $\eta = A_V - 1 + A_S$, where A_V is the number of arrivals during the vacation period V, more precisely, during C_n, we have $E[z^\eta | e_{2,n}]$ as follows:

$$E\left[z^\eta | e_{2,n}\right] = E\left[z^{A_V - 1 + A_S} | e_{2,n}\right] = \frac{1}{z} E\left[z^{A_V} | e_{2,n}\right] S(\bar{\lambda} + \lambda z). \qquad (2.13)$$

Note that

$$E\left[z^{A_V} | e_{2,n}\right] = \sum_{j=1}^{\infty} z^j \Pr\left\{A_V = j | e_{2,n}\right\} = \sum_{j=1}^{\infty} z^j \frac{\Pr\left\{A_V = j, e_{2,n}\right\}}{1 - \bar{\lambda}^{C_n}}, \qquad (2.14)$$

$$\Pr\left\{A_V = j, e_{2,n}\right\} = \binom{C_n}{j} \lambda^j \bar{\lambda}^{C_n - j}, \quad j = 1, 2, 3, \ldots, C_n. \qquad (2.15)$$

Therefore, $E\left[z^{A_V} | e_{2,n}\right]$ can be written by using the PGF of Binomial distribution as follows:

$$E\left[z^{A_V} | e_{2,n}\right] = \frac{1}{1 - \bar{\lambda}^{C_n}} \sum_{j=1}^{C_n} \binom{C_n}{j} (\lambda z)^j \bar{\lambda}^{C_n - j}$$

$$= \frac{1}{1 - \bar{\lambda}^{C_n}} \left((\bar{\lambda} + \lambda z)^{C_n} - \bar{\lambda}^{C_n}\right). \qquad (2.16)$$

Then,

$$E[z^\eta | e_{2,n}] = \frac{1}{z} \frac{(\bar{\lambda} + \lambda z)^{C_n} - \bar{\lambda}^{C_n}}{1 - \bar{\lambda}^{C_n}} S(\bar{\lambda} + \lambda z). \qquad (2.17)$$

By using the conditioning argument, we have

$$E[z^\eta] = \Pr\{e_1\} E[z^\eta | e_1] + \sum_{n=1}^{\infty} \Pr\{e_{2,n}\} E\left[z^\eta | e_{2,n}\right]$$

$$= \left(1 - D(\bar{\lambda})\right) S(\bar{\lambda} + \lambda z)$$

$$+ \frac{D(\bar{\lambda})}{z} \sum_{n=1}^{\infty} \bar{\lambda}^{\tau_n - 1} \left((\bar{\lambda} + \lambda z)^{C_n} - \bar{\lambda}^{C_n}\right) S(\bar{\lambda} + \lambda z). \qquad (2.18)$$

Let N_d be the number of data packets at the transmission completion instants. From Eq. (2.3), the PGF $N_d(z)$ of N_d satisfies the following equation:

$$N_d(z) = \Pr\{N_d \geq 1\} E\left[z^{N_d - 1 + A_S} | N_d \geq 1\right] + \Pr\{N_d = 0\} E[z^\eta]. \qquad (2.19)$$

Substituting Eqs. (2.6) and (2.18) into Eq. (2.19), we give that

$$N_d(z) = \frac{\Pr\{N_d = 0\}S(\bar{\lambda} + \lambda z)}{S(\bar{\lambda} + \lambda z) - z}$$

$$\times \left(1 - z + D(\bar{\lambda})z - D(\bar{\lambda})\sum_{n=1}^{\infty}\bar{\lambda}^{\tau_{n-1}}\left((\bar{\lambda} + \lambda z)^{C_n} - \bar{\lambda}^{C_n}\right)\right). \quad (2.20)$$

By using the normalization condition $N_d(1) = 1$ and the L'Hôspital rule, we obtain that

$$\Pr\{N_d = 0\} = \frac{1 - \rho}{1 - D(\bar{\lambda}) + \lambda D(\bar{\lambda})\sum_{n=1}^{\infty}\bar{\lambda}^{\tau_{n-1}}C_n} = \frac{1 - \rho}{H} \quad (2.21)$$

where

$$H = 1 - D(\bar{\lambda}) + \lambda D(\bar{\lambda})\sum_{n=1}^{\infty}\bar{\lambda}^{\tau_{n-1}}C_n.$$

Since N_d can be decomposed into the sum of two independent random variables. One is the queue length of the classical queueing system without vacations, the other is the additional queue length due to vacation periods. The formula of $N_d(z)$ can be given by

$$N_d(z) = \frac{(1 - \rho)(1 - z)S(\bar{\lambda} + \lambda z)}{S(\bar{\lambda} + \lambda z) - z}$$

$$\times \frac{1 - z + D(\bar{\lambda})z - D(\bar{\lambda})\sum_{n=1}^{\infty}\bar{\lambda}^{\tau_{n-1}}\left((\bar{\lambda} + \lambda z)^{C_n} - \bar{\lambda}^{C_n}\right)}{(1 - z)H}. \quad (2.22)$$

From Eq. (2.22), the average number $E[N_d]$ of data packets at the transmission completion instants is given as follows:

$$E[N_d] = \rho + \frac{\lambda^2}{2(1 - \rho)}E[S(S - 1)] + \frac{\lambda^2 D(\bar{\lambda})\sum_{n=1}^{\infty}\bar{\lambda}^{\tau_{n-1}}C_nC_{n-1}}{2H}. \quad (2.23)$$

2.4 Performance Measures

In this section, by using the performance analysis presented in Sect. 2.3, we derive performance measures of the system in terms of the energy saving rate of the system and the average response time of data packets, respectively. Then, we develop a system cost function to trade off these two performance measures to carry out the system's parameter optimization numerically in the system model.

2.4.1 System Energy

Now, we define a busy cycle R as the period between the ending instants of two consecutive busy periods. Let T_R be the time length of the busy cycle R. If a data packet arrival occurs during a close-down period D, then R consists of a close-down period D and a busy period B. If no arrival occurs during a close-down period D, R is the sum of a close-down period D, a busy period B and a vacation period V.

The actual values of D and V are denoted by T_D and T_V, respectively. Note that with probability $D(\bar{\lambda})$, $T_D = D$ and with probability $1 - D(\bar{\lambda})$, T_D equals the conditional length given that $T_D < D$. Thus, the probability distribution, the PGF $T_D(z)$ and the average $E[T_D]$ of T_D can be given as follows:

$$\Pr\{T_D = k\} = \begin{cases} \lambda \bar{\lambda}^{k-1}, & k = 1, 2, 3, \ldots, D-1 \\ \bar{\lambda}^{D-1}, & k = D, \end{cases} \tag{2.24}$$

$$T_D(z) = \frac{\lambda z + (1-z)(\bar{\lambda}z)^D}{1 - \bar{\lambda}z}, \tag{2.25}$$

$$E[T_D] = \frac{1}{\lambda}\left(1 - D(\bar{\lambda})\right). \tag{2.26}$$

Since a vacation period V can consist of multiple sleep cycles, we have the probability distribution and the average $E[T_V]$ of T_V as follows:

$$\Pr\{T_V = \tau_n\} = D(\bar{\lambda})\bar{\lambda}^{\tau_n - 1}\left(1 - \bar{\lambda}^{C_n}\right), \quad n = 1, 2, 3, \ldots, \tag{2.27}$$

$$E[T_V] = \sum_{n=1}^{\infty} \tau_n \Pr\{T_V = \tau_n\}. \tag{2.28}$$

In addition, the average number $E[N_V]$ of sleep cycles is such that

$$E[N_V] = \sum_{n=1}^{\infty} n\Pr\{T_V = \tau_n\}. \tag{2.29}$$

To compute the average busy cycle, we need to determine the probability distribution of the number of data packets denoted by Q_B in the system at the beginning of a busy period B. Considering the two possible cases e_1 and $e_{2,n}$ described above, we have the conditional PGF of Q_B as follows:

$$E\left[z^{Q_B}|e_1\right] = z, \tag{2.30}$$

$$E\left[z^{Q_B}|e_{2,n}\right] = \frac{(\bar{\lambda} + \lambda z)^{C_n} - \bar{\lambda}^{C_n}}{1 - \bar{\lambda}^{C_n}}. \tag{2.31}$$

Combining Eqs. (2.11) and (2.12), we obtain the PGF $Q_B(z)$ and the average $E[Q_B]$ of Q_B as follows:

$$Q_B(z) = z(1 - D(\bar{\lambda})) + D(\bar{\lambda}) \sum_{n=1}^{\infty} \bar{\lambda}^{\tau_{n-1}} \left((\bar{\lambda} + \lambda z)^{C_n} - \bar{\lambda}^{C_n}\right), \tag{2.32}$$

$$E[Q_B] = 1 - D(\bar{\lambda}) + \lambda D(\bar{\lambda}) \sum_{n=1}^{\infty} \bar{\lambda}^{\tau_{n-1}} C_n = H. \tag{2.33}$$

It is well known that the average busy period $E[B]$ for a standard Geom/G/1 queue is $(1 - \rho)^{-1} E[S]$. Therefore, the average length $E[B]$ of the busy period B for the system described above is such that as follows:

$$E[B] = E[Q_B](1 - \rho)^{-1} E[S] = H(1 - \rho)^{-1} E[S]. \tag{2.34}$$

Now, the average value $E[T_R]$ of T_R is obtained as follows:

$$E[T_R] = E[T_D] + E[T_V] + E[B]. \tag{2.35}$$

As stated earlier, the goal of the sleep mode operation is to reduce the energy consumption of a MS in listening periods, where the energy consumption is given by $E[N_V] \times T_L$. For the analysis of the energy saving rate of the system, we assume that the main energy cost of the MS is from maintaining the system in an awake and the listening state. We also assume that a listening slot has the same energy cost as an arbitrary awake slot.

The energy saving rate γ of the system is defined as the probability for the MS being in the state of sleep mode, which is an important measure to evaluate the efficiency of power saving, we have that

$$\gamma = \frac{E[T_V] - E[N_V] \times T_L}{E[T_R]} = \frac{E[T_V] - E[N_V] \times T_L}{E[B] + E[T_D] + E[T_V]}. \tag{2.36}$$

2.4.2 Average Response Time of Data Packets

We define the response time Y_d of a data packet as the duration in slots elapsed from the arrival of a data packet to the end of the transmission of that data packet. By using Eq. (2.23), we can get the average response time $E[Y_d]$ of data packets as follows:

$$E[Y_d] = \frac{E[N_d]}{\lambda}. \tag{2.37}$$

2.4.3 System Cost

To obtain the optimum length D^* of the close-down time D by minimizing the total average system cost, we develop a system cost function $F(D)$ as follows:

$$F(D) = f_1 E[T_D] + f_2 E[N_V] \times T_L + \frac{f_3}{E[T_V] - E[N_V] \times T_L} + f_4 E[B] \tag{2.38}$$

where f_1 is the cost per slot when the system is in close-down times, f_2 is the cost per slot when the system is on the listening windows, f_3 is the reward per slot due to the system being on the sleep-windows and f_4 is the cost per slot for transmissions of data packets, namely, when the system is in a busy period.

Differentiating the system cost function $F(D)$ with respect to D, and letting $F'(D) = 0$, we can obtain the optimum length D^* of the close-down time D.

2.5 Numerical Results

The system parameters are fixed as follows: $E[S] = 4$ slots, $T_L = 1$ slot as an example for all the numerical results. According to IEEE 802.16e standard, in the numerical results of this section, an exponential increase strategy is used for updating the sizes of the sleep windows. In fact, we can use any increase strategy, it is because we assumed the deterministic lengths of the first, second, third, etc., sleep windows to be free parameters of the system model.

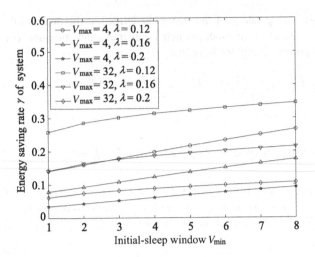

Fig. 2.2 Energy saving rate of system versus initial-sleep window

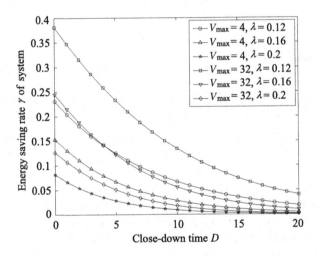

Fig. 2.3 Energy saving rate of system versus close-down time

The arrival rate $\lambda = 0.12, 0.16, 0.2$, and the size of the final-sleep window is 8 slots and 32 slots, respectively. The dependency relationships between the performance measures and the system parameters are shown in Figs. 2.2, 2.3, 2.4, and 2.5.

In Fig. 2.2, we show how the energy saving rate γ of the system obtained from Eq. (2.36) changes with the size of the initial-sleep window V_{min} for the close-down time $D = 2$ slots.

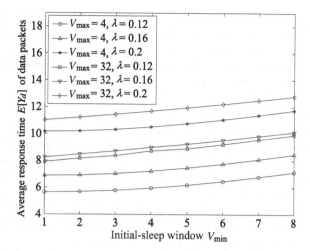

Fig. 2.4 Average response time of data packets versus initial-sleep window

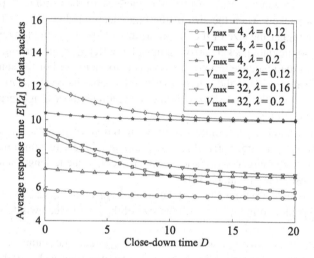

Fig. 2.5 Average response time of data packets versus close-down time

From Fig. 2.2, we can conclude that with the same arrival rate λ and the size of the final-sleep window V_{max}, an increase of the size of the initial-sleep window V_{min} leads to an increase of the energy saving rate γ of the system. This is because the bigger the size of the initial-sleep window is, the longer a MS will be in a sleep mode state, and the more energy will be saved. For the same reason, while the arrival rate λ and the size of the initial-sleep window V_{min} take the same value, bigger size of the final-sleep window V_{max} will result in a more energy saving rate of the system.

On the other hand, for the same size of the initial-sleep window V_{min} and final-sleep window V_{max}, a larger arrival rate λ results in a smaller energy saving rate γ of the system. This is because the larger the arrival rate λ is, the smaller the probability that a MS will enter sleep mode will be. Therefore, the smaller the energy saving rate γ of the system will be. From Fig. 2.2, we can say that the larger the value of energy saving rate is, the longer the MS's lifetime will be, which can improve the system performance.

Figure 2.3 shows the energy saving rate γ of the system presented in Eq. (2.36) versus the close-down time D with the size of the initial-sleep window $V_{min} = 2$ slots.

In Fig. 2.3, we can see that while λ and V_{max} take the same value, with an increase of the close-down time D, the energy saving rate γ of the system decreases sharply. This is because the longer the close-down time D is, the longer a MS will be in awake mode, so the energy saving will decrease. On the other hand, at the same close-down time D, the change trends of the energy saving rate γ of the system will be the same as shown in Fig. 2.2.

In Fig. 2.4, we show the dependency relationship for the average response time $E[Y_d]$ of data packets presented in Eq. (2.37) and the size of the initial-sleep window V_{min} with close-down time $D = 2$ slots.

From Fig. 2.4, we know that if the arrival rate λ is given, an increase of the size of the initial-sleep window V_{min} results in an increase of the average response time $E[Y_d]$ of data packets when the size of the final-sleep window V_{max} is fixed. This is because the bigger the size of the initial-sleep window V_{min}, the longer the sleep time will be, so the data packets arriving must be buffered in the serving BS buffer and this causes the response time of data packets to increase. For the same reason, when the arrival rate λ and the size of the initial-sleep window V_{min} are the same, the bigger the size of the final-sleep window is, the longer the average response time of data packets will be.

On the other hand, for the same size of the initial-sleep window V_{min} and final-sleep window V_{max}, a larger arrival rate λ results in a longer response time of data packets, because the larger λ leads to a longer average queue length and a longer average response time of data packets.

The average response time $E[Y_d]$ of data packets given in Eq. (2.37) versus close-down time D is plotted in Fig. 2.5 with the size of the initial-sleep window $V_{min} = 2$ slots.

In Fig. 2.5, it is shown that while λ and V_{max} take the same value, as the increase of the close-down time D, the average response time $E[Y_d]$ of data packets decreases smoothly. This is because the data packets arriving during the close-down time can be transmitted directly without going through a sleep-window. Therefore, the average response time of data packets will be smaller. On the other hand, at the same close-down time, the change trends of the average response time of data packets will be the same as shown in Fig. 2.4.

To show the optimum length D^* of the close-down time D by minimizing the system cost in the actual busy cycle length T_R of the busy cycle R, we give a numerical example of the system cost function $F(D)$ in Fig. 2.6. In Fig. 2.6, we assume that $S = 4$ slots, $V_{min} = 1$ slot, $T_L = 1$ slot, $f_3 = 1$, $f_1 = f_2 = f_4 = 10$.

From Fig. 2.6, we can see that at the same size of the final-sleep window, with the decrease of the arrival rate λ, the optimum value D^* of the close-down time D having the minimum average cost is smaller, such as: when the size of the final-

Fig. 2.6 System cost function versus close-down time

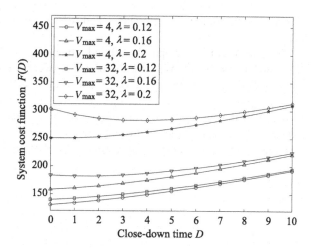

sleep window is 32 slots, $\lambda = 0.2$, $D^* = 4$ slots, $\lambda = 0.16$, $D^* = 1$ slot, and $\lambda = 0.12$, $D^* = 0$ slot. On the other hand, bigger size of the final-sleep window leads to a higher average cost. It is because bigger size of final-sleep window will result in more data packets waiting in the buffer of the serving BS when a new busy period begins. Therefore, a longer busy period and a higher average cost. In fact, we can select the parameters f_1-f_4 as in any other cases for an actual mobile wireless broadband access system and get the optimum length D^* of the close-down time D.

Moreover, by using the method of calculation presented in Eq. (2.38), we can also optimize other system parameters such as the initial-sleep window V_{min} and the final-sleep window V_{max} in the similar way.

2.6 Conclusion

We analyzed the working principle of the sleep mode for downlink traffic in power saving class type I based on IEEE 802.16e standard. In order to assess the dependency relationship of the energy saving rate of the system and the average response time of data packets on the close-down time and the sleep windows, we built a discrete-time Geom/G/1 queueing model with close-down time and multiple vacations. The contributions of this system model were taking close-down time into account. Also instead of being restricted to the exponential increase strategy of IEEE 802.16e standard, we assumed the deterministic lengths of the first, second, third, etc., sleep windows to be free parameters of the system model. By using the embedded Markov chain method, we gave the analysis procedure of this system model and derived the expressions for the energy saving rate of the system and the average response time of data packets. Finally, we produced numerical results to explain the nature of the dependency relationships between the system configuration parameters and the performance measures, as well as developed a system cost function to determine the optimum length of the close-down time under certain conditions.

The purpose of this chapter is to build a queueing model for the sleep mode in IEEE 802.16e standard. We took the period between the end of transmitting data and the beginning of entering the sleep state as close-down time, regard the time of sleep periods as vacations. Under these assumptions, we derived the performance measures of the energy saving rate of the system and the average response time of data packets in terms of the lengths of close-down time and sleep cycles, which have potential applications in network control and the design of optimal systems.

Chapter 3
Sleep Mode for Power Saving Class Type II

In order to investigate mathematically the inherent relationships between the performance measures and the system parameters, in this chapter, we propose a method for modeling the sleep mode with the power saving class type II in IEEE 802.16 and analyzing the performance of this sleep mode. Considering the attractive feature that some data packets can be transmitted during the listening state, we build a queueing model with two types of busy periods to capture the working principle of the sleep mode operations with the power saving class type II. We present methods for assessing the performance measures of the system in terms of the handover rate, the energy saving rate of the system and the average response time of data packets. Moreover, we present numerical results to show the influence of the system parameters on the system performance and the system cost function with different traffic loads. Finally, by constructing a system cost function, we give an optimal design for the time length of the sleep window.

3.1 Introduction

In recent years, the telecommunication industry has been driving the development of new metrics, especially for evaluating energy efficiency in WCNs [Min07, Wu10]. On the other hand, researchers have been striving to find approaches, simulation and optimization methods, and numerical methods for reducing the energy consumption of communication. This includes methods such as energy-efficient network capacity design, power-aware infrastructure planning, etc. [Khai10, Knop09].

Energy is a scarce resource in mobile BWA networks. It is therefore critical to design energy efficient techniques, to control the energy consumption of the Mobile Station (MS), and to extend the lifetime of the battery in the MS for the application of the BWA networks. As a way of conserving the energy of the MS, IEEE 802.16 offers three types of sleep mode operations called power saving classes type I, type II and type III, respectively.

© Springer Nature Singapore Pte Ltd. 2021
S. Jin, W. Yue, *Resource Management and Performance Analysis of Wireless Communication Networks*, https://doi.org/10.1007/978-981-15-7756-7_3

Power saving class type II is mainly used for UGS and RT-VR traffic. IEEE 802.16 supersedes and makes obsolete IEEE 802.16-2004 as well as IEEE 802.16e-2005, IEEE 802.16-2004/Cor1-2005, IEEE 802.16f-2005, and IEEE 802.16g-2007 [IEEE09]. Unlike power saving class type I, the MS can transmit or receive data packets during listening windows without deactivating power saving class type II. When the sleep mode is activated, sleep windows are interleaved with listening windows.

Several authors have shown interest in the performance of the sleep mode operation, either in the case of IEEE 802.16e or IEEE 802.16m. In [Xiao05], based on the old version of IEEE 802.16e standard, for the sleep mode of power saving class type I, the energy saving efficiency, the energy consumption and mean delay were obtained. In [Lee06], a cumulative-TIM method was provided to improve the energy efficiency of the sleep mode with power saving class type I. The MS selects the length of the sleep window in determining the trade-off function between the energy saving efficiency and the data delay by the method of simulation.

The buffer of the Base Station (BS) was modeled as a continuous-time finite-capacity queue with a Poisson arrival process and a deterministic service time, and the expressions for the average energy consumption and the average data packet delay were derived with a semi-Markov chain in [Han06]. In [Jin11a], the performance of power saving class type III initiated by the BS with unsolicited MOB_SLP-RSP or DL sleep control extended sub-header in self-similar traffic was analyzed. To capture the self-similar property in multimedia WiMAX, a batch arrival queueing model with a Pareto(c, α) distributed batch size was built. The averages and the standard deviations for the system performance using the diffusion approximation for the operating process of the system were given.

Conclusively, most of the performance research mentioned above have been focused on power saving class type I or III.

For power saving class type II in Voice over Internet Protocol (VoIP) traffic, the problem of allocation representation of VoIP packets was addressed, and an efficient uplink mapping scheme was proposed by using a simulation method in [Lee08]. To improve the energy efficiency for power saving class type II in IEEE 802.16e, [Chen09a, Chen09b] proposed an energy conservation scheme called Maximum Unavailability Interval (MUI) as well as a systematic method to determine the start frame number in MUI.

To define multiple power saving classes and their listen-and-sleep-related param-eters and packet-scheduling policy, a fold-and-demultiplex method for an IEEE 802.16 network with power saving class types I and II together with an earliest-next-bandwidth-first packet scheduler was presented in [Tsen11].

Also, a new sleep mode scheme called power saving mechanism with binary traffic indication was proposed for IEEE 802.16m, and a mathematical model for the proposed scheme to evaluate the system performance was provided in [Hwan09a].

The first published research work we can find in the literature that considers the performance of the standard power saving class type II was carried out by authors of [Kong06]. They focused on power saving class type II for its capability to potentially avoid unnecessary handover costs, and they included this feature in their analytical model. For both the conventional power saving class type II and the modified type II, the key feature is that data packets could be transmitted and received during the listening state. On the one hand, the support of this key feature means that the transmission and reception of data packets can be carried out without interruption of sleep mode.

Additional energy consumed by those data packets transmitted or received during the listening state is non-trivial. However, in order to simplify the analysis of the system performance, the additional energy consumed is neglected by previous works. In reality, it is very important to evaluate the power saving efficiency of the mobile terminals with power saving class type II.

In this chapter, we are firstly considering the comprehensive performance of power saving class type II by taking into account the energy consumption during the listening state. We then present an efficiency analysis of the system in order to evaluate the system performance. The analysis in this chapter differs considerably from former analyses.

In order to investigate the inherent relationships between the performance measures and the system parameters, to evaluate the system performance of the power saving class type II, and to improve the energy saving efficiency, an effective analytical method must be provided. On the other hand, it would be more accurate and efficient using discrete-time models than continuous-time counterparts when analyzing and designing digital transmitting systems [Jin11a, Jin12b, Tian06]. Moreover, a vacation queueing model is naturally more suitable for the system model and for performance analysis of the sleep mode in power saving schemes. Constructing an effective power saving scheme is seen as the best way to extend the lifetime of the MS in mobile BWA networks.

In this chapter, we propose an effective analysis method to evaluate the system performance of power saving class type II in IEEE 802.16 for a wireless Metropolitan Area Network (MAN). Taking into account the memoryless character of the data packet arrivals and the digital nature in power saving class type II applied in UGS and RT-VR traffic, and considering the fact that some data packets can be transmitted during the listening state, we model the system as a Geom/G/1 queueing system with two types of busy periods. One type is the normal busy period representing the awake state. The other type is a special busy period representing the listening sate, during which a limited number of data packets can be transmitted. By using an embedded Markov chain method and the boundary state variable theory, we give the performance measures of the system to illustrate the influence of the system parameters on the system performance. Moreover, we present analysis results and simulation results to show the influence of the system parameters on the system performance and the system cost function with different traffic loads. Finally, we give an optimal design for the time length of the sleep window.

The chapter is organized as follows. In Sect. 3.2, we describe the working principle for the sleep mode of power saving class type II in IEEE 802.16e. Then, we present the system model built in this chapter. In Sect. 3.3, we present the analysis of busy cycle, including the busy period in listening state, the busy period in awake state, and the time length of a busy cycle. In Sect. 3.4, we present the analysis of waiting time, including waiting time in listening state and waiting time in awake state. In Sect. 3.5, we obtain performance measures and present numerical results to evaluate the system performance. In Sect. 3.6, we obtain the optimal design of the sleep window length. Our conclusions are drawn in Sect. 3.7.

3.2   Working Principle and System Model

In this section, we first explain the working principle for the sleep mode of power saving class type II in IEEE 802.16e. Then, we establish the system model accordingly.

3.2.1   Working Principle

For all the sleep modes, the MS operates in three states: the awake state, the sleep state and the listening state. The time lengths for the system being in the sleep state and the listening state are controlled by the sleep window and the listening window, respectively. In the awake state, the MS or the BS can transmit data packets normally. In the sleep state, the MS conducts a pre-negotiated period of absence from the air interface of its serving BS. During the listening state, the MS senses the channel constantly to see if there are any data packets to be transmitted.

Different from other two power saving schemes, the time lengths of the sleep window and the listening window in power saving class type II are fixed, and a certain number of data packets can be transmitted during the listening state. If all the data packets buffered in the BS can be transmitted within a listening window, the system will return to the sleep state after the listening window is over, otherwise, the system will enter into the awake state to transmit the remaining data packets when the listening window expires. The time lengths of the sleep window and the listening window in power saving class type II are fixed, and a certain number of data packets can be transmitted during the listening state. If all the data packets buffered in the BS can be transmitted within a listening window, the system will return to the sleep state after the listening window is over, otherwise, the system will enter into the awake state to transmit the remaining data packets when the listening window expires.

It is illustrated in IEEE 802.16 that the MS will wake up any time when uplink traffic arrives. It means that the delay of uplink traffic is independent of the sleep mode, so, in this chapter, we focus on the downlink traffic only. The working principle of the sleep mode operation in power saving class type II with only downlink traffic is illustrated in Fig. 3.1.

In Fig. 3.1, SLP-REQ means the sleep-request message, and SLP-RSP means the sleep-response message.

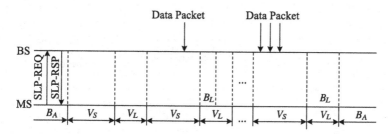

Fig. 3.1 Working principle of power saving class type II in IEEE 802.16e

3.2.2 System Model

We regard the sleep state as one vacation period V_S with a time length T_S, and the whole listening state as another special vacation period V_L with a time length T_L. The system vacation V is the sum of V_S and V_L, so obviously, $T_V = T_S + T_L$. Note that during the vacation period V_L, some data packets can be transmitted.

The time period for transmitting data packets in the listening state (the vacation period V_L) is seen as one busy period B_L with a time length of T_{B_L}. The time interval for transmitting data packets normally in the awake state is regarded as another busy period B_A with a time length of T_{B_A}.

If all the data packets arrived both in the vacation period V_S, and the vacation period V_L can be transmitted within the vacation period V_L, the system will begin a new vacation period V_S after the vacation period V_L is over. Otherwise, when the vacation period V_L expires, the system will initiate a busy periods B_A in the awake state to transmit the remaining data packets and all the subsequent data packets arriving during the sub-busy periods introduced by these remaining data packets. The system busy period B is composed of one or more busy periods B_L in the listening state and only one busy period B_A in the awake state.

For power saving class types I and III, the listening state can be regarded as a normal vacation period. However, for power saving class type II, the listening state cannot be regarded as a normal vacation state, because some data packets could be transmitted in this state. On the other hand, the listening state cannot be seen as a normal busy period, because it is possible that no data packets need to be transmitted during this stage, or that there are not enough data packets to be transmitted to occupy the whole listening state.

Therefore, the listening state in power saving class type II must be seen as a special period, some of which with transmission are regarded as busy periods, and others of which without transmission are seen as vacation periods. Analysis of this model becomes very difficult when we consider these technical key points. This chapter also proposes the effective analysis method.

Therefore, we should build a queueing model with two types of busy periods. This means that the probability behavior of the system model will more closely resemble that of the actual system. One of the busy periods is the normal busy period

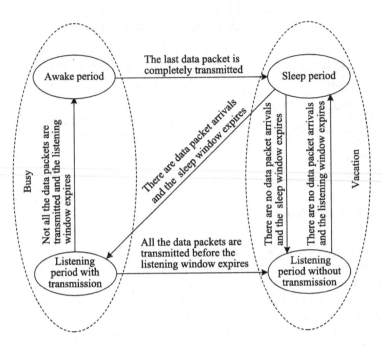

Fig. 3.2 State transition of system model

representing the awake state, and the other is a special busy period representing the listening state with transmission. The state transition in this queueing model is shown in Fig. 3.2.

From Fig. 3.2, we can observe that the listening state is classified into two parts: one belongs to the vacation period without transmission and the other belongs to the busy period with transmission.

We consider this system model as a discrete-time queueing model. The time axis is segmented into a series of equal intervals, called slots. A Late Arrival System (LAS) with immediate entrance is considered in this chapter. We suppose that the departures occur at the moment immediately before the slot boundaries and the arrivals occur at the moment immediately after the slot boundaries. It is assumed that data packets are transmitted according to a First-Come First-Served (FCFS) discipline. Moreover, the buffer capacity in the BS is assumed to be infinite. An embedded Markov chain is constructed at the end of slots where the data packet transmissions are completed. We define the system state by the number of data packets at the embedded Markov points.

Taking into account the memoryless nature of user-initiated data packet arrivals, we can assume that the arrival process of data packets follows a Bernoulli distribution. Under the assumption, in a slot, a data packet arrives with probability λ $(0 < \lambda < 1, \bar{\lambda} = 1 - \lambda)$. We call probability λ the arrival rate of data packets. The transmission time of a data packet is assumed to be an independent and identically distributed random variable denoted as S (in slots). Therefore, the probability distribution s_k, the probability generating function (PGF) $S(z)$ and the average value

$E[S]$ of S are given as follows:

$$s_k = \Pr\{S = k\}, \quad k \geq 1, \tag{3.1}$$

$$S(z) = \sum_{k=1}^{\infty} z^k s_k, \tag{3.2}$$

$$E[S] = \sum_{k=1}^{\infty} k s_k. \tag{3.3}$$

The sufficient and necessary condition for this embedded Markov chain to be positive recurrent is $\rho = \lambda E[S] < 1$, where ρ is the system load.

3.3   Analysis of Busy Cycle

In this section, we present the analysis of busy cycle, including the busy period in listening state, the busy period in awake state, and the time length of a busy cycle.

For power saving class type II in IEEE 802.16, there are two types of busy periods:

(1) The busy period B_L in the listening state.
(2) The busy period B_A in the awake state.

3.3.1   Busy Period in Listening State

Suppose that the maximal number of data packets can be transmitted within the listening window is d. Given that there is at least one data packet arrival in the sleep window, let the number of data packets transmitted during the listening window be Q_{B_L}, the PGF $Q_{B_L}(z)$ of Q are given as follows:

$$Q_{B_L}(z) = \frac{\displaystyle\sum_{j=1}^{d-1} \binom{T_V}{j}(\lambda z)^j \bar{\lambda}^{T_V-j} + \sum_{j=d}^{T_V} z^d \binom{T_V}{j}\lambda^j \bar{\lambda}^{T_V-j}}{1 - \bar{\lambda}^{T_V}}. \tag{3.4}$$

Differentiating Eq. (3.4) with respect to z at $z = 1$, the average value $E[Q_{B_L}]$ of Q_{B_L} is then obtained as follows:

$$E[Q_{B_L}] = \frac{\displaystyle\sum_{j=1}^{d-1} j\binom{T_V}{j}\lambda^j \bar{\lambda}^{T_V-j} + \sum_{j=d}^{T_V} d\binom{T_V}{j}\lambda^j \bar{\lambda}^{T_V-j}}{1 - \bar{\lambda}^{T_V}} \tag{3.5}$$

where T_V is the time length of the system vacation period V defined in Sect. 3.2, j is the number of data packets arriving both in the vacation period V_S and the vacation period T_L.

For mathematical clarity, we assume that the data packets arriving during a listening window arrive at the end of the sleep window just before the listening window. Therefore, the average time length $E[T_{B_L}]$ of the busy period B_L in the listening state can be given as follows:

$$E[T_{B_L}] = E[Q_{B_L}]E[S]$$

$$= \frac{\sum_{j=1}^{d-1} j\binom{T_V}{j}\lambda^j\bar{\lambda}^{T_V-j} + \sum_{j=d}^{T_V} d\binom{T_V}{j}\lambda^j\bar{\lambda}^{T_V-j}}{1 - \bar{\lambda}^{T_V}} \times E[S]. \tag{3.6}$$

3.3.2   Busy Period in Awake State

Given that the number of the data packets arriving within a sleep window exceeds d, the previous d data packets will be transmitted during the listening state. When the listening window expires, the system will switch to the awake state, and transmit the remaining data packets and all the subsequent data packets arriving during the sub-busy periods introduced by these remaining data packets.

Letting the number of data packets at the beginning instant of an awake state be Q_{B_A}, the PGF $Q_{B_A}(z)$ of Q_{B_A} can be obtained as follows:

$$Q_{B_A}(z) = \frac{\sum_{j=d+1}^{T_V} \binom{T_V}{j}(\lambda z)^j\bar{\lambda}^{T_V-j}}{z^d \sum_{j=d+1}^{T_V} \binom{T_V}{j}\lambda^j\bar{\lambda}^{T_V-j}}. \tag{3.7}$$

Differentiating Eq. (3.7) with respect to z at $z = 1$, the average number $E[Q_{B_A}]$ of data packets at the beginning instant of an awake state is then obtained as follows:

$$E[Q_{B_A}] = \frac{\sum_{j=d+1}^{T_V} j\binom{T_V}{j}\lambda^j\bar{\lambda}^{T_V-j}}{\sum_{j=d+1}^{T_V} \binom{T_V}{j}\lambda^j\bar{\lambda}^{T_V-j}} - d. \tag{3.8}$$

We get the average time length $E[T_{B_A}]$ for the busy period length T_{B_A} in the awake state as follows:

$$E[T_{B_A}] = E[Q_{B_A}] \times \frac{E[S]}{1-\rho} = \left(\frac{\displaystyle\sum_{j=d+1}^{T_V} j \binom{T_V}{j} \lambda^j \bar{\lambda}^{T_V - j}}{\displaystyle\sum_{j=d+1}^{T_V} \binom{T_V}{j} \lambda^j \bar{\lambda}^{T_V - j}} - d \right) \times \frac{E[S]}{1-\rho}. \quad (3.9)$$

3.3.3 Time Length of Busy Cycle

We define a busy cycle R as the time period from the instant in which a busy period B_A in the awake state ends to the instant in which the next busy period B_A in the awake state ends.

Letting N_{B_L} be the number of busy periods B_L in the listening state in a busy cycle R, the probability distribution and PGF of N_{B_L} can be given as follows:

$$\Pr\{N_{B_L} = n\} = \left(\sum_{j=0}^{d} \binom{T_V}{j} \lambda^j \bar{\lambda}^{T_V - j} \right)^{n-1} \sum_{j=d+1}^{T_V} \binom{T_V}{j} \lambda^j \bar{\lambda}^{T_V - j}, \quad n \geq 1,$$

$$(3.10)$$

$$N_{B_L}(z) = \sum_{n=1}^{\infty} \Pr\{N_{B_L} = n\} z^n = \frac{z \displaystyle\sum_{j=d+1}^{T_V} \binom{T_V}{j} \lambda^j \bar{\lambda}^{T_V - j}}{1 - z \displaystyle\sum_{j=0}^{d} \binom{T_V}{j} \lambda^j \bar{\lambda}^{T_V - j}}. \quad (3.11)$$

Differentiating Eq. (3.11) with respect to z at $z = 1$, the average value $E[N_{B_L}]$ of N_{B_L} is then obtained as follows:

$$E[N_{B_L}] = \frac{1}{\displaystyle\sum_{j=d+1}^{T_V} \binom{T_V}{j} \lambda^j \bar{\lambda}^{T_V - j}}. \quad (3.12)$$

Letting T_B be the time length of the system busy period B. By using Eqs. (3.6), (3.9) and (3.12), we can obtain the average value $E[T_B]$ of T_B as follows:

$$E[T_B] = (1 - \bar{\lambda}^{T_V})E[N_{B_L}]E[T_{B_L}] + E[T_{B_A}]$$

$$= \frac{\rho T_V - \rho E[S] \sum_{j=0}^{d} j \binom{T_V}{j} \lambda^j \bar{\lambda}^{T_V-j} - \rho T_L \sum_{j=d+1}^{T_V} \binom{T_V}{j} \lambda^j \bar{\lambda}^{T_V-j}}{\sum_{j=d+1}^{T_V} \binom{T_V}{j} \lambda^j \bar{\lambda}^{T_V-j}(1-\rho)}.$$

$$(3.13)$$

Letting T_R be the time length of a busy cycle R, the average value $E[T_R]$ of T_R is then given as follows:

$$E[T_R] = E[N_{B_L}]T_V + E[T_{B_A}]$$

$$= \frac{T_V - E[S] \sum_{j=0}^{d} j \binom{T_V}{j} \lambda^j \bar{\lambda}^{T_V-j} - T_L \sum_{j=d+1}^{T_V} \binom{T_V}{j} \lambda^j \bar{\lambda}^{T_V-j}}{\sum_{j=d+1}^{T_V} \binom{T_V}{j} \lambda^j \bar{\lambda}^{T_V-j}(1-\rho)}.$$

$$(3.14)$$

3.4 Analysis of Waiting Time

In this section, we present the analysis of waiting time, including the waiting time in listening state, the waiting time in awake state, and system waiting time.

We perform the analysis of the waiting time in two cases:

(1) Waiting time W_L for the data packets transmitted in the listening state L.
(2) Waiting time W_A for the data packets transmitted in the awake state A.

3.4.1 Waiting Time in Listening State

The waiting time W_L for the data packets transmitted in the listening state can be further divided into two parts: The residual time of a sleep window denoted as T_S^+ and the time elapsed during the listening state denoted as W_{L0}.

The residual time T_S^+ of a sleep window is the time period from the instant of a data packet arriving at a sleep window to the end of the sleep window. Note that all the data packets arriving in a listening window are seen as those data packets arriving at the end of the previous sleep window, the average value $E[T_S^+]$ of T_S^+ is then given as follows:

$$E[T_S^+] = \frac{T_S}{T_V} \times \frac{T_S - 1}{2}.$$

(3.15)

On the other hand, by using the boundary state variable theory, we can get the average value $E[W_{L0}]$ of W_{L0} as follows:

$$E[W_{L0}] = \frac{\sum_{j=1}^{d-1} j(j-1)\binom{T_V}{j}\lambda^j \bar{\lambda}^{T_V - j} + \sum_{j=d}^{T_V} d(d-1)\binom{T_V}{j}\lambda^j \bar{\lambda}^{T_V - j}}{2\left(\sum_{j=1}^{d-1} j\binom{T_V}{j}\lambda^j \bar{\lambda}^{T_V - j} + \sum_{j=d}^{T_V} d\binom{T_V}{j}\lambda^j \bar{\lambda}^{T_V - j}\right)} \times E[S].$$

(3.16)

Combining Eqs. (3.15) and (3.16), the average value $E[W_L]$ for the waiting time W_L is given as follows:

$$E[W_L] = E[W_{L0}] + E[T_S^+]$$

$$= \frac{\sum_{j=1}^{d-1} j(j-1)\binom{T_V}{j}\lambda^j \bar{\lambda}^{T_V - j} + \sum_{j=d}^{T_V} d(d-1)\binom{T_V}{j}\lambda^j \bar{\lambda}^{T_V - j}}{2\left(\sum_{j=1}^{d-1} j\binom{T_V}{j}\lambda^j \bar{\lambda}^{T_V - j} + \sum_{j=d}^{T_V} d\binom{T_V}{j}\lambda^j \bar{\lambda}^{T_V - j}\right)} \times E[S]$$

$$+ \frac{T_S}{T_V} \times \frac{T_S - 1}{2}.$$

(3.17)

3.4.2 Waiting Time in Awake State

The waiting time W_A for the data packets transmitted in the awake state can be obtained by the summing two independent random variables, namely, $W_A = W_0 + W_1$, where W_0 is the waiting time for the classical Geom/G/1 queue, and W_1 is the additional waiting time caused by the vacations introduced in this system.

We obtain the PGF $W_0(z)$ and the average value $E[W_0]$ of W_0 as follows:

$$W_0(z) = \frac{(1 - \rho)(1 - z)}{(1 - z) - \rho(1 - S(\bar{\lambda} + \lambda z))},$$

(3.18)

$$E[W_0] = \frac{\lambda}{2(1 - \rho)} \times E[S(S - 1)].$$

(3.19)

Applying the boundary state variable theory, we can get the PGF $W_1(z)$ and the average $E[W_1]$ of W_1 as follows:

$$W_1(z) = \frac{\lambda\left(1 - Q_{B_A}\left(\frac{z - \bar{\lambda}}{\lambda}\right)\right)}{E[Q_{B_A}](1 - z)},$$ (3.20)

$$E[W_1] = \frac{\displaystyle\sum_{j=d+1}^{T_V} (j - d)(j - d - 1)\binom{T_V}{j}\lambda^j \bar{\lambda}^{T_V-j}}{2\lambda \displaystyle\sum_{j=d+1}^{T_V} (j - d)\binom{T_V}{j}\lambda^j \bar{\lambda}^{T_V-j}}.$$ (3.21)

The data packets that are arriving in the awake state can be classified into two categories:

(1) Data packets that arrive in the listening state and are transmitted in the awake state. The probability of this case is $1 - \rho$, where ρ is the system load defined in Sect. 3.2. Denote the waiting time for these types of data packets as W_{A1}.
(2) Both the arrival and the transmission of a data packet occur in the awake state. The probability of this case is ρ. Denote the waiting time for these types of data packets as W_{A2}.

The data packets arriving in the listening state and being transmitted in the awake state will go through a whole listening period before their transmission, so the expression of the waiting time W_{A1} for these types of data packets is as follows:

$$W_{A1} = T_L + W_0 + W_1$$ (3.22)

where T_L is the time length of the listening window defined also in Sect. 3.2.

For the data packets both arriving and being transmitted in the awake state, the waiting time W_{A2} is the sum of W_0 and W_1. We then have that

$$W_{A2} = W_0 + W_1.$$ (3.23)

Combining Eqs. (3.22) and (3.23), the average value $E[W_A]$ of the waiting time W_A is given as follows:

$$E[W_A] = (1 - \rho)E[W_{A1}] + \rho E[W_{A2}].$$ (3.24)

3.4.3 System Waiting Time

Note that data packets will be transmitted either in the listening state, or in the awake state. Let P_L be the probability that a data packet is transmitted in the listening state, and P_A be the probability that a data packet is transmitted in the awake state. The average value $E[W]$ of the system waiting time W is given as follows:

$$E[W] = P_L E[W_L] + P_A E[W_A] \tag{3.25}$$

where the expressions of P_L and P_A are given as follows:

$$P_L = \frac{\sum_{j=1}^{d} \binom{T_V}{j} \lambda^j \bar{\lambda}^{T_V-j} + \sum_{j=d+1}^{T_V} \binom{T_V}{j} \frac{d}{j} \lambda^j \bar{\lambda}^{T_V-j}}{1 - \bar{\lambda}^{T_V}}, \tag{3.26}$$

$$P_A = \frac{\sum_{j=d+1}^{T_V} \binom{T_V}{j} \left(1 - \frac{d}{j}\right) \lambda^j \bar{\lambda}^{T_V-j}}{1 - \bar{\lambda}^{T_V}}. \tag{3.27}$$

3.5 Performance Measures and Numerical Results

In this section, we first derive performance measures of the system in terms of the handover rate, the energy saving rate of the system and the average response time of data packets, respectively. Then, we present numerical results to evaluate the performance of the system presented in this chapter.

3.5.1 Performance Measures

We define the handover rate ζ_h as the number of handovers for the system changing to an awake state from a sleep state in a slot. It is a performance measure for evaluating the additional energy consumption caused by the sleep mode of the power saving mechanism in IEEE 802.16.

From Sect. 3.3.3, we know that the number of handovers for the system entering the awake state from the sleep state in a busy cycle is N_{B_L}. Therefore, by differentiating N_{B_L} in Eq. (3.11) at $z = 1$, we can obtain the handover rate ζ_h as follows:

$$\zeta_h = \left. \frac{dN_{B_L}(z)}{dz} \right|_{z=1} = \frac{1}{\sum_{j=d+1}^{T_V} \binom{T_V}{j} \lambda^j \bar{\lambda}^{T_V-j}}. \tag{3.28}$$

We define the energy saving rate γ of the system as the ratio of the average time length for the system being in the sleep state to the average total time length of a busy cycle R. With this performance measure, we can evaluate the energy saving efficiency of power saving class type II. For this, we give the energy saving rate γ of the system as follows:

$$\gamma = \frac{E[N_{B_L}]}{E[T_R]} \times T_S. \tag{3.29}$$

We then define the response time Y_d of a data packet as the duration in slots elapsed from the arrival of a data packet to the end of the transmission of that data packet. This is a performance measure for evaluating the user's QoS. The average response time $E[Y_d]$ of data packets is actually equal to the average sojourn time of data packets, namely $E[Y_d]$ is equal to the sum of the average transmission time $E[S]$ of data packets given by Eq. (3.3) and the average waiting time $E[W]$ of the system given by Eq. (3.25). Therefore, we obtain the average response time $E[Y_d]$ of data packets as follows:

$$E[Y_d] = E[S] + E[W]. \tag{3.30}$$

3.5.2   Numerical Results

We set the system parameters as follows: $E[S] = 2$ slots, $T_L = 4$ slots. Moreover, a slot is regarded as one ms. The dependency relationships for the performance measures on the system parameters are illustrated in Figs. 3.3, 3.4, 3.5. The analytical results are compared with simulation results in Figs. 3.3, 3.4, 3.5. The results show good agreements between the analysis results and the simulation results.

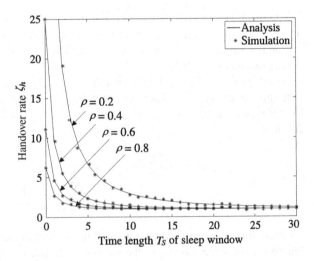

Fig. 3.3 Handover rate versus time length of sleep window

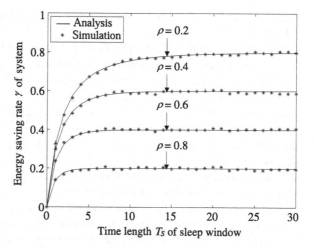

Fig. 3.4 Energy saving rate of system versus time length of sleep window

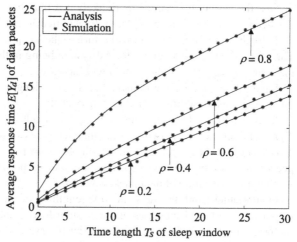

Fig. 3.5 Average response time of data packets versus time length of sleep window

Figure 3.3 shows how the handover rate ζ_h changes with the time length T_S of the sleep window for the different system loads ρ.

For all the system loads ρ, the handover rate ζ_h decreases as the time length T_S of the sleep window increases. This is because the longer the time length of the sleep window is, the lower the possibility is that the data packets arriving during the sleep state could be completely transmitted in the listening window. Thus there will be fewer handovers from the sleep state to the listening state. Therefore, the less the handover rate will be.

On the other hand, for the same time length T_S of the sleep window, the handover rate ζ_h decreases as the system load ρ increases. The reason is the larger the system load is, the more data packets will arrive during the sleep window, then the greater the possibility is that the data packets arriving during the sleep state will not be completely transmitted in the listening window. Therefore, the fewer handovers there are from the sleep state to the listening state, and the less the handover rate will be.

The influence of the time length T_S of the sleep window on the energy saving rate γ of the system for the different system loads ρ is plotted in Fig. 3.4.

It can be observed that for the same system loads ρ, with the increase of the time length T_S of the sleep window, the energy saving rate γ increases sharply and then tend to a nearly fixed value. This is because the longer the time length of the sleep window is, the longer the system is in the sleep state, so the larger the energy saving rate of the system will be. However, when the time length T_S is large enough, the data packets arriving during the sleep window cannot be transmitted in the listening state, and system will be more likely to enter into the busy period. This will result in a fewer number of handovers from the listening state to the sleep state. Therefore, the energy saving rate of the system will tend towards being fixed.

On the other hand, for a same time length T_S of the sleep window, a decrease of the system load ρ results in an increase of the energy saving rate γ. The reason is that the less the system load is, the more likely it is that the data packets arriving during the sleep window will be completely transmitted in the listening window, then the more likely it is that the system will return to the sleep state from the listening state. Therefore, the total time length for the system being in the sleep state is longer, and the energy saving rate of the system will increase.

Figure 3.5 examines the influence of the time length T_S of the sleep window on the average response time $E[Y_d]$ of data packets.

It can be noticed that for the same system loads ρ, the average response time $E[Y_d]$ of data packets increases as the time length T_S of the sleep window increases. This is because the longer the time length of the sleep window is, the longer the data packets arriving during the sleep state will wait in the sleep state, so the longer the average response time of data packets will be.

On the other hand, for the same time length T_S of the sleep window, the average response time $E[Y_d]$ increases as the system load ρ increases. The reason is that the larger the system load is, the far busier the system will be, then the longer the average response time of data packets will be.

From the numerical results presented above, we can conclude that with the sleep mode of power saving class type II provided in IEEE 802.16:

(1) Energy will be conserved.
(2) The system overhead in terms of the handover rate will be introduced.
(3) The user QoS with the average response time will be degraded.

3.6 Optimal Sleep Window Length

From the numerical results shown in Figs. 3.3, 3.4, 3.5, we can conclude that there is a trade-off between the handover rate, the energy saving rate of the system and the average response time of data packets when setting the sleep window size.

Fig. 3.6 System cost function versus time length of sleep window

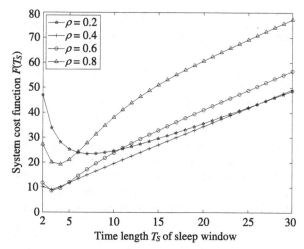

Moreover, we can observe that all the performance measures considered in this chapter are monotone functions versus the time length T_S of the sleep window for different system loads. The optimal value for the time length T_S of the sleep window can be obtained with a straightforward approach. We therefore develop a system cost function $F(T_S)$ as follows:

$$F(T_S) = f_1 \times \zeta_h + f_2 \times \frac{1}{\gamma} + f_3 \times E[Y_d] \tag{3.31}$$

where f_1 is the cost introduced by the handover from the sleep state to the awake state in a busy cycle, f_2 is the reward per slot due to the energy saving when the MS is in the sleep state, and f_3 is the cost resulting from the response time of data packets.

The system parameters of f_1, f_2 and f_3 can be set as needed in practice. For example, if we pay more attention to the cost introduced by the handover rate, the value of f_1 will be larger; if the energy saving rate of the system is the main factor to be considered among all these performance measures, the value of f_2 will be greater; if the average response time of data packets is an important constraint for user QoS, the value of f_3 must not be too small.

In this chapter, we let $f_1=3$, $f_2 = 4$, and $f_3 = 2$. We show how the system cost function $F(T_S)$ changes with the time length T_S of the sleep window for different system loads ρ in Fig. 3.6.

From Fig. 3.6, we can conclude that the system cost function experiences two stages. In the first stage, the system cost function $F(T_S)$ decreases along with an increase in the time length T_S of the sleep window. During this stage, the longer the time length of the sleep window is, the more energy will be saved and the less system cost will be. In the second stage, the system cost function $F(T_S)$ increases with an increase in the time length T_S of the sleep window. During this period, the

Table 3.1 Optimum time length of sleep window

System loads ρ	Optimal time lengths T_S^* of sleep window	Minimal costs $F(T_S^*)$
0.2	7	24.27
0.4	3	7.73
0.6	3	7.81
0.8	4	19.17

longer the time length of the sleep window is, the longer the average response time of data packets is, so the larger the system cost function will be.

Conclusively, there is a minimal system cost function $F(T_S^*)$ for all the system loads when the time length T_S of the sleep window is set to an optimal value T_S^*. The optimal time length of the sleep window for the different system loads are shown in Table 3.1.

3.7 Conclusion

The efficiency of the energy saving scheme used in battery powered MS is one of most important issues in the application of mobile BWA networks. In this chapter, we proposed an effective method to analyze the sleep mode of power saving class type II in IEEE 802.16e. We built a discrete-time queueing model with two types of busy periods: busy period in the listening state and busy period in the awake state, to capture the working principle for the sleep mode of power saving class type II. By using an embedded Markov chain and the boundary state variable theory, we obtained the performance measures of power saving class type II in terms of the handover rate, the energy saving rate of the system and the average response time of data packets. Finally, we presented analysis results and simulation results to explain the nature of the dependency relationships between the performance measures and the system parameters, as well as developed a system cost function to optimize the time length of the sleep window under certain conditions.

Chapter 4
Sleep Mode for Power Saving Class Type III

Considering the property of self-similar traffic shown widely in the networks with multimedia transmission, in this chapter, we present a method to analyze the performance of the enhanced power saving class type III with self-similar traffic. According to the operating principle of the sleep mode in the enhanced power saving class type III, considering the self-similar nature of massive multimedia packets in wireless mobile networks, we build a discrete-time batch arrival multiple vacation queueing model with vacation-delay, in which the batch size is supposed to be Pareto distributed. We present the boundary state variable theory for the batch arrival vacation queueing model to show queue length, waiting time and busy cycle in the steady state. Moreover, we derive performance measures of the system in terms of the handover rate, the energy saving rate of the system, the system utilization and the average response time of data packets, respectively. Finally, we present numerical results to demonstrate the influence of the system parameters on the system performance with different offered loads and different degrees of self-similar traffic.

4.1 Introduction

IEEE 802.16e is an emerging BWA network standard designed to support mobility [IEEE06b, Hwan09b]. How to control the energy consumption of the Mobile Station (MS) is one of the most important issues for the application of the BWA network systems in WCNs. IEEE 802.16e provides a power saving mechanism named a "sleep mode" for conserving the energy consumption of the MS. There are three types of power saving classes based on the sleep mode operations, namely, types I, II and III.

Many authors have been paying attentions to the performance evaluation for the sleep mode operations of power saving class types I and II in recent years.

© Springer Nature Singapore Pte Ltd. 2021
S. Jin, W. Yue, *Resource Management and Performance Analysis of Wireless Communication Networks*, https://doi.org/10.1007/978-981-15-7756-7_4

In [Xiao05], the author proposed an analytical model and investigated the energy consumption in the case of downlink traffic for type I. In [Kong06], the authors investigated and compared the sleep mode operations for power saving class types I and II by using the method of an embedded Markov chain. In [Niu07], the authors developed a PHase type (PH) based Markov chain model for power saving class type I, and proposed a simple utilization function to quantify the efficiency of the sleep mode operation. In [Hwan09a], a new sleep mode scheme called the power saving mechanism with binary traffic indication was proposed, and a mathematical model for the proposed scheme to evaluate the system performance was provided. In [Huo09], the authors analyzed the performance of the sleep mode for power saving class type I by using a discrete-time queueing model with multiple vacations. Moreover, in [Jin09], an enhanced power saving class type III was proposed, and the system performance with user initiated traffic was evaluated.

As we know, the three power saving class types I, II and III differ with each other by their system parameter sets, procedures of activation/deactivation, and policies of availability for data transmission. To our knowledge, there is no work on the performance analysis of the system for the enhanced power saving class type III with self-similar traffic until now. On the other hand, self-similar traffic is shown in many applications, such as the multimedia applications, which are widely found in the Internet [Jin07, Jin13]. In order to improve the energy saving efficiency of the enhanced power saving class type III with self-similar traffic, an improved analytical method must be provided. It is indicated that it would be more accurate and efficient using discrete-time models than continuous-time counterparts when analyzing and designing digital transmitting systems in [Hu05, Hu06, Ma17, Rese06].

In this chapter, we present a new method to analyze the system performance of the enhanced power saving class type III with self-similar traffic. According to the working principle of the sleep mode for the enhanced power saving class type III and considering the self-similar nature of massive multimedia packets in wireless networks, we build a discrete-time batch arrival queueing model with multiple vacations and vacation-delay, and suppose that the batch size is a random variable following a Pareto(c, α) distribution to capture the self-similar property of the network traffic. We present a boundary state variable theory for the batch arrival vacation queueing model, and analyze the queueing model built in this chapter. We investigate the influence of the sleep-delay timer on the system performance with different offered loads and different self-similar degrees for the enhanced power saving class type III with sleep-delay.

The chapter is organized as follows. In Sect. 4.2, we describe the working principle for the sleep mode of the enhanced power saving class type III. Then, we present the system model with multiple vacations and vacation-delay built in this chapter. In Sect. 4.3, we present a performance analysis of the system model in the steady state, including the number of data packets and batches, the queue length, the waiting time and the busy cycle. In Sect. 4.4, we obtain performance measures in terms of the handover rate, the energy saving rate of the system, the system utilization and the average response time of data packets. In Sect. 4.5, we present numerical results to evaluate the system performance. Our conclusions are drawn in Sect. 4.6.

4.2 Working Principle and System Model

In this section, we first describe the working principle for the sleep mode of the enhanced power saving class type III. Then, we establish a discrete-time batch arrival queueing model with multiple vacations and vacation-delay.

4.2.1 Working Principle

System parameters for power saving class type III are as follows: Final-sleep window base, final-sleep window exponent and start frame number for sleep window. Duration of the sleep window is specified as base/exponent.

In the conventional power saving class type III, if there is no packet to be received or to be transmitted, the MS will transmit a MOB_SLP-REQ message to its serving Base Station (BS). After the MS receives the MOB_SLP-RSP message from the BS, the MS will switch to the sleep state in order to save the energy consumption. After the expiration of a sleep window, the MS will automatically return to the awake state to do some assistant operation. If there is no packet to be sent in the buffer of the BS when the assistant operation period is over, the MS will return to the sleep state, otherwise, the MS will enter another awake period to transmit packets until the buffer of the BS becomes empty.

In the enhanced power saving class type III, when there is no packet to be sent to the MS in the buffer of the serving BS, a timer called sleep-delay timer with time length T will be trigged and a sleep-delay period will begin. If there is a data packet arrival in the serving BS within the time length T of the sleep-delay timer, the system will return to the awake state immediately without going through the sleep state. Only when there is no packet to be sent within the time length T of the sleep-delay timer, the system will enter the sleep state after the sleep-delay timer expires.

The setting of the sleep-delay timer will reduce the packet response time, because a data packet arrived within the time length T of the sleep-delay timer can be sent immediately without entering the sleep state. However, too long a time length of the sleep-delay timer will consume too much energy, while too short a time length of the sleep-delay timer will result in an excessive packet delay. Therefore, the setting of the time length T of the sleep-delay timer is an important issue in the enhanced power saving class type III.

In IEEE 802.16e, the MS will wake at any time when an uplink traffic arrives, so the uplink traffic is independent of the sleep mode mechanism. Thus, we focus on the downlink traffic only in this chapter. To simplify the analysis procedure, we take the periodic ranging operation as an example, and suppose a fixed size for the sleep window.

4.2.2   System Model

In the enhanced power saving class type III mentioned above in Sect. 4.2.1, the sleep period is abstracted as a vacation period denoted by V_S; the time spent on the assistant operation when the system return to awake state periodically is abstracted as another vacation period denoted by V_A. The system vacation V is the sum of V_S and V_A.

We assume the buffer capacity in the BS to be infinite. Moreover, to consider the self-similar nature of massive multimedia packets in wireless mobile networks, we assume the input process to be a batch arrival and suppose the batch size to be Pareto distributed. With these assumptions, we can build a discrete-time batch arrival Geom^X/G/1 queueing model with multiple vacations and vacation-delay for this system. The analysis is based upon a boundary state variable theory.

We assume that V_S and V_A have the fixed lengths of T_{V_S} and T_{V_A}, respectively, and V has the length of T_V, obviously, $T_V = T_{V_S} + T_{V_A}$. We regard the sleep-delay period as a vacation-delay period denoted by D. Let the actual length of the vacation-delay period be T_d, the time length T of the sleep-delay timer is obviously the maximal time length of the vacation-delay period, namely, $T_D \le T$. The state transition in this system model is shown in Fig. 4.1.

We denote by Λ the number of data packets in a batch, and call the number of data packets in a batch as a batch size Λ, which is a random variable. The probability

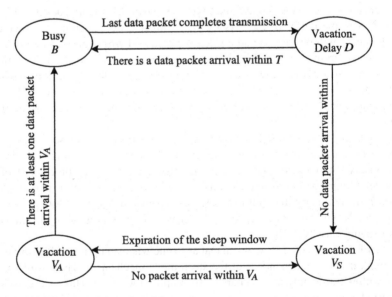

Fig. 4.1 State transition of system model

distribution λ_k, the probability generating function (PGF) $\Lambda(z)$ and the average value $E[\Lambda]$ of the batch size Λ can be given as follows:

$$\lambda_k = \Pr\{\Lambda = k\}, \quad k = 0, 1, 2, \ldots, \tag{4.1}$$

$$\Lambda(z) = \sum_{k=0}^{\infty} z^k \lambda_k, \tag{4.2}$$

$$E[\Lambda] = \sum_{k=0}^{\infty} k \lambda_k. \tag{4.3}$$

Specifically, $\lambda_0 = \Pr\{\Lambda = 0\}$ is equivalent to the probability that there are no data packet arrivals at all in a slot.

Using the first and higher moments of the PGF, we can give the averages and the standard deviation for the system performance in the diffusion approximation of the operation process in the system.

Let the transmission time of a data packet be a random variable denoted by S (in slots) following a general distribution. The probability distribution s_k, the PGF $S(z)$ and the average value $E[S]$ of the transmission time S of a data packet are given as follows:

$$s_k = \Pr\{S = k\}, \quad k = 1, 2, 3, \ldots, \tag{4.4}$$

$$S(z) = \sum_{k=1}^{\infty} z^k s_k, \tag{4.5}$$

$$E[S] = \sum_{k=1}^{\infty} k s_k. \tag{4.6}$$

Let A_S be the number of data packets arrived during the transmission time of a data packet. The PGF $A_S(z)$ of A_S is obtained as follows:

$$A_S(z) = \sum_{k=0}^{\infty} z^k \Pr\{A_S = k\} = S(\Lambda(z)). \tag{4.7}$$

In addition, it is assumed that the inter-arrival time, the batch size, the transmission time of a data packet and the sleep-delay period are mutually independent. Obviously, when the offered load is $\rho = E[\Lambda]E[S] < 1$, the system will arrive at a state of equilibrium. In Sect. 4.3, we will present the performance analysis of the queueing model in the steady state.

4.3   Performance Analysis

In this section, we carry out the performance analysis of the system model in the steady state by addressing the number of data packets and batches, the queue length and waiting time, and the time length of the busy cycle.

4.3.1   Number of Data Packets and Batches

The time interval during which the packets are transmitted continuously is called a busy period denoted by B. Let Q_B be the number of data packets in the buffer of the serving BS when a busy period begins. The PGF $Q_B(z)$ and the average value $E[Q_B]$ of Q_B are given as follows:

$$Q_B(z) = \frac{(1 - \lambda_0^T)(\Lambda(z) - \lambda_0)}{1 - \lambda_0} + \frac{\lambda_0^T\left(\Lambda(z)^{T_V} - \lambda_0^{T_V}\right)}{1 - \lambda_0^{T_V}}, \tag{4.8}$$

$$E[Q_B] = \frac{(1 - \lambda_0^T)E[\Lambda]}{1 - \lambda_0} + \frac{\lambda_0^T T_V E[\Lambda]}{1 - \lambda_0^{T_V}}. \tag{4.9}$$

Let Q_{Bg} be the number of batches in the buffer of the serving BS when a busy period begins, the PGF $Q_{Bg}(z)$ and the average value $E[Q_{Bg}]$ of Q_{Bg} are given by

$$Q_{Bg}(z) = \left(1 - \lambda_0^T\right)z + \frac{\lambda_0^T(\lambda_0 + (1 - \lambda_0)z)^{T_V} - \lambda_0^{T_V}}{1 - \lambda_0^{T_V}}, \tag{4.10}$$

$$E[Q_{Bg}] = 1 - \lambda_0^T + \frac{\lambda_0^T T_V(1 - \lambda_0)}{1 - \lambda_0^{T_V}}. \tag{4.11}$$

4.3.2   Queue Length and Waiting Time

The time axis is divided into segments of equal length called slots in a discrete-time queueing model. For mathematical clarity, we suppose that the departures occur at the moment immediately before the slot boundaries and the arrivals occur at the moment immediately after the slot boundaries. An embedded Markov chain is constructed at the transmission completion instants, and the state of the system is defined by the number of data packets at the embedded Markov points. Let L_d be the queue length at the embedded Markov chain in the discrete-time batch arrival with multiple vacations and vacation-delay built in this chapter.

In the discrete-time batch arrival $Geom^X/G/1$ queueing system with multiple vacations and vacation-delay that we built in this chapter, the stationary queue length L_d can be decomposed into the sum of two independent random variables, namely, $L_d = L_{d0} + L_{d1}$. L_{d0} and L_{d1} are two independent random variables, L_{d0} is the queue length of a classical $Geom^X/G/1$ queue without vacation, and L_{d1} is the additional queue length introduced by multiple vacations, the PGF $L_{d1}(z)$ of L_{d1} is given as follows:

$$L_{d1}(z) = \frac{E[\Lambda](1 - Q_B(z))}{E[Q_B](1 - \Lambda(z))}. \tag{4.12}$$

Let l_j be the probability that the number of data packets remained in the buffer is j after the first packet departures when a busy period started. Let k_j be the probability that the number of data packets arrived during the transmission time of a data packet is j. We can give the transition probability matrix at the embedded Markov points as follows:

$$P = \begin{pmatrix} l_0 & l_1 & l_2 & l_3 & \cdots \\ k_0 & k_1 & k_2 & k_3 & \cdots \\ & k_0 & k_1 & k_2 & \cdots \\ & & k_0 & k_1 & \cdots \\ & & & \vdots & \ddots \end{pmatrix} \tag{4.13}$$

where $l_j = \sum_{i=1}^{j} \Pr\{Q_B = i\} k_{j-i+1}$.

Let the steady-state distribution of the queue length L_d be $\Pi = (\pi_0, \pi_1, \pi_2, \ldots)$. By solving the following set of linear equations:

$$\begin{cases} \Pi P = \Pi \\ \Pi e = 1, \end{cases} \tag{4.14}$$

we can obtain π_j as follows:

$$\pi_j = \pi_0 l_j + \sum_{i=1}^{j+1} \pi_i k_{j+1-i}, \quad j \geq 0 \tag{4.15}$$

where e is a column vector with infinite elements and all elements of the vector are equal to 1.

Then, we can give the PGF $L_d(z)$ of L_d as follows:

$$L_d(z) = \frac{\pi_0 (1 - Q_B(z)) A_S(z)}{A_S(z) - z}. \tag{4.16}$$

By using the normalization condition $L_d(1) = 1$ and the L'Hôspital rule in Eq. (4.16), we can give π_0 as follows:

$$\pi_0 = \frac{1-\rho}{E[Q_B]}. \tag{4.17}$$

Substituting Eq. (4.17) to Eq. (4.16), we have the following equation:

$$
\begin{aligned}
L_d(z) &= \frac{(1-\rho)(1-Q_B(z))A_S(z)}{E[Q_B](A_S(z)-z)} \\
&= \frac{(1-\rho)(1-\Lambda(z))A_S(z)}{E[\Lambda](A_S(z)-z)} \times \frac{E[\Lambda](1-Q_B(z))}{E[Q_B](1-\Lambda(z))}.
\end{aligned} \tag{4.18}
$$

We know that the PGF $L_{d0}(z)$ of L_{d0} can be given as follows:

$$L_{d0}(z) = \frac{(1-\rho)(1-\Lambda(z))A_S(z)}{E[\Lambda](A_S(z)-z)}. \tag{4.19}$$

Therefore, we can decompose the stationary queue length L_d in Eq. (4.16) into the sum of two independent random variables, and we can give the PGF $L_{d1}(z)$ of L_{d1} as follows:

$$L_{d1}(z) = \frac{E[\Lambda](1-Q_B(z))}{E[Q_B](1-\Lambda(z))}. \tag{4.20}$$

Substituting Eqs. (4.3), (4.8) and (4.9) to Eq. (4.12), we can give the PGF $L_{d1}(z)$ of L_{d1} as follows:

$$L_{d1}(z) = \frac{1-\left(\dfrac{(1-\lambda_0^T)(\Lambda(z)-\lambda_0)}{1-\lambda_0} + \dfrac{\lambda_0^T\left(\Lambda(z)^{T_V}-\lambda_0^{T_V}\right)}{1-\lambda_0^{T_V}}\right)}{\left(\dfrac{1-\lambda_0^T}{1-\lambda_0} + \dfrac{\lambda_0^T T_V}{1-\lambda_0^{T_V}}\right)(1-\Lambda(z))}. \tag{4.21}$$

Combining Eqs. (4.19) and (4.21), we can obtain the PGF $L_d(z)$ of L_d in this system as follows:

$$
\begin{aligned}
L_d(z) &= \frac{(1-\rho)A_S(z)}{E[\Lambda](A_S(z)-z)} \\
&\quad \times \frac{1-\left(\dfrac{(1-\lambda_0^T)(\Lambda(z)-\lambda_0)}{1-\lambda_0} + \dfrac{\lambda_0^T\left(\Lambda(z)^{T_V}-\lambda_0^{T_V}\right)}{1-\lambda_0^{T_V}}\right)}{\left(\dfrac{1-\lambda_0^T}{1-\lambda_0} + \dfrac{\lambda_0^T T_V}{1-\lambda_0^{T_V}}\right)(1-\Lambda(z))}.
\end{aligned} \tag{4.22}
$$

Now, we begin to analyze the waiting time of data packets.

We focus on an arbitrary packet in the system called "tagged packet M". We note that the waiting time W of the tagged packet M can be divided into two parts as follows: One is the waiting time W_g of a batch that the tagged packet M belongs to. The other is the total transmission time J of the packets before the tagged packet M in the same batch. W_g and J are two independent random variables, so we have the PGF $W(z)$ for the waiting time W of the tagged packet M as $W(z) = W_g(z)J(z)$.

W_g can be decomposed as the sum of two independent random variables, namely, $W_g = W_{g0} + W_{g1}$, where W_{g0} is the waiting time of a classical $Geom^X/G/1$ queueing model, and W_{g1} is the additional waiting time due to multiple vacations and vacation-delay.

Applying the analysis of the single arrival $Geom/G/1$ queueing model with multiple vacations and vacation-delay, we have

$$W_{g1}(z) = \frac{1 - \lambda_0 - \left(1 - \lambda_0^T\right)(z - \lambda_0)}{E[Q_B](1 - z)} - \frac{(1 - \lambda_0)\lambda_0^T \left(z^{T_V} - \lambda_0^{T_V}\right)}{E[Q_B](1 - z)\left(1 - \lambda_0^{T_V}\right)}.$$

We also have that

$$W_{g0}(z) = \frac{(1 - \rho)(1 - z)}{(\Lambda(S(z)) - z)}, \tag{4.23}$$

$$J(z) = \frac{1 - \Lambda(S(z))}{E[\Lambda](1 - S(z))}. \tag{4.24}$$

Combining Eqs. (4.23) and (4.24), we can obtain the PGF $W(z)$ of W as follows:

$$W(z) = W_{g0}(z)W_{g1}(z)J(z). \tag{4.25}$$

Differentiating Eq. (4.25) with respect to z at $z = 1$, we can obtain the average $E[W]$ of W as follows:

$$E[W] = \frac{E[\Lambda]E[S(S - 1)] + E[\Lambda(\Lambda - 1)](E[S])^2}{2(1 - \rho)}$$

$$+ \frac{(1 - \lambda_0)\lambda_0^T T_V(T_V - 1)}{2E[Q_B]\left(1 - \lambda_0^{T_V}\right)} + \frac{E[\Lambda(\Lambda - 1)]E[S]}{2E[\Lambda]}. \tag{4.26}$$

4.3.3 Busy Cycle

We define the busy cycle R as the time interval between the ending instants of two consecutive busy periods. Let T_R be the time length of the busy cycle R.

If a data packet arrival occurs during the sleep-delay period D, a busy cycle R consists of a sleep-delay period D and a busy period B; If no arrival occurs during the sleep-delay period D, a busy cycle R is the sum of the time length T of the sleep-delay timer, one or more vacation periods V and a busy period B.

The time length of a sleep-delay D is T_D defined in Sect. 4.2.2. Note that $T_D = T$ with probability $\lambda_0{}^T$, and with probability $1 - \lambda_0^T$, T_D equals the conditional arrival interval given that $T_D < T$. Thus, the probability distribution, the PGF $T_D(z)$ and the average $E[T_D]$ of T_D can be given as follows:

$$\Pr\{T_D = j\} = \begin{cases} \lambda_0{}^{T-1}, & j = T \\ \lambda_0^{j-1}(1 - \lambda_0), & j < T, \end{cases} \tag{4.27}$$

$$T_D(z) = \frac{(\lambda_0 z)^T (1 - z) + (1 - \lambda_0)z}{1 - \lambda_0 z}, \tag{4.28}$$

$$E[T_D] = \frac{1 - \lambda_0^T}{1 - \lambda_0}. \tag{4.29}$$

Let N_V be the number of switches between the sleep state and the awake state in a busy cycle R. We get the probability distribution, the PGF $N_V(z)$ and the average $E[N_V]$ of N_V as follows:

$$\Pr\{N_V = j\} = \begin{cases} \lambda_0{}^T \lambda_0{}^{T_V(j-1)}\left(1 - \lambda_0^{T_V}\right), & j \geq 1 \\ 1 - \lambda_0^T, & j = 0, \end{cases} \tag{4.30}$$

$$N_V(z) = 1 - \lambda_0^T + \frac{\lambda_0^T \left(1 - \lambda_0^{T_V}\right) z}{1 - \lambda_0^{T_V} z}, \tag{4.31}$$

$$E[N_V] = \frac{\lambda_0^T}{1 - \lambda_0^{T_V}}. \tag{4.32}$$

Each packet at the beginning of a busy period B will introduce a sub-busy period Θ. A sub-busy period Θ of a data packet is composed of the transmission time S of this packet and the sum of all the sub-busy periods Θ incurred by all the packets arrived during the transmission time S of this packet.

All the sub-busy periods brought by the packets at the beginning of the busy period combine to make a system busy period B, we have that

$$\Theta = S + \underbrace{\Theta + \Theta + \Theta + \cdots + \Theta}_{A_S}, \quad B = \underbrace{\Theta + \Theta + \Theta + \cdots + \Theta}_{Q_B}$$

where A_S is the number of data packets arrived during the transmission time S of a data packet presented in Sect. 4.2.2, Q_B is the number of data packets in the buffer of the serving BS when a busy period begins defined in Sect. 4.3.1.

Considering the Bernoulli arrival process in this system, the PGF $\Theta(z)$ of Θ can be obtained as follows:

$$\Theta(z) = S(z(\Lambda(\Theta(z)))), \tag{4.33}$$

which yields the average $E[\Theta]$ of Θ as follows:

$$E[\Theta] = \frac{E[S]}{1-\rho}. \tag{4.34}$$

Therefore, we can obtain the PGF $B(z)$ of B as follows:

$$
\begin{aligned}
B(z) &= Q_B(z)|_{z=\Theta(z)} \\
&= \frac{\left(1 - \lambda_0^T\right)(\Lambda(\Theta(z)) - \lambda_0)}{1 - \lambda_0} + \frac{\lambda_0^T\left(\Lambda(\Theta(z))^{Tv} - \lambda_0^{Tv}\right)}{1 - \lambda_0^{Tv}}.
\end{aligned}
\tag{4.35}
$$

Differentiating Eq. (4.35) with respect to z at $z = 1$, the average $E[T_B]$ of T_B is then obtained as follows:

$$
\begin{aligned}
E[T_B] &= \frac{\rho(1 - \lambda_0^T)}{(1 - \lambda_0)(1 - \rho)} + \frac{\rho T_V \lambda_0^T}{\left(1 - \lambda_0^{Tv}\right)(1 - \rho)} \\
&= \frac{E[Q_B]E[S]}{1 - \rho}.
\end{aligned}
\tag{4.36}
$$

Conclusively, we can give the average $E[T_R]$ of the busy cycle T_R as follows:

$$
\begin{aligned}
E[T_R] &= E[T_d] + E[N_V]T_V + E[T_B] \\
&= \frac{\left(1 - \lambda_0^T\right)\left(1 - \lambda_0^{Tv}\right) + T_V \lambda_0^T(1 - \lambda_0)}{(1 - \lambda_0)(1 - \rho)\left(1 - \lambda_0^{Tv}\right)}.
\end{aligned}
\tag{4.37}
$$

4.4 Performance Measures

In this section, by using the performance analysis presented in Sect. 4.3, we derive performance measures of the system in terms of the handover rate, the energy saving rate of the system, the system utilization and the average response time of data packets, respectively. They are important performance measures to analyze and evaluate numerically the performance of the system using the enhanced power saving class type III.

4.4.1 Handover Rate

We define the handover rate ζ_h as the number of the switches from the sleep state to the awake state per slot. ζ_h is a measure for evaluating the additional power consumption due to the introduction of the sleep mode. The average number of switches from the sleep state to the awake state is N_V in a busy cycle R. Therefore, we give the handover rate ζ_h as follows:

$$
\begin{aligned}
\zeta_h &= \frac{E[N_V]}{E[T_R]} \\
&= \frac{\lambda_0^T (1 - \lambda_0)(1 - \rho)}{\left(1 - \lambda_0^T\right)\left(1 - \lambda_0^{T_V}\right) + T_V \lambda_0^T (1 - \lambda_0)}.
\end{aligned}
\tag{4.38}
$$

4.4.2 Energy Saving Rate

We define the energy saving rate γ of the system as the energy conserved per slot, by which we can evaluate the efficiency of the sleep mode in the enhanced power saving class type III. Note that the energy consumption can be lowered in the sleep state, let g_1 and g_2 be the energy consumption per slot in the awake state and the sleep state, respectively, we can obtain the energy saving rate γ of the system as follows:

$$
\begin{aligned}
\gamma &= \frac{(g_1 - g_2)E[N_V]T_{V_S}}{E[T_R]} \\
&= \frac{(g_1 - g_2)T_{V_S}(1 - \lambda_0)(1 - \rho)\lambda_0^T}{\left(1 - \lambda_0^T\right)\left(1 - \lambda_0^{T_V}\right) + T_V \lambda_0^T (1 - \lambda_0)}
\end{aligned}
\tag{4.39}
$$

where $E[N_V]T_{V_S}$ reflects the average time length for the system being in the sleep state for a busy cycle R.

4.4.3 System Utilization

The system utilization U_s is defined as the ratio of the transmission time of data packets to the total awake time in a busy cycle for the enhanced power saving class type III. Therefore, we can obtain the system utilization U_s as follows:

$$
\begin{aligned}
U_s &= \frac{E[T_B]}{E[N_V]T_{V_A} + E[T_B] + E[T_D]} \\
&= \frac{E[S]E[Q_B]}{\dfrac{\lambda_0^T T_{V_A}(1-\rho)}{1-\lambda_0^{T_V}} + E[S]E[Q_B] + \dfrac{(1-\rho)\left(1-\lambda_0^T\right)}{1-\lambda_0}} .
\end{aligned}
\tag{4.40}
$$

4.4.4 Average Response Time

We define the response time Y_d of a data packet as the duration in slots elapsed from the arrival of a data packet to the end of the transmission of that packet. This is a measure for evaluating the user's QoS. The average response time $E[Y_d]$ of data packets is the sum of the average transmission time $E[S]$ of data packets given by Eq. (4.6) and the average waiting time $E[W]$ given by Eq. (4.26). Therefore, we obtain the average response time $E[Y_d]$ of data packets as follows:

$$
\begin{aligned}
E[Y_d] &= E[S] + E[W] \\
&= E[S] + \frac{E[\Lambda]E[S(S-1)] + E[\Lambda(\Lambda-1)](E[S])^2}{2(1-\rho)} \\
&\quad + \frac{(1-\lambda_0)\lambda_0^T T_V(T_V-1)}{2E[Q_B]\left(1-\lambda_0^{T_V}\right)} + \frac{E[\Lambda(\Lambda-1)]E[S]}{2E[\Lambda]} .
\end{aligned}
\tag{4.41}
$$

4.5 Numerical Results

Considering the self-similar property of the massive multimedia packets shown in wireless mobile networks, we introduce a batch arrival process in the system model built in this chapter. Let ξ be the batch size under the condition that there is at least one packet arrival in a batch, ξ is supposed to be Pareto(c, α) distributed with that

$$
\xi_k = \Pr\{\xi = k\} = ck^{-(\alpha+1)}, \quad k = 1, 2, 3, \ldots
\tag{4.42}
$$

where c is a normalization factor for $\sum_{k=1}^{\infty} \xi_k = 1$. The parameter α is related to the Hurst factor H by

$$H = \frac{3 - \alpha}{2}, \quad 1 < \alpha < 2.$$

It is obvious that the smaller the parameter α in the Pareto distribution is, the larger the Hurst factor H is, and the greater the degree of self-similarity will be shown in network traffic.

For all the numerical results, we assume the mean length of the transmission time for a data packet is $E[S] = 4$ slots, the sleep window size is $T_{V_S} = 20$ slots, and the value of the assistant operation period size is $T_{V_A} = 3$ slots. Let $g_1 = 1.5$ mW and $g_2 = 0.6$ mW, respectively. A slot is regarded as one ms. The dependency relationships between the performance measures and the system parameters are shown in Figs. 4.2, 4.3, 4.4 and Table 4.1.

Figure 4.2 shows how the handover rate ζ_h changes with the time length T of the sleep-delay timer. It can be found that when the offered load ρ and the Pareto distribution parameter α take the same values, the handover rate ζ_h decreases as the time length T of the sleep-delay timer increases. This is because that the longer the time length T of the sleep-delay timer is, the higher the possibility is that the system will go back to the busy period directly from the sleep-delay state without the switching procedure, so the smaller the handover rate ζ_h will be.

On the other hand, for the same time length T of the sleep-delay timer and the same Pareto distribution parameter α, the handover rate ζ_h increases as the system load ρ decreases. This is because that the less the system load ρ is, the shorter the busy period B and the busy cycle R will be. Therefore, the larger the handover rate ζ_h will be.

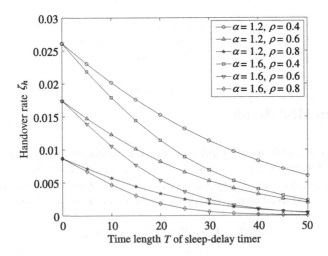

Fig. 4.2 Handover rate versus time length of sleep-delay timer

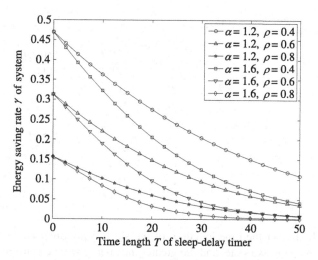

Fig. 4.3 Energy saving rate of system versus time length of sleep-delay timer

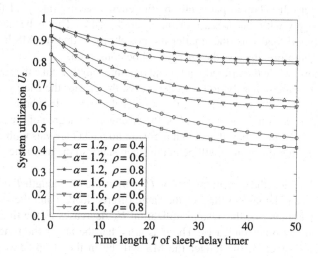

Fig. 4.4 System utilization versus time length of sleep-delay timer

Furthermore, while the offered load ρ and the time length T are the same, the smaller the Pareto distribution parameter α is, the greater the handover rate ζ_h is. The reason is that the smaller the Pareto distribution parameter α is, the bigger the Hurst parameter H is, the stronger the degree of the self-similar of the traffic is, the more possible is that there are no packet arrivals within the sleep window, so the smaller the handover rate ζ_h will be.

Figure 4.3 shows the influence of the time length T of the sleep-delay timer on the energy saving rate γ of the system. We can see that while the offered load ρ and the Pareto distribution parameter α are the same, the longer the time length T of the sleep-delay timer is, and the smaller the energy saving rate γ of the system is. This

Table 4.1 Average response time of data packets

$E[Y_d]$		ρ		
α	T	0.4	0.6	0.8
1.2	10	2641.6	12169.9	13890
1.2	20	1393.6	7192	8864
1.2	30	816.19	4799.1	7881.2
1.2	40	309.45	2222.1	6816.8
1.2	50	126.78	1118	4774.1
1.6	10	289.25	508.59	1014.7
1.6	20	97.06	170.58	770.1
1.6	30	45.87	62.99	202.9
1.6	40	37	58.08	186.22
1.6	50	28.84	50.04	54.04

is because the longer the time length T of the sleep-delay timer is, the longer the MS will be in the awake state, and the greater the energy will be consumed.

On the other hand, for the same time length T of the sleep-delay timer and the same Pareto distribution parameter α, the energy saving rate γ of the system decreases sharply with the increase of the offered load ρ. This is because that the larger the offered load ρ is, the smaller the possibility is that the MS being in the sleep state. Therefore, the less the energy will be conserved.

Furthermore, while the offered load ρ and the time length T take the same values, the smaller the Pareto distribution parameter α is, the greater the energy saving rate γ of the system is. The reason is that the smaller the Pareto distribution parameter α is, the bigger the Hurst parameter H is, the stronger the degree of the self-similar of the traffic is, the more possible is that there are no packet arrivals within the sleep window, the longer the system will be in the sleep state, so the more the energy will be saved.

Figure 4.4 depicts the system utilization U_s versus the time length T of the sleep-delay timer. It can be shown that for the same offered load ρ and the same Pareto distribution parameter α, the system utilization U_s decreases as the time length T of the sleep-delay timer increases. The reason is that the larger the time length of the sleep-delay timer is, the longer the MS will be in the sleep-delay period D. Therefore, the smaller the system utilization U_s will be.

On the other hand, for the same the time length T of the sleep-delay timer and the same Pareto distribution parameter α, the system utilization U_s increases as the system load ρ increases. It is because that the larger the system load ρ is, the shorter the time length T_d of the sleep-delay period and the less the number of switching procedures in a busy cycle will be, so the greater the system utilization U_s will be.

Moreover, while the offered load ρ and the time length T take the same values, the smaller the Pareto distribution parameter α is, the greater the system utilization U_s is. This is because that the smaller the Pareto distribution parameter α is, the larger the number of data packets at the beginning of a busy period is, and the longer the system will be in the awake state, so the greater the system utilization U_s will be.

Due to the finite first factorial moment and the infinite second factorial moment of a Pareto distributed stochastic variable, the average response time $E[Y_d]$ in

Eq. (4.41) is difficult to calculate analytically. Therefore, we investigate the average response time $E[Y_d]$ by using a simulation experiment.

The change trend for the average response time $E[Y_d]$ of data packets versus the Pareto distribution parameter α and the time length T of the sleep-delay timer is presented in Table 4.1 with the length $T_{V_s} = 20$ slots of the sleep window.

From Table 4.1, we know that if the offered load ρ and the Pareto distribution parameter α take the same values, as the increase of the time length T of the sleep-delay timer, the average packet response time $E[Y_d]$ decreases smoothly. This is because the packets arrived during the time length T of the sleep-delay timer can be transmitted directly without going through the sleep window, the waiting time for these packets are short. Therefore, the average packet response time $E[Y_d]$ will be smaller.

On the other hand, for the same the time length T of the sleep-delay timer and the same the Pareto distribution parameter α, a larger offered load ρ results in a longer average packet response time. This is bacause a larger ρ will lead to a longer queue length and a longer waiting time. Moreover, when the offered load ρ and the time length T of the sleep-delay timer are the same, the larger the Pareto distribution parameter α is, the shorter the average packet response time $E[Y_d]$ will be. The reason is that a larger parameter α means a less self-similar degree and a shorter average packet response time.

Conclusively, the introduce of the sleep-delay timer will lower the handover rate and the average response time. On the other hand, the sleep-delay timer will also lower the system utilization and energy saving rate of the system. There is a trade-off between these performance measures. In practice, we can set the time length of the sleep-delay timer as needed.

4.6 Conclusion

We presented a new method to analyze the performance of the enhanced power saving class type III with self-similar traffic. Considering the working principle of the sleep mode in the enhanced power saving class type III and the self-similar nature of massive-scale multimedia packets in the Internet, a batch arrival queueing model with multiple vacations and vacation-delay was built in this chapter. For the analysis of the system performance, we proposed an extended boundary state variable theory for the batch arrival vacation queueing model and analyzed the queue length, the waiting time and the busy cycle in the system model presented in this chapter. Correspondingly, the formula for the performance measures in terms of the handover rate, the energy saving rate of the system, the system utilization and the response time of data packets were given. From numerical results, we investigated the influence of the system configuration parameters on the system performance measures. The theory proposed in this chapter has potential applications in the effective improvement of the sleep mode mechanisms and optimal setting of the system parameters for the enhanced power saving class type III.

Chapter 5
Bernoulli Arrival-Based Sleep Mode in WiMAX 2

As an enhancement of IEEE 802.16e, IEEE 802.16m can save more energy. Considering the digital nature of the networks and regarding the initial sleep window as one half of the subsequent sleep window, in this chapter, we build a discrete-time queueing model with multiple heterogeneous vacations to analyze communication networks using the IEEE 802.16m protocol. We first describe the working principle of this system model, and then present an analytical approach to analyze the sleep mode in the steady state. By using a discrete-time embedded Markov chain, we derive performance measures of the system in terms of the average response time of data packets and the energy saving rate of the system, respectively. Finally, we present numerical results to investigate the influence of the sleep cycle and the arrival rate of data packets on the performance of the system using the sleep mode in IEEE 802.16m.

5.1 Introduction

With the development of communication industry, how to conserve the energy consumption and to extend the lifetime of the battery in Mobile Station (MS) is becoming one of the pressing issues in WiMAX [Xue11]. IEEE 802.16 standard has been designed for WiMAX. As an enhancement of IEEE 802.16 standard, IEEE 802.16e has enhanced the original standard with mobility function added so that the MS can move during services. IEEE 802.16m [IEEE11] is an evolution of mobile WiMAX and is currently being processed for standardization as an amendment of IEEE 802.16e.

IEEE 802.16m is being drafted to meet the 4G network requirements. For a next-generation mobile WiMAX network, IEEE 802.16m can provide the performance improvement required to support future advanced services. In order to improve the power saving efficiency and reduce the signal overhead required for mode switches,

© Springer Nature Singapore Pte Ltd. 2021
S. Jin, W. Yue, *Resource Management and Performance Analysis of Wireless Communication Networks*, https://doi.org/10.1007/978-981-15-7756-7_5

IEEE 802.16m defines a new sleep mode operation. The sleep mode operation in IEEE 802.16m will be explained in detail in Sect. 5.2.

Recently, there have been many studies on the performance analysis of the sleep mode operations in IEEE 802.16e. In [Kong06], the authors evaluated and compared the sleep mode operations for the power saving class types I and II by using the method of an embedded Markov chain. In order to avoid too frequent switching between the sleep state and the awake state, an enhanced power saving class type III was provided in [Jin10], and the system performances were analyzed for user initiated data packet arrivals.

IEEE 802.16m has attracted a lot of research interest. In order to reduce the power consumption and to extend the lifetime of a batter-powered MS, IEEE 802.16m provides a sleep mode scheme. In this sleep mode, the Base Station (BS) negotiates with the MS by the traffic indication. The traffic indication is periodically sent to MS at the beginning instant of every listening window.

In [Hwan09c], for the periodic traffic indication, the sleep cycle is supposed to be fixed, then the performance of the sleep mode in IEEE 802.16m is mathematically analyzed. Moreover, the optimal traffic indication interval is given to minimize the average power consumption of the MS while satisfying the QoS for the mean delay.

In [Jin11b], taking into account that the listening window can be extended and the sleep cycle length can be adjusted, the authors conducted an analytical study on the power consumption and the average data packet delay to minimize the power consumption while satisfying a user-specified packet delay constraint.

In [Baek11a], an efficient sleep mode operation was proposed by using the T_AMS timer. Also, the authors analyzed the proposed scheme by using an embedded Markov chain. The optimal system parameters are given to minimize the power consumption while satisfying the QoS requirement for the average message delay.

We note that the time length of each first sleep window continuing after an extended listening window is the remaining time length of this extended listening window. However, the sum of the average time length of all these first sleep windows continuing after all extended listening windows and the average of all these extended listening windows does not equal the integral times of the sleep cycle.

However, in the research mentioned above, the authors have assumed that the average time length of all the first sleep windows continuing after the extended listening windows is equal to the average of all remaining time lengths of these extended listening windows in their system models. In other words, the key point of the analysis was lost: that being the inter-dependence between these sleep windows and extended listening windows in the system models. Consequently, the description of the sleep mode operation in the system models does not fit IEEE 802.16m standard. Moreover, the energy consumption of traffic indication is neglected, which results in an over-evaluation of the energy conservation.

In this chapter, we disregard these assumptions and give a comprehensive performance analysis of the sleep mode for IEEE 802.16m with a heterogeneous multiple-vacation queueing model.

The chapter is organized as follows. In Sect. 5.2, we describe the working principle of the sleep mode in IEEE 802.16m. In Sect. 5.3, we present a system model with heterogeneous multiple-vacation queue built in this chapter. Then, we present a performance analysis of the system model in the steady state. In Sect. 5.4, we present numerical results to evaluate the system performance. Our conclusions are drawn in Sect. 5.5.

5.2 Working Principle of Sleep Mode in IEEE 802.16m

The aim of IEEE 802.16m is to reduce energy consumption and to improve the system performance. Different from IEEE 802.16e, in the sleep mode of IEEE 802.16m, the BS will negotiate with the MS by using traffic indication. In this way, messages for the sleep request and sleep response, which are used in IEEE 802.16e, are omitted, and the state transition overhead between the listening window and the sleep window is therefore minimized.

For the sleep mode in IEEE 802.16m, the MS is provided with a series of sleep cycles. In order to maintain synchronization between the BS and the MS, traffic indication is sent out at the beginning instant of every sleep cycle. A sleep cycle is the sum of a sleep window and a listening window.

The length of the sleep cycle remains either constant or adaptive, depending on the traffic conditions. To attain a Best Effort (BE) traffic scenario, the time length of the sleep cycle exponentially doubles until the final sleep cycle can be reached. For real-time traffic only, or a real-time and BE mixed traffic scenario, the final sleep cycle is equal to the initial sleep cycle, and the time length of sleep cycle is fixed.

The MS wakes up at every beginning instant of the sleep cycle, and listens to the traffic indication to decide whether to enter a sleep window or not. If there is no pending data packet, the MS will receive a negative traffic indication from the BS, and then the MS will enter a sleep window and power down immediately to save energy. The time length of the sleep cycle will be either doubled or fixed when the MS again receives a negative traffic indication at the beginning instant of the next sleep cycle. If there are pending data packets in the buffer for the MS, the MS will receive a positive traffic indication message from the BS, and the MS is able to receive and send data packets as long as there are data packets in the buffer. When the buffer becomes empty after the transmission of all the buffered data packets and subsequent arrivals, the BS informs the MS of this fact through the last transmission frame.

The sleep windows and the listening windows may be dynamically adjusted. The time length of the listening window may be extended due to the transmission of the data packets (called a listening window extension in IEEE 802.16m). Therefore, the time length of the initial sleep window is the time remaining to the end of the present sleep cycle.

5.3   System Model and Performance Analysis

In this section, we first establish a heterogeneous multiple-vacation queueing model to capture the sleep mode in IEEE 802.16m. Then, we present a performance analysis of the system model in the steady state.

5.3.1   System Model

Considering the "real-time traffic-only" and "real-time and BE-traffic mixed" scenarios in this model, we set a fixed time length for the sleep cycle, and assume the time length of the sleep cycle to be $2T_C$, where T_C is a parameter of the time length that will be set as required by the system, $T_C = 1, 2, 3, \ldots$.

A listening period without data packet transmission and its subsequent sleep window is regarded as a vacation period. The initial sleep window is denoted as V_1, and the subsequent sleep window is denoted as V_2. Let the time length of V_1 and V_2 be T_{V_1} and T_{V_2}, respectively. Note that the initial vacation begins with the end instant of an extended listening window, and a listening window can end at any instant during a sleep cycle. The time length T_{V_1} of the first vacation is assumed to be half of a sleep cycle, namely, $T_{V_1} = T_C$. The time length T_{V_2} of the subsequent vacation is just a sleep cycle length in slots, namely, $T_{V_2} = 2T_C$. An extended listening period with data packet transmission is regarded as a busy period B. A single channel is supposed and the data packets are transmitted according to a First-Come First-Served (FCFS) discipline. A heterogeneous multiple-vacation queueing model is constructed.

In this system model, the time axis is segmented into a sequence of equal intervals of unit duration, called slots. We assume that data packets arrive only just before the end of a slot $t = n^-$ ($n = 1, 2, 3, \ldots$) and depart just after the end of a slot $t = n^+$ ($n = 2, 3, 4, \ldots$). This is called a Late Arrival System (LAS) with delay access.

The arrival process of data packets is assumed to be a Bernoulli process. In a slot, a data packet arrives with probability λ, and no data packet arrives with probability $\bar{\lambda} = 1 - \lambda$, where $0 < \lambda < 1$. Below, we call probability λ the arrival rate of data packets.

The transmission time of a data packet is denoted by S (in slots). The probability distribution s_k, the probability generating function (PGF) $S(z)$ and the average value $E[S]$ of S are given by

$$s_k = \Pr\{S = k\}, \quad k = 1, 2, 3, \ldots, \tag{5.1}$$

$$S(z) = \sum_{k=1}^{\infty} z^k s_k, \tag{5.2}$$

$$E[S] = \sum_{k=1}^{\infty} k s_k. \tag{5.3}$$

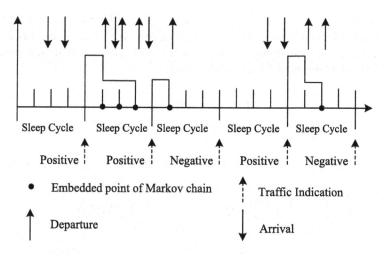

Fig. 5.1 Embedded Markov points chosen in system model

We define the system state of the MS as the number of data packets at the embedded Markov points chosen at the departure instant of every data packet. The embedded Markov points chosen in this system model are shown in Fig. 5.1.

The sufficient and necessary condition for the embedded Markov chain to be positive recurrent is $\rho = \lambda E[S] < 1$. By using a discrete-time embedded Markov chain, we can derive the average response time of data packets and the energy saving rate of the system for the sleep mode of IEEE 802.16m.

5.3.2 Performance Analysis

Let Q_B be the number of data packets in the buffer of the serving BS when a busy period B begins. For the system model presented in this chapter, a busy period B begins in one of the following two cases:

(1) If there is at least one data packet arrival within the first vacation period V_1, when V_1 is over, a busy period will begin. The probability of this case is $1 - \tilde{\lambda}^{T_C}$. For this case, the number Q_B of data packets at the beginning instant of a busy period is given as follows:

$$\Pr\{Q_B = i\} = \frac{\binom{T_C}{i} \lambda^i \tilde{\lambda}^{T_C - i}}{1 - \tilde{\lambda}^{T_C}}, \quad 1 \leq i \leq T_C. \tag{5.4}$$

(2) If there is no arrival within the first vacation period V_1, there must be at least one arrival in one of the subsequent vacation periods V_2. The probability of this event is $\bar{\lambda}^{T_C}$. For this situation, the number Q_B of data packets at the beginning instant of a busy period is given as follows:

$$\Pr\{Q_B = i\} = \frac{\binom{2T_C}{i} \lambda^i \bar{\lambda}^{2T_C - i}}{1 - \bar{\lambda}^{2T_C}}, \quad 1 \leq i \leq 2T_C. \tag{5.5}$$

Combining Eqs. (5.4) and (5.5), we can obtain the PGF $Q_B(z)$ of Q_B as follows:

$$Q_B(z) = \left(1 - \bar{\lambda}^{T_C}\right) \sum_{i=1}^{T_C} \Pr\{Q_{B1} = i\} z^i + \bar{\lambda}^{T_C} \sum_{i=1}^{2T_C} \Pr\{Q_{B2} = i\} z^i$$

$$= \left(\bar{\lambda} + \lambda z\right)^{T_C} - \bar{\lambda}^{T_C} + \frac{\bar{\lambda}^{T_C}}{1 - \bar{\lambda}^{2T_C}} \left((\bar{\lambda} + \lambda z)^{2T_C} - \bar{\lambda}^{2T_C}\right). \tag{5.6}$$

Differentiating Eq. (5.6) with respect to z at $z = 1$, we can give the average $E[Q_B]$ of Q_B as follows:

$$E[Q_B] = \frac{\lambda T_C \left(1 + 2\bar{\lambda}^{T_C} - \bar{\lambda}^{2T_C}\right)}{1 - \bar{\lambda}^{2T_C}}. \tag{5.7}$$

We define the waiting time W of a data packet as the time period from the arrival instant of a data packet to the instant that the transmission of this data packet begins.

The stationary waiting time W of a data packet can be obtained by the sum of two independent random variables, namely, $W = W_0 + W_1$. W_0 is the waiting time for the classical Geom/G/1 queueing model, and W_1 is the additional waiting time due to the multiple vacations introduced in the system.

The PGF $W_0(z)$ of W_0 is given as follows:

$$W_0(z) = \frac{(1 - \rho)(1 - z)}{(1 - z) - \rho (1 - S(z))}. \tag{5.8}$$

Differentiating Eq. (5.8) with respect to z at $z = 1$, the average $E[W_0]$ of W_0 is given as follows:

$$E[W_0] = \frac{\lambda E[S(S - 1)]}{2(1 - \rho)}. \tag{5.9}$$

By applying the boundary state variable theory, we can get the PGF $W_1(z)$ of W_1 in this system as follows:

$$
\begin{aligned}
W_1(z) &= \frac{\lambda\left(1 - Q_B\left(1 - \dfrac{1-z}{\lambda}\right)\right)}{E[Q_B](1-z)} \\
&= \frac{\left(1 - \bar{\lambda}^{2T_C}\right)\left(1 - z^{T_C} + \bar{\lambda}^{T_C}\right)}{T_C\left(1 + 2\bar{\lambda}^{T_C} - \bar{\lambda}^{2T_C}\right)(1-z)} - \frac{\bar{\lambda}^{T_C}\left(z^{2T_C} - \bar{\lambda}^{2T_C}\right)}{T_C\left(1 + 2\bar{\lambda}^{T_C} - \bar{\lambda}^{2T_C}\right)(1-z)}.
\end{aligned}
\tag{5.10}
$$

Differentiating Eq. (5.10) with respect to z at $z = 1$, the average $E[W_1]$ of W_1 is given as follows:

$$
E[W_1] = \frac{(T_C - 1)\left(1 - \bar{\lambda}^{2T_C}\right) + 2(2T_C - 1)\bar{\lambda}^{T_C}}{2\left(1 + 2\bar{\lambda}^{T_C} - \bar{\lambda}^{2T_C}\right)}.
\tag{5.11}
$$

Combining Eqs. (5.9) and (5.11), the average $E[W]$ of W is then given as follows:

$$
\begin{aligned}
E[W] &= E[W_0] + E[W_1] \\
&= \frac{pE[S(S-1)]}{2(1-\rho)} + \frac{E[Q_B(Q_B - 1)]}{2\lambda E[Q_B]} \\
&= \frac{\lambda E[S(S-1)]}{2(1-\rho)} + \frac{(T_C - 1)\left(1 - \bar{\lambda}^{2T_C}\right) + 2(2T_C - 1)\bar{\lambda}^{T_C}}{2\left(1 + 2\bar{\lambda}^{T_C} - \bar{\lambda}^{2T_C}\right)}.
\end{aligned}
\tag{5.12}
$$

Letting T_B be the length of a busy period B for the system model presented in this chapter, the average $E[T_B]$ of T_B can be obtained as follows:

$$
E[T_B] = \frac{E[S]E[Q_B]}{1 - \rho}.
\tag{5.13}
$$

Substituting Eq. (5.7) to Eq. (5.13), we can give the average $E[T_B]$ as follows:

$$
E[T_B] = \frac{\rho T_C}{1 - \rho} \times \frac{1 + 2\bar{\lambda}^{T_C} - \bar{\lambda}^{2T_C}}{1 - \bar{\lambda}^{2T_C}}.
\tag{5.14}
$$

We define a busy cycle R as the time period between the ending instants of two consecutive busy periods. Let T_R be the time length of the busy cycle R.

Also, we define a system vacation V as the time period from the ending instant of a busy period to the beginning instant of the next busy period. The first vacation V_1 and one or more subsequent vacations V_2 (if any) in a busy cycle R combine to produce a system vacation V. Letting T_V be the time length of the system vacation period V, the average $E[T_V]$ of T_V is then given as follows:

$$E[T_V] = T_C \left(1 - \bar{\lambda}^{T_C}\right) + \sum_{m=2}^{\infty} (2m - 1) T_C \bar{\lambda}^{(2m-3)T_C} \left(1 - \bar{\lambda}^{2T_C}\right)$$

$$= T_C \left(1 - \bar{\lambda}^{T_C}\right) + T_C \times \frac{3\bar{\lambda}^{T_C} - \bar{\lambda}^{3T_C}}{1 - \bar{\lambda}^{2T_C}}. \tag{5.15}$$

A busy period B and a system vacation period V will produce a busy cycle R. Combining Eqs. (5.14) and (5.15), the average $E[T_R]$ of T_R can be given as follows:

$$E[T_R] = \frac{T_C \rho}{1 - \rho} \times \frac{1 + 2\bar{\lambda}^{T_C} - \bar{\lambda}^{2T_C}}{1 - \bar{\lambda}^{2T_C}} T_C \left(1 - \bar{\lambda}^{T_C}\right) + T_C \times \frac{3\bar{\lambda}^{T_C} - \bar{\lambda}^{3T_C}}{1 - \bar{\lambda}^{2T_C}}. \tag{5.16}$$

We define the response time Y_d of a data packet as the duration in slots that has elapsed from the arrival of a data packet to the end of the transmission of that data packet. The average response time $E[Y_d]$ of data packets is actually equal to the average sojourn time of data packets, namely $E[Y_d]$ is equal to the sum of the average transmission time $E[S]$ of data packets given by Eq. (5.3) and the average waiting time $E[W]$ of the system given by Eq. (5.12). Therefore, we obtain the average response time $E[Y_d]$ of data packets as follows:

$$E[Y_d] = E[S] + E[W]$$

$$= E[S] + \frac{\lambda \left(E[S](E[S] - 1)\right)}{2(1 - \rho)} + \frac{(T_C - 1)\left(1 - \bar{\lambda}^{2T_C}\right)}{2\left(1 + 2\bar{\lambda}^{T_C} - \bar{\lambda}^{2T_C}\right)}$$

$$+ \frac{2(2T_C - 1)\bar{\lambda}^{T_C}}{2\left(1 + 2\bar{\lambda}^{T_C} - \bar{\lambda}^{2T_C}\right)}. \tag{5.17}$$

The energy saving rate γ of the system is defined as the amount of energy saved per slot. This is one of the most important performance measures for evaluating the energy saving efficiency of the sleep mode in IEEE 802.16m. Therefore, we obtain the energy saving rate γ of the system as follows:

$$\gamma = g_1 \times \frac{E[T_V]}{E[T_R]} - g_2 \times \frac{1}{2T_C} = g_1 \times (1 - \rho) - g_2 \times \frac{1}{2T_C} \tag{5.18}$$

where g_1 is the energy saved per slot for the system being in the sleep window, g_2 is the additional energy consumption per slot due to the traffic indication introduced in IEEE 802.16m.

5.4 Numerical Results

In this section, we numerically evaluate the system performance of the sleep mode of IEEE 802.16m.

By using the system parameters of $g_1 = 0.8$, $g_2 = 0.1$ and $E[S] = 1$, we present numerical results with analysis and simulation to investigate the influence of arrival rate λ of data packets on the system performance. Good agreements between the analysis results and the simulation results are observed.

The average response time $E[Y_d]$ of data packets versus the arrival rate of data packets with different sleep cycles is depicted in Fig. 5.2.

It is illustrated that for the same arrival rate λ of data packets, the longer the time length $2T_C$ of the sleep cycle is, the longer the average response time $E[Y_d]$ of data packets is. The reason is that the longer the sleep cycle is, the longer the MS will be in the sleep window. Since data packets are transmitted in the extended listening window, but not in the sleep window, the average response time of data packets will increase. On the other hand, for the same sleep cycle length $2T_C$, the larger the arrival rate λ of data packets is, the shorter the average response time $E[Y_d]$ of data packets is. This is because the higher the arrival rate of data packets is, the higher the possibility is that the system is in the listening window rather than in the sleep window, so the shorter the average response time of data packets will be.

The influence of the arrival rate λ of data packets on the energy saving rate γ of the system with different sleep cycles is plotted in Fig. 5.3.

It can be observed that for the fixed arrival rate λ of data packets, the energy saving rate γ of the system increases along with the sleep cycle $2T_C$. The reason is that the longer the sleep cycle is, the longer the MS will be in the sleep window, and the more energy will be saved, resulting in a higher energy saving rate of the system. It can also be found that, when the sleep cycle is longer, for example, $2T_C \geq 10$, the

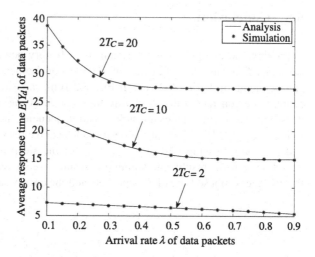

Fig. 5.2 Average response time of data packets versus arrival rate of data packets

Fig. 5.3 Energy saving rate
of system versus arrival rate
of data packets

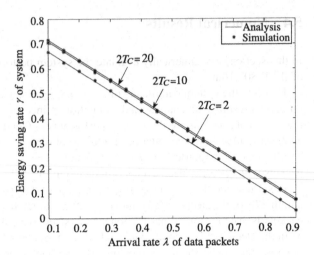

energy saving rate of the system increases slightly as the sleep cycle increases. This
is because the longer the sleep cycle is, the more data packets will arrive during that
sleep cycle, and the longer the subsequent busy period will be, so the energy saving
rate of the system will increase slightly. On the other hand, it can be found that for
the same sleep cycle $2T_C$, the energy saving rate γ of the system decreases as the
arrival rate λ of data packets, as expected, increases. This is because the bigger the
arrival rate of data packets is, the longer the MS will be in the extended listening
window with transmissions. Since more energy will be consumed in an extended
listening window than in a sleep window, the energy saving rate of the system will
be less.

5.5 Conclusion

Improving the energy saving mechanism in battery-powered MSs is one of the
keys in advancing the efficiency of mobile communications technology. As an
enhancement of IEEE 802.16e, IEEE 802.16m can save more energy. In this
chapter, we considered the stochastic time length of the initial sleep window and
constructed a heterogeneous multiple-vacation queueing model. By using a discrete-
time embedded Markov chain, we derived the average response time of data packets
and the energy saving rate of the system for the sleep mode of IEEE 802.16m.
Based on analysis results and simulation results, we showed the trade-off between
the average response time of data packets and the energy saving rate of the system.

Chapter 6
Markovian Arrival-Based Sleep Mode in WiMAX 2

We consider the sleep mode with multimedia applications in WiMAX 2 networks, where the real-time traffic includes the real-time and the Best Effort (BE) traffic mixed. In this chapter, we build a queueing model with multiple heterogeneous vacations to capture the system probability behavior in the networks with multimedia applications. Taking into account the correlation of the real-time traffic, we assume the arrival process to be a Discrete Time Markovian Arrival Process (D-MAP), and analyze the queueing model by using the method of an embedded Markov chain. Then, we give the steady-state distribution for the number of data packets. Accordingly, we derive performance measures of the system in terms of the average response time of data packets, the energy saving rate of the system, and the standard deviation for the number of data packets. We also construct a system cost function to determine the optimal length of the sleep cycle in order to maximize the energy saving rate of the system while satisfying the Quality of Service (QoS) constraint on the average response time of data packets. Finally, we present numerical results to investigate the influence of the system parameters on the system performance.

6.1 Introduction

WCNs are undergoing rapid development. They are intended to deliver wireless services for a large variety of applications in personal, local, campus, and metropolitan areas. Some protocol designs for wireless mesh networks have focused on power efficiency mechanisms.

IEEE 802.16 standard has been designed for fixed subscriber stations. As an enhancement of IEEE 802.16 standard, IEEE 802.16e, called WiMAX [Juso13], has improved the original standard supporting mobility so that the Mobile Station (MS) can move during services. Aiming at the next generation mobile WiMAX, called WiMAX 2, IEEE 802.16m [IEEE11] is currently being processed for standardization.

© Springer Nature Singapore Pte Ltd. 2021
S. Jin, W. Yue, *Resource Management and Performance Analysis of Wireless Communication Networks*, https://doi.org/10.1007/978-981-15-7756-7_6

There have been many studies analyzing the performance of the sleep mode operations for the Types I-III in WiMAX [Jin10, Jin12c, Jin15b, Kong06, Lee11, Li99, Niu07]. In [Kong06], the authors evaluated and compared the sleep mode operations for the power saving class types I and II by using the method of an embedded Markov chain. In order to avoid too frequent switching between the sleep state and the awake state, an enhanced power saving class type III was provided in [Jin10], and the system performances were analyzed for user initiated data packet arrivals. In [Lee11], by increasing the unavailability interval, the authors proposed an Enhanced Power Saving Mechanism (EPSM) where both activated power saving class types I and II exist in an MS. The performance evaluation confirms that an EPSM can save more energy than conventional schemes.

Moreover, in [Saff10], the authors used a two-phase Markovian Arrival Process (MAP) to investigate the effect of the traffic parameters, such as the correlation parameter in the performance evaluation of a system for WiMAX. The optimal sleep mode strategy was obtained by minimizing the total average system cost.

However, an IEEE 802.16m amendment has been drafted to meet the requirements of WiMAX 2. IEEE 802.16m provides the enhanced performance required to support future advanced services, and offers a sleep mode scheme which can reduce the power consumption and extend the lifetime of a battery-powered MS for a multimedia scenario.

WiMAX 2 has recently attracted a lot of research interest. In [Hwan09c], the performance of the sleep mode in WiMAX 2 was mathematically analyzed, where the sleep cycle was supposedly fixed and the traffic indication was periodic. Moreover, the optimal traffic indication interval was given to minimize the average power consumption of the MS while satisfying the QoS on the mean delay. In [Hwan09a], the authors proposed a power-saving mechanism with binary exponential traffic indication in IEEE 802.16e/m. In [Baek11a], an efficient sleep mode operation was proposed by using a T_AMS timer. Also, the authors analyzed the proposed scheme by using an embedded Markov chain. The optimal parameters were given to minimize the power consumption while satisfying the QoS requirement on the average message delay.

In [Jin11b], taking into account that the listening window can be extended and the sleep cycle length can be adjusted, the authors conducted an analytical study on the power consumption and the average data packet delay to minimize the power consumption while satisfying a user-specified packet delay constraint. In [Chen10], a concise analytical model for the sleep mode operation of WiMAX 2 was proposed. Considering both downlink and uplink traffic, the performance measures such as mean waiting time were derived, and the performance comparisons between the WiMAX 2 and WiMAX were conducted with simulations. In [Huo11], considering both downlink traffic and uplink traffic, a queueing model of two servers sharing vacations with close down time and multiple vacations was built for WiMAX 2. The arrival process was assumed to follow a Bernoulli arrival process. By employing a two-dimensional embedded Markov chain, the authors derived the performance measures.

However, in the research on the sleep mode of WiMAX 2 mentioned above, the authors assumed that a sleep cycle was constituted under the condition that the time length of the extended listening window was to be equal to or less than a sleep cycle. The authors also assumed that the average remainder time of the extended listening window was equal to the difference between the time length of the sleep cycle and the average time length of the first sleep window. Moreover, in order to simplify the analysis procedure, the authors assumed uncorrelated traffic, and modeled the incoming traffic using a Poisson process or a Bernoulli process. We know that in multimedia applications, two subsequent slots' states are correlated, and the assumption of a Poisson process or a Bernoulli process is not reasonable.

In this chapter, we present a comprehensive performance analysis by releasing those assumptions introduced in past research mentioned above for actual systems by considering the sleep mode of WiMAX 2 in a scenario where the real-time traffic includes the real-time traffic and the BE traffic mixed. Considering the correlation coefficient between the two subsequent slots' states for the real-time traffic, we assume the arrival process to be a D-MAP. The motivation for using D-MAP is that it can capture the digital nature of modern communication and model correlated traffic. Then, we build a heterogeneous multiple-vacation queue to model the stochastic behavior of the sleep mode in the network system. Note that D-MAP can represent a variety of arrival processes, including the Bernoulli arrival process and the Markov Modulated Bernoulli Process (MMBP). Therefore, the model presented in this chapter is a generalization of the conventional models for analyzing the system performance of the sleep mode in WiMAX 2 with the real-time traffic includes the real-time traffic and the BE traffic mixed.

By using the method of an embedded Markov chain, we present the probability distribution for the number of data packets. Considering the trade-off between the average response time of data packets and the energy saving rate of the system, we then develop a system cost function to determine the optimal length of the sleep cycle in order to maximize the energy saving rate of the system while satisfying the QoS constraint on the average response time of data packets. With numerical results, we investigate the influence of the system parameters on the system performance. Finally, concerning the trade-off between different performance measures, we optimize the sleep cycle with the system cost function.

The chapter is organized as follows. In Sect. 6.2, we describe the system model for the multimedia application in which the real-time traffic includes the real-time and the BE traffic mixed. Then, we present a performance analysis of the system model in the steady state by addressing the number of data packets and the busy cycle. In Sect. 6.3, we obtain performance measures and give the optimal design by constructing a system cost function for the sleep cycle. In Sect. 6.4, we present numerical results to evaluate the system performance. Finally, we draw our conclusions in Sect. 6.5.

6.2 System Model and Performance Analysis

In this section, we first describe the system model for the multimedia application in which the real-time traffic includes the real-time and the BE traffic mixed. Then, we derive the probability distribution for the number of data packets, and analyze the time length of the busy cycle.

6.2.1 System Model

In the sleep mode of WiMAX 2 networks, MS is provided with a set of sleep cycles no matter the system being at the awake period or the sleep period. Except for the initial sleep window, each subsequent sleep cycle consists of both a listening window and a sleep window. As specified in the standard of IEEE 802.16m, the real-time traffic includes the real-time traffic and the BE traffic mixed, and the length of the sleep cycle is fixed. Moreover, the value of the default listening window remains no change. Therefore, in this chapter, we present the system model to describe the sleep mode in WiMAX 2 networks by setting a fixed time length for the sleep cycle.

During the listening window, data packets (if any) will be transmitted in the same way as in the awake period, and the listening window will be extended if necessary. Therefore, in this chapter, we regard the listening window with data packet transmission as an extended listening window, the awake period following an extended listening window as a part of that extended listening window. Then, the extended listening window is defined as a busy period B. Here, the extended listening window is perhaps followed by an awake period. A listening window will terminate on reaching the end of the current listening window including any extensions of the listening window, or reaching the end of the sleep cycle.

We assume the initial sleep window following an extended window to be a vacation period V_1 with the time length T_{V_1}, the subsequent sleep cycle without data packet transmission to be another vacation period V_2 with the time length T_{V_2}, respectively. We note that the extended listening window will end at any instant during a sleep cycle, in other words, the initial sleep window will begin at any instant of an extended listening window. For this, the time length T_{V_1} of the vacation period V_1 is supposed to be one half of the time length T_{V_2} of V_2 as $T_{V_1} = 1/2 \times T_{V_2}$, where T_{V_2} is in fact the time length of the sleep cycle that will be set as required by the system.

We also define a busy cycle R as a time period between the ending instants of two consecutive busy periods. Let T_R be the time length of the busy cycle R. We further define a system vacation V as a time period from the instant where a busy period ends to the instant where the next busy period begins. We can see that the first vacation V_1 and one or more subsequent vacations V_2 (if any) in a busy cycle R combine to produce a system vacation V.

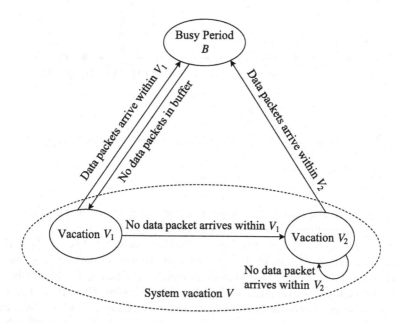

Fig. 6.1 State transition of system model

With these assumptions stated above, we can model the system as a heteroge-neous multiple-vacation queue. The state transition of the system model is illustrated in Fig. 6.1.

Figure 6.1 shows that:

(1) If there is no data packet to be transmitted in the buffer of the serving Base Station (BS), the busy period B will be over, and the vacation V_1 will begin.
(2) If there is at least one data packet arrival within the time length T_{V_1} of the vacation V_1, the MS will begin a new busy period B when V_1 is over. Otherwise, the vacation V_2 will begin.
(3) If there is at least one data packet arrival within the time length T_{V_2} of the vacation V_2, the MS will enter a new busy period B when V_2 is over. Otherwise, the MS will begin another vacation V_2 again.

In this system model, we assume that the time axis is segmented into a sequence of equal intervals of unit duration, called slots. We also assume that data packets arrive only just before the end of a slot $t = \tau^-$ ($\tau = 1, 2, 3, \ldots$), and depart only just after the end of a slot $t = \tau^+$ ($\tau = 2, 3, 4, \ldots$). This is called a Late Arrival System (LAS) with delayed access. Various time epochs at which events occur are depicted in Fig. 6.2.

The system consists of a single channel and a finite buffer size N. Data packets are supposed to be transmitted according to a First-Come First-Served (FCFS) discipline.

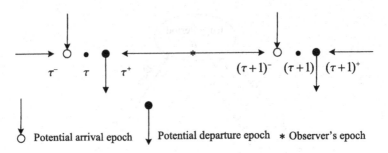

Fig. 6.2 Various time epochs in LAS with delayed access

We define the embedded Markov point as the instant at which a data packet is completely transmitted. We order these embedded points by r $(r = 1, 2, 3, \ldots)$.

Moreover, we define the arrival phase by the stochastic state of the arrival process, and define the state of the system by the number of data packets and the arrival phase at these embedded Markov points. Let L_r and J_r be the number of data packets and the arrival phase at the rth embedded point. Thus, $\{(L_r, J_r),\ r = 1, 2, 3, \ldots\}$ forms a two-dimensional discrete-time embedded Markov chain in a semi-Markov process with state space $\{(l, j) : 0 \leq l \leq N, 1 \leq j \leq m\}$, where N is the system capacity and m is the number of phases in the Underlying Markov Chain (UMC). N and m are the system parameters that we will give in the numerical results.

It should be noted that, in this chapter, the embedded Markov points are chosen at the instant at which a data packet is completely transmitted and the condition of this system model being the stationary state is the number L of data packets in real-time traffic at the embedded Markov points must be less than the system capacity N. With the assumptions and the conditions, in this system model, data packets would not be lost at the embedded Markov points.

Considering the correlation of the real-time traffic, the arrival process is assumed to be a D-MAP with representation $\{D_r,\ r = 0, 1\}$, where D_r is an $m \times m$ matrix, m is the number of phases in the UMC, where $m = 1, 2, 3, \ldots$. The element $[D_0]_{ij}$ is the probability that there is a phase transition from i to j without any data packet arrivals, and the element $[D_1]_{ij}$ is the probability that there is a phase transition from i to j with a data packet arrival, where $1 \leq i, j \leq m$.

The matrix $D = D_0 + D_1$ is the transition probability matrix of the UMC. The matrix D is assumed to be irreducible.

Let $\tilde{\pi}$ be the stationary vector of size $1 \times m$ for the UMC, namely, $\tilde{\pi}$ is a solution for the set of linear equations as follows:

$$\begin{cases} \tilde{\pi} D = \tilde{\pi} \\ \tilde{\pi} e = 1 \end{cases} \tag{6.1}$$

where e is a column vector with m elements and all elements of the vector are equal to 1.

The average arrival rate of the D-MAP is given as follows:

$$\lambda = \tilde{\pi} D_1 e. \tag{6.2}$$

The matrix probability generating function (PGF) of D is defined as follows:

$$D(z) = D_0 + z D_1. \tag{6.3}$$

The transmission time of a data packet is denoted by S (in slots). S is supposed to follow a general distribution. The probability distribution s_k, PGF $S(z)$ and the average value $E[S]$ of S are given as follows:

$$s_k = \Pr\{S = k\}, \quad k = 1, 2, 3, \ldots, \tag{6.4}$$

$$S(z) = \sum_{k=1}^{\infty} z^k s_k, \tag{6.5}$$

$$E[S] = \sum_{k=1}^{\infty} k s_k. \tag{6.6}$$

6.2.2 Number of Data Packets

Let $\pi_{l,j}$ ($0 \leq l \leq N, 1 \leq j \leq m$) be the probability distribution of the two-dimensional embedded Markov chain in the steady state. Then, we have that

$$\pi_{l,j} = \lim_{r \to \infty} \Pr\{L_r = l, J_r = j\}. \tag{6.7}$$

Moreover, let π_l be the probability vector of data packet numbers for the embedded Markov chain in the steady state. π_l can be given as follows:

$$\pi_l = (\pi_{l,1}, \pi_{l,2}, \pi_{l,3}, \ldots, \pi_{l,m}). \tag{6.8}$$

In order to compute π_l, some notations are introduced as follows.

Let $A_S(k, n)$ be a matrix of size $m \times m$, its element $[A_S(k, n)]_{ij}$ ($1 \leq i, j \leq m$) is the conditional probability that there are n data packet arrivals during the transmission time (lasting k slots) of a data packet, given that the transmission started with arrival phase i and ended with arrival phase j.

Let $A_S(n)$ denote a matrix of size $m \times m$ representing that there are n data packet arrivals within the transmission time S of a data packet. $A_S(n)$ can be given as follows:

$$A_S(n) = \sum_{k=1}^{\infty} s_k A_S(k, n), \quad n \geq 0. \tag{6.9}$$

Let $A_{V_1}(T_{V_1}, n)$ be a matrix of size $m \times m$, its element $[A_{V_1}(T_{V_1}, n)]_{ij}$ $(1 \leq i, j \leq m)$ is the conditional probability that there are n data packet arrivals during the vacation V_1 (lasting T_{V_1} slots), given that the arrival process transfer is from phase i at the beginning instant of the vacation V_1 to phase j at the end instant of the vacation V_1.

Let $A_{V_2}(T_{V_2}, n)$ be a matrix of size $m \times m$, its element $[A_{V_2}(T_{V_2}, n)]_{ij}$ $(1 \leq i, j \leq m)$ is the conditional probability that there are n data packet arrivals during the vacation V_2 (lasting T_{V_2} slots), given that the arrival process transfer is from phase i at the beginning instant of the vacation V_2 to phase j at the end instant of the vacation V_2.

Considering the multiple vacations in this chapter, let $A_V(n)$ be a matrix of size $m \times m$ with element $[A_V(n)]_{ij}$ representing the probability that there are n data packet arrivals within the system vacation, given that the system vacation V started with the arrival process of phase i and ended with the arrival process of phase j.

We calculate $A_V(n)$ in the following three possible cases.

(1) For the case of $N > T_{V_2}$:
　　If $1 \leq n \leq T_{V_1}$,

$$A_V(n) = A_{V_1}(T_{V_1}, n) + A_{V_1}(T_{V_1}, 0) \times (I - A_{V_0}(T_{V_2}, 0))^{-1} A_{V_2}(T_{V_2}, n).$$

　　If $T_{V_1} + 1 \leq n \leq T_{V_2}$,

$$A_V(n) = A_{V_1}(T_{V_1}, 0)(I - A_{V_2}(T_{V_2}, 0))^{-1} \times A_{V_2}(T_{V_2}, n).$$

　　If $n > 2T_{V_1}$,

$$A_V(n) = \mathbf{0}.$$

(2) For the case of $T_{V_1} \leq N \leq T_{V_2}$:
　　If $1 \leq n \leq T_{V_1}$,

$$A_V(n) = A_{V_1}(T_{V_1}, n) + A_{V_1}(T_{V_1}, 0) \times (I - A_{V_2}(T_{V_2}, 0))^{-1} A_{V_2}(T_{V_2}, n).$$

　　If $T_{V_1} + 1 \leq n \leq T_{V_2}$,

$$A_V(n) = A_{V_1}(T_{V_1}, 0)(I - A_{V_2}(T_{V_2}, 0))^{-1} \times A_{V_2}(T_{V_2}, n).$$

If $n = N$,

$$A_V(n) = A_{V_1}(T_{V_1}, 0)(I - A_{V_2}(T_{V_2}, 0))^{-1} \times (A_{V_2}(T_{V_2}, N) + A_{V_2}(T_{V_2}, N+1)$$

$$+ A_{V_2}(T_{V_2}, N+2) + \cdots + A_{V_2}(T_{V_2}, T_{V_2})).$$

If $n > N$,

$$A_V(n) = 0.$$

(3) For the case of $T_{V_1} > N$:
   If $n < N$,

$$A_V(n) = A_{V_1}(T_{V_1}, n) + A_{V_1}(T_{V_1}, 0) \times (I - A_{V_2}(T_{V_2}, 0))^{-1} A_{V_2}(T_{V_2}, n).$$

If $n = N$,

$$A_V(n) = (A_{V_1}(T_{V_1}, N) + A_{V_1}(T_{V_1}, N+1) + A_{V_1}(T_{V_1}, N+2) + \cdots$$

$$+ A_{V_1}(T_{V_1}, T_{V_1})) + A_{V_1}(T_{V_1}, 0) \times (I - A_{V_2}(T_{V_2}, 0))^{-1}(A_{V_2}(T_{V_2}, N)$$

$$+ A_{V_2}(T_{V_2}, N+1) + A_{V_2}(T_{V_2}, N+2) + \cdots + A_{V_2}(T_{V_2}, T_{V_2})).$$

If $n > N$,

$$A_V(n) = 0.$$

Among which, I and 0 are the Identity and Zero matrices of size $m \times m$, respectively.

It may be noted here that in order to know $A_S(n)$ and $A_V(n)$, we need to compute $A_S(k, n)$, $A_{V_1}(T_{V_1}, n)$ and $A_{V_2}(T_{V_2}, n)$ efficiently.

It can be seen that $A_S(k, n)$, $A_{V_1}(T_{V_1}, n)$ and $A_{V_2}(T_{V_2}, n)$ satisfy the following equations:

$$\begin{cases} A_S(k, n) = D_0 A_S(k-1, n) + D_1 A_S(k-1, n-1), \quad 1 \le k, 0 \le n \\ A_{V_1}(T_{V_1}, n) = D_0 A_{V_1}(T_{V_1} - 1, n) + D_1 A_{V_1}(T_{V_1} - 1, n-1), \\ \qquad\qquad 0 \le n \le T_{V_1} \\ A_{V_2}(T_{V_2}, n) = D_0 A_{V_2}(T_{V_2} - 1, n) + D_1 A_{V_2}(T_{V_2} - 1, n-1), \\ \qquad\qquad 0 \le n \le 2 T_{V_1}. \end{cases}$$

$$(6.10)$$

Specifically,

$$
\begin{cases}
A_S(0,0) = A_{V_1}(0,0) = I \\
A_S(k,-1) = A_{V_1}(T_{V_1},-1) = A_{V_2}(T_{V_2},-1) = 0 \\
A_S(k,n) = A_{V_1}(T_{V_1},n) = A_{V_2}(T_{V_2},n) = 0
\end{cases}
\tag{6.11}
$$

where $0 \le k < n, 0 \le T_{V_1} < T_{V_2} < n$.

The transition probability matrix P of the two-dimensional Markov chain $\{(L(r), J(r)), r \ge 0\}$ is given as follows:

$$
P = \begin{pmatrix}
F_0 & F_1 & \cdots & F_{N-2} & F_{N-1} & F_N \\
A_S(0) & A_S(1) & \cdots & A_S(N-2) & A_S(N-1) & \hat{A}_S(N) \\
 & A_S(0) & \cdots & A_S(N-3) & A_S(N-2) & \hat{A}_S(N-1) \\
 & & \ddots & \vdots & \vdots & \vdots \\
 & & & A_S(0) & A_S(1) & \hat{A}_S(2) \\
 & & & & A_S(0) & \hat{A}_S(1)
\end{pmatrix}
\tag{6.12}
$$

where elements of the first row are obtained as follows:

$$
\begin{cases}
F_j = \displaystyle\sum_{k=1}^{j+1} A_V(k) A_S(j+1-k), \quad 0 \le j \le N-2 \\
F_{N-1} = \displaystyle\sum_{k=1}^{N-1} A_V(k) A_S(N-k) + A_V(N) A_S(0) \\
F_N = \displaystyle\sum_{k=1}^{N-1} A_V(k) \hat{A}_S(N+1-k) + A_V(N) \hat{A}_S(1).
\end{cases}
\tag{6.13}
$$

The last element for each row, excepting the first row, is given as follows:

$$
\hat{A}_S(n) = \sum_{l=n}^{\infty} A_S(l).
$$

Let $\Pi = (\pi_0, \pi_1, \pi_2, \ldots, \pi_N)$ be the stationary probability vector of size $1 \times (N+1)m$ for the transition probability matrix P. Then, we can obtain π_l by solving a set of linear equations as follows:

$$
\begin{cases}
\Pi P = \Pi \\
\Pi e = 1
\end{cases}
\tag{6.14}
$$

where e is a column vector with $(N+1) \times m$ elements and all elements of the vector are equal to 1.

Let N_d be the number of data packets in the system at the embedded Markov points. The average value $E[N_d]$ of N_d can be given as follows:

$$E[N_d] = \sum_{l=0}^{N} l \pi_l e. \tag{6.15}$$

6.2.3 Busy Cycle

Let T_B be the time length of a busy period B and $E[T_B]$ be the corresponding average value of T_B. By using the analysis results of the classical Geom/G/1 queueing model, $E[T_B]$ can be given as follows:

$$E[T_B] = \frac{E[S]}{\pi_0 e}. \tag{6.16}$$

Let N_V be the number of sleep windows in a busy cycle R. Then, we can get the probability distribution of N_V as follows:

$$\Pr\{N_V = x\} = \begin{cases} 1 - \bar{\lambda}^{T_{V_1}}, & x = 1 \\ \bar{\lambda}^{(2x-3)T_{V_1}}(1 - \bar{\lambda}^{T_{V_2}}), & x \geq 2 \end{cases} \tag{6.17}$$

where $\bar{\lambda} = 1 - \lambda = 1 - \Pi D_1 e$.

Let T_V be the time length in slots of the system vacation period V. The average value $E[T_V]$ of T_V is then given as follows:

$$E[T_V] = T_{V_1}(1 - \bar{\lambda}^{T_{V_1}}) + T_{V_1} \bar{\lambda}^{T_{V_1}} \frac{3 - \bar{\lambda}^{T_{V_2}}}{1 - \bar{\lambda}^{T_{V_2}}}. \tag{6.18}$$

T_R is the sum of the time length T_B of a busy period B and the time length T_V of the system vacation period V. The average value $E[T_R]$ of T_R is then given as follows:

$$E[T_R] = E[T_B] + E[T_V]. \tag{6.19}$$

6.3 Performance Measures and Optimization

In this section, we first derive performance measures of the system in terms of the average response time of data packets, the energy saving rate γ of the system and the standard deviation for the number of data packets in the system, respectively.

Then, we construct a system cost function to trade off the energy saving rate of the system and the average response time of data packets to optimize these values to improve the system performance.

6.3.1 Performance Measures

We define the response time Y_d of a data packet as the duration in slots that has elapsed from the arrival instant to the end of the transmission of that data packet. This is a very important performance measure for evaluating the user QoS for the system. Based on the analysis presented in Sect. 6.2.2, we can obtain the average response time $E[Y_d]$ of data packets as follows:

$$E[Y_d] = \frac{E[N_d]}{\lambda} = \frac{\sum_{l=0}^{N} l\pi_l e}{\lambda}. \tag{6.20}$$

The energy saving rate γ of the system is defined as the amount of energy conserved per slot, which is one of the most important performance measures for evaluating the energy saving efficiency of the sleep mode in IEEE 802.16m. For this, we obtain the energy saving rate γ of the system as follows:

$$\gamma = g_1 \times \left(1 - \frac{E[T_B]}{E[T_R]}\right) - g_2 \times \frac{1}{T_{V_2}} \tag{6.21}$$

where g_1 is the energy saved per slot for the system being in the sleep period, and g_2 is the additional energy consumption per slot due to the traffic indication introduced in IEEE 802.16m.

We know that the design and development of the system require not only average performance measures but also higher moments, because the output stream from one system often forms the input stream to another.

The diffusion degree for the number of data packets in the system can reflect the delay jitter of the system in the steady state. Here, we give the standard deviation $\sigma(N_d)$ for the number of data packets in the system as follows:

$$\sigma(N_d) = \sqrt{E[N_d^2] - (E[N_d])^2} \tag{6.22}$$

where $E[N_d]$ is given by Eq. (6.15), and $E[N_d^2]$ is the second moment for the number of data packets at the embedded Markov points given as follows:

$$E[N_d^2] = \sqrt{\sum_{l=0}^{N} l^2 \pi_l e}. \tag{6.23}$$

Since D-MAP can represent a variety of arrival processes, we decide on a special case of D-MAP, namely, a Bernoulli modulated ON-OFF traffic model, as an example of the system input. The ON-OFF traffic model refers to the arrival process modulated by a two-state Markov chain. When the system is in the ON state, the arrival process follows a Bernoulli process with probability β. If the system is in the OFF state, no arrival occurs.

We assume the system is in the ON state with the probability ω. The correlation coefficient between the two subsequent slot's states is $1-1/M$, where M is the correlation parameter for the average time lengths for the system being in the ON and OFF state, respectively. The mean time length for the system being in the ON state is $M/(1-\omega)$, and correspondingly, M/ω is the average time length for the system being in the OFF state.

Therefore, the matrix PGF $\boldsymbol{D}(z)$ for the traffic model in this chapter can be given as follows:

$$\boldsymbol{D}(z) = \begin{pmatrix} 1-\beta+\beta z & 0 \\ 0 & 1 \end{pmatrix} \begin{pmatrix} 1-\dfrac{1-\omega}{M} & \dfrac{1-\omega}{M} \\ \dfrac{\omega}{M} & 1-\dfrac{\omega}{M} \end{pmatrix}. \tag{6.24}$$

Letting $z=0$ in Eq. (6.24), we can derive the formula of $\boldsymbol{D}_0$ as follows:

$$\boldsymbol{D}_0 = \begin{pmatrix} (1-\beta)\left(1-\dfrac{1-\omega}{M}\right) & (1-\beta)\dfrac{1-\omega}{M} \\ \dfrac{\omega}{M} & 1-\dfrac{\omega}{M} \end{pmatrix}. \tag{6.25}$$

Letting $z=1$ in Eq. (6.24), we can obtain the formula of $\boldsymbol{D}$ as follows:

$$\boldsymbol{D} = \begin{pmatrix} 1-\dfrac{1-\omega}{M} & \dfrac{1-\omega}{M} \\ \dfrac{\omega}{M} & 1-\dfrac{\omega}{M} \end{pmatrix}. \tag{6.26}$$

Combining Eqs. (6.25) and (6.26), we can obtain the formula of $\boldsymbol{D}_1$ as follows:

$$\boldsymbol{D}_1 = \boldsymbol{D} - \boldsymbol{D}_0 = \begin{pmatrix} \beta\left(1-\dfrac{1-\omega}{M}\right) & \beta\dfrac{1-\omega}{M} \\ 0 & 0 \end{pmatrix}. \tag{6.27}$$

Meanwhile, we assume that the transmission time S of a data packet follows a geometric distribution with probability μ called the service rate, $0 \leq \mu \leq 1$. Then, the PGF $S(z)$ and the average value $E[S]$ of S are given as follows:

$$S(z) = \sum_{k=1}^{\infty} \bar{\mu}^{k-1} \mu z^k = \frac{\mu z}{1 - \bar{\mu} z}, \tag{6.28}$$

$$E[S] = \sum_{k=1}^{\infty} k \bar{\mu}^{k-1} \mu = \frac{1}{\mu} \tag{6.29}$$

where $\bar{\mu} = 1 - \mu$.

Now, we will discuss the influence of the sleep cycle on the system performance measures for different correlation parameters M with numerical results. Obviously, the smaller the value of correlation parameters M is, the less the correlation degree of the real-time traffic. Specifically, $M = 1$ means that there is only the BE traffic in the system and $M = \infty$ means that there is only the real-time traffic when the traffic is mixed. The ratio of the BE traffic to the real-time traffic in the mixed traffic is given as $1/M$. When we let B_Q denote the quantity of the BE traffic, R_Q denotes the quantity of the real-time traffic, and M_Q denotes the quantity of the mixed traffic, we can have

$$B_Q = M_Q \times \frac{1}{M}, \tag{6.30}$$

$$R_Q = M_Q \times \left(1 - \frac{1}{M}\right). \tag{6.31}$$

6.3.2 Performance Optimization

Obviously, a longer sleep cycle can enhance the energy saving rate of the system while increasing the average response time of data packets. On the other hand, a shorter sleep cycle can decrease the average response time of data packets while reducing the energy saving rate of the system. Therefore, it is important to determine an optimal value for the sleep cycle in order to maximize the energy saving rate of the system while satisfying the QoS constraint on the average response time of data packets.

To this end, we introduce a system cost function $F(T_{V_2})$, which is proportional to the average response time $E[Y_d]$ of data packets given in Eq. (6.20), and is inversely proportional to the energy saving rate γ of the system given in Eq. (6.21). Thus, we construction the system cost function $F(T_{V_2})$ as follows:

$$F(T_{V_2}) = f_1 \times E[Y_d] + f_2 \times \frac{1}{\gamma} \tag{6.32}$$

where f_1 is the factor of the average response time of data packets to the system cost, and f_2 is the factor of the energy saving rate of the system to the system cost.

6.4 Numerical Results

Now, we present numerical results of the analysis and simulation results to evaluate the system performance for the sleep mode in WiMAX 2.

We list the system parameters settings in Table 6.1 as an example for all the numerical results.

Moreover, the system parameters are fixed as follows: $N = 15$, $E[S] = 1$, $\lambda = 0.2$. From numerical results shown in the following figures, good agreements between the analysis results and the simulation results are observed.

In Fig. 6.3, we show the average response time $E[Y_d]$ of data packets versus the sleep cycle T_{V_2} with different correlation parameters M.

It is illustrated that for the same correlation parameter M, the longer the time length T_{V_2} of the sleep cycle is, the greater the average response time $E[Y_d]$ of data packets is. The reason is that when the time length of the sleep cycle is longer, the system is more likely to be in the sleep period, so the average response time of data packets is longer. On the other hand, for all the time lengths T_{V_2} of the sleep

Table 6.1 Parameter settings in numerical results

Parameters	Values
1 slot	5 ms
Mean transmission time $E[S]$ of a data packet	5 ms
Probability ω being at the ON state	0.8
Power conservation g_1 during sleep period	100 mW
Energy consumption g_2 for sending a traffic indication	15 mJ

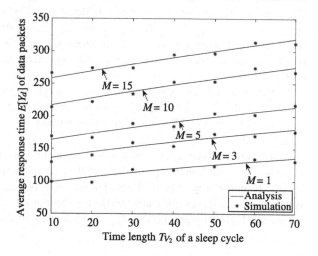

Fig. 6.3 Average response time of data packets versus time length of sleep cycle

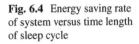

Fig. 6.4 Energy saving rate of system versus time length of sleep cycle

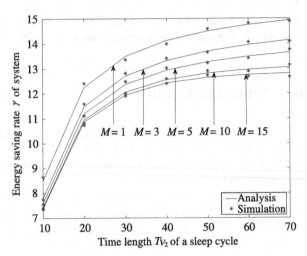

cycle, the larger the correlation parameter M is, the longer the average response time $E[Y_d]$ of data packets is. That is because the greater the correlation parameter is, the denser the data packet arrivals are, and the longer the queueing length is, so the longer the average response time of data packets will be. Moreover, we observe that if the BE traffic only is considered in multimedia application, the average response time of data packets will be under-evaluated.

The energy saving rate γ of the system versus the time length T_{V_2} of the sleep cycle with different correlation parameters M is demonstrated in Fig. 6.4.

It can be observed that for the same correlation parameter M, the energy saving rate γ of the system increases along with the time length T_{V_2} of the sleep cycle. The reason is that when the time length of the sleep cycle is longer, the system is more likely to be in a sleep period. Energy will be saved in the sleep period, so the energy saving rate of the system will be greater. It can also be found that for all the time lengths T_{V_2} of the sleep cycle, the energy saving rate γ of the system decreases as the correlation parameter M decreases. This is because the bigger the correlation parameter M is, the denser the data packet arrivals are, the busier the MS is, and the less energy will be saved. Moreover, we can find that if the BE traffic is only considered for a multimedia scenario, the energy saving effect will be over-evaluated.

In Figs. 6.5 and 6.6, we show how the average response time $E[Y_d]$ of data packets and the energy saving rate γ of the system change versus the correlation parameter M from 1 to 25 with different time lengths T_{V_1} of the initial sleep window from 5 to 35.

Moreover, we observe that for the same correlation parameter M, with a smaller value for the time length T_{V_1} of the initial sleep window, such as $T_{V_1} = 5$, the average response time $E[Y_d]$ of data packets and the energy saving rate γ of the system have smaller results. But when the time length T_{V_1} of the initial sleep window becomes a higher value, such as $T_{V_1} = 35$, results of the average response time

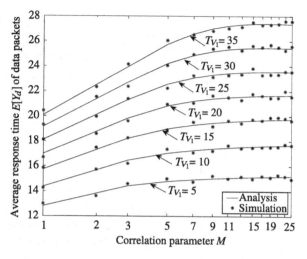

Fig. 6.5 Average response time of data packets versus correlation parameter

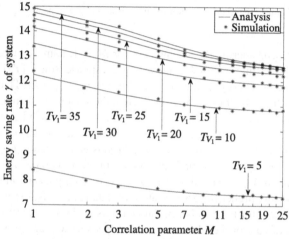

Fig. 6.6 Energy saving rate of system versus correlation parameter

$E[Y_d]$ of data packets and the energy saving rate γ of the system are greater. The reason is that when the time length T_{V_1} of the initial sleep window is longer, the system is more likely to be in the sleep period, so the longer the response time of data packets and the greater the energy saving rate of the system will be.

We also find that the differences among values of the energy saving rate γ of the system are bigger when the time length T_{V_1} is smaller. This is because for a smaller value of the time length T_{V_1}, there will be more sleep windows in a busy cycle. Therefore, a slight increase in the time length T_{V_1} will result in a huge increase in the value of the energy saving rate γ of the system. On the other hand, when the time length T_{V_1} is greater, the system is more likely to transfer from a sleep state to an awake state, and the fewer sleep windows there will be in a busy cycle. Therefore, an increase in the time length T_{V_1} will result in a gradual increase in the energy saving rate γ of the system.

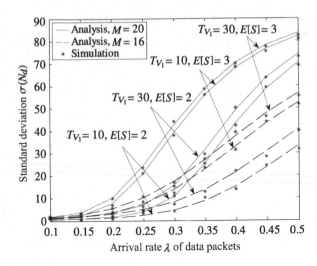

Fig. 6.7 Standard deviation versus arrival rate of data packets

Considering the special case for $M = 1$, namely there is only the BE traffic, for all the time length T_{V_1}, the average response time $E[Y_d]$ of data packets are shorter and the energy saving rate γ of the system are bigger. That is to say, if the correlation of the real-time traffic is neglected, the system performance of the sleep mode with multimedia application in WiMAX 2 networks will be over-optimistically.

The change trend of the standard deviation $\sigma(N_d)$ of the number of data packets versus the arrival rate λ of data packets is depicted in Fig. 6.7.

In Fig. 6.7, we find that for all the time lengths T_{V_1} of the initial sleep cycle, the average transmission time $E[S]$ of data packets and the correlation parameter M, the higher the arrival rate λ of data packets is, the larger the standard deviation $\sigma(N_d)$ for the number of data packets is. This is because the higher the arrival rate λ of data packets is, the more data packets will arrive within a certain time period. Note that the minimal number of data packets is 1, a greater average number of data packets in the buffer during the extended listening period means a maximal number of data packets, thus the standard deviation $\sigma(N_d)$ for the number of data packets will be greater.

We notice that for the same time length T_{V_1} of the initial sleep cycle, the correlation parameter M and the arrival rate λ of data packets, the standard deviation $\sigma(N_d)$ of the number of data packets increases with an increase in the average transmission time $E[S]$ of data packets. The reason is that the longer the average transmission time $E[S]$ of data packets is, the maximal value of the queueing length in the system will be during in the extended listening period. Therefore, the standard deviation $\sigma(N_d)$ of the number of data packets will be greater.

In Fig. 6.7, we also find that for the same time length T_{V_1} of the initial sleep cycle, the same average transmission time $E[S]$ of data packets and the same arrival rate λ of data packets, the larger the correlation parameter M is, and the greater the standard deviation $\sigma(N_d)$ for the number of data packets is. For example, the standard deviation $\sigma(N_d)$ of the number of data packets for the case of $M = 20$ is much greater than that for the case of $M = 16$ for each case of T_{V_1}. This is because in the $M = 20$ case, the quantity of the real-time traffic is much greater than that of the BE traffic in the mixed traffic, so it is more likely that some of the real-time traffic happens to be waiting in the buffer. This results in the standard deviation $\sigma(N_d)$ for the number of data packets being greater.

In Fig. 6.7, we also observe that for the same correlation parameter M, the same average transmission time $E[S]$ of data packets and the same arrival rate λ of data packets, the longer the time length T_{V_1} is, and the greater the standard deviation $\sigma(N_d)$ for the number of data packets is. This is because the longer the time length T_{V_1} is, the greater the maximal number of data packets arriving within the time length T_{V_1} will be, so that the number range of the data packets will become wide at the beginning instant of the data packets' transmission. It causes the standard deviation $\sigma(N_d)$ of the data packets' number to be greater.

Figure 6.7 clearly shows that the standard deviation $\sigma(N_d)$ for the number of data packets is heavily dependent on the correlation of the real-time traffic. If we underestimate the correlation degree of the real-time traffic in multimedia application, as in the system models presented in the available literature, the standard deviation $\sigma(N_d)$ for the number of data packets will be under-evaluated. It means that the results from analyzing the system model will be considerably separated from actual systems.

To maximize the energy saving rate γ of the system while satisfying the QoS constraint on the average response time $E[Y_d]$ of data packets, we investigate the changing trend for the system cost function $F(T_{V_2})$ given by Eq. (6.32) versus the time length T_{V_2} of the sleep cycle with different correlation parameters M. Taking $f_1 = 0.3$ and $f_2 = 0.5$ as an example, the results are shown in Fig. 6.8.

From Fig. 6.8, we see that two stages are experienced by the system cost function $F(T_{V_2})$. In the first stage, the system cost function $F(T_{V_2})$ decreases along with an increase in the time length T_{V_2} of the sleep cycle. During this stage, the longer the time length of the sleep cycle is, the longer the MS will stay in the sleep period, and the more energy will be saved, so the less the system cost function will be. In the second stage, the system cost function $F(T_{V_2})$ increases along with the time length T_{V_2} of the sleep cycle. During this period, the longer the time length of the sleep cycle is, the longer the MS will stay in the sleep period, and the longer the average response time $E[Y_d]$ of data packets is, and thus the greater the system cost function will be.

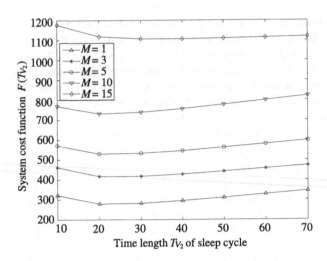

Fig. 6.8 System cost function versus time length of sleep cycle

Table 6.2 Optimum time length of sleep cycle

Correlation parameters	Optimum time lengths $T_{V_2}^*$ of the sleep cycle	Minimum costs $F(T_{V_2}^*)$
$M = 1$	20	279.4107
$M = 3$	20	417.1222
$M = 5$	20	529.5695
$M = 10$	20	731.3965
$M = 15$	30	1106.5756

Conclusively, there is a minimal cost $F(T_{V_2}^*)$ when the time length T_{V_2} of the sleep cycle is set to an optimal value. The optimal time length of the sleep cycle and the minimal cost for the different correlation parameters are summarized in Table 6.2.

6.5   Conclusion

How to improve the energy saving mechanism in battery powered MSs is one of the most pressing questions facing researchers of communication networks. As an enhancement of WiMAX, WiMAX 2 can save more energy. In this chapter, concerning the real-time traffic including the real-time and the BE traffic mixed, by setting the initial sleep window as one half of the fixed sleep cycle, we built a heterogeneous multiple-vacation queueing model. Considering the correlation of the data packets shown in real-time traffic with multimedia application, the arrival process was assumed to be D-MAP. The steady-state distribution of the queueing model was derived by using the method of an embedded Markov chain. For the performance measures, the average response time of data packets, the energy saving rate of the system and the standard deviation for the number of data packets were given to evaluate the system performance with different correlation parameters. Moreover, an optimal design with a system cost function for the sleep cycle was given to maximize the energy efficiency with the constraint of user QoS in term of the average response time of data packets. Finally, by using numerical results obtained from analysis and simulation, the influence of the system parameters on the system performance was studied and the trade-off between different performance measures was investigated. The research in this chapter provides a theoretical basis for improving the power saving schemes in WiMAX 2 and has potential applications for solving other energy related problems in modern communication networks.

Chapter 7
Two-Stage Vacation Queue-Based Active DRX Mechanism in an LTE System

When using a Discontinuous Reception (DRX) mechanism in Long Term Evolution (LTE) for wireless communication of high-speed data, two different operational modes are employed: Idle DRX and Active DRX. In this chapter, we propose an enhanced energy saving strategy based on the Active DRX mechanism in an LTE system to improve the sleep strategy for a better balance between response performance and energy efficiency by introducing a sleep-delay timer. We build a discrete-time multiple-vacation queueing model with a vacation-delay period and a set-up period by addressing the busy period, the queue length, the waiting time and the busy cycle. We derive performance measures of the system in terms of the handover rate, the energy saving rate of the system and the average response time of data packets, respectively. We present numerical results to show the impact of the thresholds of the short DRX stages, the time lengths of the sleep-delay timer, the short DRX stage and the long DRX stage on the system performance so that we can evaluate the influence of the configuration parameters on the system performance. Finally, by considering the trade-off between different performance measures, we optimize the enhanced energy saving strategy for the Active DRX.

7.1 Introduction

With the development of WCNs, a raft of new applications, such as instant message services, multimedia services and social network services, have imposed higher demands than before for high data transmission rates, greater energy conservation and larger system capacities [Jin11a, Jin14]. The LTE project for UMTs has been initiated by the 3GPP [Wiga09]. The purpose of LTE is to accommodate more users in every cell, accelerate the data transmission rate, and reduce the energy consumption and the cost of the network. Many telecom operators have deployed LTE networks and concentrated their research into LTE productions

© Springer Nature Singapore Pte Ltd. 2021
S. Jin, W. Yue, *Resource Management and Performance Analysis of Wireless Communication Networks*, https://doi.org/10.1007/978-981-15-7756-7_7

[Abet10, Nga11]. Recently, techniques in LTE projects have become a focus of research.

As an energy saving strategy, a DRX mechanism is defined in the Medium Access Control (MAC) specification of LTE system [Miho10]. There are two operational modes in the DRX mechanism, namely, Idle DRX mechanism and Active DRX mechanism [Ting11]. For an Idle DRX mechanism, the system stays in the Radio Resource Control IDLE (RRC_IDLE) state when there is no data packet in the user's buffer. If a data packet arrives, the User Equipment (UE) will send a RRC request and rebuild the port connection to an evolved Node B (eNodeB). For an Active DRX mechanism, the system stays in the state of RRC_CONNECTED even if there is no data packet to be transmitted or to be received. But the UE will shut down the transmitter-receiver unit temporarily to save energy. When there is a data packet to be transmitted or received, the UE will switch into the working state from the sleep state directly without rebuilding the RRC connection with an eNodeB and the signaling overhead will be decreased [Luo11].

Recently, many energy saving strategies in DRX mechanisms have been investigated. For example, in [Bont09], the authors explained the energy saving methods for both a network attached mode and a network idle mode in LTE system. They then defined the optimum criteria for different applications in the RRC_CONNECTED and RRC_IDLE states. They found that through the reasonable setting of the DRX parameters, energy could be saved. In [Yin12], the author proposed a simple but efficient application aware DRX mechanism to optimize the system performance of LTE-Advanced (LTE-A) networks with Carrier Aggregation (CA). With CA technology, the UE could support high transmission rates over a wider bandwidth, and the enhanced DRX mechanism was able to achieve nearly 50% energy saving compared to the conventional DRX mechanism. In [Li10], the authors provided an algorithm in which the cycle length of the DRX was adjusted dynamically to realize the variable growth of the sleep cycle. The result of simulation experiments in [Li10] indicated that the algorithm had an improved effect on the energy saving.

Conclusively, most available research into DRX mechanisms has been focused on improving energy savings to the system or optimizing the system parameters. Moreover, performance evaluations of the DRX mechanism have relied solely on simulation experiments. In pursuing greater energy conservation in the DRX mechanism, a lesser latency is another important constraint on the QoS of wireless networks. However, in order to improve the system performance of the DRX mechanism, very accurate models that faithfully reproduce the stochastic behavior must be produced.

In this chapter, we propose an enhanced energy saving strategy based on the Active DRX mechanism in an LTE system to improve the sleep strategy for a better balance between response performance and energy efficiency by introducing a sleep-delay timer. We call this enhanced energy saving strategy based the Active DRX mechanism an "enhanced energy saving strategy". For this, we build a discrete-time queueing model with two stages of vacations, taking into account the sleep-delay period and the waking-up procedure in order to model the system operation. Moreover, we exactly analyze the system model by using a DTMC and numerically evaluate the performance measures with regard to the handover rate, the energy consumption and the average response time of data packets. Finally, we optimize the enhanced energy saving strategy by considering a trade-off between different performance measures. The numerical results show that the overall performance of the enhanced energy saving strategy proposed in this chapter is superior to the conventional one in LTE system.

The chapter is organized as follows. In Sect. 7.2, we describe the enhanced energy saving strategy proposed in this chapter. In Sect. 7.3, we present the system model with vacation-delay period and set-up period to capture the system operation. Then we present a performance analysis of the system model in the steady state by addressing the busy period, the queue length, the waiting time and the busy cycle. And then, we obtain performance measures. In Sect. 7.4, we present numerical results to evaluate and optimize the system performance based on the enhanced energy saving strategy proposed in this chapter. Our conclusions are drawn in Sect. 7.5.

7.2   Enhanced Energy Saving Strategy

In this section, we present the enhanced energy saving strategy proposed in this chapter. In this enhanced energy saving strategy, we introduce a sleep-delay timer and set a threshold to restrict the number of the short DRX stages in a DRX cycle, denoted by K as a system parameter. Obviously, a larger number of the short DRX stages in a DRX cycle will shorten the average response time of data packets. On the other hand, a lesser value of K corresponds to a higher number of long DRX stages in a DRX cycle, which will be beneficial for terminal energy conservation. Therefore, a reasonable threshold K will make the system more effective.

In order to guarantee a low latency while reducing the energy consumption, we introduce a sleep-delay timer before the system enters into a sleep state from a working state, and we adopt a procedure where the system switches to a sleep state from a working state, resulting in the enhanced energy saving strategy. The time period within the time length of the sleep-delay timer is called a sleep-delay period and the maximal length of the sleep-delay period is denoted as T. During the procedure where the system switches to a sleep state from a working state, the transmitter-receiver unit should be activated. We call this procedure a waking-up procedure.

The working principle of the enhanced energy saving strategy proposed in this chapter is given as follows:

(1) When all the data packets in the buffer are completely transmitted, a sleep-delay timer will be activated and the system will enter into the sleep-delay period from the working state. If there is a data packet arrival within the sleep-delay period, the sleep-delay period will end at once and the data packet will be transmitted immediately. However, if there is no data packet arrival within the sleep-delay period, the system will switch into a short DRX stage after the sleep-delay timer expires.

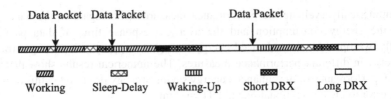

Fig. 7.1 Time sequence of proposed enhanced energy saving strategy

(2) If there is at least one data packet arrival in a short DRX stage, the system will initiate a waking-up procedure and then enter into the working state. If there are no data packets arriving during a short DRX stage, and the number of short DRX stages is less than the threshold K, the system will continue with another short DRX stage when the previous short DRX stage is over. If the number of short DRX stages reaches the threshold K and there has still been no data packet arrival, a long DRX stage will begin after the last short DRX stage is finished.

(3) If a data packet arrives during a long DRX stage, when the long DRX stage is over, the system will initiate a waking-up procedure and then enter into a working state. Otherwise, the system will start a new long DRX stage until there is a data packet arrival.

This process will be repeated.

The time sequence of the enhanced energy saving strategy proposed in this chapter is illustrated in Fig. 7.1.

In this enhanced energy saving strategy, a shorter time length for the sleep-delay timer will make the system to enter into the sleep state quickly. This will reduce the average response time of data packets, but will increase the energy consumption. However, a shorter time length will also cause the system to switch between the sleep state and the working state frequently, and the system handover rate will increase. On the other hand, for a larger threshold for the short DRX stages, the system will spend more time in the short DRX stages. This will decrease the energy conservation and increase the system handover. A larger threshold for the short DRX stages has the advantage of reducing the average response time of data packets. Therefore, the values of the time length of the sleep-delay timer and threshold for the short DRX stages play an important role in the application of the enhanced energy saving strategy based Active DRX mechanism in LTE system.

7.3 System Model and Performance Analysis

In this section, we build a discrete-time multiple-vacation queueing model with vacation-delay period and set-up period to model the system operation. Then, we analyze the system model in the steady state and derive performance measures of the system.

7.3.1   System Model

In order to evaluate the system performance mathematically, the system model should be constructed reasonably and the values for the system parameters should be set optimally.

The sleep-delay period is seen as a vacation-delay period D, and the time length of D is denoted as T_D. The short DRX stage and the long DRX stage are regarded as the short vacation period V_1 and the long vacation period V_2, respectively. The time lengths of the short vacation period V_1 and the long vacation period V_2 are denoted as T_{V_1} and T_{V_2}, respectively. The time period from the instant that the first short vacation period V_1 begins to the instant that the waking-up procedure begins is regarded as a system vacation V. One or more short vacation periods V_1 and long vacation periods V_2 combine to constitute a system vacation V, and the time length of V is denoted as T_V.

The sleep-delay period is seen as a vacation-delay period D, and the time length of D is denoted as T_D. The short DRX stage and the long DRX stage are regarded as the short vacation period V_1 and the long vacation period V_2, respectively. The time lengths of the short vacation period V_1 and the long vacation period V_2 are denoted as T_{V_1} and T_{V_2}, respectively. The time period from the instant that the first short vacation period V_1 begins to the instant that the waking-up procedure begins is regarded as a system vacation V. One or more short vacation periods V_1 and long vacation periods V_2 combine to constitute a system vacation V, and the time length of V is denoted as T_V.

We regard the waking-up procedure as a set-up period Δ. Let T_Δ denote the time length of Δ. The time period for the data packets being transmitted continuously is seen as a busy period B, and its length is denoted as T_B. A busy cycle R is defined as the time period from the instant that a busy period B ends to the instant that the next busy period B ends. The busy cycle can be regarded as the DRX cycle in LTE system.

A queueing model with two stages of vacations is built considering the vacation-delay period and the set-up period in this chapter. The state transition of this queueing model is illustrated in Fig. 7.2, where K is the maximum number of short vacation periods V_1 in a busy cycle defined as the threshold in this chapter.

To comply with the digital nature of modern communication, we evaluate the system performance by using a discrete-time queueing model.

In this system model, the time axis is segmented into a series of equal intervals called slots. The data packet arrivals or departures are supposed to happen only at the boundary of a slot. A single channel and an infinite capacity are considered in this model. The transmissions of data packets are supposed to follow a First-Come First-Served (FCFS) discipline.

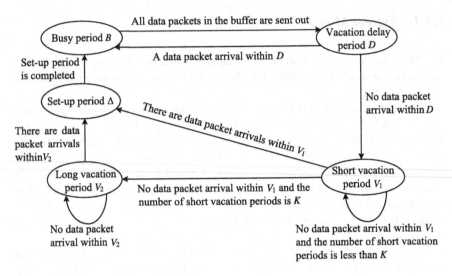

Fig. 7.2 State transition of system model

Moreover, we assume that the arrival process of data packets follows a Bernoulli distribution, such that a data packet arrives in a slot with probability λ $(0 < \lambda < 1)$, and no data packet arrives with probability $\bar{\lambda}$, $\bar{\lambda} = 1 - \lambda$. The transmission time S (in slots) of a data packet is supposed to follow a general distribution. Let the transmission time S of a data packet be an independent and identically distributed random variable. The probability distribution s_k, the probability generating function (PGF) $S(z)$ and the average value $E[S]$ of S can be expressed as follows:

$$\Pr\{S = k\} = s_k, \quad k = 1, 2, 3, \ldots, \tag{7.1}$$

$$S(z) = \sum_{k=1}^{\infty} z^k s_k, \tag{7.2}$$

$$E[S] = \sum_{k=1}^{\infty} k s_k \tag{7.3}$$

We select the departure instant of every data packet as the embedded Markov point and define the state of the system by the number of data packets at these embedded Markov points. Therefore, the system structure at the embedded points constitutes an embedded Markov chain. The sufficient and necessary condition for the embedded Markov chain to be stable is $\rho = \lambda E[S] < 1$.

7.3.2   Busy Period

In this system model, a busy period B begins either at the completion moment of a vacation-delay period D because there is a data packet arrival, or at the instant that a set-up period Δ ends. The set-up period Δ begins with a short vacation period V_1 or a long vacation period V_2, during which there is at least one data packet arrival.

Let P_D be the probability that there is a data packet arrival in the vacation-delay period D. For this case, the system will enter into a busy period B directly from the vacation-delay period D. P_D can be given as follows:

$$P_D = 1 - \bar{\lambda}^T. \tag{7.4}$$

Let P_{V_1} be the probability that the system will switch into the set-up period Δ from a short vacation period V_1, and P_{V_2} be the probability that the system will enter into the set-up period Δ from a long vacation period V_2. P_{V_1} and P_{V_2} can be given as follows:

$$P_{V_1} = 1 - \bar{\lambda}^{KT_{V_1}}, \tag{7.5}$$

$$P_{V_2} = \bar{\lambda}^{KT_{V_1}}. \tag{7.6}$$

Let $A_j^{V_1}$ be the probability that there are j data packet arrivals during a short vacation period V_1 given that there is at least one data packet arrival in this short vacation period. $A_j^{V_1}$ is given as follows:

$$A_j^{V_1} = \frac{\lambda^j \bar{\lambda}^{T_{V_1} - j}}{1 - \bar{\lambda}^{T_{V_1}}}, \quad j \geq 1. \tag{7.7}$$

Let $A_j^{V_2}$ be the probability that there are j data packet arrivals during a long vacation period V_2 given that there is at least one data packet arrival in this long vacation period. $A_j^{V_2}$ is given as follows:

$$A_j^{V_2} = \frac{\lambda^j \bar{\lambda}^{T_{V_2} - j}}{1 - \bar{\lambda}^{T_{V_2}}}, \quad j \geq 1. \tag{7.8}$$

Let A_V be the number of data packets that have arrived at the beginning instant of a set-up period Δ. The probability distribution of A_V can be given as follows:

$$\Pr\{A_V = j\} = \begin{cases} P_{V_1} A_j^{V_1} + P_{V_2} A_j^{V_2}, & 1 \leq j \leq T_{V_1} \\ P_{V_2} A_j^{V_2}, & T_{V_1} + 1 \leq j \leq T_{V_2}. \end{cases} \tag{7.9}$$

Accordingly, the PGF $A_V(z)$ of A_V can be given as follows:

$$A_V(z) = \sum_{j=1}^{T_{V_1}} \Pr\{A_V = j\}z^j + \sum_{j=T_{V_1}+1}^{T_{V_2}} \Pr\{A_V = j\}z^j$$

$$= P_{V_1} \times \frac{(\bar{\lambda}+\lambda z)^{T_{V_1}} - \bar{\lambda}^{T_{V_1}}}{1 - \bar{\lambda}^{T_{V_1}}} + P_{V_2} \times \frac{(\bar{\lambda}+\lambda z)^{T_{V_2}} - \bar{\lambda}^{T_{V_2}}}{1 - \bar{\lambda}^{T_{V_2}}}. \tag{7.10}$$

Differentiating Eq. (7.10) with respect to z at $z = 1$, the average value $E[A_V]$ of A_V can be obtained as follows:

$$E[A_V] = P_{V_1} \times \frac{\lambda T_{V_1}}{1 - \bar{\lambda}^{T_{V_1}}} + P_{V_2} \times \frac{\lambda T_{V_1}}{1 - \bar{\lambda}^{T_{V_2}}}. \tag{7.11}$$

Supposing the time length T_Δ of a set-up period Δ follows a general distribution, the probability distribution u_k, PGF $T_\Delta(z)$ and average value $E[T_\Delta]$ of T_Δ will be obtained as follows:

$$u_k = \Pr\{T_\Delta = k\}, \quad k = 1, 2, 3, \ldots, \tag{7.12}$$

$$T_\Delta(z) = \sum_{k=1}^{\infty} z^k u_k, \tag{7.13}$$

$$E[T_\Delta] = \sum_{k=1}^{\infty} k u_k. \tag{7.14}$$

By letting A_Δ be the number of data packets that have arrived during a set-up period Δ, the PGF $A_\Delta(z)$ can be obtained as follows:

$$A_\Delta(z) = \sum_{j=0}^{k} \sum_{k=1}^{\infty} u_k \binom{k}{j} \lambda^j \bar{\lambda}^{k-j} z^j = T_\Delta(\bar{\lambda} + \lambda z). \tag{7.15}$$

Differentiating Eq. (7.15) with respect to z at $z = 1$, the average value $E[A_\Delta]$ of A_Δ is given as follows:

$$E[A_\Delta] = \lambda E[T_\Delta]. \tag{7.16}$$

A busy period B begins given one of the following two cases:

(1) If there is a data packet arrival within the maximum length T of a vacation-delay period D, the arriving data packet will trigger a busy period B immediately. The probability of this event is P_D, which is given in Eq. (7.4). Letting Q_{B_1} be the

number of data packets at the beginning instant of a busy period B in this case, the PGF $Q_{B_1}(z)$ of the Q_{B_1} is given as follows:

$$Q_{B_1}(z) = z. \tag{7.17}$$

(2) If there is no data packet arrival in the vacation-delay period D, the system vacation period V will begin when a vacation-delay period D is over. The probability that there is no data packet arrival during the vacation-delay period D is $\bar{\lambda}^T$. When the vacation period is over, the system will firstly go through a set-up period, and then enter into a busy period. For this case, letting Q_{B_2} be the number of data packets at the beginning instant of a busy period B, the PGF $Q_{B_2}(z)$ of Q_{B_2} can be obtained as follows:

$$Q_{B_2}(z) = A_V(z)A_\Delta(z). \tag{7.18}$$

Let Q_B be the number of data packets at the beginning instant of a busy period B. Taking into account both cases mentioned above, we can obtain the PGF $Q_B(z)$ of Q_B as follows:

$$Q_B(z) = (1 - \bar{\lambda}^T)Q_{B_1}(z) + \bar{\lambda}^T Q_{B_2}(z). \tag{7.19}$$

Differentiating Eq. (7.19) with respect to z at $z = 1$, the average value $E[Q_B]$ of Q_B can be given by

$$E[Q_B] = 1 - \bar{\lambda}^T + \bar{\lambda}^T E[A_V] + \bar{\lambda}^T E[A_\Delta] = \lambda H \tag{7.20}$$

where H is given as follows:

$$H = \frac{1 - \bar{\lambda}^T}{\lambda} + \frac{\bar{\lambda}^T P_{V_1} T_{V_1}}{1 - \bar{\lambda}^{T_{V_1}}} + \frac{\bar{\lambda}^T P_{V_2} T_{V_2}}{1 - \bar{\lambda}^{T_{V_2}}} + \bar{\lambda}^T E[T_\Delta].$$

7.3.3 Queue Length and Waiting Time

At the embedded point, the queue length L_d can be decomposed into two parts, namely, $L_d = L_{d0} + L_{d1}$. L_{d0} is the queue length of the classical Gemo/G/1 model, and L_{d1} is the additional queue length introduced by the multiple vacations.

The average value $E[L_{d0}]$ of L_{d0} can be given as follows:

$$E[L_{d0}] = \frac{\lambda^2 E[S(S - 1)]}{2(1 - \rho)}. \tag{7.21}$$

By using the boundary state variable theory, the PGF $L_{d1}(z)$ of L_{d1} is given as follows:

$$L_{d1}(z) = \frac{1 - Q_B(z)}{E[Q_B](1 - z)}.$$

(7.22)

Differentiating Eq. (7.22) with respect to z at $z = 1$, the average value $E[L_{d1}]$ of L_{d1} is obtained as follows:

$$E[L_{d1}] = \frac{E[Q_B(Q_B - 1)]}{2E[Q_B]}$$

$$= \frac{\lambda \bar{\lambda}^T}{2H} \left(P_{V_1} T_{V_1} \frac{(T_{V_1} - 1) + 2E[T_\Delta]}{1 - \bar{\lambda}^{T_{V_1}}} + P_{V_2} T_{V_2} \frac{(T_{V_2} - 1) + 2E[T_\Delta]}{1 - \bar{\lambda}^{T_{V_2}}} \right).$$

(7.23)

Combining Eqs. (7.21) and (7.23), we can give the average value $E[L_d]$ of the queue length L_d as follows:

$$E[L_d] = E[L_{d0}] + E[L_{d1}].$$

(7.24)

We can also obtain the average waiting time $E[W]$ of data packets as follows:

$$E[W] = \frac{E[L_d]}{\lambda}$$

$$= \frac{\bar{\lambda}^T}{2H} \left(P_{V_1} T_{V_1} \frac{(T_{V_1} - 1) + 2E[T_\Delta]}{1 - \bar{\lambda}^{T_{V_1}}} + P_{V_2} T_{V_2} \frac{(T_{V_2} - 1) + 2E[T_\Delta]}{1 - \bar{\lambda}^{T_{V_2}}} \right).$$

(7.25)

7.3.4 Busy Cycle

Note that a busy cycle is regarded as the DRX cycle in LTE system. If there is a data packet arrival within the vacation-delay period D, the busy cycle R will consist of a vacation-delay period D and a busy period B. Otherwise, the busy cycle R will include a vacation-delay period D, a system vacation period V, a set-up period Δ and a busy period B.

Note that T is the maximum length of a vacation-delay period D, and T_D is the actual length of D, so we have $T_D \leq T$. The probability distribution and the average value $E[T_D]$ of the actual length T_D are given as follows:

$$\Pr\{T_D = j\} = \begin{cases} \bar{\lambda}^{j-1}\lambda, & 0 < j < T \\ \bar{\lambda}^{T-1}, & j = T, \end{cases}$$

(7.26)

$$E[T_D] = \sum_{j=1}^{T} j\Pr\{T_D = j\} = \frac{1 - \bar{\lambda}^T}{\lambda}. \tag{7.27}$$

By applying the boundary state variable theory, the average value $E[T_B]$ of the busy period B is given as follows:

$$E[T_B] = E[Q_B]\frac{E[S]}{1 - \rho} = \frac{\rho H}{1 - \rho}. \tag{7.28}$$

Let N_V be the number of vacation periods in a busy cycle, including the short vacation periods V_1 and long vacation periods V_2. The probability distribution of N_V can be obtained as follows:

$$\Pr\{N_V = j\} = \begin{cases} \left(\bar{\lambda}^{T_{V_1}}\right)^{j-1}\left(1 - \bar{\lambda}^{T_{V_1}}\right), & 1 \le j \le K \\ \left(\bar{\lambda}^{T_{V_1}}\right)^{K}\left(\bar{\lambda}^{T_{V_2}}\right)^{j-K-1}\left(1 - \bar{\lambda}^{T_{V_2}}\right), & j > K \end{cases} \tag{7.29}$$

where K is the maximum number of short vacation periods V_1 in a busy cycle defined in Sect. 7.3.1.

The average length $E[T_V]$ of the system vacation V is then given as follows:

$$E[T_V] = \sum_{j=1}^{K} jT_{V_1}\Pr\{N_V = j\} + \sum_{j=K+1}^{\infty} (KT_{V_1} + (j - K)T_{V_2})\Pr\{N_V = j\}$$

$$= \frac{P_{V_1}T_{V_1}}{1 - \bar{\lambda}^{T_{V_1}}} + \frac{P_{V_2}T_{V_2}}{1 - \bar{\lambda}^{T_{V_2}}} \tag{7.30}$$

where P_{V_1} and P_{V_2} are given in Eqs. (7.5) and (7.6), respectively.

Let T_R be the time length of a busy cycle period R, and $E[T_R]$ be the average value of T_R. Combining Eqs. (7.14), (7.27)–(7.30), $E[T_R]$ can be calculated as follows:

$$E[T_R] = E[T_D] + E[T_B] + \bar{\lambda}^T E[T_V] + \bar{\lambda}^T E[T_\Delta] = \frac{H}{1 - \rho}. \tag{7.31}$$

7.3.5 Performance Measures

We define the handover rate ζ_h as the number of switches between the working state and the sleep state per slot. The handovers will occur only when there are no data packet arrivals within the time length of the sleep-delay timer in a DRX cycle. Therefore, we obtain the handover rate ζ_h as follows:

$$\zeta_h = \frac{\bar{\lambda}^T}{E[T_R]}. \tag{7.32}$$

We define the energy saving rate γ of the system as the energy conservation per slot due to the introduction of the sleep mode. Energy is consumed normally in the working state, and is saved in the DRX stages. The system also loses some energy when the system switches between the working state and the sleep state. We express the energy saving rate γ of the system using the following equation:

$$\gamma = \frac{(g_1 - g_2)E[T_V]}{E[T_R]} - g_3 \zeta_h \tag{7.33}$$

where g_1 and g_2 are the energy consumption per slot in the working state and the DRX stage, respectively, g_3 is the energy consumption for each switch from the sleep state to the working state.

We define the response time Y_d of a data packet as the duration from the arrival moment of a data packet to the end moment of the transmission of that data packet. Obviously, the average response time $E[Y_d]$ of data packets is the sum of the average transmission time $E[S]$ of data packets given by Eq. (7.3) and the average waiting time $E[W]$ of data packets given by Eq. (7.25). Therefore, we obtain the average response time $E[Y_d]$ of data packets as follows:

$$E[Y_d] = E[S] + E[W]. \tag{7.34}$$

7.4 Numerical Results and Performance Optimization

In this section, we present numerical results to evaluate the performance of the system using the enhanced energy saving strategy proposed in this chapter. Then, we optimize the threshold of the short DRX stages and the time length of the sleep-delay timer in this strategy to improve the system performance.

7.4.1 Numerical Results

In practice, the system parameters, such as the time length for one slot, the arrival rate of data packets, the average transmission time of a data packet, the time length of the waking-up procedure, the energy consumption per second during the working state and the sleep state, and the energy consumption for each switch from the sleep state to the working state, are set as needed.

The system parameters are fixed as follows: 1 slot $= 1$ ms, $\lambda = 0.05$, $E[S] = 4$ ms, $E[T_\Delta] = 1$ ms, $g_1 = 80$ mW, $g_2 = 5$ mW, $g_3 = 5$ mJ as an example for all the numerical results.

Figure 7.3 examines how the handover rate ζ_h changes in relation to the threshold K of the short stages with different time lengths T of the sleep-delay timer, T_{V_1} of the short DRX stage, and T_{V_2} of the long DRX stage.

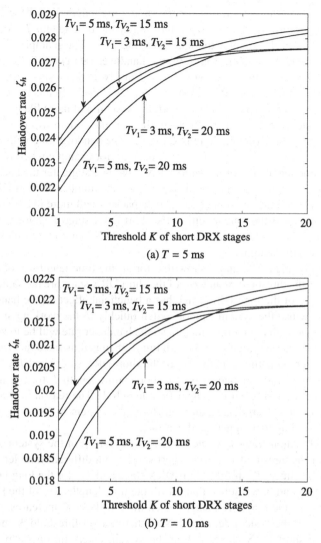

Fig. 7.3 Handover rate versus threshold of short DRX stages

In both Figs. 7.3a and b, we find that for all the time lengths T_{V_1} of the short DRX stage and T_{V_2} of the long DRX stage, when the threshold K increases, the handover rate ζ_h also increases. The reason is that as the threshold K increases, the number of the short DRX stages in a DRX cycle will increase. This increases the likelihood that there will be more data packet arrivals in the short DRX stage, so the system will switch into the working state from the sleep state earlier, meaning the handover rate will inevitably increase.

In Figs. 7.3a and b, we see that for the same time length T_{V_2} of the long DRX stages, when the threshold K is small, the shorter the time length T_{V_1} of the short DRX stage is, and the smaller the handover rate ζ_h is. For example, for the case $T_{V_2} = 15$ ms and $K < 10$ in Fig. 7.3a, the handover rate ζ_h with $T_{V_1} = 5$ ms is greater than the handover rate ζ_h with $T_{V_1} = 3$ ms. Note that a smaller threshold K means a lesser number of short DRX stages in a DRX cycle. In this case, if the time length of the short DRX stage is also short, the system will more likely enter into a long DRX stage from the short DRX stage. Because the system will not easily enter into the working state from the long DRX stages, the handover rate will be lower.

However, when the threshold K is high enough, for example when $K \geq 10$, the shorter the time length T_{V_1} of the short DRX stage is, the greater the handover rate ζ_h is. In this case, when the threshold K is higher, the number of short DRX stages in a DRX cycle will also be high, and the data packets will more likely arrive in the short DRX stages. If the time length of the short DRX stage is shorter, the system will enter into the working state from the short DRX stage more quickly, so the handover rate will be higher.

Additionally, Figs. 7.3a and b show that for all the time lengths T of the sleep-delay timer, if the threshold K and T_{V_1} of the short DRX stage are the same, a longer time length T_{V_2} of the long DRX stage has a beneficial effect on the handover rate ζ_h. This is because the system cannot switch quickly to the working state from a long DRX stage with a long time length, so the handover rate will be lower.

Moreover, by comparing Fig. 7.3a with Fig. 7.3b, it is illustrated that for the same threshold K, the same time lengths T_{V_1} of the short DRX stages and T_{V_2} of the long DRX stages, the handover rate ζ_h with a shorter time length T for the sleep-delay timer is higher than that with a longer time length T. The reason is that a shorter time length for the sleep-delay timer will force the system to enter the sleep state more easily, leading to a higher handover rate.

Figure 7.4 demonstrates how the energy saving rate γ of the system changes as a function of the threshold K for the short stages with different time lengths T for the sleep-delay timer; T_{V_1} for the short DRX stage and T_{V_2} for the long DRX stage.

In Figs. 7.4a and b, we notice that for all the time lengths T_{V_1} of the short DRX stage and T_{V_2} of the long DRX stage, when the threshold K increases, the energy saving rate γ of the system decreases. Note that a larger threshold K means there will be more short DRX stages before the system enters into a sleep state. The energy conservation mainly focuses on the DRX stages. The short DRX stages conserve less energy than the long DRX stages for energy conservation. Therefore, less energy will be saved when there is a greater number of short DRX stages, and the energy saving rate of the system will decrease.

In Figs. 7.4a and b, we also observe that for the same time length T_{V_2} of the long DRX stages, when the threshold K is small, the shorter the time length T_{V_1} of the short DRX stage is, and the higher the energy saving rate γ of the system is. For example, in the case where $T_{V_2} = 15$ ms and $K < 10$ in Fig. 7.4a, the energy saving rate γ of the system with $T_{V_1} = 3$ ms is higher than the energy saving rate γ of the system with $T_{V_1} = 5$ ms. Note that a smaller threshold K means a lesser number of short DRX stages in a DRX cycle. In this case, if the time length of the short DRX

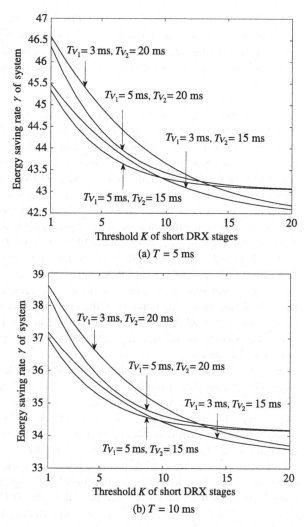

Fig. 7.4 Energy saving rate of system versus threshold of short DRX stages

stage is shorter, the system will more likely enter into the long DRX stage from the short DRX stage. As more energy will be saved in the long DRX stages, this will result in a greater energy saving rate of the system.

However, when the threshold K is high enough, for example $K \geq 10$, the shorter the time length T_{V_1} of the short DRX stage is, the lower the energy saving rate γ of the system is. The reason is that when the threshold K is higher, the number of short DRX stages in a DRX cycle will be greater, so the more likely it will be that there will be a data packet arrival during the short DRX stages. In this case, the smaller the time length T_{V_1} of the short DRX stage is, the more quickly the system will switch to the working state, the less energy will be saved, and the energy saving rate of the system will be lower.

Additionally, Figs. 7.4a and b illustrate that for the same threshold K and same time length T_{V_1} of the short DRX stages, a longer time length T_{V_2} results in a higher energy saving rate γ of the system. This is because the long DRX stages are beneficial to energy conservation, so the energy saving rate of the system will be higher.

Moreover, by comparing Fig. 7.4a with Fig. 7.4b, it is illustrated that for all the same thresholds K, the same time lengths T_{V_1} of the short DRX stage and T_{V_2} of the long DRX stage, the energy saving rate γ of the system where time length T of the sleep-delay timer is short, is greater than that one with a long T. The reason is that a shorter time length of the sleep-delay timer will force the system to enter the DRX stage more easily. Note that energy is saved in DRX stages. Obviously, a shorter sleep-delay timer length will result in a higher energy saving rate of the system.

Figure 7.5 illustrates the change trend for the average response time $E[Y_d]$ of data packets versus the threshold K of the short stages with different time lengths T of the sleep-delay timer, T_{V_1} of the short DRX stage and T_{V_2} of the long DRX stage.

From Figs. 7.5a and b, we notice that for the same time lengths T_{V_1} of the short DRX stage and T_{V_2} of the long DRX stage, when the threshold K increases, the average response time $E[Y_d]$ of data packets decreases. This is because as the threshold K increases, the number of the short DRX stages in a DRX cycle will increase, the data packets will more likely arrive in the short DRX stages, and the system will switch into the working state early, so the average response time of data packets will decrease.

In Figs. 7.5a and b, we observe that for the same time length T_{V_2} of the long DRX stage, when the threshold K is smaller, the shorter the time length T_{V_1} of the short DRX stage is, and the greater the average response time $E[Y_d]$ of data packets is. For example, for the case of $T_{V_2} = 15$ ms and $K < 11$ in Fig. 7.5a, the average response time $E[Y_d]$ of data packets with $T_{V_1} = 3$ ms is greater than the average response time $E[Y_d]$ of data packets with $T_{V_1} = 5$ ms.

Note that a lower threshold K means a fewer short DRX stages in a DRX cycle. In this case, if the time length T_{V_1} of the short DRX stage is shorter, the system will more likely enter the long DRX stage from the short DRX stage. If there is a data packet arrival in a long DRX stage, the system will have to wait for the end of that long DRX stage and then initiate the wake-up procedure, so the average response time will be greater.

However, when the threshold K is big enough, for example $K \geq 11$, the shorter the time length T_{V_1} of the short DRX stage is, and the lower the average response time $E[Y_d]$ of data packets is. Note that when the threshold K is higher, the number of the short DRX stages will be larger, and the data packets will more likely arrive in the short DRX stages. If the time length T_{V_1} of the short DRX stage is shorter, the system will respond more quickly when there is a data packet arrival in the short DRX stage, so the average response time of data packets will be lower.

Additionally, Figs. 7.5a and b illustrate that for the same threshold K, the same time lengths T of the sleep-delay timer and T_{V_1} of the short DRX stage, a shorter time length T_{V_2} of the long DRX stage leads to a lower average response time $E[Y_d]$ of data packets. This is because that if a data packet arrives during the long DRX

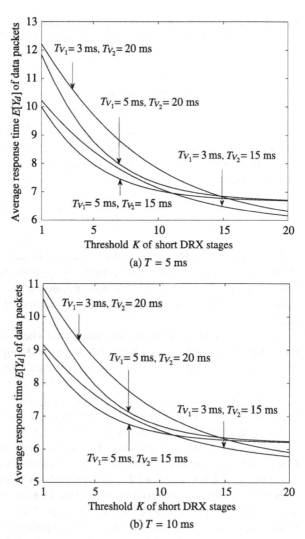

Fig. 7.5 Average response time versus threshold of short DRX stages

stage, the system will have to wait a long time for the end of that long DRX stage, and then initiate the wake-up procedure and enter into the working state. Therefore, the shorter the long DRX stage is, the lower the average response time of data packets will be.

Moreover, by comparing Fig. 7.5a with Fig. 7.5b, we conclude that for the same threshold K, and the same time lengths T_{V_1} of the short DRX stage and T_{V_2} of the long DRX stage, the average response time $E[Y_d]$ of data packets with a longer time length T of the sleep-delay timer is lower than that one with a shorter time length T. The reason is that the system will return to the working state at once if there is a data packet arrival in the sleep-delay period. Therefore, when the time length T is greater, the average response time of data packets will be lower.

7.4.2 Performance Optimization

From Figs. 7.3, 7.4, 7.5, we conclude that there is a trade-off between different performance measures. Aiming to optimize the threshold K of the short DRX stages and the time length T of the sleep-delay timer, we construct a system cost function $F(X)$ as follows:

$$F(X) = f_1 \times \zeta_h + f_2 \times \frac{1}{\gamma} + f_3 \times E[Y_d] \tag{7.35}$$

where f_1 and f_3 are the factors of the handover rate and the average response time of data packets to the system cost, respectively. f_2 is the factor of the energy saving rate of the system in relation to the system reward. f_1, f_2 and f_3 are all system parameters and they are variable in different numerical examples as required in practice.

We can obtain the system cost function $F(K)$ when we set X of Eq. (7.35) to K versus the threshold K of the short DRX stages by fixing the time length T of the sleep-delay timer, and the system cost function $F(T)$ when we set X of Eq. (7.35) to T versus the time length T of the sleep-delay timer by fixing the threshold K of the short DRX stages, where ζ_h, γ and $E[Y_d]$ are given by Eqs. (7.32)–(7.34), respectively. ζ_h, γ and $E[Y_d]$ are all the functions of T and K.

The parameters are fixed as follows: $f_1 = 1.3$, $f_2 = 1$ and $f_3 = 1.9$ as an example for all the numerical results. The change trend of the system cost function $F(K)$ versus the threshold K of the short DRX stages with different time lengths T_{V_1} of the short DRX stage and T_{V_2} of the long DRX stage is shown in Fig. 7.6.

In Fig. 7.6, we show the system cost function $F(K)$ versus the threshold K for the time length $T = 10$ ms as an example. From Fig. 7.6, we find that as the threshold K increases, the system cost function $F(K)$ experiences two stages. Note that as the threshold K increases, the number of the short DRX stages in a DRX cycle will be greater. During the first stage, the system cost function $F(K)$ decreases as the threshold K increases. If there is a data packet arrival in the short DRX stage, the system will enter into a working state quickly and respond rapidly, and the average response time of data packets will be lower. On the other hand, if no data packets arrive in the short DRX stages, the system will enter the long DRX stage easily after a number of the short DRX stages reach the threshold K. Consequently, more energy will be saved. In this case, the system cost function $F(K)$ will decrease.

During the next stage, the system cost function $F(K)$ increases as the threshold K continues increasing. This means that the data packets will more likely arrive in the short DRX stages and the system will switch into the working state quickly, so the handover rate will inevitably increase. If there are no data packet arrivals during

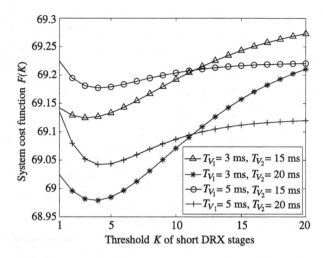

Fig. 7.6 System cost function versus threshold of short DRX stages

Table 7.1 Optimum threshold of short DRX stages

Time lengths T_{V_1} of short DRX stage	Time lengths T_{V_2} of long DRX stage	Optimum thresholds K^*	Minimum costs $F(K^*)$
3	15	3	69.1246
3	20	4	68.9792
5	15	4	69.1773
5	20	4	69.0424

the short DRX stages, the system will not enter the long DRX stages easily, not until the number of the short DRX stages reaches K. This means there will be less energy conservation. In this case, the system cost function $F(K)$ will be greater. This shows conclusively that when the threshold is set to an optimal value, there is a minimum cost for all the time lengths of the short DRX stage and the long DRX stage.

We set the time length T of the sleep-delay timer as $T = 10$ ms. With different time lengths T_{V_1} of the short DRX stage and T_{V_2} of the long DRX stage, the optimal thresholds K^* and the minimum costs $F(K^*)$ are presented in Table 7.1.

The change trend of the system cost function $F(T)$ versus the time lengths T of the sleep-delay timer with different time lengths T_{V_1} of the short DRX stage and T_{V_2} of the long DRX stage is shown in Fig. 7.7.

In Fig. 7.7, we show the system cost function $F(T)$ versus the time length T for the threshold $K = 5$ as an example. From Fig. 7.7, we find that as the time length T of the sleep-delay timer increases, the system cost function $F(T)$ experiences two stages. During the first stage, the system cost function $F(T)$ decreases as the time length T increases. The reason is that when the time length T of the sleep-delay timer is shorter, and as the time length T increases, the more likely it is that a data packet will arrive within the time length of the sleep-delay timer. The system will

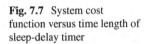

Fig. 7.7 System cost function versus time length of sleep-delay timer

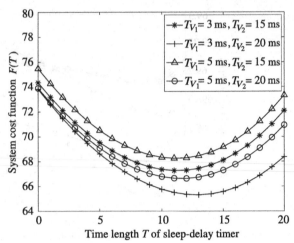

more likely enter the working state at once, and the average response time of data packets will be lower. Therefore, the system cost function $F(T)$ will decrease.

During the next stage, the system cost function $F(T)$ increases as the time length T becomes longer and longer. This is because when the time length T of the sleep-delay timer is long, if there is no data packet arrival within T, the system will have to wait for the sleep-delay timer to expire before it can enter the DRX stage, thus reducing the level of energy conservation. Therefore, the system cost function $F(T)$ will be greater. This leads to the conclusion that when the time length of the sleep-delay timer is set to an optimal value, there will be a minimum cost for all the time lengths of the short DRX stage and the long DRX stage.

Note that when the time length of the sleep-delay timer is 0 ms, the enhanced energy saving strategy based Active DRX mechanism proposed in this chapter will be downgraded to a conventional energy saving strategy in LTE system. From the change trend shown in Fig. 7.7, we see that the cost of the conventional energy saving strategy is higher than that of the enhanced energy saving strategy when the time length of the sleep-delay timer T is set reasonably.

When the threshold K of the short DRX stages is set as $K = 5$, with different time lengths T_{V_1} of the short DRX stage and T_{V_2} of the long DRX stage, the optimal time lengths T^* of the sleep-delay timer and the minimum costs $F(T^*)$ are shown in Table 7.2.

Table 7.2 Optimum time length of sleep-delay timer

Time lengths T_{V_1} of short DRX stage	Time lengths T_{V_2} of long DRX stage	Optimum time lengths T^* of sleep-delay timer	Minimum costs $F(T^*)$
3	15	12	67.2612
3	20	13	65.3125
5	15	12	68.2427
5	20	12	66.6293

7.5 Conclusion

In this chapter, we proposed an enhanced energy saving strategy based on the Active DRX mechanism in an LTE system to improve the sleep strategy for a better balance between response performance and energy efficiency by introducing a sleep-delay timer. Accordingly, we built a discrete-time multiple-vacation queueing model with a vacation-delay period and a set-up period by addressing the busy period, the queue length, the waiting time and the busy cycle. Moreover, we analyzed the system model in the steady state by using an embedded Markov chain and calculated the formulas for the performance measures, including the handover rate, the energy saving rate and the average response time of data packets. Using numerical results, we demonstrated the impact of the different thresholds of the short DRX stages, the different time lengths of the sleep-delay timer, the short DRX stage and the long DRX stage on the system performance. We also investigated the trade-off between different performance measures. Finally, we optimized the threshold of the short DRX stages and the time length of the sleep-delay timer while minimizing the system cost function. This chapter has potential applications in the improvement of energy saving strategies for WCNs.

7.5 Conclusion

Chapter 8
Multiple-Vacation Queue-Based Active DRX Mechanism in an LTE System

In order to reduce the average response time of data packets while guaranteeing a greater energy saving rate of the system, in this chapter, we propose an enhanced Active Discontinuous Reception (DRX) mechanism with a sleep-delay strategy in the Long Term Evolution (LTE) system to influence the downlink transmission at the User Equipment (UE). Utilizing several logical channels for one connection, we build a multiple synchronous vacation queueing system with a wake-up period and a sleep-delay. We derive performance measures of the system in terms of the energy saving rate of the system, the blocking rate and the average response time of data packets. We present numerical results to show the validity of the proposed enhanced Active DRX mechanism with a sleep-delay strategy. Finally, by constructing a system profit function, we optimize the number of the logical channels for one connection, the time lengths of the sleep-delay timer and the sleep period.

8.1 Introduction

LTE technology is one of the 4G standards being employed in many advanced communication technologies. Compared with 3G technology, LTE has the ability to operate at a higher transmission rate [Abet10]. However, the improvement of the transmission rate will lead to excessive energy consumption at the mobile terminal. In order to reduce the energy consumption and to achieve more efficient and greener communication, a DRX mechanism has been introduced into the LTE technology [Koc13]. This mechanism influences the downlink transmission at the UE.

Nowadays, some scholars concentrate their research on the sleep mode in the DRX mechanism. In [Baek11b], the authors considered the downlink packet arrivals at the UE and the uplink packet arrivals at an evolved node with the help of a properly constructed discrete-time two-dimensional embedded Markov chain. They optimized some parameters of the DRX mechanism by analyzing the Markov chain

© Springer Nature Singapore Pte Ltd. 2021
S. Jin, W. Yue, *Resource Management and Performance Analysis of Wireless Communication Networks*, https://doi.org/10.1007/978-981-15-7756-7_8

in the steady state. In [Yin12], the author introduced Carrier Aggregation (CA) technology to the DRX mechanism and proposed an enhanced DRX mechanism in which the mobile terminals could choose different sleeping cycles in different channels. They validated the mechanism by simulation, which showed that the introduced mechanism was both flexible and efficient.

In the conventional DRX mechanism, UEs will switch to a sleep mode whenever it is possible. As a result, the energy consumption will be decreased. However, the average response time of data packets will be increased. In [Fowl12], the authors modeled the DRX mechanism with adjustable and fixed DRX cycles as a semi-Markov process. They provided numerical results to investigate the trade-off between the power saving and the wake-up period. In [Kall12], the authors presented an architecture for a DRX mechanism in the case of video streaming over Real-time Transport Protocol/User Datagram Protocol (RTP/UDP) transport in the Evolved Packet System (EPS). By predicting the number of the pending video frames, the UE can determine whether it enters into the sleep mode or not. Simulation revealed that the DRX mechanism can improve energy savings by 30%–80%. In [Jha12], the authors proposed an algorithm to set DRX parameters efficiently and to ensure a balanced trade-off between latency and power saving. In [Kuo11], the authors proposed a light sleeping mode and evaluated the performance of this mechanism by considering the drawbacks and deficiencies of the sleep mode in conventional DRX mechanisms.

In [Fowl11], the author evaluated and compared the influences of a DRX light sleep mode and a DRX deep sleep mode on power consumption for Voice and Web traffic in relation to its level of dependency on the Transmission Time Interval (TTI) size. They found that the combination of the light sleep mode and the deep sleep mode might be a desirable method for achieving maximal power efficiency with minimum delay.

In [Gao11], the authors demonstrated that Single-threshold Automated config-uration DRX (S-ADRX) will reduce the energy consumption but it will increase the delay of data packets. Furthermore, they showed that Proportional Fair (PF) will increase the system utility, whereas the energy consumption will be lower. By considering the trade-off between S-ADRX and PF, they proposed a Multi-threshold Automated configuration DRX (M-ADRX) mechanism. Their simulation results showed that the M-ADRX mechanism can better reduce energy consumption than the conventional DRX mechanism.

In [Bont09], the authors optimized the DRX parameters by means of a numerical procedure by introducing an energy saving method in both RRC-CONNECTED and RRC-IDLE states. In [Yu12, Zhan13b], the authors proposed traffic-based DRX cycles in LTE systems. By conjecturing the traffic status with a partially observable Markov decision process, they also proposed a method for setting the DRX parameters.

In most of the above-mentioned research, a single channel was considered and the switching procedure from a sleep state to an awake state was neglected. For a 4G network with LTE technology, a mobile intelligent terminal provides good support for multiple logical channels. Therefore, it is more challenging to model several logical channels for one connection and also account for the above switching procedure. In this way, we address more realistic modeling of the real-time network traffic in 4G networks.

In order to reduce the average response time of data packets while guaranteeing a greater energy saving rate of the system, in this chapter, we propose an enhanced Active DRX mechanism with a sleep-delay strategy influencing the downlink transmission at the UE. We call this Active DRX mechanism an "enhanced Active DRX mechanism". We model the network using the enhanced Active DRX mechanism as a discrete-time synchronous multiple-vacation queueing system with a wake-up period and a sleep-delay. Then we evaluate the system performance of the enhanced Active DRX mechanism by constructing a Markov chain. We present numerical result to evaluate the system performance. Finally, we optimize the system parameters by constructing a system profit function.

The chapter is organized as follows. In Sect. 8.2, we describe the enhanced Active DRX mechanism proposed in this chapter. In Sect. 8.3, we present the system model with wake-up and sleep-delay to capture the proposed mechanism by studying the transition probability matrix in two cases: Case I and Case II. We also obtain performance measures in this section. In Sect. 8.4, we present numerical results to evaluate the system performance, and construct a system profit function to optimize the enhanced Active DRX mechanism. Finally, we draw our conclusions in Sect. 8.5.

8.2 Enhanced Active DRX Mechanism

In this section, we present the enhanced Active DRX mechanism proposed in this chapter to improve the downlink transmission at the UE.

For the purpose of energy saving, a DRX mechanism is introduced into an LTE technology, which affects the control of the downlink transmission at the UE with a Physical Downlink Control CHannel (PDCCH) frame. In the DRX mechanism there are two types of DRX patterns, namely, the Idle DRX and the Active DRX. In the Idle DRX pattern, the UE stays in an idle state accounting for the situation, in which there is no data packet to be transmitted. In the Active DRX pattern, the UE stays in a connected state even though there is no data packet to be transmitted.

In the conventional DRX mechanism, when the transmission of a data packet is completed, the system will change immediately to the sleep state from its working condition. The main goal of the new strategy is to reduce the average response time of data packets with an energy consumption constraint. In this Active DRX mechanism there are four different system periods for a UE: An awake period, a sleep period, a sleep-delay period and a wake-up period. During the awake period, data packets are transmitted normally. In the sleep-delay period, the UE will listen to the channel all the time. During the sleep period, the UE will periodically listen to the channel. During the wake-up period, the system will listen to channel, but will not transmit data packets. Additionally, in the mechanism, there are three different timers: An inactivity timer denoted as T_1 operating in the awake period, a sleep-delay timer denoted as T_2 operating in the sleep-delay period, and an on-duration timer denoted as T operating in the sleep period.

When considering a connection composed of logical channels with C ($C = 1, 2, 3, \ldots, C_{max}$), the main steps of the improved algorithm for the enhanced Active DRX mechanism are given as follows, where C is a system parameter and C_{max} is the maximum value of C. C_{max} is related to the bandwidth of the network under consideration:

Step 1: Input timers T_1, T_2, T, input the number C of logical channels.
Step 2: System being at awake period.
       while timer T_1 doesn't expire
         transmit data packets and listen PDCCH
         if PDCCH control frame arrives
            if number of data packets in system < C
              reset timer T_1
            endif
         endif
       endwhile
       go into sleep-delay period
       reset timer T_2
Step 3: System being at sleep-delay period.
       while timer T_2 doesn't expire
         listen PDCCH
         if PDCCH control frame arrives
            reset timer T_1
            go into awake period
         endif
       endwhile
       go into sleep period
Step 4: System being at sleep period.
       counter = 0
       while true
         if sleep interval is over
           reset timer T
           listen PDCCH
           if PDCCH control frame arrives before timer T expires
             if counter < C
               counter = counter + 1
             endif
           endif
           if timer T expires
             if counter > 0
               break;
             else

            begin next sleep interval
        endif
        endif
        endif
    endwhile
    go into wake-up period
Step 5: System being at wake-up period.
    while wake-up period is not over
        listen PDCCH
        if counter < C
            counter = counter + 1
        endif
    endwhile
    go into awake period.

The time sequence of the enhanced Active DRX mechanism with a sleep-delay strategy proposed in this chapter is illustrated in Fig. 8.1.

From Fig. 8.1, we observe that at instant t_1, the UE begins the awake period, and the timer T_1 is activated. The timer T_1 would expire at instant t_3. However, the UE receives a PDCCH frame at instant t_2. Therefore, the data packets will be transmitted immediately and the timer T_1 will be reset at the same time. There is no PDCCH control frame arrival before the expiration instant t_4 of the timer T_1, so the UE will enter into a sleep-delay period, and the timer T_2 will be activated.

Before the expiration instant t_6 of the timer T_2, a PDCCH control frame arrives at the system at instant t_5. Then the UE returns to the awake period from the sleep-delay period, and the timer T_1 is restarted. At instant t_7, the awake period is over, the UE will enter into a sleep-delay period again, and the timer T_2 will be activated.

During the sleep-delay period, there is no PDCCH control frame arrival, so the UE enters into the sleep period at instant t_8. After experiencing one sleep interval, the UE activates the timer T at instant t_9. There is no PDCCH control frame arrival before the expiry of the timer T, so the UE begins a new sleep interval at instant t_{10}.

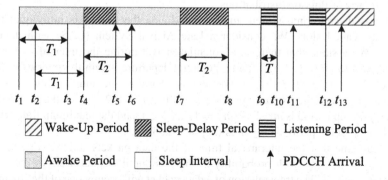

Fig. 8.1 Time sequence of proposed enhanced Active DRX mechanism

There is a PDCCH control frame arrival during the sleep interval at instant t_{11}, hence the UE will switch to a wake-up period, after the next listening period, namely, at instant t_{12} when the timer T next expires.

In the enhanced Active DRX mechanism proposed in this chapter, the sleep interval, the number of logical channels, the channel bandwidth, and also the time length of the timer T_2 have significant influences on the system performance.

8.3 System Model and Performance Analysis

In this section, we first build a discrete-time synchronous multiple-vacation queue with wake-up period and sleep-delay to capture the enhanced Active DRX mechanism proposed in this chapter. Then, we analyze the system model in the steady state and derive the performance measures of the system.

8.3.1 System Model

The time period consisting of one sleep interval and one listening period can be seen as a vacation period. The system consists of a finite number of logical channels having the awake period, the sleep-delay period and the wake-up period for the data packets' transmissions.

Therefore, the enhanced Active DRX mechanism proposed in this chapter can be treated as a synchronous multiple-vacation queueing model with a set-up period and a vacation-delay period.

In alignment with the digital nature and the synchronization transmission technology of modern communication, we treat the above synchronous multiple-vacation queueing model as a discrete-time system. We notice that a geometric or Poisson distribution will reduce the complexity of the analysis procedure and will lead to simple analysis results. We also notice that the user-initiated arrivals, such as remote-login and file-transfer, are well-modeled as memoryless processes of geometric or Poisson distributions.

In this discrete-time system model, we divide the time axis into equal time intervals, called slots. We consider a Late Arrival System (LAS) with delayed access. We assume that there is a potential arrival during the interval (n^-, n) of the nth slot ($n = 0, 1, 2, \ldots$), and a potential departure within the interval (n, n^+) of the nth slot ($n = 1, 2, 3, \ldots$). We suppose that the time length of the sum of a sleep interval and one listening period is m slots ($m = 1, 2, 3, \ldots$), the time length of the wake-up period is m_1 slots ($m_1 = 1, 2, 3, \ldots$), and the maximum time length of the sleep-delay period is m_2 slots ($m_2 = 1, 2, 3, \ldots$).

We assume that the interarrival time of the data packets follows a geometric distribution, namely, with probability λ, there is one data packet arrival in a slot, where $0 < \lambda < 1$. The transmission of a data packet will occupy one of the channels. The transmission time of a data packet follows a geometric distribution, namely, with probability μ, the data packet occupying one of the channels departs the system in a slot, where $0 < \mu < 1$. Throughout this chapter, λ is called the arrival rate of data packets, and μ is called the service rate of data packets (packets/slot).

We define the number $x(n^+)$ of data packets in the system at the instant n^+ as the system level, namely, $x(n^+) = i$ $(i = 0, 1, 2, \ldots)$. Note that there are no buffers for any of the logical channels, that meaning the maximum system level is $C_{\max}$.

We define the system period $y(n^+)$ at the instant n^+ as the system phase, namely, $y(n^+) = j$, where $j = 0$ indicates the system being in a sleep period; $j = 1$ indicates the system being in an awake period; $j = 2$ indicates the system being in a sleep-delay period; $j = 3$ indicates the system being in a wake-up period.

We also define the sequence number $z(n^+)$ of the nth slot in a specific system period as the system stage. If the system is in an awake period, the $z(n^+) = 0$; if the system is in a sleep period, then $z(n^+) = k$ $(k = 1, 2, 3, \ldots, m)$; if the system is in a wake-up period, then $z(n^+) = k$ $(k = 1, 2, 3, \ldots, m_1)$; if the system is in a sleep-delay period, then $z(n^+) = k$ $(k = 1, 2, 3, \ldots, m_2)$.

Therefore, $\{(x(n^+), y(n^+), z(n^+)), \ n \geq 0\}$ constitutes a three-dimensional DTMC.

Let $A_{i,j}$ be the one-step transition probability sub-matrix of the system transferring from level i to level j. According to different system levels, the one-step transition probability matrix P of the three-dimensional Markov chain can be given in a block form as follows:

$$
P = \begin{pmatrix}
A_{0,0} & A_{0,1} & & & & \\
A_{1,0} & A_{1,1} & A_{1,2} & & & \\
A_{2,0} & A_{2,1} & A_{2,2} & A_{2,3} & & \\
\vdots & \vdots & \vdots & \ddots & \ddots & \\
A_{C-1,0} & A_{C-1,1} & A_{C-1,2} & \cdots & A_{C-1,C-1} & A_{C-1,C} \\
A_{C,0} & A_{C,1} & A_{C,2} & \cdots & A_{C,C-1} & A_{C,C}
\end{pmatrix}
\tag{8.1}
$$

Let $\eta_1(i)$ be the total number of possible stages in level i. Let $\eta_2(i)$ indicate whether the level i is allowed when the system stays in the awake period. If the level i is allowed in the awake period, then $\eta_2(i) = 1$, otherwise, $\eta_2(i) = 0$. The size of the sub-matrix $A_{i,j}$ in Eq. (8.1) is $(\eta_1(i) + \eta_2(i)) \times (\eta_1(j) + \eta_2(j))$. The indexing in the sub-matrix $A_{i,j}$ is ordered in increasing system stages and in the order of possible system periods in level i as awake, sleep, sleep-delay and wake-up. Furthermore, the indexing of the sub-matrices starts with 1.

Let LS be the sum of m slots of the sleep interval and m_1 slots of the wake-up period, namely, $LS = m + m_1$ in slots. On the other hand, since only one data packet can be submitted in one slot, this LS can also represent the number of data packets that can be submitted during the time length of LS slots.

For a practical application, C, the number of system logical channels, can be considered as a maximum system level, where $C \leq C_{\max}$. That is to say that when $LS > C$, then at most C data packets among LS packets that have arrived at the system during the time length of LS slots can be transmitted by the system, namely,

$x(n^+) \leq C$. When $LS \leq C$, all the data packets that have arrived at the system during the time length of LS slots will be submitted by the system, namely, $x(n^+) \leq LS$. In this case, the number of the data packets submitted by the system during the time length of LS slots is LS.

According to the relationship between LS and C, we study the transition probability matrix P in the following two cases:

Case I: $LS > C$.
Case II: $LS \leq C$.

8.3.2 Transition Probability Sub-Matrices for Case I

To give the transition probability matrix P of case I, $LS > C$, here we present the transition probability sub-matrices $A_{i,j}$ $(0 \leq i, j \leq C)$ in P as follows.

(1) $A_{0,0}$ is the one-step probability sub-matrix describing the transitions at level 0. Note that level 0 exists only in the sleep period and in the sleep-delay period. During the sleep period, the system stage increases by 1 (or returns to stage 1 from stage m) with probability $(1 - \lambda)$, while the system level remains at 0. During the sleep-delay period, the system stage increases by 1 (or transfers from stage m_2 of the sleep-delay period to stage 1 of the sleep period) with probability $(1 - \lambda)$, while the system level remains at 0. Thus $A_{0,0}$ is a square matrix with the size $(m + m_2) \times (m + m_2)$, since $\eta_1(0) = m + m_2$, $\eta_2(0) = 0$. Therefore, the sub-matrix $A_{0,0}$ is given as follows:

$$A_{0,0} = \begin{pmatrix} 0 & 1-\lambda & & & & & & & \\ 0 & 0 & 1-\lambda & & & & & & \\ \vdots & \vdots & \ddots & \ddots & & & & & \\ 0 & 0 & & 0 & 1-\lambda & & & & \\ 1-\lambda & 0 & & & 0 & 0 & & & \\ 0 & 0 & & & & 0 & 1-\lambda & & \\ \vdots & \vdots & & & & & \ddots & \ddots & \\ 0 & 0 & & & & & & 0 & 1-\lambda \\ 1-\lambda & 0 & \ldots & \ldots & \ldots & \ldots & \ldots & \ldots & 0 \end{pmatrix}. \qquad (8.2)$$

(2) $A_{0,1}$ is the one-step probability sub-matrix describing the transition from level 0 to level 1. Level 0 belongs to the sleep period or the sleep-delay period, level 1 belongs to the awake period or the wake-up period, and the transition

probability is λ. Therefore, $\eta_1(0) = m + m_2$, $\eta_2(0) = 0$, $\eta_1(1) = m_1$, $\eta_2(1) = 1$, and $A_{0,1}$ is an $(m + m_2) \times (m_1 + 1)$ matrix, which is given as follows:

$$
A_{0,1} = \begin{pmatrix}
0 & \lambda & & & & \\
0 & 0 & \lambda & & & \\
\vdots & \vdots & \ddots & \ddots & & \\
0 & 0 & & 0 & \lambda & \\
\lambda & 0 & & & 0 & 0 \\
\vdots & \vdots & & & & 0 & 0 \\
\lambda & 0 & \cdots & \cdots & \cdots & & 0
\end{pmatrix}.
\tag{8.3}
$$

(3) $A_{i,0}$ ($0 < i < C$) is the one-step transition probability sub-matrix describing the transition to level 0 of the sleep-delay period from level i of the awake period. This means that all the data packets in the system are completely transmitted and there are no new data packet arrivals. Therefore, $\eta_1(i) = m + m_1 - i$, $\eta_2(i) = 1$, $\eta_1(0) = m + m_2$, $\eta_2(0) = 0$, and $A_{i,0}$ is an $(m + m_1 - i + 1) \times (m + m_2)$ matrix. In $A_{i,0}$, the only non-zero element $a_{1,m+1}$ is given as follows:

$$
a_{1,m+1} = (1 - \lambda)\mu^i.
\tag{8.4}
$$

(4) $A_{C,0}$ is the one-step transition probability sub-matrix describing the transition to level 0 of the sleep-delay period from level C of the awake period. This means that all the data packets in the system are completely transmitted. Since the number of data packets in the system is saturated before the transition, the system will no longer receive any new data packets. Therefore, $\eta_1(C) = m + m_1 - C$, $\eta_2(C) = 1$, $\eta_1(0) = m + m_2$, $\eta_2(0) = 0$, and $A_{C,0}$ is an $(m + m_1 - C + 1) \times (m + m_2)$ matrix. In $A_{C,0}$, the only non-zero element $a_{1,m+1}$ is given as follows:

$$
a_{1,m+1} = \mu^C.
\tag{8.5}
$$

(5) $A_{i,j}$ ($i < C$ and $0 < j < i$) is the one-step transition probability sub-matrix describing the transition to level j of the awake period from level i of the awake period. This means that the system remains in the awake period and it receives at most one new data packet. Therefore, $\eta_1(i) = m + m_1 - i$, $\eta_2(i) = 1$, $\eta_1(j) = m + m_1 - j$, $\eta_2(j) = 1$, and $A_{i,j}$ is an $(m + m_1 - i + 1) \times (m + m_1 - j + 1)$ matrix. In $A_{i,j}$, the only non-zero element $a_{1,1}$ is given as follows:

$$
a_{1,1} = (1 - \lambda)\binom{i}{i - j}\mu^{i-j}(1 - \mu)^j + \lambda\binom{i}{i - j + 1}\mu^{i-j+1}(1 - \mu)^{j-1}.
\tag{8.6}
$$

(6) $A_{C,j}$ ($0 < j < C$) is the one-step transition probability sub-matrix describing the transition to level j in the awake period from level C in the awake period. This means that the system remains at the awake period and cannot receive any new data packets. Therefore, $\eta_1(C) = m + m_1 - C$, $\eta_2(C) = 1$, $\eta_1(j) = m + m_1 - j$, $\eta_2(j) = 1$, and $A_{i,j}$ is an $(m + m_1 - C + 1) \times (m + m_1 - j + 1)$ matrix. In $A_{C,j}$, the only non-zero element $a_{1,1}$ is given as follows:

$$a_{1,1} = \binom{C}{C-j} \mu^{C-j}(1-\mu)^j. \tag{8.7}$$

(7) $A_{i,i}$ ($0 < i < C$) is the one-step transition probability sub-matrix describing the transition from level i to level i. Given that the system level remains unchanged, the system phase may remain unchanged, or change to the wake-up period from the sleep period, or change to the awake period from the wake-up period. Similar to the explanation in Item (4), $A_{i,i}$ is an $(m+m_1-i+1) \times (m+m_1-i+1)$ square matrix and it is given by

$$A_{i,i} = \begin{pmatrix} (1-\lambda)(1-\mu)^i + i\lambda\mu(1-\mu)^{i-1} & 0 & & & \\ 0 & & 0 & 1-\lambda & \\ \vdots & & \vdots & \ddots & \ddots \\ 0 & & 0 & \cdots & 0 & 1-\lambda \\ 1-\lambda & & 0 & \cdots & \cdots & 0 \end{pmatrix}. \tag{8.8}$$

(8) $A_{C,C}$ is the one-step transition probability sub-matrix describing the transition from level C to level C. This means that during the awake period, none of the data packets in the system complete the transmission in a slot, the system remains at level C. In this case, the transition of the system phase is similar to that in Item (7). $A_{C,C}$ is an $(m+m_1-C+1) \times (m+m_1-C+1)$ square matrix and it is given as follows:

$$A_{i,i} = \begin{pmatrix} (1-\mu)^C & 0 & & & \\ 0 & & 0 & 1 & \\ \vdots & & \vdots & \ddots & \ddots \\ 0 & & 0 & \cdots & 0 & 1 \\ 1 & & 0 & \cdots & \cdots & 0 \end{pmatrix}. \tag{8.9}$$

(9) $A_{i,i+1}$ ($0 < i < C$) is the one-step transition probability sub-matrix describing the transition to level $(i+1)$ from level i. In the awake period, none of the data packets in the system are completely transmitted, and there is a new data packet arrival. The system level changes to $(i+1)$ from i with probability $\lambda(1 - \mu)^i$. In the sleep and wake-up periods, the transition probability is λ. The system may change to the awake period from stage m_1 of the wake-up period, or remain in the wake-up period with an increased stage, or change to the first stage of

the wake-up period from stage m of the sleep period, or remain in the sleep period with an increased stage. Similar to the explanation in Item (4), $A_{i,i+1}$ is an $(m + m_1 - i + 1) \times (m + m_1 - i)$ matrix given as follows:

$$
A_{i,i+1} =
\begin{pmatrix}
\lambda(1-\mu)^i & & & \\
0 & \lambda & & \\
0 & 0 & \lambda & \\
\vdots & \vdots & \ddots & \ddots \\
0 & 0 & \cdots & 0 & \lambda \\
\lambda & 0 & \cdots & \cdots & 0
\end{pmatrix}.
\tag{8.10}
$$

8.3.3 Transition Probability Sub-Matrices for Case II

To give the transition probability matrix P of case II, $LS \leq C$, here we present the transition probability sub-matrices $A_{i,j}$ ($LS \leq i \leq C$ or $0 \leq j \leq C$) in P as follows. Note that when $LS \leq i \leq C$ or $LS \leq j \leq C$, the sub-matrix $A_{i,j}$ in the transition probability matrix P will be degenerated into a vector or a value.

(1) When $LS \leq i < C$ and $j = 0$, there is no level i in either the sleep period or the wake-up period. However, level 0 exists in the sleep-delay period after a one-step transition. In this case, $\eta_2(i) = 1$, $\eta_1(0) = m_2$, $\eta_2(0) = 0$, and $A_{i,0}$ is degenerated into a $1 \times m_2$ row vector. The only non-zero element $a_{1,1}$ in the row vector is obtained as follows:

$$
a_{1,1} = (1-\lambda)\mu^i.
\tag{8.11}
$$

(2) When $i = C$ and $j = 0$, there is no level C in either the sleep period or the wake-up period. However, level 0 exists in the sleep-delay period after a one-step transition. In this case, $\eta_2(i) = 1$, $\eta_1(0) = m_2$, $\eta_2(0) = 0$, and $A_{C,0}$ is degenerated into a $1 \times m_2$ row vector. The only non-zero element $a_{1,1}$ in this vector is obtained as follows:

$$
a_{1,1} = \mu^C.
\tag{8.12}
$$

(3) When $LS \leq i < C$ and $1 \leq j \leq LS - 1$, there is no level i in either the sleep period or the wake-up period. However, level j exists also in the sleep period or the wake-up period after a one-step transition. In this case, $\eta_2(i) = 1$, $\eta_1(j) =$

$m + m_1 - j$, $\eta_2(j) = 1$, and $A_{i,j}$ is degenerated into a $1 \times (m + m_1 - j + 1)$ row vector. The only non-zero element $a_{1,1}$ in this row rector is given as follows:

$$a_{1,1} = \binom{i}{i-j} \mu^{i-j}(1-\mu)^j(1-\lambda) + \lambda \binom{i}{i-j+1} \mu^{i-j+1}(1-\mu)^{j-1}.$$

(8.13)

(4) When $i = C$ and $1 \leq j \leq LS - 1$, there is no level C in either the sleep period or the wake-up period. However, level j exists in both the sleep period and the wake-up period after a one-step transition. In this case, $\eta_2(C) = 1$, $\eta_1(j) = m + m_1 - j$, $\eta_2(j) = 1$, and $A_{i,j}$ is degenerated into a $1 \times (m + m_1 - j + 1)$ row vector. The only non-zero element $a_{1,1}$ in this vector is given as follows:

$$a_{1,1} = \binom{C}{C-j} \mu^{C-j}(1-\mu)^j.$$

(8.14)

(5) When $LS \leq i < C$, $LS \leq j < C - 1$ and $i > j$, there is neither level i nor level j in either the sleep period or the wake-up period before or after a one-step transition. Then, $A_{i,j}$ is degenerated into a value $a_{1,1}$ given as follows:

$$a_{1,1} = \binom{i}{i-j} \mu^{i-j}(1-\mu)^j(1-\lambda) + \lambda \binom{i}{i-j+1} \mu^{i-j+1}(1-\mu)^{j-1}.$$

(8.15)

(6) When $i = C$ and $LS \leq j < C$, there is neither level C nor level j in either the sleep period or the wake-up period before or after a one-step transition. Then, $A_{i,j}$ is degenerated into a value $a_{1,1}$ given as follows:

$$a_{1,1} = \binom{C}{C-j} \mu^{C-j}(1-\mu)^j.$$

(8.16)

(7) When $LS \leq i < C$, $LS \leq j < C$ and $i = j$, the system level remains fixed after a one-step transition. There is no level i in either the sleep period or the wake-up period before or after a one-step transition. Then, $A_{i,i}$ is degenerated into a value $a_{1,1}$ given as follows:

$$a_{1,1} = (1-\lambda)(1-\mu)^i + i\lambda\mu(1-\mu)^{i-1}.$$

(8.17)

(8) When $i = C$ and $j = C$, after a one-step transition, the system level remains at C. There is no level C in either the sleep period or the wake-up period. Then, $A_{C,C}$ is degenerated into a value $a_{1,1}$ given as follows:

$$a_{1,1} = (1-\mu)^C.$$

(8.18)

(9) When $i = LS - 1$ and $j = LS$, level $(m + m_1 - 1)$ exists in the wake-up period. However, level $(m + m_1)$ does not exist in the wake-up period after a one-step transition. In this case, $\eta_1(m + m_1 - 1) = 1$, $\eta_2(m + m_1 - 1) = 1$, $\eta_2(m+m_1) = 1$, and $A_{m+m_1-1,m+m_1}$ is degenerated into a 2×1 column vector given as follows:

$$A_{m+m_1-1,m+m_1} = \begin{pmatrix} \lambda(1 - \mu)^{m+m_1-1} \\ \lambda \end{pmatrix}. \tag{8.19}$$

(10) When $LS \leq i < C$, $LS < j \leq C$ and $j = i + 1$, there is neither level i nor level $(i + 1)$ in either the sleep period or the wake-up period. Then, $A_{i,j+1}$ is degenerated into a value $a_{1,1}$ given as follows:

$$a_{1,1} = \lambda(1 - \mu)^i. \tag{8.20}$$

In this way, we have obtained all the elements in the transition probability matrix P. We can also give the state space of the Markov chain as follows:
$\Omega = \Omega_1 \cup \Omega_2 \cup \Omega_3 \cup \Omega_4$, where $\Omega_1 = \{(i, 0, k) : 0 \leq i \leq \min(C, m - 1), 1 \leq k \leq m\}$, $\Omega_2 = \{(i, 1, 0) : 1 \leq i \leq C\}$, $\Omega_3 = \{(0, 2, k) : 1 \leq k \leq m_2\}$, $\Omega_4 = \{(i, 3, k) : 1 \leq i \leq \min(C, m + m_1 - 1), 1 \leq k \leq m_1\}$.

Let $\pi_{i,j,k}$ be the probability distribution of the three-dimensional DTMC $\{(x(n^+), (y(n^+), (z(n^+)), n \geq 0\}$ in the steady state. $\pi_{i,j,k}$ is defined as follows:

$$\pi_{i,j,k} = \lim_{n \to +\infty} \Pr\{x(n^+) = i, y(n^+) = j, z(n^+) = k\}, \quad (i, j, k) \in \Omega. \tag{8.21}$$

Let π_i be the steady-state probability vector of the system being at level i. When $i = 0$:
$\pi_0 = (\pi_{0,0,1}, \pi_{0,0,2}, \pi_{0,0,3}, \ldots, \pi_{0,0,m}, \pi_{0,2,1}, \pi_{0,2,2}, \ldots, \pi_{0,2,m_2})$.
When $0 < i < m$:
$\pi_i = (\pi_{i,1,0}, \pi_{i,0,i+1}, \pi_{i,0,i+2}, \ldots, \pi_{i,0,m}, \pi_{i,3,1}, \pi_{i,3,2}, \ldots, \pi_{i,3,m_1})$.
When $m \leq i < m + m_1$:
$\pi_i = (\pi_{i,1,0}, \pi_{i,3,i-m+1}, \pi_{i,3,i-m+2}, \ldots, \pi_{i,3,m_1})$.
When $m + m_1 \leq i \leq C$, especially when $LS \leq C$:
$\pi_i = \pi_{i,1,0}$.

Let Π be the steady-state distribution of the system. Π can be given as follows:

$$\Pi = (\pi_0, \pi_1, \pi_2, \ldots, \pi_C). \tag{8.22}$$

Combining the steady-state equation with the normalization condition, we have

$$\begin{cases} \Pi P = \Pi \\ \Pi e = 1 \end{cases} \tag{8.23}$$

where e is a three-dimensional column vector and all elements of the vector are equal to 1.

This system of linear equations uniquely determines the steady-state distribution Π, which can be computed numerically.

8.3.4 Performance Measures

The main factors influencing the UE's networking experience are the standby time of the mobile terminal and the transmission quality. In order to evaluate the standby time, we introduce the measure of the energy saving rate of the system, and in order to evaluate the transmission quality, we introduce the measures of the blocking rate of data packets and the average response time of data packets.

The energy saving rate γ of the system is defined as the energy conservation per slot. Obviously, energy will be consumed normally during the awake period. We also note that less energy will be saved during the sleep-delay period than the sleep period. Therefore, we give the system energy saving rate γ of the system as follows:

$$
\gamma =
\begin{cases}
g_1 \displaystyle\sum_{k=1}^{m_2} \pi_{0,2,k} + g_2 \sum_{i=0}^{C} \sum_{k=i+1}^{m} \pi_{i,0,k}, & m > C \\[4ex]
g_1 \displaystyle\sum_{k=1}^{m_2} \pi_{0,2,k} + g_2 \sum_{i=0}^{m-1} \sum_{k=i+1}^{m} \pi_{i,0,k}, & m \le C
\end{cases}
\tag{8.24}
$$

where g_1 is the energy conservation per slot during the sleep-delay period, g_2 is the energy conservation per slot during the sleep period. Obviously, $g_1 < g_2$.

The blocking rate B_d of data packets is defined as the probability that the transmission request of a newly arriving data packet is blocked. The transmission request of a newly arriving data packet will be rejected when all the logical channels are occupied, namely, the number of data packets in the system is C. We note that a blocking event may occur during all the system periods except the sleep-delay period. Therefore, we give the blocking rate B_d of data packets as follows:

$$
B_d =
\begin{cases}
\lambda \left(\pi_{C,1,0}(1-\mu)^C + \displaystyle\sum_{k=C+1}^{m} \pi_{C,0,k} + \sum_{k=1}^{m_1} \pi_{C,3,k} \right), & m > C \\[4ex]
\lambda \left(\pi_{C,1,0}(1-\mu)^C + \displaystyle\sum_{k=C-m+1}^{m_1} \pi_{C,3,k} \right), & m \le C < m + m_1 \\[3ex]
\lambda \pi_{C,1,0}(1-\mu)^C, & m + m_1 \le C.
\end{cases}
\tag{8.25}
$$

The response time Y_d of a data packet is defined as the duration that has elapsed from the arrival instant of a data packet to the departure instant of that data packet. The data packets arriving during the sleep period and the wake-up period can only be transmitted in the awake period. The data packets arriving at the awake period and the sleep-delay period can be transmitted immediately.

Based on the analysis presented in Sects. 8.3.2 and 8.3.3, we can obtain the average response time $E[Y_d]$ of data packets as follows:

$$E[Y_d] = \frac{1}{\lambda - B_d} \times \sum_{i=0}^{C} i \pi_i e \qquad (8.26)$$

where e is a three-dimensional column vector and all elements of the vector equal to 1, and B_d is the blocking rate of data packets given in Eq. (8.25).

8.4 Numerical Results and Performance Optimization

In this section, we present numerical results to evaluate the performance of the system using the enhanced Active DRX mechanism in LTE system. Then, we optimize the enhanced Active DRX mechanism by constructing a system profit function to improve the system performance.

8.4.1 Numerical Results

By taking examples in numerical results, we set the system parameters as follows: The arrival rate of data packets $\lambda = 0.5$, the time length of the wake-up period $m_1 = 3$ slots, and the service rate of the logical channel is either $\mu = 0.5$ or $\mu = 0.9$. Moreover, we set one slot as 1 s, the energy conservation per slot during the sleep-delay period $g_1 = 0.3$ mW, and the energy conservation per slot during the sleep period $g_2 = 0.9$ mW.

The dependency of the energy saving rate γ of the system on several parameters is shown in Fig. 8.2.

From Fig. 8.2, we observe that for the same values of the service rate μ of the logical channel, the sleep interval m and the number C of the logical channels, the energy saving rate γ of the system decreases as the time length m_2 of the sleep-delay timer increases. As the time length of the sleep-delay timer increases, the probability of the system switching to the sleep period from the sleep-delay period is lower. Therefore, γ will be lower.

We also see that when the service rate μ of the logical channel, the sleep interval m and the time length m_2 of the sleep-delay timer are given, the energy saving rate γ of the system decreases with an increase in the number C of logical channels.

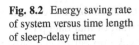

Fig. 8.2 Energy saving rate of system versus time length of sleep-delay timer

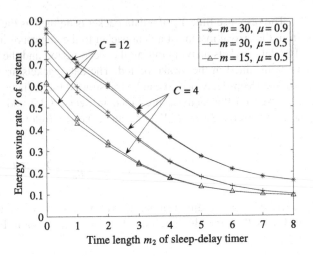

Only when all the logical channels are idle within the sleep-delay timer will the system enter into the sleep period. The greater the number of the logical channels is, the lower the probability of the system switching into the sleep period is. Then, the system will spend less time being in the sleep period, and γ will be lower.

Moreover, we conclude that for the same time length m of the sleep interval, the time length m_2 of the sleep-delay timer and the number C of the logical channels, the energy saving rate γ of the system increases with any increase in the service rate μ of the logical channel. The higher the service rate of the logical channel is, the higher the possibility is that all the logical channels are idle. This means that the system is more likely to switch into a sleep period from the sleep-delay period. Thus, γ increases.

Additionally, we also find that when the service rate μ of the logical channel, the time length m_2 of the sleep-delay timer and the number C of the logical channels are the same, the energy saving rate γ of the system increases with any increase in the sleep interval m. With an increase of the sleep interval, the time length for the system being in the sleep period is longer, so γ will be higher.

The dependency of the blocking rate B_d of data packets on several parameters is shown in Fig. 8.3.

From Fig. 8.3, we also observe that when we take the same values of the service rate μ of the logical channel, the sleep interval m and the number C of the logical channels, the blocking rate B_d of data packets decreases as the time length m_2 of the sleep-delay timer increases. The increase in the time length of the sleep-delay timer leads to a decrease in the time length of the system being in the sleep period and a decrease in the number of the data packets blocked in the sleep period. Therefore, B_d will decrease.

We also see that the service rate μ of the logical channel, the sleep interval m and the time length m_2 of the sleep-delay timer are provided, the blocking rate B_d of data packets decreases with an increase in the number C of logical channels. If

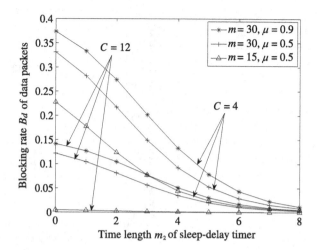

Fig. 8.3 Blocking rate of data packets versus time length of sleep-delay timer

the number of logical channels increases, the system can accommodate more data packets, the number of data packets blocked by the system is fewer, so B_d will decrease.

Moreover, we conclude that for the same time length m of the sleep interval, the time length m_2 of the sleep-delay timer, and the number C of the logical channels, the blocking rate of data packets increases with an increase in the service rate μ of the logical channel. As the service rate of the logical channel increases, the probability of the system switching into the sleep period increases. Data packets arriving during the sleep period cannot be transmitted immediately, so fewer data packets are able to be transmitted. Therefore, B_d increases.

Additionally, we find that when the service rate μ of the logical channel, the time length m_2 of the sleep-delay timer, and the number C of the logical channels are provided, the blocking rate B_d of data packets increases with an increase in the sleep interval m. With an increase in the sleep interval m, the time length for the system being in the sleep period is longer. Data packets will not be transmitted in the sleep period, so B_d will be higher.

The dependency of the average response time $E[Y_d]$ of data packets on several parameters is shown in Fig. 8.4.

From Fig. 8.4, we observe that when we take the same values of the service rate μ of the logical channel, the sleep interval m and the number C of the logical channels, the average response time $E[Y_d]$ of data packets decreases as the time length m_2 of the sleep-delay timer increases. The longer the time length of the sleep-delay timer is, the lower the probability is that the system will enter into a sleep period. The number of data packets arriving during the sleep period decreases, and the average response time of data packets decreases.

We also see that when the service rate μ of the logical channel, the sleep interval m and the time length m_2 of the sleep-delay timer are fixed, with an increase in the number C of logical channels, the average response time $E[Y_d]$ of data packets

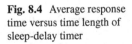

Fig. 8.4 Average response
time versus time length of
sleep-delay timer

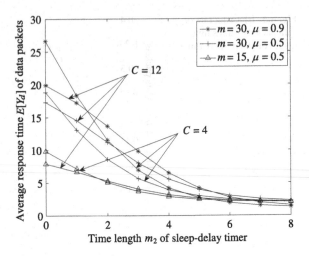

shows two types of trends. One is that when the time length m_2 of the sleep-delay
timer is shorter, $E[Y_d]$ with a larger number C, such as $C = 12$, is lower than that
with a smaller number C, such as $C = 4$. This is because in this case, the main factor
influencing $E[Y_d]$ is the possibility of the system entering into the sleep period. The
more the logical channels C there are, the less likely it is that the system is in the
awake period. The data packets arriving at the system during the awake period will
be transmitted without a long delay. Therefore, $E[Y_d]$ is lower. The other is that
when the time length m_2 of the sleep-delay timer is greater, $E[Y_d]$ with more logical
channels, such as $C = 12$, is greater than in the case having less logical channels,
such as $C = 4$. This is because in this case, the main factor influencing $E[Y_d]$ is the
number of data packets entering into the system during the sleep period. The more
the logical channels C there are, the greater the number of data packets arriving at
system during the sleep period is. The data packets arriving at the system during the
sleep period experience longer delays before being transmitted. Therefore, $E[Y_d]$ is
higher.

Moreover, we notice that for the same time length m of the sleep interval, the
time length m_2 of the sleep-delay timer, and the number C of the logical channels,
with an increase in the service rate μ of logical channel, the average response time
$E[Y_d]$ of data packets shows two types of changing trends. One is that when the time
length of the sleep-delay timer is shorter, the greater the service rate of the logical
channel is, and the higher possibility there is that the system switches into the sleep
period. The data packets arriving during the sleep period experience long delays
before being transmitted, so $E[Y_d]$ increases. The other is that when the time length
of the sleep-delay timer is longer, the main factor affecting $E[Y_d]$ is the transmission
time of data packets in the awake period. The transmission time decreases with an
increase in the service rate of the logical channel, so $E[Y_d]$ decreases.

Additionally, we find that when the service rate μ of the logical channel, the time length m_2 of the sleep-delay timer and the number C of the logical channels are fixed, the average response time $E[Y_d]$ of data packets increases with an increase in the sleep interval m. The greater the sleep interval is, the longer the system will be in a sleep period, and the more data packets will arrive in that sleep period. These data packets will be buffered in the system until the awake period begins, then the average waiting time of data packets increases, so $E[Y_d]$ increases.

On the other hand, we can also compare the system performance of the conventional Active DRX mechanism ($m_2 = 0$) and the enhanced Active DRX mechanism ($m_2 > 0$). From the numerical results shown in Figs. 8.2, 8.3, 8.4, we conclude that the system performance for the enhanced Active DRX mechanism has improved in terms of the blocking rate of data packets and the average response time of data packets. This is in line with our expectations. However, the energy saving rate of the system will be reduced slightly in the enhanced Active DRX mechanism. This result is outside our estimation. In other words, there is a trade-off between different performance measures. Therefore, the enhanced Active DRX mechanism needs to be optimized for setting the time lengths of the sleep interval and the sleep-delay timer, as well as for setting the number of logical channels.

8.4.2 Performance Optimization

In order to optimize the enhanced Active DRX mechanism, we construct a system profit function as follows:

$$F_1(m) = F_2(m_2) = F_3(C) = f_1 \times (\lambda - B_d) + f_2 \times \gamma - f_3 \times C - f_4 \times E[Y_d] \tag{8.27}$$

where f_1 is the reward per slot due to the successful transmission of a data packet, f_2 is the benefit due to the energy saving per slot, f_3 is the cost of the maintaining a logical channel, f_4 is the cost per slot associated with the average response time of data packets. The values of f_1-f_4 should be set as needed in practice. For the network with a throughput or energy conservation sensitive application, the factor f_1 or f_2 will be set relatively higher. On the other hand, for the networks with lower tolerance on the channel maintenance cost or the average response time of data packets, the factor f_3 or f_4 will be set greater.

In this chapter, as an example, we set $f_1 = 6$, $f_2 = 2.55$, $f_3 = 0.05$, $f_4 = 0.06$ and apply other parameters used in Sect. 8.4.1. The development of the system profit function is shown in Figs. 8.5, 8.6, 8.7 dependent on different system parameters.

In Fig. 8.5, we show the system profit function $F_1(m)$ versus the sleep interval m for the time length $m_2 = 3$ and the number $C = 15$ as an example. From Fig. 8.5, we find that system profit function $F_1(m)$ shows two trends with an increase in the sleep interval m. When the sleep interval is smaller, the main factor affecting the system profit function is the time length for the system being in the sleep period

Fig. 8.5 System profit
function versus sleep interval

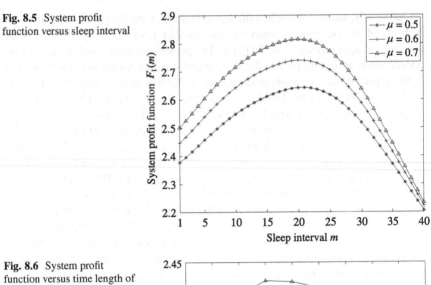

Fig. 8.6 System profit
function versus time length of
sleep-delay timer

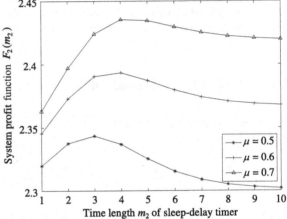

per slot. The greater the sleep interval is, the longer the time length for the system
being in the sleep period is, so the system profit will be greater. However, when the
sleep interval increases to a certain value, the main factor affecting the system profit
becomes the number of data packets transmitted successfully per slot. The greater
the sleep interval is, the more the data packets are blocked in the sleep period, so the
system profit will be lower.

In Fig. 8.6, we show the system profit function $F_2(m_2)$ versus the time length m_2
for the sleep interval $m = 36$ and the number $C = 16$ as an example. From Fig. 8.6,
we find that system profit function $F_2(m_2)$ shows two trends with an increase in the
time length m_2 of the sleep-delay timer. When the time length of the sleep-delay
timer is shorter, the main factor affecting the system profit is the number of data
packets transmitted successfully per slot. The longer the time length of the sleep-
delay timer is, the lower the probability that the system switches into the sleep period

Fig. 8.7 System profit function versus number of logical channels

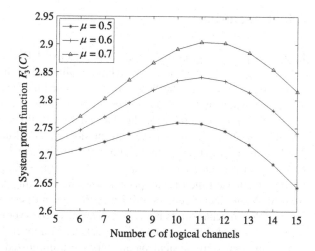

Table 8.1 Optimum parameters in proposed enhanced Active DRX mechanism

Transmission rates μ	Optimum sleep intervals m^*	Optimum time lengths m_2^* of sleep-delay timer	Optimum numbers C^* of logical channels
0.5	21	3	10
0.6	20	4	11
0.7	20	4	11

is. As a result, the number of data packets blocked during the sleep period decreases, hence the system profit will be greater. When the time length of the sleep-delay timer is long enough, the main factor affecting the system profit becomes the cost of the system listening to the logical channels in the sleep-delay period. The greater the time length of the sleep-delay period is, the higher the cost of the system listening to the channels is, so the system profit will be lower.

In Fig. 8.7, we show the system profit function $F_3(C)$ versus the number C of logical channels for the sleep interval $m = 20$ and the time length $m_2 = 3$ as an example. From Fig. 8.7, we observe that the system profit function $F_3(C)$ shows two trends with an increase in the number C of logical channels. When the number of logical channels is smaller, the main factor affecting the system profit is the number of data packets transmitted successfully per slot. With an increase in the number of logical channels, the blocking rate of data packets decreases, and the system profit increases. When the number of logical channels is greater, the main factor affecting the system profit is the cost of the system maintaining the logical channels. The greater the number of logical channels is, the higher the cost of the system maintaining the logical channels is. Therefore, the system profit will be lower.

Consequently, there will be maximum profits when the system parameters are set optimally. According to the results shown in Figs. 8.5, 8.6, 8.7, the summary of the optimal values of the sleep interval, the sleep-delay timer and the number of logical channels for different service rates μ can be given as shown in Table 8.1.

8.5 Conclusion

In this chapter, we proposed an enhanced Active DRX mechanism with a sleep-delay strategy in 4G networks, which has an influence on the control of the downlink transmission at the UE. Furthermore, we modeled the system with the mechanism as a synchronous multiple-vacation queueing system with wake-up and sleep-delay. Accounting for the number of the data packets in the system, the system period and the sequence number of the current slot, we constructed a three-dimensional Markov chain by studying the transition probability matrix in two cases: Case I and Case II. Performance measures such as the energy saving rate of the system, the average response time of data packets and the blocking rate of data packets were derived. Numerical results showed that the system performance of the enhanced Active DRX mechanism is improved in terms of the average response time of data packets and the blocking rate of data packets. Moreover, by constructing a system profit function, some system parameters were optimized in terms of the number of the logical channels for one connection, and the time lengths of the sleep-delay timer and the sleep period.

Part II
Resource Management and Performance Analysis on Cognitive Radio Networks

Part II discusses the dynamic spectrum allocation and energy saving strategy in Cognitive Radio Networks (CRNs). We present an analytic framework to evaluate the system performance by constructing priority queueing models with possible service interruptions, using multiple channels, with several types of vacation mechanisms, and possible transmission interruptions.

There are seven chapters in Part II, beginning with Chap. 9.

In Chap. 9, we propose a channel aggregation strategy in which all the channels in a spectrum are aggregated as one channel for the transmission of a Primary User (PU) packet, while each Secondary User (SU) packet occupies only one of the channels in the spectrum for its transmission. In Chap. 10, we propose an adaptive control approach to determine the reservation ratio of the licensed spectrum for SUs and present an adaptive spectrum reservation strategy to better adapt to systemic load changes in CRNs. In such a strategy, the licensed spectrum is separated into two logical channels, namely, the reserved channel and the shared channel, respectively. In Chap. 11, we establish a priority queueing model in which two types of packets, the PU packets and the SU packets, may interfere with each other. In this priority queueing model, we take into account the impatient behavior of the interrupted SU packets, the tolerance delay of an SU packet, the sensing errors of SUs and the preemptive priority of PU packets. In Chap. 12, we present a mini-slotted spectrum allocation strategy with the purpose of improving the throughput of SU packets and reducing the spectrum switching frequency in CRNs. Due to the mistake detections in practice, the PU packet and the SU packet will occupy the spectrum simultaneously, namely, a collision will occur on the spectrum. In Chap. 13, we establish a two-dimensional Continuous-Time Markov Chain (CTMC) model to record the stochastic behavior of two types of user packets, the PU packets and the SU packets, with a channel reservation strategy. In this channel reservation strategy, part licensed channels are reserved for SU packets for the purpose of properly controlling the interference between the PU packets and the SU packets. In Chap. 14, we propose an energy saving strategy in CRNs with the aim of alleviating the spectrum scarcity crisis and reducing the energy consumption. By establishing a preemptive priority queueing model with a single-vacation to capture the stochastic behavior of the proposed strategy, and by using the matrix-geometric solution method, we derive performance measures of the system in terms of the average latency of SU packets and the energy saving degree. In Chap. 15, we establish a preemptive priority queueing model with multiple vacations to capture the stochastic behavior of user packets, the PU packets and the SU packets, and present analyses to numerically evaluate the energy saving strategy using a multiple-sleep mode in CRNs.

Chapter 9
Channel Aggregation Strategy with Perfect-Sensing Results

In this chapter, we propose a channel aggregation strategy in Cognitive Radio Networks (CRNs) to improve the service rate of Primary User (PU) packets and eliminate the forced termination of Secondary User (SU) packets. In this strategy, all the channels in a spectrum are aggregated as one channel for the transmission of a PU packet, while each SU packet occupies only one of the channels in the spectrum for its transmission. Considering the stochastic behavior of SU packets, we build a discrete-time preemptive retrial queueing model with multiple servers, a retrial buffer and implement synchronous transmission interruptions. We derive performance measures of the system in terms of the blocking rates for both types of user packets, the average latency of SU packets, the channel utilization and the system cost function. We present numerical results to evaluate the system performance and optimize the channel aggregation intensity. Furthermore, we investigate the Nash equilibrium and the socially optimal behaviors of SU packets, and we propose an appropriate pricing policy to maximize the value of the social benefit function.

9.1 Introduction

With the development of the Internet of Things (IoT) and the emergence of Big Data, the number and scale of WCNs are growing dramatically. The demand for wireless spectrum is presenting a huge challenge to the network resources of these businesses. In conventional static spectrum allocation strategy, the spectrum usage is unbalanced and the spectrum utilization is low. The use of CRNs based on cognitive radio technology is being used to try and solve the problem of spectrum inefficiency [Akyi06]. Scholars are currently concentrating their research attention to dynamic spectrum allocation strategies in CRNs to maximize the value of the social benefit function [Liu20].

© Springer Nature Singapore Pte Ltd. 2021
S. Jin, W. Yue, *Resource Management and Performance Analysis of Wireless Communication Networks*, https://doi.org/10.1007/978-981-15-7756-7_9

One key technique used in dynamic spectrum allocation strategies is channel aggregation. In [Lei10], in order to achieve higher bandwidth utilization, the authors proposed a channel aggregation scheme in CRNs by assembling several channels together for SUs. In [Li09], with the aim of enhancing the throughput, fairness and latency performance, two carrier schedulers with both joint and disjoint queues for channel aggregation in an LTE-A system were proposed. In [Lu09], in order to obtain a higher service rate, SUs were enabled to sense multiple channels simultaneously, and several idle and discontinuous channels were aggregated by using a channel bonding technique.

A spectrum assignment method in cognitive Ad-hoc networks was investigated in [Chen08]. With the help of discontiguous channel access, small channel fragments could be aggregated and further utilized, which dramatically improved the channel utilization.

In some research models, the interrupted SUs were supposed to claim another idle channel rather than directly give up their transmissions. This gave rise to a new type of handoff, namely, the spectrum handoff in CRNs.

In [Wang12], aiming to choose an appropriate channel for each spectrum handoff and resume the unfinished transmission, on-demand manner spectrum sensing was performed, and the influence of the spectrum handoff on channel utilization was investigated.

In [Pham14], a proactive handoff approach was analyzed, and a spectrum handoff model with prediction to optimize the spectrum handoff scheme was proposed. It was found that although a spectrum handoff might increase the channel utilization, a handoff delay and an extra energy cost would be inadvertently introduced. Additionally, if no idle channel was available, the interrupted SUs would be dropped from the system.

Many researchers have also been concentrating on the problem of reducing the forced termination rate. The use of channel reservation has been proposed as one way to address this problem. An analytical framework of CRNs with channel reservation for PUs was designed and a trade-off between different system parameters was derived in [Tama13]. Using a channel reservation scheme, the spectrum handoff was dealt with and a fuzzy logic to detect spectrum channel priority was employed in [Liu13]. However, use of a channel reservation scheme comes at the cost of a smaller system throughput and a lower channel utilization.

Another way to reduce the forced termination rate is to set a buffer for SUs. In [Wang13a], a framework for admission control was presented by taking into account a finite buffer for interrupted SUs together with newly arriving SUs, the system performance was evaluated by considering that the SUs queueing at the buffer were able to leave the system when they became impatient. In [Peng13], the authors focused on the handoff delay of SUs, then set a finite buffer for newly arriving and interrupted SUs. They also investigated the scheme by using a Continuous-Time Markov Chain (CTMC). In [Zhao13], a channel access strategy with α-retry policy was proposed. By using a retrial policy, the forced termination rate could be reduced. Based on the research mentioned above, we can draw a conclusion that the buffer schemes work well for decreasing the forced termination rate. However, it is vital that the buffer size and schedule for the handoff of SUs are set correctly.

In this chapter, we propose a channel aggregation strategy in which all the channels in a spectrum are aggregated as one channel for the transmission of a PU packet, while each SU packet occupies only one of the channels in the spectrum for its transmission. Considering the stochastic behavior of SU packets, we build a discrete-time preemptive retrial queueing model with multiple channels, a retrial buffer and synchronous transmission interruptions. Accordingly, we evaluate the system performance under the proposed strategy and optimize the arrival rate of SU packets socially.

The chapter is organized as follows. In Sect. 9.2, we describe the channel aggregation strategy proposed in this chapter. Then, we present the system model in detail. In Sect. 9.3, we present a performance analysis of the system model, through an analysis of the steady-state distribution to obtain the performance measures and the system cost. We present numerical results to evaluate the system performance and optimize the channel aggregation intensity. In Sect. 9.4, we firstly investigate the Nash equilibrium and the socially optimal behaviors of SUs in the channel aggregation strategy proposed in this chapter. Then, we propose an appropriate pricing policy to maximize the value of the social benefit function for imposing an appropriate admission fee for SU packets. Finally, we draw our conclusions in Sect. 9.5.

9.2 Channel Aggregation Strategy and System Model

In this section, we first describe the channel aggregation strategy proposed in this chapter. Then, we present the system model in detail.

9.2.1 Channel Aggregation Strategy

In this chapter, establishing that all spectrums in the system to have the same probability nature, we focus on one spectrum, called a "tagged spectrum", and propose a channel aggregation strategy on the tagged spectrum. In order to ensure the transmission quality of PU packets, all the channels in the tagged spectrum can be aggregated together in one channel, called "PU's channel", for the transmission of a PU packet. However, the transmission of an SU packet will occupy only one of the channels in the tagged spectrum, so multiple SU packets can be transmitted concurrently if there are multiple idle channels in the spectrum at the same time. We call the number of the channels in the tagged spectrum the aggregation intensity. This is denoted by c, which is one system parameter to be optimized.

The PU packets have preemptive priority to occupy the PU's channel. SU packets, however, can only make opportunistic use of the channels. When a PU packet appears, the transmissions of all the SU packets occupying the channels in the tagged spectrum will be interrupted. All the interrupted SU packets will enter the retrial buffer to protect them from any forced termination.

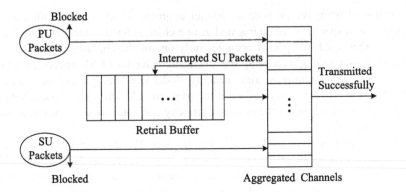

Fig. 9.1 Proposed channel aggregation strategy

When an SU packet arrives at the system, if there is no idle channel available in the tagged spectrum, the SU packet will be blocked. Generally speaking, from the view point of user preference, a forced termination is less acceptable than the blocking of a new transmission request. This is because the new transmission request might earn a chance to be switched to other spectrums and receive prompt transmission service. Obviously, the handoff overhead for a new transmission is lighter than that for an interrupted transmission.

For this reason, we should consider that the SU packets already in the retrial buffer are supposed to have a higher priority to access the channels in the tagged spectrum over the newly arriving SU packets. Once the tagged spectrum is no longer occupied by a PU packet, the SU packets in the retrial buffer will access the channels in the tagged spectrum immediately to resume their transmissions. Importantly, we note that an SU packet already in the system will never drop away, which eliminates any forced termination of SU packets.

The channel aggregation strategy proposed in this chapter is demonstrated in Fig. 9.1.

9.2.2   System Model

From Fig. 9.1, we describe the working principle of the channel aggregation strategy and present the system model as a preemptive retrial queue with multiple aggregated channels, a retrial buffer and synchronous transmission interruptions by using this strategy as follows.

(1) We assume that SUs sense the channel perfectly. If there is no PU packet in the tagged spectrum, whether or not the channels are occupied by any SU packets, a newly arriving PU packet will occupy the PU's channel. Otherwise, this PU packet will be blocked by the tagged spectrum.

(2) When an SU packet emerges, if there is at least one idle channel in the tagged spectrum, the central controller will allocate one of the available channels in the spectrum to this SU packet. Otherwise, this SU packet will be blocked by the tagged spectrum.

(3) If part or all of the channels in the tagged spectrum are being occupied by some SU packets, a newly arriving PU packet will interrupt the transmissions of these SU packets, and occupy the PU's channel with preemptive priority. In this case, all the interrupted SU packets will return to the buffer and wait for a retrial. That is to say, once an SU packet accesses a channel in the tagged spectrum, its transmission will be guaranteed. This results in an improvement in the level of satisfaction with the quality of transmission of SU packets.

(4) Once the transmission of the PU packet occupying the PU's channel is completed and there are no new arrivals of PU packets, the SU packets in the retrial buffer will access the channels in the tagged spectrum and resume their transmissions. In order to simplify the analysis of this system model, in this chapter, we omit the procedure for accumulating the transmission information related to those packets that have been forcibly terminated.

(5) Since the number of interrupted SU packets at one time is never more than the aggregation intensity c, namely, the number of channels is the tagged spectrum, we set the retrial buffer size as the aggregation intensity. We note that if the buffer size is less than the aggregation intensity, the forced termination of SU packets cannot be eliminated.

From the perspective of the SU packets, we can model the system as a preemptive retrial queue with multiple channels, a retrial buffer and synchronous transmission interruptions.

The time axis is segmented into a series of equal intervals, called slots. We consider an Early Arrival System (EAS), namely, the packets are assumed to arrive at the system immediately before the beginning instant n^+ of the nth slot ($n = 1, 2, 3, \ldots$), and depart from the system immediately after the end instant n^- of the nth slot ($n = 2, 3, 4, \ldots$).

The arriving intervals and transmission times of the packets are supposed to be i.i.d. random variables. The inter-arrival times of the SU packets and PU packets are assumed to follow a geometric distribution with arrival rate of SU packets λ_1 ($0 \le \lambda_1 \le 1$, $\bar{\lambda}_1 = 1 - \lambda_1$) and arrival rate of PU packets λ_2 ($0 \le \lambda_2 \le 1$, $\bar{\lambda}_2 = 1 - \lambda_2$), respectively. The transmission time of an SU packet is assumed to follow a geometric distribution with parameter μ_1. The transmission time of a PU packet is assumed to follow another geometric distribution with parameter μ_2. In the system model considered in this chapter, the parameters μ_1 and μ_2 are in fact the probabilities that an SU packet or a PU packet is being completely transmitted in a slot, so μ_1 and μ_2 cannot be greater than 1, namely, $0 \le \mu_1 \le 1$, $\bar{\mu}_1 = 1 - \mu_1$ and $0 \le \mu_2 \le 1$, $\bar{\mu}_2 = 1 - \mu_2$. As a result, the time length of a slot should be set appropriately short. We call μ_1 and μ_2 the service rates of SU packets and PU packets, respectively. For PU packets under the proposed strategy, the service rate μ_2 on the PU's channel is approximately the sum of the service rate μ_0 for all the channels in the tagged spectrum. Therefore, we have $\mu_2 = \min\{c\mu_0, 1\}$, where μ_0 is the relative service rate of one channel, and c is the aggregation intensity, namely, the number of channels in the spectrum.

Let $L_n^{(1)} = i$ be the number of SU packets in the system at the instant $t = n^+$, and $L_n^{(2)} = j$ be the number of PU packets in the system at the instant $t = n^+$, where $0 \le i \le c$, $j = 0, 1$. $\left\{\left(L_n^{(1)}, L_n^{(2)}\right), n \ge 0\right\}$ constitutes a two-dimensional Markov chain. The state space of this Markov chain is given as follows:

$$\Omega = \{(i, j) : 0 \le i \le c, j = 0, 1\} \tag{9.1}$$

where $(0, 0)$ denotes that there are no packets in the system; $(i, 0)$ denotes that there are i SU packets occupying i channels in the tagged spectrum and no PU packets in the system; $(i, 1)$ denotes that there is a PU packet occupying the PU's channel and i SU packets in the retrial buffer, where $1 \le i \le c$.

9.3 Performance Analysis and Numerical Results

In this section, we present a performance analysis of the system model, through an analysis of the steady-state distribution to obtain the performance measures and the system cost. We present numerical results to evaluate the performance of the system and optimize the channel aggregation intensity.

9.3.1 Steady-State Distribution

We define the system phase as the total number of SU packets in the system. Let P be the state transition probability matrix of the system phases. According to different system phases, P can be given as a $(c + 1) \times (c + 1)$ block-structured matrix as follows:

$$P = \begin{pmatrix} A_{0,0} & A_{0,1} & & & \\ A_{1,0} & A_{1,1} & A_{1,2} & & \\ \vdots & \vdots & \vdots & \ddots & \\ A_{c-1,0} & A_{c-1,1} & A_{c-1,2} & \cdots & A_{c-1,c} \\ A_{c,0} & A_{c,1} & A_{c,2} & \cdots & A_{c,c} \end{pmatrix} \tag{9.2}$$

where sub-matrix $A_{u,v}$ is the transition probability matrix from the system phase u ($u = 0, 1, 2, \ldots, c$) to the system phase v ($v = 0, 1, 2, \ldots, c$). Considering that no buffer is prepared for PU packets, there is at most one PU packet, namely, $j = 0$ or $j = 1$, in the system. It is easy to find that each sub-matrix $A_{u,v}$ has an order of 2×2 structure. $A_{u,v}$ is discussed as follows.

(1) At the instant $t = n^+$, the system phase is $u = 0$, namely, there are no SU packets in the system, and the system phase will be v ($v = 0, 1$) at the instant $t = (n + 1)^+$.

If the system phase $v = 0$, namely, there are also no packets in the system at the instant $t = (n + 1)^+$, the transition probability matrix $A_{0,0}$ is given as follows:

$$A_{0,0} = \begin{pmatrix} \bar{\lambda}_1 \bar{\lambda}_2 & \lambda_2 \\ \bar{\lambda}_1 \bar{\lambda}_2 \mu_2 & \lambda_2 \mu_2 + \bar{\mu}_2 \end{pmatrix}, \quad v = 0. \tag{9.3}$$

If the system phase $v = 1$, namely, there is an SU packet in the system at the instant $t = (n + 1)^+$, the transition probability matrix $A_{0,1}$ is given as follows:

$$A_{0,1} = \begin{pmatrix} \lambda_1 \bar{\lambda}_2 & 0 \\ \lambda_1 \bar{\lambda}_2 \mu_2 & 0 \end{pmatrix}, \quad v = 1. \tag{9.4}$$

(2) At the instant $t = n^+$, the system phase is u ($u = 1, 2, 3, \ldots, c-1$), the system phase will be v ($v = 0, 1, 2, \ldots, u + 1$) at the instant $t = (n + 1)^+$.

The system phase $v = 0$ means that there are no SU packets in the system at the instant $t = (n + 1)^+$. In this case, all the SU packets in the system complete their transmissions and leave the system together. At the same time, there are no new SU packet arrivals at the system. Therefore, the transition probability matrix $A_{u,0}$ is given as follows:

$$A_{u,0} = \begin{pmatrix} \bar{\lambda}_1 \bar{\lambda}_2 \mu_1^u & \lambda_2 \mu_1^u \\ 0 & 0 \end{pmatrix}, \quad v = 0. \tag{9.5}$$

The system phase $v > 0$ means that there are v ($v \geq 1$) SU packets in the system at the instant $t = (n + 1)^+$. One case is that $u \geq v$ and $(u - v)$ of the SU packets in the system complete their transmissions and leave the system together. At the same time, no new SU packets arrive at the system. A second case is that $u \geq v$ and $(u - v + 1)$ of the SU packets in the system complete their transmissions and leave the system. Meanwhile, a new SU packet arrives at the system. A third case ($v = u + 1$) means that all the SU packets in the system do not complete their transmissions and a new SU packet arrives at the system. Therefore, the transition probability matrix $A_{u,v}$ is given as follows:

$$A_{u,v} = \begin{cases} \begin{pmatrix} \bar{\lambda}_2\left(\bar{\lambda}_1\binom{u}{v}\mu_1^{u-v}\bar{\mu}_1^v + \lambda_1\binom{u}{v-1}\mu_1^{u-v+1}\bar{\mu}_1^{v-1}\right) & \lambda_2\binom{u}{v}\mu_1^{u-v}\bar{\mu}_1^v \\ 0 & 0 \end{pmatrix}, \\ \qquad\qquad\qquad\qquad\qquad 1 \le v < u \\[4pt] \begin{pmatrix} \bar{\lambda}_2\left(\bar{\lambda}_1\bar{\mu}_1^v + \lambda_1\binom{u}{1}\mu_1\bar{\mu}_1^{u-1}\right) & \lambda_2\bar{\mu}_1^v \\ \bar{\lambda}_1\bar{\lambda}_2\mu_2 & \lambda_2\mu_2 + \bar{\mu}_2 \end{pmatrix}, \quad v = u \\[4pt] \begin{pmatrix} \lambda_1\bar{\lambda}_2\bar{\mu}_1^u & 0 \\ \lambda_1\bar{\lambda}_2\mu_2 & 0 \end{pmatrix}, \quad v = u+1. \end{cases}$$

$$\text{(9.6)}$$

(3) At the instant $t = n^+$, the system phase is $u = c$, namely, there are u SU packets in the system, and the system phase will be v ($v = 0, 1, 2, \ldots, c$) at the instant $t = (n+1)^+$. Similar to the matrix structures shown in Eqs. (9.5) and (9.6), the transition probability matrix $A_{u,v}$ ($v = 0, 1, 2, \ldots, c-1$) can be given as follows:

$$A_{c,0} = \begin{pmatrix} \bar{\lambda}_1\bar{\lambda}_2\mu_1^c & \lambda_2\mu_1^c \\ 0 & 0 \end{pmatrix}, \quad v = 0, \tag{9.7}$$

$$A_{c,v} = \begin{pmatrix} \bar{\lambda}_2\left(\bar{\lambda}_1\binom{c}{v}\mu_1^{c-v}\bar{\mu}_1^v + \lambda_1\binom{c}{v-1}\mu_1^{c-v+1}\bar{\mu}_1^{v-1}\right) & \lambda_2\binom{c}{v}\mu_1^{c-v}\bar{\mu}_1^v \\ 0 & 0 \end{pmatrix},$$

$$1 \le v \le c-1.$$
$$\text{(9.8)}$$

The system phase $v = c$ means that there are c SU packets in the system at the instant $t = (n+1)^+$. In this phase, one possible case is that one of the SU packets leaves the system. The other possible case is that a new SU packet arrives at the system. Therefore, the transition probability matrix $A_{c,c}$ can be given as follows:

$$A_{c,c} = \begin{pmatrix} \bar{\lambda}_2\left(\bar{\mu}_1^c + \lambda_1\binom{c}{1}\mu_1\bar{\mu}_1^{c-1}\right) & \lambda_2\bar{\mu}_1^c \\ \bar{\lambda}_2\mu_2 & \lambda_2\mu_2 + \bar{\mu}_2 \end{pmatrix}, \quad v = c.$$

Now, all the sub-matrices in P have been presented.

The structure of the transition probability matrix P indicates that the two-dimensional Markov chain $\left\{ \left(L_n, L_n^{(1)} \right), n \geq 0 \right\}$ is non-periodic, irreducible and positive recurrent. Letting $\pi_{i,j}$ be the steady-state distribution of the two-dimensional Markov chain, $\pi_{i,j}$ can be given as follows:

$$\pi_{i,j} = \lim_{n \to \infty} \Pr\{L_n = i, L_n^{(1)} = j\}, \quad 0 \leq i \leq c, j = 0, 1. \tag{9.9}$$

Let π_i be the steady-state probability vector of the system being at phase i. π_i can be given as follows:

$$\pi_i = (\pi_{i,0}, \pi_{i,1}), \quad 0 \leq i \leq c. \tag{9.10}$$

Combining the system equilibrium equation and the normalization condition for the Markov chain mentioned above, we have

$$\begin{cases} (\pi_0, \pi_1, \pi_2, \ldots, \pi_c) \, P = (\pi_0, \pi_1, \pi_2, \ldots, \pi_c) \\ (\pi_0, \pi_1, \pi_2, \ldots, \pi_c) \, e = 1 \end{cases} \tag{9.11}$$

where e is a column vector with $2 \times (c + 1)$ elements and all elements of the vector are equal to 1.

Equation (9.11) is a linear system of equations with $2 \times (c + 1)$ unknowns. By using a Gaussian elimination method, we establish an iterative algorithm to calculate the steady-state distribution $\Pi = (\pi_0, \pi_1, \pi_2, \ldots, \pi_c)$. The main steps of the iterative algorithm are given as follows:

Step 1: Input state transition probability matrix P and initialize a small constant ε (for example, $\varepsilon = 10^{-6}$).

Step 2: Set $\tau = 2 \times (c + 1)$.

Step 3: Set the maximum error ε.

Step 4: Set the initial iterative time as $m = 0$ and the initial value $\Pi^{(0)}$.

Step 5: Construct a $\tau \times \tau$ matrix G by replacing an arbitrary column of matrix $(P - E)$ with column vector e. Here E is a $\tau \times \tau$ unit matrix.

Step 6: Construct a $1 \times \tau$ row vector $b = (0, 1)$, where 0 is a $1 \times (\tau - 1)$ zero row vector.

Step 7: $\tau \times \tau$ matrixes denoted as U, V and R. U is the strictly lower triangular part of G, V is the strictly upper part of G, and R is the diagonal part of G.

Step 8: $\Pi^{(m+1)} = -\Pi^{(m)} V (W + U)^{-1} + b(W + U)^{-1}$.

Step 9:
  while $\left\| \Pi^{(m+1)} - \Pi^{(m)} \right\| > \varepsilon$
    $m = m + 1$
    $\Pi^{(m+1)} = -\Pi^{(m)} V (R + U)^{-1} + b(R + U)^{-1}$
  endwhile

Step 10: Output steady-state distribution Π.

By selecting a suitably small ε, we can obtain the steady-state distribution Π with enough precision.

9.3.2 Performance Measures and Analysis of System Cost

We suppose that the transmissions for the PU packets are independent of those for the SU packets. Since there is no buffer for PU packets, the transmission process of PU packets can be considered as a simple pure losing system model with a single channel. Let p_2 be the probability that the tagged spectrum is occupied by a PU packet in the system. p_2 is then given as follows:

$$p_2 = \lim_{n \to \infty} \Pr \left\{ L_n^{(2)} = 1 \right\} = \frac{\lambda_2}{\lambda_2 + \mu_2}. \tag{9.12}$$

We define the blocking rate B_p of PU packets as the probability that a newly arriving PU packet is blocked by the system. We note that only when the PU's channel is occupied by a PU packet will the newly arriving PU packets be blocked by the system. We give the blocking rate B_p of PU packets as follows:

$$B_p = \lambda_2 p_2 \bar{\mu}_2 = \frac{\lambda_2^2 \bar{\mu}_2}{\lambda_2 + \mu_2}. \tag{9.13}$$

In CRNs, the transmission of an SU packet can be influenced by PU packets. As a result, the performance measures of SU packets are affected by PU packets' activities. This is important to note as we next mathematically derive some required performance measures for the SU packets.

We define the blocking rate B_s of SU packets as the probability that a newly arriving SU packet is blocked by the system. When an SU packet arrives at the system, if no channel is available, the newly arriving SU packet will not be allowed by the system.

In an EAS, the newly arriving SU packet will be blocked by the system in the following four cases:

(1) In the previous slot, a PU packet occupies the PU's channel and this PU packet doesn't complete its transmission at the end of the slot. Therefore, the new SU packet arriving at the current slot will be blocked.
(2) When a PU packet and an SU packet arrive at the system simultaneously, the newly arriving SU packet will be blocked due to the higher priority of the PU packets.
(3) In the previous slot, all the channels in the tagged spectrum are occupied by SU packets. If none of the SU packets complete their transmission at the end of the slot and there are no new PU packet arrivals at the current slot, the newly arriving SU packet cannot access the channel.

(4) In the previous slot, there is a PU packet occupying the PU's channel and the retrial buffer is full of interrupted SU packets. If the transmission of the PU packet is completed at the end of the previous slot and there are no PU packet arrivals, the SU packets in the retrial buffer will occupy the vacated channels in the spectrum.

Therefore, the newly arriving SU packet has to leave the system. Conclusively, we give the blocking rate B_s of SU packets as follows:

$$B_s = \lambda_1 (p_2 \bar{\mu}_2 + (p_2 \mu_2 + (1 - p_2)) \lambda_2 + \pi_{c,0} \bar{\lambda}_2 \bar{\mu}_1^c + \pi_{c,1} \bar{\lambda}_2 \mu_2)$$

$$= \frac{\lambda_1 \lambda_2 + \lambda_1 \lambda_2^2 \mu_2}{\lambda_2 + \mu_2} + \pi_{c,0} \lambda_1 \bar{\lambda}_2 \bar{\mu}_1^c + \pi_{c,1} \lambda_1 \bar{\lambda}_2 \mu_2. \tag{9.14}$$

We define the latency Y_s of an SU packet as the duration from the instant at which an SU packet joins the system to the instant that the SU packet is successfully transmitted. Considering an EAS, during the transmission period of an SU packet, possible interruption occurs at the beginning instant of every slot other than the first. We suppose that an SU packet will experience k interruptions before it is transmitted successfully. For each interruption, the SU packet has to wait for a period of time, during which several PU packets will perform their transmissions. The latency of an SU packet is the sum of the waiting time and the transmission time. This means the average latency $E[Y_s]$ of SU packets can be expressed as follows:

$$E[Y_s] = 1 + \sum_{j=1}^{\infty} \sum_{k=0}^{\infty} k \left(\frac{1}{\mu_2} + 1 \right) \lambda_2^k \bar{\lambda}_2 (j - 1) \mu_1 \bar{\mu}_1^{j-1}$$

$$= 1 + \frac{\bar{\mu}_1}{\mu_1} \left(\frac{\lambda_2}{\bar{\lambda}_2 \mu_2} + 1 \right). \tag{9.15}$$

We define the channel utilization U_c of the system as the probability that the channels in the tagged spectrum are being occupied by PU or SU packets in a slot. According to the working principle of the proposed channel aggregation strategy, the channel states are classified into three categories:

(1) With probability $\pi_{0,0}$, there are no packets in the system, namely, all the channels in the spectrum are idle. For this case, the channel utilization is 0.

(2) With probability p_2, the tagged spectrum is occupied by a PU packet. Because all the channels in the tagged spectrum are aggregated together for the transmission of one PU packet, the channel utilization for this state is up to 100%.

(3) With probability $\pi_{i,0}$, there are i ($i = 1, 2, 3, \ldots, c$) SU packets occupying part channels in the tagged spectrum. Since the transmission of each SU packet occupies only one of the channels in the tagged spectrum, the channel utilization for this case is i/c.

In summary, we give the channel utilization U_c as follows:

$$U_c = p_2 + \sum_{i=1}^{c} \frac{i\pi_{i,0}}{c}$$

$$= \frac{\lambda_2}{\lambda_2 + \mu_2} + \sum_{i=1}^{c} \frac{i\pi_{i,0}}{c}. \tag{9.16}$$

We know that with an increase in the aggregation intensity, the blocking rates for the two types of user packets, the PU packets and the SU packets, and the average latency of SU packets will decrease. However, the channel utilization will also decrease. This means that in the channel aggregation strategy proposed in this chapter there is a trade-off among different performance measures when setting the aggregation intensity. In order to improve the beneficial effects and reduce any negative effects on the system performance, it is necessary to optimize the aggregation intensity. Hence, we construct a system cost function $F(c)$ as follows:

$$F(c) = f_1 B_p + f_2 B_s + f_3 E[Y_s] + \frac{f_4}{U_c} + f_5 c \tag{9.17}$$

where f_1, f_2, f_3, f_4 and f_5 are assumed to be the cost impact factors of the blocking rate B_p of PU packets, the blocking rate B_s of SU packets, the average latency $E[Y_s]$ of SU packets, the channel utilization U_c and the channel aggregation intensity c, respectively.

The system cost function can achieve the minimum value when the channel aggregation intensity is optimized. The value of the channel aggregation intensity c minimizing the system cost function $F(c)$ is the optimal channel aggregation intensity c^*. Therefore, c^* can be given as follows:

$$c^* = \underset{c \in \{1,2,3,\ldots\}}{\text{argmin}} \ \{F(c)\} \tag{9.18}$$

where "argmin" stands for the argument of the minimum.

9.3.3 Numerical Results

In this section, the system performance for the channel aggregation strategy proposed in this chapter is investigated using numerical results. In the numerical results, we set the service rate for SU packets as $\mu_1 = 0.2$, and the service rate for PU packets as $\mu_2 = c\mu_0$ ($\mu_0 = 0.05$).

Figure 9.2 illustrates the blocking rate B_s of SU packets versus the channel aggregation intensity c for different arrival rates λ_1 of SU packets and λ_2 of PU packets.

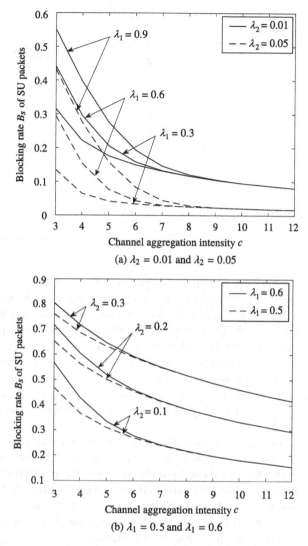

Fig. 9.2 Blocking rate of SU packets versus channel aggregation intensity

(a) $\lambda_2 = 0.01$ and $\lambda_2 = 0.05$

(b) $\lambda_1 = 0.5$ and $\lambda_1 = 0.6$

From Fig. 9.2, we observe that for the same arrival rate λ_1 of SU packets or the same arrival rate λ_2 of PU packets, the blocking rate B_s of SU packets decrease as the channel aggregation intensity c increases. This is because the higher the channel aggregation intensity is, the greater the service rate of PU packets is, the less likely it is that the PU's channel is occupied by a PU packet, and the more likely is that a newly arriving SU packet is able to access the spectrum. Therefore, the blocking rate of SU packets will be lower.

In addition, from Fig. 9.2a we find that the arrival rate λ_1 of SU packets has an impact on the blocking rate of SU packets. We see that for the same arrival rate λ_2 of PU packets, such as $\lambda_2 = 0.01$, when the channel aggregation intensity c is lower,

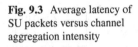
Fig. 9.3 Average latency of
SU packets versus channel
aggregation intensity

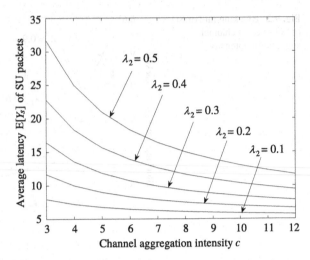

the blocking rate B_s of SU packets shows a sharp increasing trend as the arrival
rate λ_1 of SU packets increases. The reason being is that when the arrival rate of
SU packets is higher, the more SU packets will occupy the channels in the tagged
spectrum, the greater the likelihood is that all the channels will be occupied, so any
newly arriving SU packets are more likely to be blocked. As the channel aggregation
intensity c continuously increases, the impact of the arrival rate λ_1 of SU packets
on the blocking rate lessens. This is because when the channel aggregation intensity
is greater than a certain value, such as $c \geq 9$, all the channels are more likely to
be occupied by SU packets. Therefore, the blocking rate of SU packets tends to be
fixed.

On the other hand, from Fig. 9.2b we find that for the same arrival rate λ_1 of
SU packets, such as $\lambda_1 = 0.3$, and the same channel aggregation intensity c, the
higher the arrival rate λ_2 of PU packets is, the greater the blocking rate B_s of the SU
packets is. This is because as the arrival rate of PU packets increases, the possibility
that the PU's channel will be occupied by a PU packet is higher, so the possibility
of a new SU packet being able to access the channel will be lower. As a result, the
blocking rate of SU packets will increase.

In addition, we examine the influence of the channel aggregation intensity c on
the average latency $E[Y_s]$ of SU packets for different arrival rates λ_2 of PU packets
in Fig. 9.3.

In Fig. 9.3, we observe that for the same arrival rate λ_2 of PU packets, the
average latency $E[Y_s]$ of SU packets decreases as the channel aggregation intensity
c increases. The reason is that the larger the channel aggregation intensity is, the
quicker the PU packets will be transmitted. Therefore, the waiting time for an SU
packet at the system will be shorter, and this will result in a decrease in the average
latency of SU packets.

Moreover, from Fig. 9.3, we find that for the same channel aggregation intensity
c, the average latency $E[Y_s]$ of SU packets increases as the arrival rate λ_2 of PU

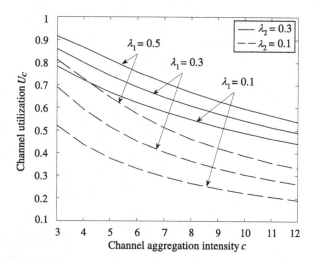

Fig. 9.4 Channel utilization versus channel aggregation intensity

packets increases. This is because as the arrival rate of PU packets increases, the possibility that the PU's channel will be occupied by a PU packet increases, so the time period for an SU packet waiting in the buffer will be longer. Therefore, the average latency of SU packets will be greater.

Figure 9.4 illustrates the channel utilization U_c versus the channel aggregation intensity c for different arrival rates λ_1 of SU packets and λ_2 of PU packets.

In Fig. 9.4, it can be observed that for the same arrival rate λ_1 of SU packets and the same arrival rate λ_2 of PU packets, the channel utilization U_c decreases as the channel aggregation intensity c increases. The reason is that the larger the channel aggregation intensity is, the quicker the PU packets are transmitted. For a certain traffic load of SU packets, there is a higher possibility that the channels will be idle. Therefore, the channel utilization will be lower.

When the channel aggregation intensity c is fixed, the channel utilization U_c increases as the arrival rate λ_1 of SU packets or the arrival rate λ_2 of PU packets increases. This is because the higher the arrival rate λ_1 of SU packets or the arrival rate λ_2 of PU packets is, the greater the probability is that the channels in the tagged spectrum will be occupied by PU or SU packets, so the channel utilization will be greater.

As an example, we set the arrival rate of SU packets as $\lambda_1 = 0.3$, and the service rate on each channel for SU packets as $\mu_1 = 0.2$. Moreover, we consider the service rate on each channel for a PU packet as $\mu_0 = 0.05$. By setting $\lambda_2 = 0.3$, $\lambda_2 = 0.4$ and $\lambda_2 = 0.5$ as an example, we plot how the system cost function $F(c)$ changes with respect to the channel aggregation intensity c for two groups of cost impact factors in Fig. 9.5. In Fig. 9.5a, we set the cost impact factors as $f_1 = f_2 = f_3 = f_4 = f_5 = 1$; in Fig. 9.5b, the cost impact factors are set as $f_1 = 5$, $f_2 = 2$, $f_3 = 3$, $f_4 = 7$, $f_5 = 2$.

From Fig. 9.5, we observe that all values of the system cost function $F(c)$ experience two stages. In the first stage, the system cost function $F(c)$ decreases

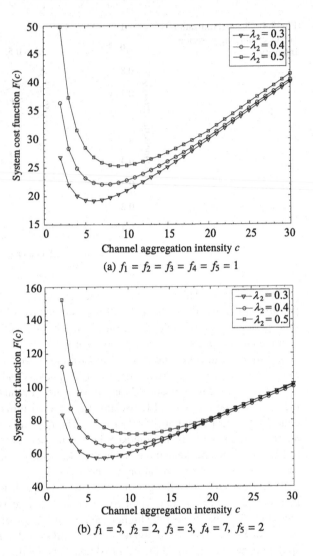

Fig. 9.5 System cost function versus channel aggregation intensity

(a) $f_1 = f_2 = f_3 = f_4 = f_5 = 1$

(b) $f_1 = 5$, $f_2 = 2$, $f_3 = 3$, $f_4 = 7$, $f_5 = 2$

along with an increase in the channel aggregation intensity c. During this stage, the greater the channel aggregation intensity is, the lower the blocking rates for both PU packets and SU packets will be. Moreover, the average latency of SU packets will decrease sharply. We note that the cost introduced by the decrease in the channel utilization changes slowly. Therefore, the system cost illustrates an overall decreasing trend in the first stage. In the second stage, the system cost function $F(c)$ increases with an increase in the channel aggregation intensity c. During this period, the higher the channel aggregation intensity is, the greater effect the channel utilization and the channel aggregation intensity will have on the system cost function. Namely, when the channel aggregation intensity exceeds a critical

Table 9.1 Optimum channel aggregation intensity

f_1	f_2	f_3	f_4	f_5	Arrival rates λ_2 of PU packets	Minimum costs $F(c^*)$	Optimum channel aggregation intensities c^*
1	1	1	1	1	0.3	19.17	6
					0.4	22.08	7
					0.5	25.33	9
5	2	3	7	2	0.3	57.41	7
					0.4	64.17	9
					0.5	71.69	11

value, the channel utilization and channel aggregation intensity will play more important roles in influencing the system cost function. A smaller channel utilization and a greater channel aggregation intensity will raise the system cost function.

In summary, the system cost function $F(c)$ invariably exists as a minimal value $F(c^*)$ for all the system parameters when the channel aggregation intensity c is set to an optimal value c^*. For two groups of cost impact factors, the optimal channel aggregation intensities with different arrival rates of PU packets are shown in Table 9.1.

9.4 Analysis of Admission Fee

In this section, we first investigate the Nash equilibrium behavior and socially optimal behavior of SU packets in the channel aggregation strategy proposed in this chapter. Then, we present a pricing policy for the SU packets to optimize the system socially. This issue can be addressed by imposing an appropriate admission fee for SU packets.

9.4.1 Nash Equilibrium Behavior

Every SU is individually selfish and tries to access the system for its own benefit. If an SU packet is admitted to the system, it will be transmitted successfully and get reward R_g. Conversely, if an SU packet tries to access the system but fails, it will not be rewarded. No matter whether an SU packet is admitted to the system or not, a trial cost C_g ($C_g < R_g$), such as the cost in channel sensing, propagation delay and sojourn time, has to be paid. Therefore, SUs will adjust their transmission requests.

We consider an SU packet's strategy with probability q. Namely, an SU packet decides to join the buffer with a probability of q ($0 \leq q \leq 1$), and leaves the system with probability $\bar{q}$ ($\bar{q} = 1 - q$). Since SU packets are allowed to make their own decisions, there will be a non-cooperative and symmetric game among the SU packets. In the presence of the joining probability of SU packets, the effective arrival rate λ_1^e deviates from the potential arrival rate λ_1 with $\lambda_1^e = q_e \lambda_1$, where q_e is the Nash equilibrium probability. The effective arrival rate λ_1^e in the Nash equilibrium state is called the Nash equilibrium arrival rate.

We define the individual benefit function $G_{ind}(\lambda_1)$ of as follows:

$$G_{ind}(\lambda_1) = \left(1 - \frac{B_s}{\lambda_1}\right) \times R_g - C_g \qquad (9.19)$$

where B_s is the blocking rate of SU packets given in Eq. (9.14).

With an increase in the arrival rate λ_1 of SU packets, the blocking rate B_s of SU packets increases monotonically. Hence the individual benefit function $G_{ind}(\lambda_1)$ is a decreasing function about the arrival rate λ_1 of SU packets. In other words, as the arrival rate of SU packets increases, the blocking rate of SU packets grows and the individual benefit function decreases. Since all SUs are individually selfish, they all try their best to access the system. Provided the benefit is positive, the arrival rate of SU packets will continue to grow. If there is at least one solution for the inequality $G_{ind}(\lambda_1) \geq 0$ within the closed interval $[\lambda_{min}, \lambda_{max}]$, the maximal value of the solutions is the Nash equilibrium arrival rate λ_1^e. Otherwise, $\lambda_1^e = \lambda_{min}$. No SU packet has any incentive to deviate unilaterally from the Nash equilibrium arrival rate. We discuss the Nash equilibrium of the proposed strategy as follows:

(1) Letting $\lambda_1 = \lambda_{min}$, if $C_g/(1 - B_s/\lambda_1) > R_g$, the individual benefit function $G_{ind}(\lambda_1)$ for one SU packet is less than zero. For this case, even if all the SU packets arrive at the system at the lowest arrival rate, the value of the individual benefit function is negative. Therefore, $\lambda_1^e = \lambda_{min}$ is a Nash equilibrium arrival rate and no other Nash equilibrium arrival rates exist. That is to say, the dominant strategy is one where SU packets arrive at the system at the lowest possible rate.

(2) Letting $\lambda_1 = \lambda_{max}$, if $C_g/(1 - B_s/\lambda_1) \leq R_g$, the value of the individual benefit function $G_{ind}(\lambda_1)$ is no less than zero. For this case, even if all the SU packets arrive at the system at the maximum arrival rate, they all enjoy non-negative individual benefits. Therefore, $\lambda_1^e = \lambda_{max}$ is a Nash equilibrium arrival rate and no other Nash equilibrium arrival rate is possible. That is to say, the dominant strategy is one where SU packets arrive at the system at the highest rate.

(3) Letting $\lambda_{min} < \lambda_1 < \lambda_{max}$, if $C_g/(1 - B_s/\lambda_1) > R_g$, some SU packets will suffer negative benefits, so this cannot be a Nash equilibrium strategy. On the other hand, if $C_g/(1 - B_s/\lambda_1) < R_g$, some SU packets will receive positive benefits, so this also cannot be a Nash equilibrium strategy. Therefore, there is a unique Nash equilibrium strategy $\lambda_{min} < \lambda_1^e < \lambda_{max}$ satisfying $C_g/(1 - B_s/\lambda_1) = R_g$. In this case, λ_1^e is a mixed Nash equilibrium arrival rate.

We investigate the monotonicity of the individual benefit function $G_{ind}(\lambda_1)$ using numerical results. The system parameters are fixed as follows: $\lambda_{min} = 0.05$, $\lambda_{max} = 0.5$, $\mu_1 = 0.2$, $\mu_0 = 0.05$, $c = 3$, $R_g = 20$ as an example for all the numerical results.

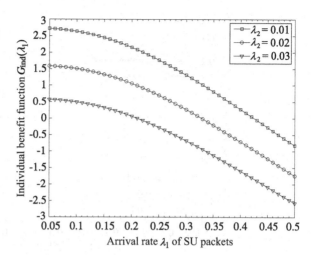

Fig. 9.6 Individual benefit function versus arrival rate of SU packets

By setting $\lambda_2 = 0.01$, 0.02 and 0.03, respectively, we show the change trend of individual benefit function $G_{\text{ind}}(\lambda_1)$ versus arrival rate λ_1 of SU packets in Fig. 9.6.

In Fig. 9.6, we find that with the parameters set above, all the individual benefit functions $G_{\text{ind}}(\lambda_1)$ show downward trends as the arrival rate λ_1 of SU packets increases. We also find that all the individual benefit functions $G_{\text{ind}}(\lambda_1)$ go through $G_{\text{ind}}(\lambda_1) = 0$, namely, there are always values of λ_1^e subject to $G_{\text{ind}}(\lambda_1) = 0$. That is to say, a Nash equilibrium behavior for our proposed strategy exists.

9.4.2 Socially Optimal Behavior

In the system design, it is necessary to consider the level of social benefit of the system derived during operation as well as the benefit to the individual users. In this subsection, we turn our attention to the socially optimal behavior of SU packets. We define the social benefit function $G_{\text{soc}}(\lambda_1)$ as follows:

$$G_{\text{soc}}(\lambda_1) = \lambda_1 \times \left(\left(1 - \frac{B_s}{\lambda_1} \right) \times R_g - C_g \right). \tag{9.20}$$

By maximizing the value of the social benefit function, we can derive the socially optimal arrival rate λ_1^*, where λ_1^* is given as follows:

$$\lambda_1^* = \underset{\lambda_1 \in [\lambda_{\min}, \lambda_{\max}]}{\operatorname{argmax}} \{ G_{\text{soc}}(\lambda_1) \} \tag{9.21}$$

where "argmax" stands for the argument of the maximum.

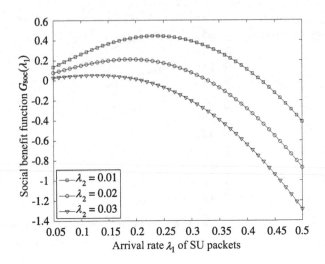

Fig. 9.7 Social benefit function versus arrival rate of SU packets

Table 9.2 Nash equilibrium and socially optimal arrival rates of SU packets

Arrival rates λ_2 of SU packets	Nash equilibrium arrival rates λ_1^e of SU packets	Socially optimal arrival rates λ_1^* of SU packets
0.01	0.428	0.24
0.02	0.33	0.188
0.03	0.215	0.126

With the same parameters as used in Fig. 9.6, we show how the social benefit function $G_{\mathrm{soc}}(\lambda_1)$ changes with respect to the arrival rate λ_1 of SU packets in Fig. 9.7.

In Fig. 9.7, we find that there is always a socially optimal arrival rate λ_1^* and a maximal value of the social benefit function $G_{\mathrm{soc}}(\lambda_1^*)$ for all the arrival rates λ_2 of PU packets.

Combining the results given in Figs. 9.6 and 9.7, we summarize the Nash equilibrium arrival rate λ_1^e and the socially optimal arrival rate λ_1^* in Table 9.2.

From Table 9.2, we conclude that optimizing the individual benefit function leads to a higher arrival rate of SU packets than that socially desired. This issue can be addressed by imposing an appropriate admission fee for SU packets.

9.4.3 Pricing Policy

One approach that would oblige the SU packets to adopt the socially optimal arrival rate is to charge a fee to the SU packets joining the system. We assume the Base Station (BS) acts as a pricing agent and imposes an admission fee on all the SU packets transmitted successfully. Therefore, for the channel aggregation strategy

Table 9.3 Numerical results for admission fee

Arrival rates λ_2 of PU packets	Socially maximum benefits $G_{soc}(\lambda_1^*)$	Admission fees f
0.01	0.4445	2.0748
0.02	0.2125	1.3196
0.03	0.0524	0.5071

proposed in this chapter, we present a pricing policy. It is worth mentioning that the admission fee f is different from the trial cost C_g. The admission fee f is only imposed on the SU packets transmitted successfully, whereas the trial fee C_g is a cost that each arriving SU packet has to pay.

When the pricing policy is implemented, the individual benefit function $G'_{ind}(\lambda_1)$ will be given as follows:

$$G'_{ind}(\lambda_1) = (R_g - f) \times \left(1 - \frac{B_s}{\lambda_1}\right) - C_g. \qquad (9.22)$$

Substituting the arrival rate λ_1 of SU packets in Eq. (9.22) with the socially optimal arrival rate λ_1^* of SU packets given in Table 9.2 and letting $G'_{ind}(\lambda_1) = 0$, we can calculate the admission fee f as follows:

$$f = R_g - \frac{C_g}{\left(1 - \dfrac{B_s}{\lambda_1}\right)}. \qquad (9.23)$$

With the socially optimal arrival rates λ_1^* given in Table 9.2, we calculate the blocking rate B_s using Eq. (9.14). Afterward, we can give the admission fee f using Eq. (9.23). For different arrival rates λ_2 of PU packets, we summarize the maximum of the social benefit $G_{soc}(\lambda_1^*)$ and the admission fee f in Table 9.3.

9.5 Conclusion

Taking into account the transmission quality for both PU packets and SU packets, we proposed a channel aggregation strategy in CRNs. We presented the performance analysis of the system model, through an analysis of the steady-state distribution to obtain the performance measures and the system cost. Then, we presented numerical results to evaluate the system performance and optimize the channel aggregation intensity. Moreover, after we investigated the Nash equilibrium and the socially optimal behaviors of SU packets, we proposed an appropriate pricing policy to maximize the value of the social benefit function. This issue can be addressed by imposing an appropriate admission fee for SU packets.

Chapter 10
Spectrum Reservation Strategy with Retrial Feedback and Perfect-Sensing Results

In order to better adapt to systemic load changes in Cognitive Radio Networks (CRNs), in this chapter, we present an adaptive control approach to determine the reservation ratio of the licensed spectrum for Secondary User (SU) packets and propose an adaptive spectrum reservation strategy with retrial feedback and perfect-sensing results. We establish a three-dimensional DTMC model to capture the stochastic behavior of users. By using a method similar to that of the matrix-geometric solution, we obtain the steady-state distribution of the system model and derive some required performance measures of the system. Numerical results show that the system performance is sensitive to system parameters like the adaptive control factor and the admission threshold. Finally, we construct a system cost function to balance different performance measures and present an intelligent searching algorithm to optimize the system parameters with the global minimum value of the system cost function.

10.1 Introduction

In the current communication resource allocation framework, the demand for efficient radio spectrum is increasing rapidly with continuing growth in wireless applications. Most spectrum bands have already been exclusively allocated to licensed service providers [Mari12], and the remaining wireless spectrum suitable for wireless communication is being exhausted. In order to improve the spectrum utilization and cope with the immense popularity of wireless devices, the concept of CRNs emerged [Fede13]. In CRNs, Secondary Users (SUs) opportunistically exploit the spectrum unused by Primary Users (PUs) [Tang06, Zhan14a]. The design of spectrum allocation strategy is a hot topic in the field of wireless communications [Li15a, Wang14a].

© Springer Nature Singapore Pte Ltd. 2021
S. Jin, W. Yue, *Resource Management and Performance Analysis of Wireless Communication Networks*, https://doi.org/10.1007/978-981-15-7756-7_10

In CRNs with multiple channels, in order to improve the utilization of the spectrum hole, channel bonding technology has been investigated, where contiguous idle channels are bonded as one logical channel for SUs [Anan14]. Channel aggregate technology has also been proposed, where non-contiguous idle channels can be aggregated as one logical channel for SUs or PUs [Bala14, Zhao15a]. Mixed aggregate/bonding technology that takes channel handoff into account has been investigated in [Liao15a]. Considering the low arrival rate of PU packets, many studies have also researched spectrum reservation strategy. With spectrum reservation strategy, part of the licensed spectrum is reserved for SUs, and the remaining spectrum is used by PUs with preemptive priority and used by SUs opportunistically. This method can decrease handoff and lower the interruption probability of SU packets so as to enhance the system throughput [Hong09]. In [Wang15a], the trade-offs for the forced termination probability and the blocking probability against the number of reserved channels were examined. However, this study ignored the retrial of the SU packets interrupted by PU packets. In [Wang13a], a finite buffer capacity and user impatience were considered in spectrum reservation strategy, but the issue of how to reduce the interference between the PU packets and the SU packets was not mentioned.

We note that reserving a fixed ratio of the licensed spectrum for SU packets is relatively conservative. On the one hand, an overly high arrival rate of PU packets will lead to an increase in the average latency of PU packets, and the QoS for PU packets will go down. On the other hand, an excessively low arrival rate of PU packets will result in a considerable wastage of spectrum resources. According to the change in the spectrum environment, it is necessary to adjust the reservation ratio of the licensed spectrum adaptively for SU packets. In addition, in CRNs, SUs have cognition, so the SU packets interrupted by PU packets can return to the buffer to wait for future transmission on the original spectrum. Furthermore, how to control the interference between the PU packets and the SU packets is also a significant problem to be solved in spectrum reservation strategy.

To overcome the limitations of previous works, in this chapter we propose a spectrum reservation strategy with an adaptive control approach for the setting the spectrum aside ratio in centralized CRNs. In addition, in order to reduce the interference between the PU packets and the SU packets, we also set an admission threshold for SU packets. The proposed adaptive spectrum reservation strategy is more flexible than the conventional strategy. Aiming at mathematically evaluating and optimizing the spectrum reservation strategy proposed in this chapter, we establish a system cost function. Performance optimization will involve complicated nonlinear equations and nonlinear optimization problems, and the conventional optimization methods, such as the steepest descent method or Newton's method, are inappropriate. We therefore turn to intelligent optimization algorithms with a strong global convergence ability. By using the Teaching-Learning-Based Optimization (TLBO) algorithm, we optimize the system parameters in terms of the adaptive control factor and the admission threshold.

The chapter is organized as follows. In Sect. 10.2, we describe the spectrum reservation strategy proposed in this chapter. Then, we present the system model in detail. In Sect. 10.3, we present a performance analysis to give the steady-state distribution of the system model and obtain performance measures. And then, we present numerical results to evaluate the system performance. In Sect. 10.4, by analyzing the system cost and using the TLBO algorithm, we optimize the system parameters in terms of the adaptive control factor and the admission threshold. Finally, we draw our conclusions in Sect. 10.5.

10.2 Spectrum Reservation Strategy and System Model

In this section, we first propose a spectrum reservation strategy with an adaptive control approach, called the adaptive spectrum reservation strategy. Then, we build a three-dimensional DTMC model accordingly.

10.2.1 Spectrum Reservation Strategy

We consider a licensed spectrum in CRNs. In such networks, the licensed spectrum is separated into two logical channels, namely, the reserved channel and the shared channel, respectively. In this chapter, we assume that the reserved channel is only used by SU packets, and the shared channel is used by both PU packets and SU packets. The PU packets have preemptive priority to use the shared channel and can reclaim the shared channel at any time, while the SU packets use the shared channel opportunistically.

The working principle of the adaptive spectrum reservation strategy proposed in this chapter is shown in Fig. 10.1.

As shown in Fig. 10.1, we also make the following assumptions:

(1) Once a PU packet arrives at the system, the transmission of the SU packet on the shared channel is forcibly interrupted, and the terminated SU packet returns to the buffer set for SU packets (called the SU packets' buffer). We assume that the capacity of the SU packets' buffer is infinite.

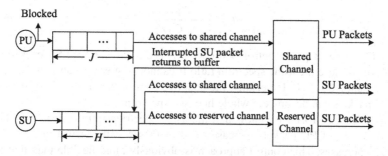

Fig. 10.1 Proposed adaptive spectrum reservation strategy

(2) Considering the low arrival rate of PU packets, a buffer (called the PU packets' buffer) with a finite capacity J ($J \geq 0$) is set for PU packets. J is the system parameter of the CRN under consideration. When a PU packet arrives at the system, if the shared channel is occupied by a PU packet and the PU packets' buffer is full, the newly arriving PU packet will be blocked.

(3) In order to reduce the interference between the PU packets and the SU packets, we set an admission threshold H ($0 \leq H \leq J$) for SU packets. That is to say, if the number of SU packets waiting in the SU packets' buffer is not greater than H, the SU packet queueing at the head of the SU packets' buffer cannot access the idle shared channel. H is a system parameter that can be adjusted to improve the system QoS.

(4) An SU packet waiting in the SU packets' buffer prefers to occupy the idle reserved channel over the idle shared channel.

(5) For the sake of clarity, the SU packets that are interrupted by PU packets and return to the SU packets' buffer are termed retrial SU packets. The retrial SU packet has a higher priority than both the newly arriving SU packet and all the SU packets waiting in the SU packets' buffer. That is to say, a retrial SU packet will queue at the head of the SU packets' buffer to wait for transmission service. In addition, the transmission of two types of packets is supposed to follow a First-Come First-Served (FCFS) discipline.

In order to describe the strategy more clearly, we introduce the ratio of the reserved channel's bandwidth to the total licensed spectrum's bandwidth, called the aside spectrum ratio ϑ ($0 \leq \vartheta \leq 1$). It is obvious the aside spectrum ratio ϑ may affect the blocking rate of PU packets, the interruption rate of SU packets, and the average latency for these two types of packets.

As usual, in order to ensure the QoS of users and achieve system stability, a higher arrival rate of users requires a greater service rate. With our proposed spectrum reservation strategy, a too small aside spectrum ratio ϑ will lead to a strong interference between the PU packets and the SU packets; Contrary to this, a too large aside spectrum ratio ϑ will lead to a decrease in the QoS of the PU packets. Considering both the priority of the PU packets and the need to better adapt to systemic load changes in CRNs, we present an adaptive control approach for the setting aside spectrum ratio ϑ as follows:

$$\vartheta = \frac{\lambda_1}{\lambda_1 + C_f \lambda_2} \tag{10.1}$$

where C_f is the adaptive control factor, λ_1 (λ_2) is the arrival rate of SU packets (PU packets). Because the aside spectrum ratio ϑ cannot be greater than 1, the adaptive control factor $C_f \geq 0$. Especially, the adaptive control factor $C_f = 0$ means that the SU packets can occupy the whole licensed spectrum.

Based on Eq. (10.1), we know that the aside spectrum ratio ϑ decreases as the arrival rate λ_2 of PU packets increases, and increases as the arrival rate λ_1 of SU packets increases. This control approach is obviously more flexible than that with a fixed aside spectrum ratio. We call this spectrum strategy the adaptive spectrum reservation strategy.

10.2.2   System Model

In this system, a shared channel and a reserved channel are used to transmit the two types of packets: the PU packets and the SU packets. There are two buffers to be set in the system. As presented in Sect. 10.2.1, one is for the PU packets called the PU packets' buffer which has a finite size J. The other is for the SU packets called the SU packets' buffer which has an infinite capacity.

The time axis is slotted into segments of equal length, called slots. The arriving intervals of the SU packets and the PU packets are independent random variables. The inter-arrival times of SU packets and PU packets are supposed to follow geometric distributions with arrival rate of SU packets λ_1 ($0 < \lambda_1 < 1, \bar{\lambda}_1 = 1 - \lambda_1$) and arrival rate of PU packets λ_2 ($0 < \lambda_2 < 1, \bar{\lambda}_2 = 1 - \lambda_2$), respectively. The service times on the shared channel and the reserved channel are geometrically distributed with probability μ_s called the service rate of the shared channel, where $0 < \mu_s < 1, \bar{\mu}_s = 1 - \mu_s$, and with probability μ_r called the service rate of the reserved channel, where $0 < \mu_r < 1, \bar{\mu}_r = 1 - \mu_r$, respectively. In this chapter, we also assume that the system is an Early Arrival System (EAS).

It is well known that if the Signal-to-Noise Ratio (SNR) in a channel is fixed, the channel capacity increases linearly with channel bandwidth. We further assume that if the shared channel and the reserved channel are homogeneous and have the same SNR, then the service rate μ_s on the shared channel is linearly decreased with the aside spectrum ratio ϑ. Conversely, the service rate μ_r on the reserved channel is linearly increased with the aside spectrum ratio ϑ. Based on the above assumptions, we obtain $\mu_s = (1 - \vartheta) \times \mu$ and $\mu_r = \vartheta \times \mu$, where μ is the service rate for the whole spectrum.

Let $X_n = i$ ($i = 0, 1, 2, \ldots$) and $Y_n = j$ ($j = 0, 1$) indicate the total number of SU packets in the system and on the reserved channel, respectively, at the instant n^+. Let $Z_n = k$ ($k = -1, 0, 1, \ldots, J + 1$) indicate the state of the shared channel at instant n^+. $k = -1$ means that the shared channel is occupied by an SU packet. $k \geq 0$ means that there are k PU packets in the system at the instant n^+. Using a three-dimensional vector $\{(X_n, Y_n, Z_n), n \geq 1\}$ to record the stochastic behavior of PU packets and SU packets, we establish a DTMC model to capture our proposed adaptive spectrum reservation strategy. The state space of the Markov chain is given as follows:

$$\Omega = \{(i, j, k) : i \geq 0, j = 0, 1, -1 \leq k \leq J + 1\}. \tag{10.2}$$

Let $\pi_{i,j,k}$ be the steady-state distribution of the three-dimensional DTMC. $\pi_{i,j,k}$ is then defined as follows:

$$\pi_{i,j,k} = \lim_{n \to \infty} \Pr\{X_n = i, Y_n = j, Z_n = k\}, \quad (i, j, k) \in \Omega. \tag{10.3}$$

10.3 Performance Analysis and Numerical Results

In this section, we obtain the steady-state distribution of the system model and derive performance measures of the system. Then, we present numerical results to evaluate the performance of the system using the spectrum reservation strategy proposed in this chapter.

10.3.1 Performance Analysis

Let $p_{i,j,k;l,m,h} = \Pr\{X_{n+1} = l, Y_{n+1} = m, Z_{n+1} = h | X_n = i, Y_n = j, Z_n = k\}$, (i, j, k), $(l, m, h) \in \Omega$. All the one-step transition probabilities from the original state (i, j, k) to the other possible state (l, m, h) are discussed accordingly as follows.

(1) When a new SU packet arrives at the system, if the reserved channel is occupied by an SU packet, and neither of the SU packets in the system departs in one slot, then all the one-step transition probabilities from the original state (i, j, k) can be written as follows:

$$p_{i,j,k;i+1,j,k-2} = \lambda_1 \bar{\lambda}_2 \bar{\mu}_r \mu_s, \quad i \geq H+2, \ j = 1, \ k = 1, \tag{10.4}$$

$$p_{i,j,k;i+1,j,k-1} = \begin{cases} \lambda_1 \bar{\lambda}_2 \bar{\mu}_r, & i \geq H+1, \ j = 1, \ k = 0 \\ \lambda_1 \bar{\lambda}_2 \bar{\mu}_r \mu_s, & 1 \leq i \leq H+1, \ j = 1, \ k = 1 \\ \lambda_1 \bar{\lambda}_2 \bar{\mu}_r \mu_s, & i \geq 1, \ j = 1, \ 2 \leq k \leq J+1, \end{cases} \tag{10.5}$$

$$p_{i,j,k;i+1,j,k} = \begin{cases} \lambda_1 \bar{\lambda}_2 \bar{\mu}_r \bar{\mu}_s, & i \geq 2, \ j = 1, \ k = -1 \\ \lambda_1 \bar{\lambda}_2 \bar{\mu}_r, & 0 < i \leq H+1, \ j = 1, \ k = 0 \\ \lambda_1 \bar{\lambda}_2 \bar{\mu}_r \bar{\mu}_s + \lambda_1 \lambda_2 \bar{\mu}_r \mu_s, & i \geq 1, \ j = 1, \ 1 \leq k \leq J \\ \lambda_1 \bar{\mu}_r \bar{\mu}_s + \lambda_1 \lambda_2 \bar{\mu}_r \mu_s, & i \geq 1, \ j = 1, \ k = J+1, \end{cases} \tag{10.6}$$

$$p_{i,j,k;i+1,j,k+1} = \begin{cases} \lambda_1 \lambda_2 \bar{\mu}_r, & i \geq 1, \ j = 1, \ k = 0 \\ \lambda_1 \lambda_2 \bar{\mu}_r \bar{\mu}_s, & i \geq 1, \ j = 1, \ 1 \leq k \leq J, \end{cases} \tag{10.7}$$

$$p_{i,j,k;i+1,j,k+2} = \lambda_1 \lambda_2 \bar{\mu}_r \bar{\mu}_s, \quad i \geq 2, \ j = 1, \ k = -1. \tag{10.8}$$

(2) When a new SU packet arrives at the system, if the reserved channel is idle, and neither of the SU packets in the system departs in one slot, then all the one-step transition probabilities from the original state (i, j, k) can be written as follows:

$$p_{i,j,k;i+1,j+1,k-1} = \bar{\lambda}_2 \lambda_1 \mu_s, \quad i = 0, \ j = 0, \ 1 \leq k \leq J+1, \tag{10.9}$$

$$
p_{i,j,k;i+1,j+1,k} = \begin{cases} \bar{\lambda}_2 \lambda_1 \bar{\mu}_s, & i = 1, \ j = 0, \ k = -1 \\ \bar{\lambda}_2 \lambda_1, & i = 0, \ j = 0, \ k = 0 \\ \bar{\lambda}_2 \lambda_1 \bar{\mu}_s + \lambda_1 \lambda_2 \mu_s, & i = 0, \ j = 0, \ 1 \le k \le J \\ \lambda_1 \bar{\mu}_s + \lambda_1 \lambda_2 \mu_s, & i = 0, \ j = 0, \ k = J+1, \end{cases}
$$
(10.10)

$$
p_{i,j,k;i+1,j+1,k+1} = \lambda_2 \lambda_1, \quad i = 0, \ j = 0, \ 0 \le k \le J,
$$
(10.11)

$$
p_{i,j,k;i+1,j+1,k+2} = \lambda_2 \lambda_1 \bar{\mu}_s, \quad i = 1, \ j = 0, \ k = -1.
$$
(10.12)

(3) If the number of SU packets is fixed in one slot, then all the one-step transition probabilities from the original state (i, j, k) can be written as follows:

$$
p_{i,j,k;i,j,k-2} = \lambda_1 \bar{\lambda}_2 \mu_r \mu_s + \bar{\lambda}_1 \bar{\lambda}_2 \bar{\mu}_r \mu_s, \quad i \ge H + 2, \ j = 1, \ k = 1,
$$
(10.13)

$$
p_{i,j,k;i,j,k-1} = \begin{cases} \bar{\lambda}_1 \bar{\lambda}_2 \bar{\mu}_r \mu_s + \lambda_1 \bar{\lambda}_2 \mu_r \mu_s, \\ \qquad\qquad\qquad 1 \le i \le H+1, \ j = 1, \ k = 1 \\ \bar{\lambda}_1 \bar{\lambda}_2 \bar{\mu}_r \mu_s + \lambda_1 \bar{\lambda}_2 \mu_r \mu_s, \\ \qquad\qquad\qquad i \ge 1, \ j = 1, \ 2 \le k \le J+1 \\ \bar{\lambda}_1 \bar{\lambda}_2 \mu_s, \quad i = 0, \ j = 0, \ k \ge 1, \end{cases}
$$
(10.14)

$$
p_{i,j,k;i,j,k+1} = \begin{cases} \lambda_1 \bar{\lambda}_2 \bar{\mu}_r \mu_s, & 1 \le i \le H+1, \ j = 1, \ k = -1 \\ \lambda_1 \lambda_2 \mu_r + \bar{\lambda}_1 \lambda_2 \bar{\mu}_r, & 1 \le i \le H+1, \ j = 1, \ k = 0 \\ \lambda_1 \lambda_2 \mu_r \bar{\mu}_s + \bar{\lambda}_1 \lambda_2 \bar{\mu}_r \bar{\mu}_s, & i \ge 1, \ j = 1, \ 1 \le k \le J \\ \bar{\lambda}_1 \lambda_2 \bar{\mu}_s, & i = 0, \ j = 1, \ k \ge 1 \\ \bar{\lambda}_1 \lambda_2, & i = 0, \ j = 0, \ k = 0, \end{cases}
$$
(10.15)

$$
p_{i,j,k;i,j,k+2} = \lambda_2 (\bar{\lambda}_1 \bar{\mu}_r \bar{\mu}_s + \lambda_1 \mu_r \bar{\mu}_s + \lambda_1 \bar{\mu}_r \mu_s),
$$
$$
i \ge 2, \ j = 1, \ k = -1,
$$
(10.16)

$$
p_{i,j,k;i,j+1,k+2} = \bar{\lambda}_1 \lambda_2 \bar{\mu}_s, \quad i = 1, \ j = 0, \ k = -1,
$$
(10.17)

$$
p_{i,j,k;i,j,k} = \begin{cases}
\bar{\lambda}_1 \bar{\lambda}_2 \bar{\mu}_s, & i = 1, \ j = 0, \ k = -1 \\[4pt]
\bar{\lambda}_1 \bar{\lambda}_2 \bar{\mu}_r \bar{\mu}_s + \lambda_1 \bar{\lambda}_2 \mu_r \bar{\mu}_s, \\
& 1 \le i \le H + 1, \ j = 1, \ k = -1 \\[4pt]
\bar{\lambda}_1 \bar{\lambda}_2 \bar{\mu}_r \bar{\mu}_s + \lambda_1 \bar{\lambda}_2 \mu_r \bar{\mu}_s + \lambda_1 \bar{\lambda}_2 \bar{\mu}_r \mu_s, \\
& i \ge H + 2, \ j = 1, \ k = -1 \\[4pt]
\bar{\lambda}_1 \bar{\lambda}_2, & i = 0, \ j = 0, \ k = 0 \\[4pt]
\bar{\lambda}_1 \bar{\lambda}_2 \bar{\mu}_r + \lambda_1 \bar{\lambda}_2 \mu_r, & 1 \le i \le H + 1, \ j = 1, \ k = 0 \\[4pt]
\bar{\lambda}_1 \bar{\lambda}_2 \bar{\mu}_s + \bar{\lambda}_1 \lambda_2 \mu_s, & i = 0, \ j = 0, \ 1 \le k \le J \\[4pt]
\bar{\lambda}_1 \bar{\mu}_s + \bar{\lambda}_1 \lambda_2 \mu_s, & i = 0, \ j = 0, \ k = J + 1 \\[4pt]
\bar{\lambda}_1 \bar{\lambda}_2 \bar{\mu}_r \bar{\mu}_s + \bar{\lambda}_1 \lambda_2 \bar{\mu}_r \mu_s + \lambda_1 \bar{\lambda}_2 \mu_r \bar{\mu}_s + \lambda_1 \lambda_2 \mu_r \mu_s, \\
& i \ge 1, \ j = 1, \ 1 \le k \le J \\[4pt]
\bar{\lambda}_1 \bar{\mu}_r \bar{\mu}_s + \bar{\lambda}_1 \lambda_2 \bar{\mu}_r \mu_s + \lambda_1 \mu_r \bar{\mu}_s + \lambda_1 \lambda_2 \mu_r \mu_s, \\
& i \ge 1, \ j = 1, \ k = J + 1.
\end{cases}
\tag{10.18}
$$

(4) If the number of SU packets departing the system is greater than the number of SU packets arriving at the system, and the reserved channel does not become idle in one slot, then all the one-step transition probabilities from the original state (i, j, k) can be written as follows:

$$
p_{i,j,k;i-1,j,k-2} = \bar{\lambda}_1 \bar{\lambda}_2 \mu_r \mu_s, \quad i \ge H + 2, \ j = 1, \ k = 1, \tag{10.19}
$$

$$
p_{i,j,k;i-1,j,k-1} = \begin{cases}
\bar{\lambda}_1 \bar{\lambda}_2 \mu_r \mu_s, & 2 \le i \le H + 2, \ j = 1, \ k = 1 \\[4pt]
\bar{\lambda}_1 \bar{\lambda}_2 \mu_r \mu_s, & i \ge 2, \ j = 1, \ 2 \le k \le J + 1,
\end{cases}
\tag{10.20}
$$

$$
p_{i,j,k;i-1,j,k} = \begin{cases}
\bar{\lambda}_1 \bar{\lambda}_2 \mu_r \bar{\mu}_s, & 2 \le i \le H + 2, \ j = 1, \ k = -1 \\[4pt]
\bar{\lambda}_1 \bar{\lambda}_2 \mu_r \bar{\mu}_s + \lambda_1 \bar{\lambda}_2 \bar{\mu}_r \mu_s + \lambda_1 \bar{\lambda}_2 \mu_r \mu_s, \\
& i \ge H + 3, \ j = 1, \ k = -1 \\[4pt]
\bar{\lambda}_1 \bar{\lambda}_2 \mu_r, \ 2 \le i \le H + 1, & j = 1, \ k = 0 \\[4pt]
\bar{\lambda}_1 \bar{\lambda}_2 \mu_r \bar{\mu}_s + \bar{\lambda}_1 \lambda_2 \mu_r \mu_s, & i \ge 2, \ j = 1, \ 1 \le k \le J \\[4pt]
\bar{\lambda}_1 \mu_r \bar{\mu}_s + \bar{\lambda}_1 \lambda_2 \mu_r \mu_s, & i \ge 2, \ j = 1, \ k = J + 1,
\end{cases}
\tag{10.21}
$$

$$P_{i,j,k;i-1,j,k+1} = \begin{cases} \bar{\lambda}_2\mu_s(\bar{\lambda}_1\bar{\mu}_r + \lambda_1\mu_r), \\ \qquad\qquad 2 \le i \le H+2, \ j=1, \ k=-1 \\ \bar{\lambda}_1\lambda_2\mu_r, \quad 2 \le i \le H+1, \ j=1, \ k=0 \\ \bar{\lambda}_1\lambda_2\mu_r\bar{\mu}_s, \quad i \ge 2, \ j=1, \ 1 \le k \le J \\ \bar{\lambda}_2\bar{\lambda}_1\mu_s, \quad i=1, \ j=0, \ k=-1, \end{cases} \tag{10.22}$$

$$P_{i,j,k;i-1,j,k+2} = \begin{cases} \lambda_2\left(\lambda_1\mu_r\mu_s + \bar{\lambda}_1\mu_r\bar{\mu}_s + \lambda_1\bar{\mu}_r\mu_s\right), \\ \qquad\qquad\qquad i \ge 2, \ j=1, \ k=-1 \\ \bar{\lambda}_2\bar{\lambda}_1\mu_s, \quad i=1, \ j=0, \ k=-1. \end{cases} \tag{10.23}$$

(5) If the number of SU packets departing the system is greater than the number of SU packets arriving at the system, and the reserved channel becomes idle in one slot, then all the one-step transition probabilities from the original state (i, j, k) can be written as follows:

$$P_{i,j,k;i-1,j-1,k-1} = \bar{\lambda}_2\bar{\lambda}_1\mu_r\mu_s, \quad i=1, \ j=1, \ 1 \le k \le J+1, \tag{10.24}$$

$$P_{i,j,k;i-1,j-1,k} = \begin{cases} \bar{\lambda}_2\bar{\lambda}_1\mu_r, \quad i=1, \ j=1, \ k=0 \\ \bar{\lambda}_2\bar{\lambda}_1\mu_r\bar{\mu}_s + \lambda_2\bar{\lambda}_1\mu_r\mu_s, \quad i=1, \ j=1, \ 1 \le k \le J \\ \bar{\lambda}_1\mu_r\bar{\mu}_s + \lambda_2\bar{\lambda}_1\mu_r\mu_s, \quad i=1, \ j=1, \ k=J+1 \\ \bar{\lambda}_2\bar{\lambda}_1\mu_r\bar{\mu}_s, \quad i=2, \ j=1, \ k=-1, \end{cases} \tag{10.25}$$

$$P_{i,j,k;i-1,j-1,k+1} = \begin{cases} \lambda_2\bar{\lambda}_1\mu_r, \quad i=1, \ j=1, \ k=0 \\ \lambda_2\bar{\lambda}_1\mu_r\bar{\mu}_s, \quad i=1, \ j=1, \ 1 \le k \le J. \end{cases} \tag{10.26}$$

(6) When two SU packets depart the system and no SU packet arrives at the system, if the state of the reserved channel does not change in one slot, then the one-step transition probabilities from the original state (i, j, k) can be written as follows:

$$P_{i,j,k;i-2,j,k} = \bar{\lambda}_1\bar{\lambda}_2\mu_r\mu_s, \quad i \ge H+4, \ j=1, \ k=-1, \tag{10.27}$$

$$P_{i,j,k;i-2,j,k+1} = \bar{\lambda}_1\bar{\lambda}_2\mu_r\mu_s, \quad 3 \le i \le H+3, \ j=1, \ k=-1, \tag{10.28}$$

$$P_{i,j,k;i-2,j,k+2} = \bar{\lambda}_1\lambda_2\mu_r\mu_s, \quad i \ge 3, \ j=1, \ k=-1. \tag{10.29}$$

(7) When two SU packets depart the system and no SU packet arrives at the system, if the state of the reserved channel changes in one slot, then all the one-step transition probabilities from the original state (i, j, k) can be written as follows:

$$p_{i,j,k;i-2,j-1,k+1} = \bar{\lambda}_2 \bar{\lambda}_1 \mu_r \mu_s, \quad i = 2, \ j = 1, \ k = -1, \qquad (10.30)$$

$$p_{i,j,k;i-2,j-1,k+2} = \lambda_2 \bar{\lambda}_1 \mu_r \mu_s, \quad i = 2, \ j = 1, \ k = -1. \qquad (10.31)$$

Let P be the state transition probability matrix of the Markov chain $\{(X_n, Y_n, Z_n), \ n \geq 0\}$. Let $A_{i,k}$ be the transition probability sub-matrix for the number of SU packets in the system changing from i $(i = 0, 1, 2, \ldots)$ to k $(k = 0, 1, 2, \ldots)$. The one-step transition probability matrix P can be written as a block matrix as follows:

$$P = \begin{pmatrix} A_{0,0} & A_{0,1} & & & & & \\ A_{1,0} & A_{1,1} & A_{1,2} & & & & \\ A_{2,0} & A_{2,1} & A_{2,2} & A_{2,3} & & & \\ & \ddots & \ddots & \ddots & \ddots & & \\ & & A_{H+4,H+2} & A_{H+4,H+3} & A_{H+4,H+4} & A_{H+4,H+5} & \\ & & & A_{H+4,H+2} & A_{H+4,H+3} & A_{H+4,H+4} & A_{H+4,H+5} \\ & & & & \ddots & \ddots & \ddots & \ddots \end{pmatrix}. \qquad (10.32)$$

Employing Eqs. (10.4)–(10.31), each sub-matrix $A_{i,j}$ can be computed. The structure of the one-step transition probability matrix P shows that the three-dimensional DTMC $\{(X_n, Y_n, Z_n), \ n \geq 0\}$ has a structure similar to that of the Quasi Birth-Death (QBD) process. If the number of SU packets is no less than $(H + 4)$, the one-step probabilities are repeatable. Therefore, we can use a method similar to that of the matrix-geometric solution to obtain the steady-state distribution $\pi_{i,j,k}$ of the system model.

For the Markov chain $\{(X_n, Y_n, Z_n), \ n \geq 0\}$ with the one-step transition probability matrix P, the necessary and sufficient condition of positive recurrence is that the 3rd order matrix equation:

$$R^3 A_{H+4,H+2} + R^2 A_{H+4,H+3} + R A_{H+4,H+4} + A_{H+4,H+5} = R \qquad (10.33)$$

has a minimal non-negative solution R, and the spectral radius $\mathrm{Sp}(R) < 1$. In order to employ a method similar to that of the matrix-geometric solution, we construct new sub-matrices $B_{0,0}$, $B_{0,1}$, B_1 and B_2 as follows:

$$B_{0,0} = \begin{pmatrix} A_{0,0} & A_{0,1} & & & & & \\ A_{1,0} & A_{1,1} & A_{1,2} & & & & \\ A_{2,0} & A_{2,1} & A_{2,2} & A_{2,3} & & & \\ & \ddots & & \ddots & & \ddots & \\ & & A_{H+3,H+1} & A_{H+3,H+2} & A_{H+3,H+3} & & \end{pmatrix}, \qquad (10.34)$$

$$B_{0,1} = (\mathbf{0}^T, \mathbf{0}^T, \mathbf{0}^T, \ldots, \mathbf{0}^T, A_{H+3,H+4}^T)^T \qquad (10.35)$$

where T is the matrix transpose.

$$B_1 = (\mathbf{0}, \mathbf{0}, \mathbf{0}, \ldots, \mathbf{0}, \ A_{H+4,H+2}, A_{H+4,H+3}), \qquad (10.36)$$

$$B_2 = (\mathbf{0}, \mathbf{0}, \ldots, \mathbf{0}, A_{H+4,H+2}). \qquad (10.37)$$

Furthermore, we construct a stochastic matrix as follows:

$$B[R] = \begin{pmatrix} B_{0,0} & B_{0,1} \\ RB_2 + B_1 & R^2 A_{H+4,H+2} + R A_{H+4,H+3} + A_{H+4,H+4} \end{pmatrix}. \qquad (10.38)$$

Letting

$$\boldsymbol{\pi}_0 = (\pi_{0,0,0}, \pi_{0,0,1}, \pi_{0,0,2}, \ldots, \pi_{0,0,J+1}),$$

$$\boldsymbol{\pi}_1 = (\pi_{1,0,-1}, \pi_{1,1,0}, \pi_{1,1,1}, \ldots, \pi_{1,1,J+1}),$$

$$\boldsymbol{\pi}_i = (\pi_{i,1,-1}, \pi_{i,1,0}, \pi_{i,1,1}, \ldots, \pi_{i,1,J+1}), \quad 2 \leq i \leq H+1,$$

and

$$\boldsymbol{\pi}_i = (\pi_{i,1,-1}, \pi_{i,1,1}, \pi_{i,1,2}, \ldots, \pi_{i,1,J+1}), \quad i \geq H+2,$$

we obtain the steady-state distribution of the Markov chain by solving the following set of linear equations:

$$\begin{cases} (\boldsymbol{\pi}_0, \boldsymbol{\pi}_1, \boldsymbol{\pi}_2, \ldots, \boldsymbol{\pi}_{H+3})B[R] = (\boldsymbol{\pi}_0, \boldsymbol{\pi}_1, \boldsymbol{\pi}_2, \ldots, \boldsymbol{\pi}_{H+3}) \\ \boldsymbol{\pi}_0 e + \boldsymbol{\pi}_1 e + \boldsymbol{\pi}_2 e + \cdots + \boldsymbol{\pi}_{H+2} e + \boldsymbol{\pi}_{H+3}(I-R)^{-1} e = 1 \\ \boldsymbol{\pi}_i = \boldsymbol{\pi}_{H+3} R^{i-H-3}, \quad i \geq H+3 \end{cases} \qquad (10.39)$$

where e is a three-dimensional column vector and all elements of the vector are equal to 1.

10.3.2 Performance Measures

We define the interruption rate β_s of SU packets as the number of SU packets which are interrupted by PU packets per slot. An SU packet which is on the shared channel will be interrupted by a newly arriving PU packet. Therefore, we give the interruption rate β_s of SU packets as follows:

$$\beta_s = \lambda_2 \times \left(\sum_{i=2}^{\infty} \pi_{i,1,-1} + \pi_{1,0,-1} \right) \bar{\mu}_s. \tag{10.40}$$

We define the blocking rate B_p of PU packets as the number of PU packets which are blocked due to the finite capacity of the PU packets' buffer per slot. A newly arriving PU packet will be blocked when the shared channel is occupied by a PU packet and the PU packets' buffer is full. In addition, since the PU packets have preemptive priority, the blocking rate B_p of PU packets is not influenced by SU packets. Therefore, we give the blocking rate B_p of PU packets as follows:

$$\begin{aligned} B_p &= \lambda_2 \times \left(\sum_{i=1}^{\infty} \pi_{i,1,J+1} + \pi_{0,0,J+1} \right) \\ &= \frac{\dfrac{\lambda_2 \bar{\mu}_r}{\mu_r} \eta^{J+1}}{\dfrac{1 - \eta^{J+1}}{1 - \eta} + \dfrac{\lambda_2 \bar{\mu}_r}{\mu_r} \eta^J} \end{aligned} \tag{10.41}$$

where $\eta = \lambda_2 \bar{\mu}_r (\bar{\lambda}_2 \mu_r)^{-1}$.

We define the latency Y_s of an SU packet as the duration from the arrival instant of an SU packet to its departure instant. Based on the analysis presented in Sect. 10.3.1, we can obtain the average latency $E[Y_s]$ of SU packets as follows:

$$E[Y_s] = \frac{\sum_{k=0}^{J+1} \sum_{j=0}^{1} \sum_{i=0}^{\infty} i \pi_{i,j,k}}{\lambda_1}. \tag{10.42}$$

We define the throughput θ as a proportion of the number of SU packets transmitted actually per slot to the number of SU packets transmitted maximally per slot across the whole spectrum. With the throughput θ, we can evaluate the efficiency of the whole spectrum.

We know from the stochastic behavior of this system model, when both the shared channel and the reserved channel are occupied by users, the efficiency of the whole spectrum is 1; When the shared channel is idle and the reserved channel

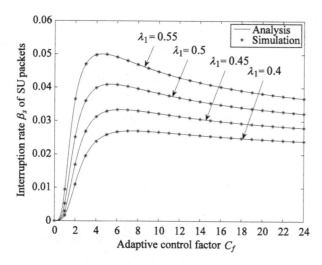

Fig. 10.2 Interruption rate of SU packets versus adaptive control factor

is busy, the efficiency of the whole spectrum is ϑ. Therefore, we give the throughput θ as follows:

$$\theta = \sum_{i=1}^{\infty} \sum_{k=-1,\ k\neq0}^{J+1} \pi_{i,1,k} + \sum_{i=1}^{\infty} \pi_{i,1,0}\vartheta + \sum_{k=1}^{J+1} \pi_{0,0,k}(1-\vartheta) \qquad (10.43)$$

where ϑ is the aside spectrum ratio defined in Sect. 10.2.1.

10.3.3 Numerical Results

In this subsection, we investigate the influences of the adaptive control factor C_f and the admission threshold H on the system performance. Unless otherwise specified, the system parameters in numerical results are set as follows: $J = 2$, $\lambda_1 = 0.4, 0.45, 0.5, 0.55, \lambda_2 = 0.2$ and $\mu = 0.8$.

Taking the admission threshold $H = 4$ as an example, for different arrival rates λ_1 of SU packets, we show the change trend of the interruption rate β_s of SU packets with respect to the adaptive control factor C_f in Fig. 10.2.

In Fig. 10.2, we find that the interruption rate β_s of SU packets exhibits two stages as the adaptive control factor C_f increases.

During the first stage, the interruption rate β_s of SU packets increases sharply as the adaptive control factor C_f increases. When the adaptive control factor is smaller, the dominant element influencing the interruption rate of SU packets is the service rate μ_r on the reserved channel. Based on Eq. (10.1), we note that the aside spectrum ratio decreases as the adaptive control factor increases. Therefore,

Fig. 10.3 Interruption rate of SU packets versus admission threshold

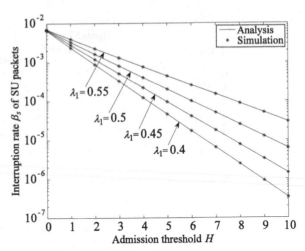

the larger the adaptive control factor is, the lower μ_r on the reserved channel is, and the more SU packets enter into the shared channel to receive transmission service. The transmission of SU packets on the shared channel may be interrupted by PU packets. When this occurs, the interruption rate of SU packets will increase.

During the second stage, the interruption rate β_s of SU packets decreases slowly as the adaptive control factor C_f increases. When the adaptive control factor is larger, the dominant element influencing the interruption rate of SU packets is the service rate μ_s on the shared channel. As the adaptive control factor increases, μ_s on the shared channel increases, the PU packets on the shared channel are transmitted quickly, and the synthetical service rate for the SU packets increases. Therefore, the transmission of SU packets on the shared channel is less likely to be interrupted by PU packets, so the interruption rate of SU packets will decrease.

By setting the adaptive control factor $C_f = 0.5$ as an example, for different arrival rates λ_1 of SU packets, we show the change trend for the interruption rate β_s of SU packets in relation to the admission threshold H in Fig. 10.3.

In Fig. 10.3, for all the arrival rates λ_1 of SU packets, we find that the interruption rate β_s of SU packets decreases as the threshold H increases. The obvious reason is that the higher the admission threshold is, the more SU packets are transmitted on the reserved channel without interruption. As a result, the interruption rate of SU packets will decrease.

Taking the admission threshold $H = 4$ as an example, for different arrival rates λ_1 of SU packets, we show the change trend for the average latency $E[Y_s]$ of SU packets in relation to the adaptive control factor C_f in Fig. 10.4.

We discuss the average latency $E[Y_s]$ of SU packets in the following two cases.

(1) For a higher arrival rate of SU packets, such as $\lambda_1 = 0.5$ and $\lambda_1 = 0.55$, the average latency $E[Y_s]$ of SU packets exhibits two stages as the adaptive control factor C_f increases.

Fig. 10.4 Average latency of SU packets versus adaptive control factor

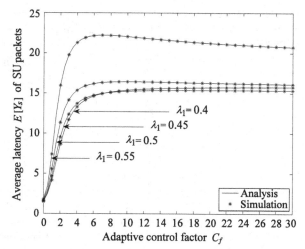

During the first stage, the average latency $E[Y_s]$ of SU packets increases sharply as the adaptive control factor C_f increases. When the adaptive control factor is smaller, the service rate μ_r on the reserved channel is the dominant element influencing the average latency of SU packets. As the adaptive control factor increases, the service rate μ_r on the reserved channel decreases. This leads to a decrease in the synthetical service rate for SU packets, and so the average latency of SU packets will be greater.

During the second stage, as the adaptive control factor C_f increases, the average latency $E[Y_s]$ of SU packets decreases slowly and tends to be fixed. When the adaptive control factor exceeds a certain value, the service rate μ_s on the shared channel is the dominant element influencing the average latency of SU packets. The bigger the adaptive control factor is, the higher μ_s on the shared channel is, and the greater the synthetical service rate for SU packets is.

Therefore, the average latency of SU packets will decrease. As the adaptive control factor continuously increases, nearly all the SU packets are transmitted on the shared channel, so the average latency of SU packets will tend to be fixed.

(2) For a lower arrival rate of SU packets, such as $\lambda_1 = 0.4$ and $\lambda_1 = 0.45$, the average latency $E[Y_s]$ of SU packets increases sharply when the adaptive control factor C_f is smaller, increases slowly and tends to be fixed when the adaptive control factor C_f is greater than a certain value. When the arrival rate of SU packets is lower, with the constraint of the admission threshold, the average latency of SU packets is mainly influenced by the service rate μ_r on the reserved channel. The bigger the adaptive control factor is, the lower μ_r on the reserved channel is. Therefore, the average latency of SU packets will increases sharply.

Fig. 10.5 Average latency of SU packets versus admission threshold

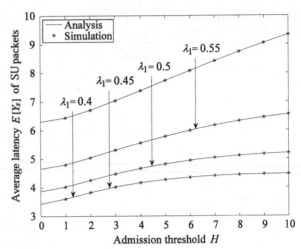

However, when the adaptive control factor exceeds a certain value, some SU packets are transmitted opportunistically on the shared channel, so the average latency of SU packets will increase slowly. As the adaptive control factor continuously increases, more and more SU packets will enter into the shared channel. Therefore, the average latency of SU packets will tend to be fixed.

By setting the adaptive control factor $C_f = 1$ as an example, for different arrival rates λ_1 of SU packets, we show the change trend for the average latency $E[Y_s]$ in relation to the admission threshold H in Fig. 10.5.

In Fig. 10.5, we find that, for all the arrival rates of SU packets, the average latency $E[Y_s]$ of SU packets increases as the admission threshold H increases. The reason is that the higher the admission threshold is, the fewer the SU packets are transmitted on the shared channel, and the lower the synthetical service rate for SU packets is. This will lead to an increase in the average latency of SU packets.

PU packets receive service on the shared channel with preemptive priority. The transmission of PU packets is only affected by the adaptive control factor and the arrival rate of PU packets, so the blocking rate of PU packets has nothing to do with the admission threshold. By setting $\lambda_2 = 0.15, 0.2, 0.25, 0.3$ and $H = 4$ as examples, we show the change trend for the blocking rate B_p of PU packets in relation to the adaptive control factor C_f in Fig. 10.6.

From Fig. 10.6, we conclude that, for all the arrival rates λ_2 of PU packets, the blocking rate B_p of PU packets decreases as the adaptive control factor C_f increases. The intuitive reason is that the larger the adaptive control factor is, the smaller the aside spectrum ratio is, and the higher the service rate μ_s for the PU packets on the shared channel is. This leads to a decrease in the blocking rate of PU packets.

By setting the admission threshold $H = 4$ as an example, for different arrival rates λ_1 of SU packets, we show the change trend for the systematic throughput θ in relation to the adaptive control factor C_f in Fig. 10.7.

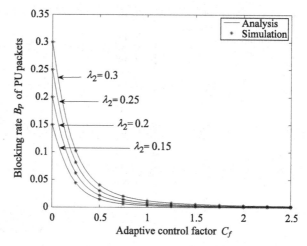

Fig. 10.6 Blocking rate of PU packets versus adaptive control factor

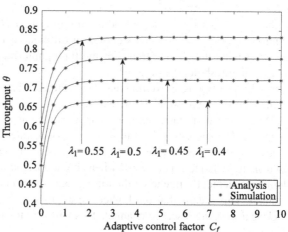

Fig. 10.7 Throughput versus adaptive control factor

From Fig. 10.7, we observe that as the adaptive control factor C_f increases, the throughput θ increases firstly when the adaptive control factor C_f is smaller; and then, when the adaptive control factor C_f exceeds a certain value, the throughput tends to be fixed. Since the capacity set for the PU packets is finite and the capacity set for SU packets is infinite, when the adaptive control factor is less than a certain value, the greater the adaptive control factor is, and the higher the service rate μ_s on the shared channel is, and the lower the blocking rate of PU packets is. As a result, throughput will increase. However, as the adaptive control factor further increases, when the adaptive control factor is greater than a certain value, nearly all the users are transmitted on the shared channel, so the normalized throughput tends to be fixed.

In order to clearly show the throughput θ in relation to the admission threshold H, we calculate the normalized throughput increment as $\eta = \theta - \theta_0$, where θ_0

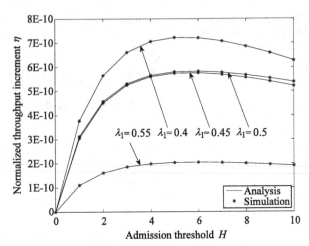

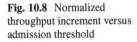

Fig. 10.8 Normalized throughput increment versus admission threshold

is the throughput by setting the admission threshold $H = 0$. Taking the adaptive control factor $C_f = 1$ as an example, for different arrival rates λ_1 of SU packets, we investigate the change trend for the throughput increment η in relation to the admission threshold H in Fig. 10.8.

Looking at Fig. 10.8, we find that the throughput increment η exhibits two stages as the admission threshold H increases.

During the first stage, the throughput increment η increases as the admission threshold H increases. We note that the service rate μ_r on the reserved channel is greater than the service rate μ_s on the shared channel with the parameters used in this chapter. Therefore, when the admission threshold H is smaller, the dominant element influencing the throughput is the service rate μ_r on the reserved channel. This means that the higher the admission threshold is, the more the SU packets access the reserved channel. This will lead to an increase in the throughput increment.

During the second stage, as the admission threshold further increases, when the admission threshold is greater than a certain value, the dominant element influencing the throughput is the service rate μ_s on the shared channel, the higher the admission threshold is, the more likely the shared channel is idle, resulting in a decrease in the throughput increment.

10.4 Performance Optimization

In this section, we first construct a system cost function to trade off different performance measures. Then, we jointly optimize the adaptive control factor and the admission threshold in the spectrum reservation strategy proposed in this chapter to improve the system performance.

10.4.1 Analysis of System Cost

From the experiment results provided in Sect. 10.3.3, we can draw the following conclusions. The interruption rate of β_s of SU packets, the average latency $E[Y_s]$ of SU packets and the throughput θ heavily depend on the adaptive control factor C_f and the admission threshold H. However, the blocking rate B_p of PU packets heavily depends on the adaptive control factor C_f.

In order to get the utmost out of the spectrum resource and meet the demands for QoS requirements of these two types of packets, considering the trade-off between the performance measures of these two types of packets, we construct a system cost function $F(C_f, H)$ as follows:

$$F(C_f, H) = f_1 B_p + f_2 E[Y_s] + \frac{f_3}{\theta} + f_4\beta_s + \frac{f_5}{\lambda_1} \tag{10.44}$$

where f_1, f_2, f_3, f_4 and f_5 are cost factors from the blocking rate of PU packets, the average latency of SU packets, the normalized throughput, the interruption rate of SU packets and the arrival rate of SU packets, respectively.

By minimizing the system cost function, the optimal combination (C_f^*, H^*) is given as follows:

$$(C_f^*, H^*) = \underset{C_f \geq 0, 0 \leq H \leq J}{\arg\min} \{F(C_f, H)\}. \tag{10.45}$$

Taking system parameters $f_1 = 20$, $f_2 = 0.6$, $f_3 = 4$, $f_4 = 3$, $f_5 = 0.2$, $M = 2$, $\mu = 0.8$, $\lambda_1 = 0.4, 0.5, 0.6$, $\lambda_2 = 0.1, 0.15$, $H = 4$ and $C_f = 1$ as an example, we investigate the change trend for the system cost function $F(C_f, H)$ in relation to the adaptive control factor C_f and the admission threshold H in Figs. 10.9 and 10.10, respectively.

Looking at Figs. 10.9 and 10.10, we conclude that there is an optimal adaptive control factor and an optimal admission threshold with local minimum $\omega = 0.05$ system costs. Based on these local minimum system costs, we can further obtain the global minimum system cost. However, it is difficult to give an analytical expression for the system cost function $F(C_f, H)$ in close form. Conventional optimization methods, such as steepest descent method or Newton's method, are inappropriate. Therefore, we turn to an intelligent optimization algorithm with a strong global convergence ability to obtain the optimal combination (C_f^*, H^*) with a global minimum system cost.

10.4.2 Optimization of System Parameters

The TLBO algorithm is a new and efficient meta-heuristic optimization method based on the philosophy of teaching and learning. This algorithm has the advantages

Fig. 10.9 System cost
function versus adaptive
control factor

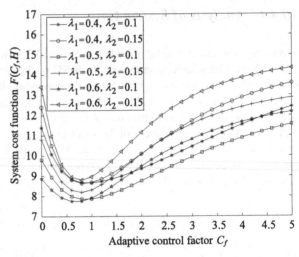

Fig. 10.10 System cost
function versus admission
threshold

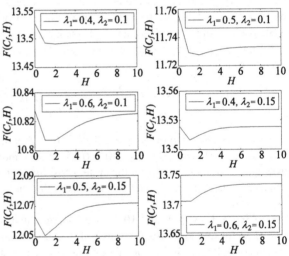

of having fewer parameters, being easy to understand and having a high degree of precision. By using the TLBO algorithm, we optimize the system parameters in terms of the adaptive control factor C_f and the admission threshold H.

Inspired by the teaching-learning process, we firstly randomly set a group of (C_f, H) as the students, and the corresponding system cost function $F(C_f, H)$ as academic records. The student who achieves the best academic record is assigned as a teacher. After a period of the teaching-learning process, we can deduce the best student. This means that we can derive the global minimum system cost function $F(C_f^*, H^*)$ and the optimize combination (C_f^*, H^*). The complexity of the TLBO algorithm depends on the maximum iterations N, the number M of students and the upper bound of the admission threshold H_1. The complexity T of this algorithm is

$T = O(N \times M \times H_1)$. The main steps for the TLBO algorithm to obtain the optimal adaptive control factor C_f^* and admission threshold H^* are given as follows:

Step 1: Set the maximum iterations N and number of students M. Initialize the admission threshold as $H = 0$, the current iterations as $d = 0$, and the number of local minimum system costs as $q = 0$.

Step 2: Set the upper bound of the adaptive control factor as $C_{f1} = (\mu - \lambda_1)/\lambda_2$ and the upper bound of the admission threshold as $H_1 = \lambda_1 \bar{\mu}_r / (\bar{\lambda}_1 \mu_r - \lambda_1 \bar{\mu}_r)$.

Step 3: Initialize each student $(C_f, H)_a$ $(a = 1, 2, 3, \ldots, M)$ within the constraint condition $C_f \in [0, C_{f1}]$, and calculate the system cost function $F((C_f, H)_a)$.

Step 4: Calculate the average value $(C_f, H)_{\text{mean}}$ for all students.
$$(C_f, H)_{\text{mean}} = \underset{a \in \{1,2,3,\ldots,M\}}{\text{mean}} \{(C_f, H)_a\}$$

Step 5: Select a teacher $(C_f, H)_{\text{teacher}}$ from all the students.
$$(C_f, H)_{\text{teacher}} = \underset{a \in \{1,2,3,\ldots,M\}}{\text{argmin}} \{F((C_f, H)_a)\}$$

Step 6:

> **for** $a = 1 : M$
> $\quad G = \text{round}(1 + \text{rand})$
> $\quad (C_f, H)_a^* = (C_f, H)_a + \text{rand} \times ((C_f, H)_{\text{teacher}} - G \times (C_f, H)_{\text{mean}})$
> $\quad$ % rand is a random number selected in the interval $(0, 1)$.
> $\quad$ **if** $F((C_f, H)_a) > F((C_f, H)_a^*)$
> $\quad\quad (C_f, H)_a = (C_f, H)_a^*$
> $\quad$ **endif**
> **endfor**

Step 7:

> **for** $a = 1 : M$
> $\quad$ randomly select the bth student $(C_f, H)_b$ $(b \neq a)$
> $\quad$ **if** $F((C_f, H)_a) > F((C_f, H)_b)$
> $\quad\quad (C_f, H)_a = (C_f, H)_a + \text{rand} \times ((C_f, H)_a - (C_f, H)_b)$
> $\quad$ **else** $(C_f, H)_a = (C_f, H)_a + \text{rand} \times ((C_f, H)_b - (C_f, H)_a)$
> $\quad$ **endif**
> **endfor**

Step 8:

> **if** $d < N$
> $\quad d = d + 1$, go to **Step 4**
> **else** $F((C_f, H)_q) = \underset{a \in \{1,2,3,\ldots,M\}}{\min} \{F((C_f, H)_a)\}, q = q + 1$
> % $(C_f, H)_q$ is a local minimum system cost.
> **endif**

Step 9:

> **if** $H \leq H_1$
> $\quad H = H + 1$
> $\quad$ go to **Step 3**

Table 10.1 Optimum combination of parameters in proposed strategy

Arrival rates λ_1 of SU packets	Arrival rates λ_2 of PU packets	Optimum combinations (C_f^*, H^*)	System costs $F(C_f^*, H^*)$
0.4	0.1	(1.04,1)	8.61
0.4	0.15	(1.14,0)	8.45
0.5	0.1	(1.03,1)	7.78
0.5	0.15	(0.99,0)	8.04
0.6	0.1	(0.77,1)	7.68
0.6	0.15	(0.58,1)	8.46

$$\textbf{else } (C_f^*, H^*) = \underset{s \in \{1,2,3,\dots,q\}}{\text{argmin}} \ \{F((C_f, H)_s)\}$$

% (C_f^*, H^*) is the optimal combination.

endif

Step 10: Output (C_f^*, H^*) as an optimal combination.

By setting the same parameters as used in Figs. 10.9 and 10.10, we obtain the optimal combination (C_f^*, H^*) with the adaptive control factor C_f and the admission threshold H as shown in Table 10.1.

10.5 Conclusion

In this chapter, we proposed a spectrum reservation strategy with an adaptive control approach for setting the spectrum aside ratio in centralized CRNs. We firstly constructed a three-dimensional DTMC model and obtained the steady-state distribution of the system model by using a method similar to that of the matrix-geometric solution. Accordingly, we evaluated the system performance mathematically. Moreover, we provided numerical results to investigate the influence of the adaptive control factor and the admission threshold on the system performance. Based on the trade-off between different performance measures, we built a system cost function. Finally, by using the TLBO algorithm, we optimized the adaptive control factor and the admission threshold with the minimum system cost. Numerical results showed that the proposed spectrum reservation strategy is effective in improving the spectrum utilization and coping with the immense demand from wireless devices.

Chapter 11
Opportunistic Spectrum Access Mechanism with Imperfect Sensing Results

In this chapter, we propose an opportunistic channel access mechanism with admission threshold and probabilistic feedback in Cognitive Radio Networks (CRNs) to reduce the average latency of Secondary User (SU) packets and adapt to various levels of tolerance for transmission interruption. Considering the preemptive priority of Primary User (PU) packets, as well as the sensing errors of missed detections and false alarms caused by SUs, we establish a priority queueing model, in which two types of packets, the PU packets and the SU packets, may interfere with each other. By employing a matrix-geometric solution, we derive performance measures of the system in terms of the blocking rate and the average latency of SU packets. We present numerical results to evaluate the performance of the system using the proposed opportunistic channel access mechanism. Moreover, we investigate the Nash equilibrium and the socially optimal behaviors of SU packets, and then, we propose an appropriate pricing policy to maximize the value of the social benefit function.

11.1 Introduction

In future wireless application, such as 5G networks, the demand for wireless spectrum resources will have a huge increase [Boch14, Wang14b]. Currently, a large portion of the assigned spectrum remains under-utilized. This is the key reason for the shortage of spectrum resources in WCNs [Wang11a]. How to improve the utilization of spectrum resources is a hot research topic. Cognitive radio is predicted to become one of the most popular wireless technologies due to its efficient spectrum utilization [Lian11]. Given this prediction, CRNs will emerge as a required technology [Kim15]. Recently, the opportunistic channel access mechanism in CRNs [Altr14, Ghos14] has been paid more attention as a method of improving the spectrum utilization.

© Springer Nature Singapore Pte Ltd. 2021
S. Jin, W. Yue, *Resource Management and Performance Analysis of Wireless Communication Networks*, https://doi.org/10.1007/978-981-15-7756-7_11

In [Kahv13], the authors proposed a dynamic channel selection approach to reduce the overhead caused by interrupted SU packets. Depending on the probability of a channel being idle and the average waiting time in the channel queue, they calculated a value for each channel to demonstrate suitable a channel was for selection when an interruption occurred. Furthermore, the arrival rate of PU packets was set optimally to reduce the average transmission delay of SU packets. In [Hu13a], the authors divided a time slot into three parts in terms of sensing time, reporting time and transmission time. They also proposed a scheme in which one SU reporting time was also used for other SUs' sensing. Then, the sensing time was optimized by minimizing the transmission delay of SU packets under the condition of sufficient protection to PU packets. In [Tran13], the authors studied the pricing mechanisms and their effects on equilibrium behaviors of SU packets' self-optimizing. With an individual optimal strategy, they showed that there was a unique equilibrium in the joining probability of SU packets. In addition, they also analyzed the relation between the revenue maximization and social benefit maximization by using pricing mechanisms. In the above analysis, the perfect sensing results of SUs were assumed. However, in practice, there are two types of errors associated with spectrum sensing, namely, mistake detections and false alarms [Lian08].

In [Tan13], considering collisions between PU packets and SU packets, and collisions among SU packets themselves, the authors proposed a channel-aware opportunistic spectrum access strategy consisting of multiple SU packets. They investigated the strategies for both cooperative and non-cooperative settings. They introduced pure threshold policies for both scenarios, where the SU packets compared channel qualities with a fixed threshold to decide whether to access the channel or not. Finally, they set the threshold optimally to obtain a maximum throughput. Based on [Hu13a], the authors performed a further work [Hu14]. By introducing the method of energy detection, they designed the sensing time so as to minimize the transmission delay of SU packets via spectrum sensing in CRNs. In [Bhow14], the authors analyzed a cooperative spectrum sensing strategy. They investigated the performance of the network in terms of maximum throughput with an optimal number of SU packets and sensing time. Unfortunately, the preemptive priority of PU packet to SU packet was neglected in these researches.

Inspired by the above observation, aiming to reduce the average delay of SU packets, in our previous work [Ge15], we proposed an opportunistic channel access mechanism in CRNs. Considering the preemptive priority of PU packets, as well as the sensing errors caused by SUs, we modeled the network system as a priority queue with two types of packets, the PU packets and the SU packets, in which these two types of packets may interfere with each other. We presented an analysis to obtain the throughput of SU packets and gave numerical results to optimize the energy sensing threshold with a maximum throughput of SU packets. In [Ge15], however, the authors did not take into account the blocking rate, the average latency of SU packets and the pricing policy for SU packets.

In this chapter, in order to investigate the performance of the system, we extend the analysis of [Ge15] with additional analyses to give furthermore the performance measures including the blocking rate and the average latency of SU packets by considering the mistake detections and the false alarms. Moreover, by building a reward function, we investigate the strategies for both the Nash equilibrium and the social optimization. Also, we provide a pricing policy for SU packets to coordinate these two strategies. With numerical results, we investigate the influence of admission threshold and feedback probability on the system performance, and verify the effectiveness of the proposed opportunistic channel access mechanism and the rationality of the proposed pricing policy.

The chapter is organized as follows. In Sect. 11.2, by addressing the activities of PU and SU packets, we describe the opportunistic channel access mechanism proposed in this chapter. Then, we present the system model in detail. In Sect. 11.3, by considering the mistake detections and the false alarms, we construct the transition probability matrix and analyze the steady-state distribution of the system model. In Sect. 11.4, we obtain performance measures and present numerical results to evaluate the system performance. In Sect. 11.5, we firstly investigate the Nash equilibrium and the socially optimal behaviors of SUs. Then, we propose an appropriate pricing policy to maximize the value of the social benefit function for imposing an appropriate admission fee for SU packets. Finally, we draw our conclusions in Sect. 11.6.

11.2 Opportunistic Spectrum Access Mechanism and System Model

We consider a CRN with a single licensed channel. The channel is used by PU packets preemptively and shared by SU packets opportunistically. Based on the sensing results using energy detection, an SU packet decides whether to occupy the channel or not. To maximize the throughput of SU packets, a buffer for SU packets is considered. However, in order to satisfy the delay requirement of PU packets to a maximum extent, no buffer is prepared for PU packets. With these considerations, in this section, we propose an opportunistic channel access mechanism.

Next, we discuss the activities for both PU packets and SU packets in this opportunistic channel access mechanism. Furthermore, we build a system model accordingly.

11.2.1 Activity of PU Packets

The transmission behavior of a PU packet in this opportunistic channel access mechanism can be characterized as a busy-idle alternate process. When the transmission of a PU packet is completed successfully or interfered by an SU packet, the channel state will change to idle from busy at the slot boundary. Conversely, when a PU packet starts to occupy the idle channel for its transmission, the idle state will change to busy.

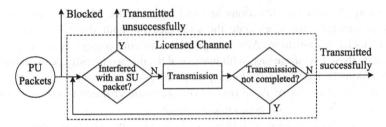

Fig. 11.1 Transmission process of PU packets

When a new PU packet arrives at the system, if there is no other PU packet being transmitted at the channel, the arriving PU packet will occupy the channel immediately. Otherwise, the newly arriving PU packet will be refused by the system; that is to say, it will be blocked. If a sensing error of mistake detection occurs, the PU packet will collide with an SU packet in the system, and the transmission of the PU packet is interfered. As a result, the expected transmission time of the interfered PU packet becomes shorter, and the transmission is unsuccessful.

The transmission process of PU packets is illustrated in Fig. 11.1.

11.2.2 Activity of SU Packets

In this subsection, we discuss the activity for an SU packet in this opportunistic channel access mechanism by dividing the transmission process into the system access, the channel sense and the interruption behavior.

(1) *System Access*: By trading off the throughput and the average latency of SU packets, we set an admission threshold H ($H \geq 0$) as a system parameter in the opportunistic channel access mechanism proposed in this chapter. When an SU packet arrives at the system, the central controller will compare the number L of SU packets in the system with H. If $L \geq H$, the SU packet will be admitted to join the system with probability r or leave the system with probability $\bar{r}$ ($\bar{r} = 1 - r$). Otherwise, the packet will be admitted to access the system with probability 1. Obviously, the larger the value of r is, the more the SU packets will be admitted to join the buffer, however, the greater the average latency of SU packets will be. The SU packets admitted to join the system will queue at the buffer.

(2) *Channel Sense:* Under the schedule of the central controller, the SU senses the channel at the boundary of each slot. Due to the effect of channel fading and the stochastic noise, the sensing errors in terms of mistake detections and false alarms are unavoidable. If a false alarm occurs, the channel will be in the idle state; If a mistake detection occurs, a PU packet and an SU packet will collide with each other and the transmissions for both the PU packet and the SU packet

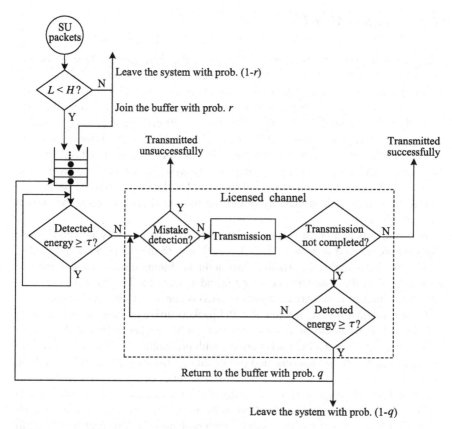

Fig. 11.2 Transmission process of SU packets

will be unsuccessful. Energy detection is one of the popular methods of channel sensing. And with this method, we introduce the energy threshold τ.

(3) *Interruption Behavior*: When the transmission of an SU packet is interrupted, the SU packet would like to return to the buffer with probability q $(0 \leq q \leq 1)$ or to leave the system with probability $\bar{q}$ $(\bar{q} = 1 - q)$. We note that the larger the value of q is, the more patient the SU packet will be. Generally speaking, the patience of the SU packets waiting in the buffer is greater than those being interrupted. Therefore, the interrupted SU packet returning to the buffer is supposed to queue at the head of the buffer.

The transmission process of SU packets is illustrated in Fig. 11.2.

11.2.3   System Model

In this network system, there are two types of the data packets. One is the PU packets having a high priority to be transmitted without a buffer, and the other is the SU packets having an unlimited buffer. A single channel is used to transmit these two types' packets. We note that the two types' packets may interfere with each other. We can model this network system as a discrete-time priority queueing system for performance analysis and numerical evaluation. In this priority queueing model, we take into account the impatient behavior of the interrupted SU packets, the tolerance delay of an SU packet, the sensing errors of SUs and the preemptive priority of PU packets. Meanwhile, we suppose that the interrupted SU packets go back to the buffer with probability q and the interfered packets leave the system with probability 1.

The time axis is divided into equal intervals, and the intervals are called slots. The slots are marked as n ($n = 1, 2, 3, \ldots$). For both the PU packets and the SU packets, we suppose that the arrival occurs at the beginning instant of a slot, marked as (n, n^+) while the departure occurs at the end instant of a slot, marked as (n^-, n). This means that an Early Arrival System (EAS) is considered in this chapter.

We assume that the SU packets and PU packets arrive at the system according to independent Bernoulli processes. In a slot, an SU packet arrives at the system with probability λ_1 and a PU packet arrives with probability λ_2. We call probability λ_1 the arrival rate of SU packets, and probability λ_2 the arrival rate of PU packets. We also assume that the necessary transmission times of an SU packet and a PU packet follow independent geometrical distributions. In a slot, the transmission of an SU packet is completed successfully with probability μ_1 and the transmission of a PU packet is completed successfully with probability μ_2, respectively. We call probability μ_1 the service rate of SU packets and probability μ_2 the service rate of PU packets.

We define the total number $X_n = i$ ($i = 0, 1, 2, \ldots$) of SU packets in the system at the instant n^+ as the system level and the channel state $Y_n = j$ ($j = 0, 1, 2, 3$) as the system stage at the instant n^+. $j = 0$ means the channel is idle; $j = 1$ means the channel is occupied by a PU packet; $j = 2$ means the channel is occupied by an SU packet; $j = 3$ means the channel is disordered, namely, a collision occurs between a PU packet and an SU packet. Therefore, $\{(X_n, Y_n), n \geq 0\}$ constitutes a two-dimensional Markov chain. The state space Ω of this Markov chain is given as follows:

$$\Omega = \{(i, j) : i \geq 0, j = 0, 1, 2, 3\}. \tag{11.1}$$

Let $\pi_{i,j}$ be the steady-state probability that the system level is i and the system stage is j. Thus $\pi_{i,j}$ can be given as follows:

$$\pi_{i,j} = \lim_{n \to \infty} \Pr\{X_n = i, Y_n = j\}, \quad i \geq 0, j = 0, 1, 2, 3. \tag{11.2}$$

11.3 Performance Analysis

Considering the influence of channel fading and the stochastic noise, we first calculate the mistake detection ratio and false alarm ratio. Then, we construct the transition probability matrix to analyze the system model in the steady state.

11.3.1 Mistake Detections and False Alarms

When SUs sense the channel via energy detection, two types of sensing errors, in terms of mistake detections and false alarms, can possibly occur.

Let t_s be the sensing time, f_s be the sensing frequency, ξ be the SNR and ϖ be the variance of noise. Let p_{md} be the mistake detection ratio and p_{fa} be the false alarm ratio. p_{md} and p_{fa} are given as follows:

$$\begin{cases} p_{md} = 1 - Q\left(\left(\frac{\tau}{\varpi^2} - \xi - 1\right)\sqrt{\frac{t_s f_s}{2\xi + 1}}\right) \\ p_{fa} = Q\left(\left(\frac{\tau}{\varpi^2} - 1\right)\sqrt{t_s f_s}\right) \end{cases} \tag{11.3}$$

where τ is defined in Sect. 11.2.2 and $Q(v)$ is the tail probability of the standard normal distribution given by

$$Q(v) = \frac{1}{\sqrt{2\pi}} \int_v^\infty \exp\left(-\frac{t^2}{2}\right) dt.$$

Taking into account the sensing results of SUs, we construct the transition probability matrix of the system model.

11.3.2 Transition Probability Matrix

We define P to be the one-step transition probability matrix of the Markov chain $\{(X_n, Y_n), n \geq 0\}$. Let P_{ik} be the transition probability sub-matrix for the number of SU packets in the system changing from i to k. The sub-matrix P_{ik} is discussed as follows.

(1) If $i = 0$ and $k = 0$, it means that there is no SU packet arrival at the system during the one-step transition. In this case, the change of the system stage only depends on whether there is a PU packet arrival or not. The one-step transition

sub-matrix P_{00} can be given as follows:

$$P_{00} = \bar{\lambda}_1 \begin{pmatrix} \bar{\lambda}_2 & \lambda_2 & 0 & 0 \\ \bar{\lambda}_2 \mu_2 & \omega_2 & 0 & 0 \end{pmatrix} \tag{11.4}$$

where ω_2 ($\omega_2 = \bar{\mu}_2 + \lambda_2 \mu_2$) is the probability that the channel is always occupied by a PU packet.

(2) If $i = 0$ and $k = 1$, it means that there is an SU packet arrival at the system during the one-step transition. In this case, the change of the system stage depends on the sensing results of SUs and the behavior of PU packets. The one-step transition probability sub-matrix P_{01} is given as follows:

$$P_{01} = \lambda_1 \begin{pmatrix} \bar{\lambda}_2 & \lambda_2 & \bar{\lambda}_2 & \lambda_2 \\ \bar{\lambda}_2 \mu_2 & \omega_2 & \bar{\lambda}_2 \mu_2 & \omega_2 \end{pmatrix} \times K \tag{11.5}$$

where

$$K = \begin{pmatrix} p_{\text{fa}} & 0 & 0 & 0 \\ 0 & \bar{p}_{\text{md}} & 0 & 0 \\ 0 & 0 & \bar{p}_{\text{fa}} & 0 \\ 0 & 0 & 0 & p_{\text{md}} \end{pmatrix}.$$

(3) If $i = 1$ and $k = 0$, it means that there is an SU packet departure and no arrival at the system during the one-step transition. The departure of an SU packet occurs in three cases:

   (i) The transmission of the SU packet is completed successfully.
   (ii) The transmission of the SU packet is interrupted by a PU packet and then the SU packet leaves the system with probability $\bar{q}$.
   (iii) The SU packet collides with a PU packet and the collided SU packet leaves the system with probability 1.

In stage 0 or stage 1, the channel is not occupied by SU packets, so there is certainly no departure of any SU packet. In stage 2, the change of the system stage depends on the sensing results of SUs, the behavior of PU packets and whether there is an SU packet departure or not. In stage 3, the interfered SU packet is forced to leave the system, and the change of the system stage depends on whether there is a new PU packet arrival or not.

Therefore, the one-step transition probability sub-matrix P_{10} of this case is given as follows:

$$P_{10} = \bar{\lambda}_1 \begin{pmatrix} 0 & 0 & 0 & 0 \\ 0 & 0 & 0 & 0 \\ \mu_1 \bar{\lambda}_2 & \mu_1 \lambda_2 & 0 & 0 \\ \bar{\lambda}_2 & \lambda_2 & 0 & 0 \end{pmatrix} + \bar{\lambda}_1 \begin{pmatrix} 0 & 0 & 0 & 0 \\ 0 & 0 & 0 & 0 \\ \bar{q}\bar{\mu}_1\bar{\lambda}_2 & \bar{q}\bar{\mu}_1\lambda_2 & 0 & 0 \\ 0 & 0 & 0 & 0 \end{pmatrix} \times K. \tag{11.6}$$

(4) Let $\boldsymbol{B}_0$ be the transition probability matrix representing that there is an SU packet arrival at the system, and no SU packet departure during the one-step transition from system levels higher than 0. In stage 0 or stage 1, the channel is not occupied by SU packets, so the departure of an SU packet is impossible. In this case, the change of the system stage depends on the sensing results of SUs and the behavior of PU packets. In stage 2, the change of the system stage also depends on the behavior of the interrupted SU packet. In stage 3, the number of SU packets in the system cannot increase during the one-step transition. $\boldsymbol{B}_0$ is given as follows:

$$
\boldsymbol{B}_0 = \lambda_1 \begin{pmatrix} \bar{\lambda}_2 & \lambda_2 & \bar{\lambda}_2 & \lambda_2 \\ \bar{\lambda}_2\mu_2 & \omega_2 & \bar{\lambda}_2\mu_2 & \omega_2 \\ q\bar{\mu}_1\bar{\lambda}_2 & q\bar{\mu}_1\lambda_2 & \bar{\mu}_1\bar{\lambda}_2 & \bar{\mu}_1\lambda_2 \\ 0 & 0 & 0 & 0 \end{pmatrix} \times \boldsymbol{K}. \tag{11.7}
$$

(i) For the case of $1 \le i < H$ and $k = i + 1$, $\boldsymbol{P}_{i,i+1}$ is given by

$$
\boldsymbol{P}_{i,i+1} = \boldsymbol{B}_0.
$$

(ii) For the case of $i \ge H$ and $k = i + 1$, the new arriving SU packet (if any) is admitted to join the buffer with probability r. By $\boldsymbol{A}_0$ we denote the one-step transition sub-matrix $\boldsymbol{P}_{i,i+1}$ for the above stochastic behavior. $\boldsymbol{A}_0$ then is given by

$$
\boldsymbol{A}_0 = r\boldsymbol{B}_0.
$$

(5) Let $\boldsymbol{B}_1$ be the transition probability matrix representing the number of SU packets to be fixed at a value greater than 0 during the one-step transition. In stage 0 or stage 1, the channel is not occupied by SU packets, so the departure of an SU packet is impossible. In stage 2, if the SU packet being transmitted leaves the system, a new SU packet will arrive. If the SU packet being transmitted does not leave, a new SU packet will not arrive at the system. In stage 3, the interfered SU packet is forced to leave the system with probability 1. For this case, the change of the system stage depends on both the sensing results of SUs and the behavior of PU packets. $\boldsymbol{B}_1$ is given as follows:

$$
\boldsymbol{B}_1 = \begin{pmatrix} \bar{\lambda}_1 & 0 & 0 & 0 \\ 0 & \bar{\lambda}_1 & 0 & 0 \\ 0 & 0 & \lambda_1(\mu_1 + \bar{q}\bar{\mu}_1) + \bar{\lambda}_1 q\bar{\mu}_1 & 0 \\ 0 & 0 & 0 & \lambda_1 \end{pmatrix} \times \begin{pmatrix} \bar{\lambda}_2 & \lambda_2 & \bar{\lambda}_2 & \lambda_2 \\ \bar{\lambda}_2\mu_2 & \omega_2 & \bar{\lambda}_2\mu_2 & \omega_2 \\ \bar{\lambda}_2 & \lambda_2 & \bar{\lambda}_2 & \lambda_2 \\ \bar{\lambda}_2 & \lambda_2 & \bar{\lambda}_2 & \lambda_2 \end{pmatrix} \times \boldsymbol{K}. \tag{11.8}
$$

(i) For the case of $1 \leq i < H$ and $k = i$, $\boldsymbol{P}_{ii}$ is given by

$$\boldsymbol{P}_{ii} = \boldsymbol{B}_1.$$

(ii) For the case of $i = H$ and $k = i$, the newly arriving SU packet (if any) is refused admittance to the buffer with probability $\bar{r}$ given that there is no SU packet being completed successfully. The number of SU packets in the system remains equal to the admission threshold H. $\boldsymbol{P}_{HH}$ is given by

$$\boldsymbol{P}_{HH} = \boldsymbol{B}_1 + \bar{r}\lambda_1 \begin{pmatrix} \bar{\lambda}_2 & \lambda_2 & \bar{\lambda}_2 & \lambda_2 \\ \bar{\lambda}_2 & \lambda_2 & \bar{\lambda}_2 & \lambda_2 \\ (2q-1)\bar{\mu}_1\bar{\lambda}_2 & (2q-1)\bar{\mu}_1\lambda_2 & \bar{\mu}_1\bar{\lambda}_2 & \bar{\mu}_1\lambda_2 \\ 0 & 0 & 0 & 0 \end{pmatrix} \times \boldsymbol{K}.$$

(iii) For the case of $i > H$ and $k = i$, the newly arriving SU packet (if any) is refused admittance to the buffer with probability $\bar{r}$. By $\boldsymbol{A}_1$ we denote the one-step transition sub-matrix $\boldsymbol{P}_{ii}$ for the above stochastic behavior. $\boldsymbol{A}_1$ then is given by

$$\boldsymbol{A}_1 = \boldsymbol{B}_1 + \bar{r}\lambda_1$$
$$\times \begin{pmatrix} \bar{\lambda}_2 & \lambda_2 & \bar{\lambda}_2 & \lambda_2 \\ \bar{\lambda}_2 & \lambda_2 & \bar{\lambda}_2 & \lambda_2 \\ (2q\bar{\mu}_1-1)\bar{\lambda}_2 & (2q\bar{\mu}_1-1)\lambda_2 & (2\bar{\mu}_1-1)\bar{\lambda}_2 & (2\bar{\mu}_1-1)\lambda_2 \\ -\bar{\lambda}_2 & -\lambda_2 & -\bar{\lambda}_2 & -\lambda_2 \end{pmatrix} \times \boldsymbol{K}.$$

(6) Let $\boldsymbol{B}_2$ be the transition probability matrix representing that there is no SU packet arrival at the system, and an SU packet departure during the one-step transition from system levels higher than 0. In stage 0 or stage 1, the channel is not occupied by SU packets, so the departure of an SU packet is impossible. For this case, the change of the system stage depends on the sensing results of SUs and whether there is a PU packet arrival or not. In stage 2, the change of the system stage depends on the sensing results of SUs, the behavior of the interrupted SU packet and whether there is a PU packet arrival or not. In stage 3, the interfered SU packet is forced to leave the system and the change of the system stage depends on the sensing results of SUs and whether there is a new PU packet arrival or not. $\boldsymbol{B}_2$ is given as follows:

$$\boldsymbol{B}_2 = \bar{\lambda}_1 \begin{pmatrix} 0 & 0 & 0 & 0 \\ 0 & 0 & 0 & 0 \\ (\mu_1 + \bar{q}\bar{\mu}_1)\bar{\lambda}_2 & (\mu_1 + \bar{q}\bar{\mu}_1)\lambda_2 & \mu_1\bar{\lambda}_2 & \mu_1\lambda_2 \\ \bar{\lambda}_2 & \lambda_2 & \bar{\lambda}_2 & \lambda_2 \end{pmatrix} \times \boldsymbol{K}. \qquad (11.9)$$

(i) For the case of $1 \leq i < H$ and $k = i - 1$, $P_{i,i-1}$ is given by

$$P_{i,i-1} = B_2.$$

(ii) For the case of $i = H$ and $k = H - 1$, the potentially arriving SU packet may be refused by the system if no SU packet is transmitted completely. The one-step transition probability sub-matrix $P_{H,H-1}$ can be given by $P_{H,H-1}$ is given by

$$P_{H,H-1} = B_2 + \bar{q}\bar{\mu}_1\bar{r}\lambda_1 \begin{pmatrix} 0 & 0 & 0 & 0 \\ 0 & 0 & 0 & 0 \\ \bar{\lambda}_2 & \lambda_2 & 0 & 0 \\ 0 & 0 & 0 & 0 \end{pmatrix} \times K.$$

(iii) For the case of $i > H$ and $k = i - 1$, the newly arriving SU packet (if any) is refused admittance to the buffer with probability $\bar{r}$. By A_2 we denote the one-step transition sub-matrix $P_{i,i-1}$ for the above stochastic behavior. A_2 then is given by

$$A_2 = B_2 + \bar{r}\lambda_1 \begin{pmatrix} 0 & 0 & 0 & 0 \\ 0 & 0 & 0 & 0 \\ (\bar{q}\bar{\mu}_1 + \mu_1)\bar{\lambda}_2 & (\bar{q}\bar{\mu}_1 + \mu_1)\lambda_2 & \mu_1\bar{\lambda}_2 & \mu_1\lambda_2 \\ \bar{\lambda}_2 & \lambda_2 & \bar{\lambda}_2 & \lambda_2 \end{pmatrix} \times K.$$

So far, we have obtained all the sub-matrices in the transition probability matrix P. Therefore,

$$P = \begin{pmatrix} P_{00} & P_{01} & & & & & \\ P_{10} & P_{11} & P_{12} & & & & \\ & \ddots & \ddots & \ddots & & & \\ & & P_{H-1,H-2} & P_{H-1,H-1} & P_{H-1,H} & & \\ & & & P_{H,H-1} & P_{HH} & A_0 & \\ & & & & A_2 & A_1 & A_0 \\ & & & & & \ddots & \ddots & \ddots \end{pmatrix}. \tag{11.10}$$

Structure of P shows that the system transition occurs only in adjacent levels. Therefore, the two-dimensional Markov chain $\{(X_n, Y_n), n \geq 0\}$ is a Quasi Birth-Death (QBD) process. Moreover, it is clear that the rows of the transition probability matrix P are repeating after the $(H + 1)$th row. Therefore, by using the matrix-geometric solution method, we can get the steady-state probability $\pi_{i,j}$ iteratively with numerical results.

11.4   Performance Measures and Numerical Results

In this section, we first derive performance measures of the system in terms of the throughput, the block rate and the average latency of SU packets, respectively. Then, we present numerical results to evaluate the performance of the system using the opportunistic channel access mechanism proposed in this chapter.

11.4.1   Performance Measures

In this subsection, we derive the formula for the throughput, the block rate and the average latency of SU packets.

(1) The throughput θ of SU packets is defined as the number of SU packets transmitted successfully per slot. If the transmission of an SU packet is completed successfully at the end boundary of a slot, the channel must be occupied by this SU packet normally during this slot. We give the throughput θ of SU packets as follows:

$$\theta = \sum_{i=0}^{\infty} \pi_{i,2} \mu_1. \tag{11.11}$$

(2) The blocking rate B_s of SU packets is defined as the number of SU packets blocked by the system per slot. Given that the number L of SU packets in the system is no less than the admission threshold H, namely, $L \geq H$, the central controller will refuse the arriving SU packet admittance to the buffer with probability $\bar{r}$. Then, we give the blocking rate B_s as follows:

$$B_s = \bar{r}\lambda_1 \left(\sum_{i=H+1}^{\infty} \sum_{j=0}^{3} \pi_{i,j} + \pi_{H,0} + \pi_{H,1} + \pi_{H,2}\bar{\mu}_1 \right). \tag{11.12}$$

(3) The latency Y_s of an SU packet is defined as the duration from the arrival instant of a SU packet to its departure instant. Based on the analysis presented in Sect. 11.3.2, we can obtain the average latency $E[Y_s]$ of SU packets as follows:

$$E[Y_s] = \frac{1}{\lambda_1 (1 - B_s)} \sum_{i=0}^{\infty} \sum_{j=0}^{3} i \pi_{i,j}. \tag{11.13}$$

11.4.2   Numerical Results

In order to illustrate the influence of the system parameters on the opportunistic channel access mechanism proposed in this chapter, we present numerical results with analysis. Considering practical application, we set the mistake detection ratio $p_{md} \leq 0.5$ and the false alarm ratio $p_{fa} \leq 0.07$ in Eq. (11.3), then we get the energy threshold $0 < \tau < 7$. Moreover, we present numerical results with simulation to validate iteration accuracy of the model analysis. Good agreements between the analysis results and the simulation results are observed.

Based on IEEE 802.11 standard operating in the 2.4 GHz band, we list the system parameter settings in Table 11.1 as an example.

Figure 11.3 illustrates the change trend of the throughput θ of SU packets along with the energy threshold τ for different admission thresholds H, admission probabilities r and feedback probabilities q.

Table 11.1 Parameter settings in numerical results

Parameters	Values
Slot	1 ms
Transmission rate in physical layer	11 Mbps
Arrival rate λ_1 of SU packets	0.3
Mean size of an SU packet	1760 Byte
Arrival rate λ_2 of PU packets	0.05
Mean size of a PU packet	2010 Byte
Feedback probability q	0, 0.4, 0.7
Simulation scale	3 million slots
Sensing time t_s	0.1 ms
Sensing frequency f_s	10 times/ms

Fig. 11.3 Throughput of SU packets versus energy threshold

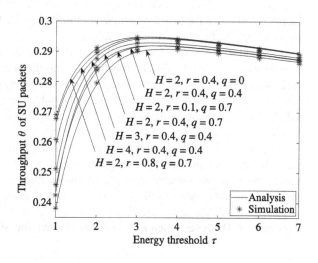

$H = 2, r = 0.4, q = 0$
$H = 2, r = 0.4, q = 0.4$
$H = 2, r = 0.1, q = 0.7$
$H = 2, r = 0.4, q = 0.7$
$H = 3, r = 0.4, q = 0.4$
$H = 4, r = 0.4, q = 0.4$
$H = 2, r = 0.8, q = 0.7$

— Analysis
* Simulation

Throughput θ of SU packets

Energy threshold τ

In Fig. 11.3, we notice that for all the admission thresholds H, the admission probability r and the feedback probability q, as the energy threshold τ increases, the throughput θ of SU packets increases initially and then decreases slowly. When the energy threshold is smaller, the false alarm ratio p_{fa} defined in Sect. 11.3.1 is greater, so the false alarm is the main factor impacting the throughput of SU packets. As the energy threshold increases, the false alarm ratio becomes lesser, which will result in fewer SU packets leaving the system due to their transmission interruption. On the other hand, when the energy threshold increases continuously, the throughput of SU packets decreases slowly. The reason is that when the energy threshold is greater, the mistake detection ratio p_{md} defined in Sect. 11.3.1 becomes the main factor impacting the throughput of SU packets. A greater mistake detection ratio will lead to more SU packets and PU packets colliding with each other, so the throughput of SU packets will decrease.

We also observe that for all the energy thresholds τ, if the admission probability r and the feedback probability q are given, e.g., $r = 0.4$ and $q = 0.4$, the throughput θ of SU packets increases as the admission threshold H increases. Note that a larger admission threshold means more SU packets queueing in the buffer, so the number of SU packets transmitted successfully will increase accordingly.

Moreover, we find that for all the energy thresholds τ, if the admission threshold H and the feedback probability q are given, e.g., $H = 2$ and $q = 0.7$, the throughput θ of SU packets increases along with the admission probability r. The reason is that as the admission probability increases, an SU packet is more likely to be admitted to the buffer, so more SU packets will be transmitted successfully.

Furthermore, we see that for all the energy thresholds τ, if the admission threshold H and the admission probability r are given, e.g., $H = 2$ and $r = 0.4$, the throughput θ of SU packets increases with the feedback probability q. An interrupted SU packet will return to the buffer with probability q waiting for future re-transmission. The greater the feedback probability is, the more likely it is that an interrupted SU packet will return to the buffer and will be transmitted successfully.

Figure 11.4 demonstrates how the Blocking rate B_s of SU packets changes as the energy threshold τ for different admission thresholds H, admission probabilities r and feedback probabilities q of SU packets.

In Fig. 11.4, we notice that for all the admission thresholds H, the admission probabilities r and the feedback probabilities q, as the energy threshold τ increases, the blocking rate B_s of SU packets decreases. As the energy threshold increases, the false alarm ratio will decrease and the mistake detection ratio will increase, which will lead to a decrease in the number of SU packets in the system. Therefore, the arriving SU packets are less likely to be refused by the system.

We also observe that for all the energy thresholds τ, if the admission probability r and the feedback probability q are given, e.g., $r = 0.4$ and $q = 0.4$, the Blocking rate B_s of SU packets decreases as the admission threshold H increases. When the number L of SU packets in the system is no less than the admission threshold H, namely, $L \geq H$, the newly arriving SU packet will be refused by the system with probability $\bar{r}$. However, when $L < H$ the newly arriving SU packet will be admitted

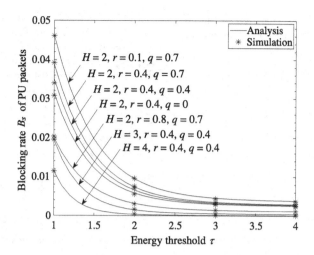

Fig. 11.4 Blocking rate of SU packets versus energy threshold

to the buffer. Therefore, the greater the admission threshold is, the less likely it is that the newly arriving SU packet will be refused by the system.

Moreover, we find that for all the energy thresholds τ, if the admission threshold H and the feedback probability q are given, e.g., $H = 2$ and $q = 0.7$, the Blocking rate B_s of SU packets decreases as the admission probability r increases. When $L \geq H$, the newly arriving SU packet will be admitted to the buffer with probability r. A smaller admission probability will certainly lead to an increase in the number of the blocked SU packets.

Furthermore, we see that for all the energy thresholds τ, if the admission threshold H and the admission probability r are given, e.g., $H = 2$ and $r = 0.4$, the Blocking rate B_s of SU packets increases as the feedback probability q increases. A greater feedback probability will lead more interrupted SU packets returning to the buffer, then the number of SU packets in the system is less likely to be smaller than the admission threshold. Therefore, the newly arriving SU packet is more likely to be refused by the system.

Figure 11.5 reveals how the average latency $E[Y_s]$ of SU packets changes with the energy threshold τ for different admission thresholds H, admission probabilities r and feedback probabilities q of SU packets.

In Fig. 11.5, we notice that for all the admission thresholds H, the admission probabilities r and the feedback probabilities q, the average latency $E[Y_s]$ of SU packets decreases as the energy threshold τ increases. When the energy threshold is lower, as the energy threshold increases, the false alarm ratio becomes lower, which will reduce the possibility that the channel is in the idle state. On the other hand, when the energy threshold continuously increases, the mistake detection ratio becomes greater than before, which will improve the possibility that a PU packet

Fig. 11.5 Average latency of
SU packets versus energy
threshold

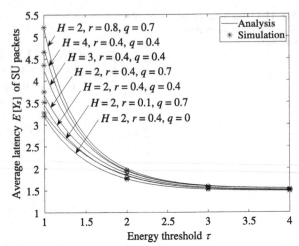

and an SU packet to collide with each other. Finally, the interfered SU packets will
leave the system ahead of time.

We also observe that for all the energy thresholds τ, if the admission probability
r and the feedback probability q are given, e.g., $r = 0.4$ and $q = 0.4$, the average
latency $E[Y_s]$ of SU packets increases along with the admission threshold H. A
greater admission threshold leads more SU packets to be admitted to the buffer.
Therefore, an SU packet queueing at the buffer will wait for a longer time before
accessing the channel.

Moreover, we find that for all the energy thresholds τ, if the admission threshold
H and the feedback probability q are given, e.g., $H = 2$ and $q = 0.7$, the average
latency $E[Y_s]$ of SU packets increases as the admission probability r increases.
When $L \geq H$, the newly arriving SU packet will be admitted to the buffer with
probability r. The greater the admission probability is, the more the SU packets will
join the buffer. As a result, the number of SU packets in the buffer will be greater,
and the average delay $E[Y_s]$ will be correspondingly larger.

Furthermore, we see that for all the energy thresholds τ, if the admission
threshold H and the admission probability r are given, e.g., $H = 2$ and $r = 0.4$, the
average latency $E[Y_s]$ decreases as the feedback probability q decreases. Recalling
that an interrupted SU packet will leave the system with probability $\bar{q}$, a smaller
feedback probability means that an interrupted SU packet is less likely to return
to the buffer for future transmission, which will make the SU packets stay in the
system for a shorter time.

In addition, from the numerical results shown in Figs. 11.3, 11.4, 11.5, we also
find that the analysis results match well with the simulation results.

11.5 Analysis of Admission Fee

In this section, we first investigate the Nash equilibrium behavior and socially optimal behavior of SU packets in the opportunistic channel access mechanism proposed in this chapter. Then, we present a pricing policy for the SU packets to optimize the system socially. This issue can be addressed by imposing an appropriate admission fee for SU packets.

11.5.1 Behaviors of Nash Equilibrium and Social Optimization

An SU packet queueing in the buffer leaves the system in two ways: either the transmission is completed successfully; or the transmission is interrupted and the SU packet leaves the system with probability $\bar{r}$. Based on the system model built in Sect. 11.2.3, we give some hypothesis as follows:

(1) Before an SU packet joins the buffer, it does not have any information about the system, such as the queue length or the channel state.
(2) The reward for an SU packet to be transmitted successfully is R_g.
(3) The cost of an SU packet staying in the system is C_g per slot.
(4) The benefits for all the SU packets can be added together.

With the hypothesis mentioned above, the individual benefit function is given as follows:

$$G_{\text{ind}}(\lambda_1) = R_g \frac{\theta}{\lambda_1(1 - B_s)} - C_g E[Y_s] \tag{11.14}$$

where $E[Y_s]$ is the average latency of SU packets given in Eq. (11.13).

By aggregating the individual benefits, the social benefit function $G_{\text{soc}}(\lambda_1)$ can be given as follows:

$$G_{\text{soc}}(\lambda_1) = \lambda_1(1 - B_s)G_{\text{ind}}(\lambda_1)$$
$$= \lambda_1(1 - B_s)\left(R_g \frac{\theta}{\lambda_1(1 - B_s)} - C_g E[Y_s]\right). \tag{11.15}$$

Maximizing the value of the social benefit function $G_{\text{ind}}(\lambda_1)$ in Eq. (11.15), we can get the socially optimal arrival rate λ_1^* as follows:

$$\lambda_1^* = \underset{0 \leq \lambda_1 < 1}{\text{argmax}} \left\{\lambda_1(1 - B_s)\left(R_g \frac{\theta}{\lambda_1(1 - B_s)} - C_g E[Y_s]\right)\right\}. \tag{11.16}$$

In CRNs, all the SU packets want to occupy the licensed channel to get benefit socially. However, when the number of SU packets in the system increases, the

Fig. 11.6 Individual benefit
function versus arrival rate of
SU packets

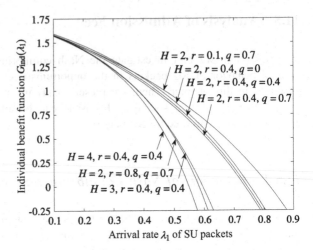

Fig. 11.7 Social benefit
function versus arrival rate of
SU packets

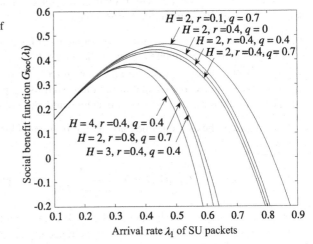

average latency of SU packets will be greater and the value of the individual benefit
function will be diminished. Therefore, there is an equilibrium behavior for the
arrival rate λ_1 of SU packets. Considering the complexity for the individual benefit
function $G_{\text{ind}}(\lambda_1)$ given in Eq. (11.14) and the social benefit function $G_{\text{soc}}(\lambda_1)$ given
in Eq. (11.15), we explore the monotonic property of $G_{\text{ind}}(\lambda_1)$ and $G_{\text{soc}}(\lambda_1)$ with
numerical results.

Using the system parameters given in Table 11.1, setting $R_g = 4.5$, $C_g = 2$,
$\tau = 3$, we illustrate the change trends for the individual benefit function $G_{\text{ind}}(\lambda_1)$
and the social benefit function $G_{\text{soc}}(\lambda_1)$ in Figs. 11.6 and 11.7, respectively.

In Fig. 11.6, we observe that the individual benefit function $G_{\text{ind}}(\lambda_1)$ shows a
decrease trend along with the arrival rate λ_1 of SU packets where there is a value of
λ_1 subjected to $G_{\text{ind}}(\lambda_1) = 0$. That is to say, there is a Nash equilibrium strategy for
the arrival rate λ_1 of SU packets, and the equilibrium arrival rate λ_1^e of SU packets

is unique. Letting $G_{\text{ind}}(\lambda_1) = 0$, we can obtain the equilibrium arrival λ_1^e of SU packets.

In Fig. 11.7, we find that the social benefit function $G_{\text{soc}}(\lambda_1)$ increases firstly and then decreases as the arrival rate λ_1 of SU packets increases. Therefore, an optimal arrival rate λ_1^* exists with the maximum value of the social benefit function.

For the same admission threshold H, admission probability r and feedback probability q, we compare the equilibrium arrival rate λ_1^e in Fig. 11.6 with the socially optimal arriving rate λ_1^* in Fig. 11.7, and we find $\lambda_1^e > \lambda_1^*$. That is to say, there are more SU packets joining the buffer under a Nash equilibrium strategy.

11.5.2 Pricing Policy

By charging an admission fee f to SU packets, we obtain the modified individual benefit function $G_{\text{ind}}'(\lambda_1)$ as follows:

$$G_{\text{ind}}'(\lambda_1) = R_g \frac{\theta}{\lambda_1(1 - B_s)} - C_g E[Y_s] - f. \tag{11.17}$$

In order to realize social optimization, the Nash equilibrium arrival rate λ_1^e of SU packets should be restrained against the socially optimal arrival rate λ_1^*. Letting $G_{\text{ind}}'(\lambda_1) = 0$ and $\lambda_1 = \lambda_1^*$ we can obtain the admission fee f.

With the same parameters used in Figs. 11.3, 11.4, 11.5, we present numerical results for the admission fee f in Table 11.2.

From Table 11.2 we find that for the same admission probability r and feedback probability q, the greater the admission threshold H is, the more SU packets will be admitted to the system. Therefore, the admission fee f should be set higher. We also observe that for the same admission threshold H and feedback probability q, the greater the admission probability r is, the more likely it is that an SU packet will be admitted to the buffer even though the number of SU packets already in the system is larger than the admission threshold. This is another reason why the admission fee f should be set higher.

Table 11.2 Numerical results for admission fee

Admission thresholds H	Admission probabilities r	Feedback probabilities q	Admission fees f
4	0.4	0.4	1.0873
3	0.4	0.4	1.0687
2	0.4	0.4	1.0272
2	0.4	0	1.0341
2	0.4	0.7	1.0163
2	0.8	0.7	1.064
2	0.1	0.7	0.9938

In addition, we see that for the same admission threshold H and admission probability r, the admission fee f increases as the feedback probability q decreases. A smaller feedback probability means that an interrupted SU packet is more likely to leave the system resulting in the number of SU packets in the system becoming less. As an incentive for more SU packets to join the buffer, a lower admission fee f should be set.

11.6   Conclusion

In this chapter, for the purpose of reducing the average latency of SU packets, we proposed an opportunistic channel access mechanism with admission threshold and probabilistic feedback. Based on the imperfect sensing results of SUs, we established a one-step transition probability matrix and evaluated the system performance quantitatively. Moreover, we investigated the behaviors of Nash equilibrium and social optimization of the proposed opportunistic channel access mechanism. Numerical results showed that the equilibrium arrival rate is greater than the socially optimal arrival rate. Moreover, after we investigated the Nash equilibrium and the socially optimal behaviors of SU packets, we proposed an appropriate pricing policy to maximize the value of the social benefit function. This issue can be addressed by imposing an appropriate admission fee for SU packets.

Chapter 12
Mini-Slotted Spectrum Allocation Strategy with Imperfect Sensing Results

In order to improve the throughput of Secondary User (SU) packets and reduce the spectrum switching frequency in Cognitive Radio Networks (CRNs), in this chapter, we propose a mini-slotted spectrum allocation strategy. Due to the mistake detection in practice, the Primary User (PU) packets and the SU packets will occupy the spectrum simultaneously, namely, a collision will occur on the spectrum. We build a heterogeneous discrete-time queueing model with possible collisions to model the system operation. Taking into account the imperfect sensing results, we construct the transition probability matrix. Applying the method of a matrix-geometric solution, we derive performance measures of the system in terms of the interruption rate of PU packets, the throughput of SU packets, the switching rate of SU packets and the average latency of SU packets, respectively. We present numerical results to verify the effectiveness of the proposed mini-slotted spectrum strategy. Finally, by trading off different performance measures, we construct a system profit function to optimize the slot size.

12.1 Introduction

Nowadays, with the rapid development of WCNs, the demand for wireless spectrum increases gradually, the spectrum resource has become scarcer than ever before. However, recent studies show that as much as 90% of the time, most of the allocated spectrum is not used under the static spectrum assignment policy. There are two types of users, namely, PUs and SUs in CRNs, the PUs have higher priority to the licensed spectrum, while the SUs are capable of sensing spectrum holes and utilizing them in an opportunistic way without causing harmful interference with the PUs [Nguy14, Wang11a, Zhao07].

As an important function of CRNs, spectrum allocation strategy has drawn more and more attentions. Recently, a lot of researches on the performance of

© Springer Nature Singapore Pte Ltd. 2021
S. Jin, W. Yue, *Resource Management and Performance Analysis of Wireless Communication Networks*, https://doi.org/10.1007/978-981-15-7756-7_12

spectrum allocation strategies have been carried out from different view points. In [Sole13], for the sake of decreasing the spectrum switches, with a Hidden Markov Model (HMM), the authors investigated the handoff procedure based on prediction approach. In [Tang13], aiming to adapt the spectrum access behavior of PUs, the authors proposed an access strategy for SUs and analyzed the impact of the access strategy on the spectrum switches. In [Do12], taking into account the CRN with a single SU and multiple PUs, the authors applied the M/G/1 preemptive priority queueing model to analyze the average latency of SUs. They also proposed an adaptive algorithm with the delay constraint. The researches mentioned above are based on the assumption of perfect spectrum sensing. However, due to the channel fading in a spectrum and the interference at the physical layer [Akyi06], the sensing errors are inevitable in practice.

Some related works have been studied by considering the mistake detections and false alarms. In [Altr14], the authors applied the Continuous-Time Markov Chain (CTMC) model to analyze the performance of opportunistic spectrum access under the imperfect sensing conditions. In [Ko14], the authors provided a framework for IEEE-802.11 MAC, and studied the trade-off between the sensing time and throughput of SUs. Moreover, they also investigated the optimal sensing time by considering both the unsaturated and the saturated traffic conditions. Unfortunately, in these researches, the PUs and SUs are supposed to operate on the homogeneous unit of time.

For the purpose of making full use of the spectrum, some researches have been carried out based on a heterogeneous structure. In [Bae10], in order to improve the throughput, the authors proposed a modified spectrum allocated strategy, where the SUs were supposed to access the spectrum with a binary exponential back-off algorithm. According to the remaining time of the current slot, the winning SU packets with appropriate length will be transmitted. The main drawback of this work is that if the remaining time of the current slot is smaller than the length of the shortest SU packets, the remaining time of the current slot will be idle. In [Atma13], considering that the PUs employed TDMA to access the spectrum, the SUs utilized slotted CSMA and accessed the spectrum when the slot was not occupied by PUs, the author divided the idle time slot into contention period and data transmission period. They also analyzed the throughput for the two types of users and the overall network to evaluate the spectrum utilization. But if the SU queue is empty and there is an SU packet arrival after the contention period, the newly arriving SU packet will be not transmitted after the contention period.

To reuse the spectrum with higher throughput and decrease the expense of the spectrum switching in CRNs, in this chapter, we propose a mini-slotted spectrum allocation strategy for SUs. Considering the imperfect spectrum sensing results in practice, we build a heterogeneous discrete-time queueing model with possible collisions. We also investigate and optimize the performance measures of the interruption rate of PU packets, the throughput of SU packets, the spectrum switching rate and the average latency of SU packets. In [Zhan14b], the authors presented a little result of this research in an early stage as a lecture note. However, in this chapter we present an analysis framework to derive the steady-state distribution of the queueing model. We also give some important performance measures in terms of the interruption rate of PU packets, the normalized throughput of SU packets, the switching rate of SU packets and the average latency of SU packets to investigate the stochastic behavior of the system. In addition, we present numerical results with analysis and simulation to show the change trends of the performance measures. Moreover, we establish a reward-cost structure-based function to optimize the slot size.

The chapter is organized as follows. In Sect. 12.2, we describe the mini-slotted spectrum allocation strategy proposed in this chapter. Then, we present the system model in detail. In Sect. 12.3, we present a performance analysis based on the analysis of the transition probability matrix and the steady-state distribution of the system model. In Sect. 12.4, we derive performance measures and present numerical results to evaluate the system performance. In Sect. 12.5, performance optimization is carried out with a system profit function. Finally, we draw our conclusions in Sect. 12.6.

12.2   Mini-Slotted Spectrum Allocation Strategy and System Model

In this section, we propose a mini-slotted spectrum allocation strategy in CRNs. Then, we build a discrete-time queueing model with possible collisions accordingly.

12.2.1   Mini-Slotted Spectrum Allocation Strategy

We propose a mini-slotted spectrum allocation strategy in CRNs. Firstly, we present the mini-slotted allocation strategy used in the system as follows: Time axis is divided into mini slots with fixed length, and several mini slots combine to constitute a slot. Then, we assume that the transmission of an SU packet is based on the mini slot, while the transmission of a PU packet is based on the slot. Moreover, we suppose that there are multiple PUs, multiple SUs and adequate licensed spectrums in a CRN.

In order to maximum the throughput of the SUs, we introduce a buffer for the SU packets, and the buffer is implemented with infinite length for simplicity. The SU packets are transmitted on a First-Come First-Served (FCFS) discipline. However, for the purpose of minimizing the average latency of the PUs, no buffer is set for the PU packets.

Because a PU packet can access the spectrum at any slot boundary, so at the beginning instant of each slot, the SUs will sense the PUs' activity and then send the sensing results to the central controller. By synthesizing these sensing results, the central controller will allocate one of the idle spectrums for the SU packet queueing

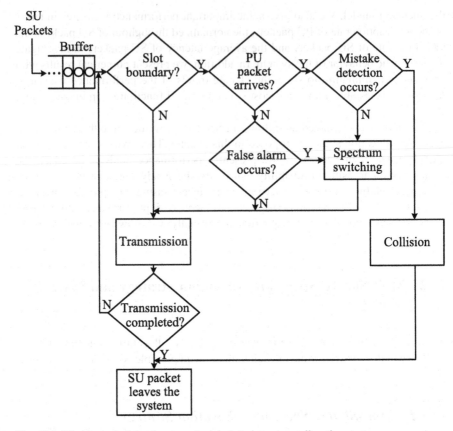

Fig. 12.1 Working principle of proposed mini-slotted spectrum allocation strategy

at the head of the SU buffer, then this SU packet will be transmitted on this spectrum. In this chapter, we call the spectrum on which the SU packets are being transmitted as a tagged spectrum.

The working principle of the mini-slotted spectrum allocation strategy proposed in this chapter is presented in Fig. 12.1.

As can be seen in Fig. 12.1, throughout the transmission procedure of an SU packet, the SU will check whether the current mini slot is a slot boundary or not. If the current mini slot is not a slot boundary, the SU will ignore the PUs' activity and proceed the transmission. Otherwise, the SU will perform the spectrum sensing to find out whether the PU is active or not.

We note that the sensing errors are inevitable in practice. When a PU packet arrives at the system, but the SU is not aware this arrival, namely, a mistake detection occurs, the arriving PU packet and the SU packet will occupy the spectrum simultaneously, it means that the arriving PU packet will be collided with the SU packet. After a mini slot, both of the collided packets will be dropped out of the system, namely, the collided packets are disrupted. The remaining SU packets in the buffer will start the transmission from the next mini slot.

When SUs sense the channel via energy detection, two types of sensing errors, in terms of mistake detections and false alarms, can possibly occur. Let t_s be the sensing time, f_s be the sensing frequency, ξ be the SNR and ϖ be the variance of noise. Let p_{md} be the mistake detection ratio and p_{fa} be the false alarm ratio. In discrete-time field, the mistake detection ratio p_{md} and the false alarm ratio p_{fa} are also called the mistake detection probability and the false alarm probability, respectively. p_{md} and p_{fa} are given as follows:

$$
\begin{cases}
p_{md} = 1 - Q\left(\left(\dfrac{\tau}{\varpi^2} - \xi - 1\right)\sqrt{\dfrac{t_s f_s}{2\xi + 1}}\right) \\
p_{fa} = Q\left(\left(\dfrac{\tau}{\varpi^2} - 1\right)\sqrt{t_s f_s}\right)
\end{cases}
\tag{12.1}
$$

where τ is the energy threshold used in channel sensing, and $Q(v)$ is the tail probability of the standard normal distribution given by

$$
Q(v) = \frac{1}{\sqrt{2\pi}} \int_v^\infty \exp\left(-\frac{t^2}{2}\right) dt.
$$

When a PU packet arrives at the system, and the SU being transmitted senses this arrival, the transmission of the SU packet will be preempted. Based on the spectrum condition in the CRN, the central controller will allocate one of idle spectrums for the preempted SU packet. We suppose that there are always available spectrums for use. The dispatch process that the central controller allocates idle spectrums is not considered in this chapter. The preempted SU packet together with other SU packets in the buffer will be switched to the allocated available spectrum, and the preempted SU packet will queue at the head of the buffer. After the spectrum switching, the transmission of the preempted SU packet will be continued.

By contrary, when a PU packet does not arrive at the system, but the SU misjudges that there is a PU packet arrival, namely, a false alarm occurs, the transmission of the SU packet will be preempted too. In this case all the SU packets in the system will perform a spectrum switching, then the transmission of the preempted SU packet will be continued on the new spectrum. When a PU packet does not arrive at the system, and the SU being transmitted senses the spectrum condition correctly, the transmission of the SU packet will be continued on the current spectrum.

12.2.2 System Model

In this network system, there are two types of the data packets. One is the PU packets having a high priority without a buffer, and the other is the SU packets having an unlimited buffer. A tagged spectrum is used to transmit these two types' packets. Considering the imperfect spectrum sensing results of SUs, the mini-slotted spectrum allocation strategy proposed in this chapter can be described as a heterogeneous discrete-time queueing model with possible collisions.

We suppose that there are m mini slots in a slot, the mini slot boundaries in a slot are numbered as n ($n = 1, 2, 3, \ldots, m$), while the slot boundaries are numbered as N ($N = 1, m + 1, 2m + 1, \ldots$). For the mathematical clarify, an SU packet is supposed to arrive at the beginning instant of a mini slot, and depart at the end instant of a mini slot; a PU packet is supposed to arrive at the beginning instant of a slot, and departure at the end instant of a slot. In other words, an Early Arrival System (EAS) with a heterogeneous time structure is considered.

Considering the digital nature of modern communications, the following assumptions are made in order to develop our analytical model.

(1) The arrival of an SU packet is assumed to follow Bernoulli process with probability λ ($0 < \lambda < 1$, $\bar{\lambda} = 1 - \lambda$), that is to say, in a mini slot, an SU packet arrives at the system with probability λ, no SU packet arrives at the system with probability $\bar{\lambda}$. We call probability λ the arrival rate of SU packets.

(2) The transmission time of an SU packet is assumed to follow geometric distribution with probability μ ($0 < \mu < 1$, $\bar{\mu} = 1 - \mu$), it means that in a mini slot, the transmission of an SU packet will be completed successfully with probability μ, and continued with probability $\bar{\mu}$.

   (i) The arrival of a PU packet is assumed to follow Bernoulli process with probability α ($0 < \alpha < 1$, $\bar{\alpha} = 1 - \alpha$), that is to say, in a slot, a PU packet arrives at the system with probability α, no PU packet arrives at the system with probability $\bar{\alpha}$. We call probability α the arrival rate of PU packets.

   (ii) On the other hand, the arrivals and departures for these two types of packets are supposed to be independent. Moreover, we assume that the switching procedure is neglected.

We define the total number of SU packets in the system as the system level, the mini slot number in a slot as the system phase, and the tagged spectrum condition as the system stage. Let X_n denote the system level at the instant n^+, Y_n denote the system phase at the instant n^+, and Z_n denote the system stage at the instant n^+. $Z_n = 0$ represents the tagged spectrum is idle or being used by an SU packet, namely, the spectrum is in a "normal" condition. $Z_n = 1$ represents that a PU packet and an SU packet appear at the tagged spectrum simultaneously, namely, there is a collision on the spectrum and the spectrum is in a "disorder" condition. $\{(X_n, Y_n, Z_n), n \geq 0\}$ constitutes a three-dimensional Markov chain. The state space of this Markov chain is given as follows:

$$\mathbf{\Omega} = \{(i, j, l) : i \geq 0, 1 \leq j \leq m, l = 0, 1\}. \tag{12.2}$$

12.3 Performance Analysis

In this section, we present a performance analysis of the system model, through the analysis of the transition probability matrix and the steady-state distribution.

12.3.1 Transition Probability Matrix

Since the SUs sense the spectrum at the boundary of a slot, the imperfect sensing results occur only at the beginning instant of a slot. We note that for the system level $X_n > 0$ and the system phase $Y_n = 1$, there are two stages: $Z_n = 0$ and $Z_n = 1$; otherwise, there is only one stage: $Z_n = 0$. We also note that whether or not there is an SU packet arrival in a mini slot, if the current system phase is $1 \leq Y_n \leq m - 1$, the next system phase will be increased to $Y_{n+1} = Y_n + 1$; if the current system phase is $Y_n = m$, the next system phase will be $Y_{n+1} = 1$.

Let P be the one-step transition probability matrix of the $\{(X_n, Y_n, Z_n),\ n \geq 0\}$, $P(u, v)$ be the one-step transition probability sub-matrix from the system level u to v. $P(u, v)$ is all on the order of $(m+1) \times (m+1)$ and can be discussed as follows.

(1) If $u = 0$ and $v = 0$, it means that there is no SU packet arrival at the system with probability $\bar{\lambda}$ during the one-step transition. Therefore, the one-step transition probability sub-matrix $P(0, 0)$ is given as follows:

$$P(0, 0) = \begin{pmatrix} 0 & 0 & & & & \\ & 0 & \bar{\lambda} & & & \\ & & 0 & \bar{\lambda} & & \\ & & & \ddots & \ddots & \\ & & & & 0 & \bar{\lambda} \\ 0 & \bar{\lambda} & 0 & \cdots & 0 & 0 \end{pmatrix}. \tag{12.3}$$

(2) If $u = 0$ and $v = 1$, it means that there is an SU packet arrival at the system with probability λ during the one-step transition. Firstly, we discuss the transition for the system phase changing to $Y_{n+1} = 1$ from $Y_n = m$. If a PU packet arrives at the beginning instant of a slot with probability α, and a mistake detection occurs with probability p_{md}, the system stage $Z_n = 0$ will be changed to the system stage $Z_{n+1} = 1$ with probability αp_{md}. There are two cases will make the system stage being fixed at $Z_{n+1} = 0$: no PU packet arrives at the beginning instant of a slot with probability $\bar{\alpha}$; a PU packet arrives at the beginning instant of a slot with probability α, but no mistake detection occurs with probability $\bar{p}_{\mathrm{md}}$ ($\bar{p}_{\mathrm{md}} = 1 - p_{\mathrm{md}}$), so the system stage will be fixed at $Z_{n+1} = 0$ with probability $\bar{\alpha} + \alpha \bar{p}_{\mathrm{md}}$.

Therefore, the one-step transition probability sub-matrix $P(0, 1)$ is given as follows:

$$P(0, 1) = \begin{pmatrix} 0 & 0 & & & & \\ & 0 & \lambda & & & \\ & & 0 & \lambda & & \\ & & & \ddots & \ddots & \\ & & & & 0 & \lambda \\ \lambda\alpha p_{md} & \lambda(\alpha\bar{p}_{md} + \bar{\alpha}) & 0 & \cdots & 0 & 0 \end{pmatrix}. \tag{12.4}$$

(3) If $u = 1$ and $v = 0$, there is no SU packet arrival at the system with probability $\bar{\lambda}$. For the system stage $Z_n = 1$, the SU packet being collided with a PU packet has to leave the system during the one-step transition. For the system stage $Z_n = 0$, the transmission of the SU packet occupying the spectrum is completed successfully with probability μ during one-step transition.

Therefore, the one-step transition probability sub-matrix $P(1, 0)$ is given by

$$P(1, 0) = \begin{pmatrix} 0 & 0 & \bar{\lambda} & & & \\ & 0 & \bar{\lambda}\mu & & & \\ & & 0 & \bar{\lambda}\mu & & \\ & & & \ddots & \ddots & \\ & & & & 0 & \bar{\lambda}\mu \\ 0 & \bar{\lambda}\mu & 0 & \cdots & 0 & 0 \end{pmatrix}. \tag{12.5}$$

(4) If $u = v \geq 1$, for the system stage $Z_n = 1$, the collided SU packet has to leave the system and an SU packet arrives at the system with probability λ during the one-step transition. For the system stage $Z_n = 0$, there are two types of cases to be addressed: an SU packet arrives at the system with probability λ and the transmission of the SU packet occupying the spectrum is completed successfully with probability μ; no SU packet arrives at the system with probability $\bar{\lambda}$ and no SU packet departs the system with probability $\bar{\mu}$. Similar to Item (2), when the system phase changes to $Y_{n+1} = 1$ from $Y_n = m$, the stage $Z_n = 0$ will change to $Z_{n+1} = 1$ with probability αp_{md}, the stage $Z_n = 0$ will be fixed at $Z_{n+1} = 0$ with probability $\alpha\bar{p}_{md} + \bar{\alpha}$. By A_1 we denote the one-step transition probability sub-matrix $P(u, u)$, A_1 is then given as follows:

$$A_1 = \begin{pmatrix} 0 & 0 & \lambda & & & \\ & 0 & \lambda\mu + \bar{\lambda}\bar{\mu} & & & \\ & & 0 & \lambda\mu + \bar{\lambda}\bar{\mu} & & \\ & & & \ddots & \ddots & \\ & & & & 0 & \lambda\mu + \bar{\lambda}\bar{\mu} \\ (\lambda\mu + \bar{\lambda}\bar{\mu})\alpha p_{md} & (\lambda\mu + \bar{\lambda}\bar{\mu})(\alpha\bar{p}_{md} + \bar{\alpha}) & 0 & \cdots & 0 & 0 \end{pmatrix}. \tag{12.6}$$

(5) If u ($u \geq 1$) and $v = u+1$, there is at most one SU packet arrival in a mini slot. When the i system stage changes to $Z_{n+1} = 0$ from $Z_n = 1$, it is impossible for the system level transferring from u to $(u+1)$. For the system stage $Z_n = 0$, there is an SU packet arrival at the system with probability λ and no SU packet departure from the system with probability $\bar{\mu}$ during the one-step transition. Similar to Item (2), when the system phase changes to $Y_{n+1} = 1$ from $Y_n = m$, the system stage will change to $Z_{n+1} = 1$ from $Z_n = 0$ with probability αp_{md}, while the system stage will be fixed at 0 with probability $\alpha \bar{p}_{md} + \bar{\alpha}$. Let A_0 be the one-step transition probability sub-matrix $P(u, u+1)$. A_0 is given as follows:

$$A_0 = \begin{pmatrix} 0 & 0 & & & & \\ & 0 & \lambda\bar{\mu} & & & \\ & & 0 & \lambda\bar{\mu} & & \\ & & & \ddots & \ddots & \\ & & & & 0 & \lambda\bar{\mu} \\ \lambda\bar{\mu}\alpha p_{md} & \lambda\bar{\mu}(\alpha\bar{p}_{md} + \bar{\alpha}) & 0 & \cdots & 0 & 0 \end{pmatrix}. \tag{12.7}$$

(6) For the case of u ($u \geq 2$) and $v = u - 1$, we denote the one-step transition probability sub-matrix $P(u, u-1)$ as A_2. Similar to Items (2) and (3), A_2 is obtained as follows:

$$A_2 = \begin{pmatrix} 0 & 0 & \bar{\lambda} & & & \\ & 0 & \bar{\lambda}\mu & & & \\ & & 0 & \bar{\lambda}\mu & & \\ & & & \ddots & \ddots & \\ & & & & 0 & \bar{\lambda}\mu \\ \bar{\lambda}\mu\alpha p_{md} & \bar{\lambda}\mu(\alpha\bar{p}_{md} + \bar{\alpha}) & 0 & \cdots & 0 & 0 \end{pmatrix}. \tag{12.8}$$

Combining Eqs. (12.3)–(12.8), the one-step transition probability P of the system is given by

$$P = \begin{pmatrix} P(0,0) & P(0,1) & & & \\ P(1,0) & A_1 & A_0 & & \\ & A_2 & A_1 & A_0 & \\ & & A_2 & A_1 & A_0 \\ & & & \ddots & \ddots & \ddots \end{pmatrix}. \tag{12.9}$$

From the structure of the one-step transition probability P, we know that the stochastic process of $\{(X_n, Y_n, Z_n), n \geq 0\}$ is a Quasi Birth-Death (QBD) process.

12.3.2 Steady-State Distribution

Let $\pi_{i,j,l}$ be the steady-state distribution of the three-dimensional Markov chain $\{(X_n, Y_n, Z_n),\ n \geq 0\}$ being at the state (i, j, l), where $(i, j, l) \in \Omega$. $\pi_{i,j,l}$ can be given as follows:

$$\pi_{i,j,l} = \lim_{n \to \infty} \Pr\{X_n = i, Y_n = j, Z_n = l\}, \quad i \geq 0,\ 1 \leq j \leq m,\ l = 0, 1.$$

(12.10)

Let π_i be the steady-state probability vector of the system being at level i. π_i can be given as follows:

$$\pi_i = (\pi_{i,1,0}, \pi_{i,1,1}, \pi_{i,2,0}, \ldots, \pi_{i,m,1}).$$

(12.11)

Then, the steady-state distribution Π of the system can be given as follows:

$$\Pi = (\pi_0, \pi_1, \pi_2, \ldots).$$

(12.12)

Let matrix R be the minimum nonnegative solution of the following matrix quadratic equation:

$$R^2 A_2 + R A_1 + A_0 = R.$$

(12.13)

From Eq. (12.13), we have

$$R = (R^2 A_2 + A_0)(I - A_1)^{-1}$$

(12.14)

where I denotes an $(m + 1) \times (m + 1)$ unit matrix.

The reasonable approximation of R is obtained iteratively. When the spectral radius of the matrix R is less than 1, the system will achieve the stationary state.

Combining the system equilibrium equation and the normalization condition, we have

$$\begin{cases} (\pi_0, \pi_1) B[R] = (\pi_0, \pi_1) \\ \pi_0 e + \pi_1 (I - R)^{-1} e = 1 \end{cases}$$

(12.15)

where e is a column vector with $m+1$ elements and all elements of the vector equal to 1. Matrix $B[R]$ has the structure as follows:

$$B[R] = \begin{pmatrix} P(0, 0) & P(0, 1) \\ P(1, 0) & A_1 + R A_2 \end{pmatrix}.$$

(12.16)

By using the matrix-geometric solution method, we have

$$\boldsymbol{\pi}_i = \boldsymbol{\pi}_1 \boldsymbol{R}^{i-1}, \quad i \geq 1. \tag{12.17}$$

Combining Eqs. (12.15)–(12.17), we can get the steady-state distribution $\boldsymbol{\Pi}$ with numerical results.

12.4 Performance Measures and Numerical Results

In this section, we first derive performance measures of the system in terms of the interruption rate of PU packets, the throughput of SU packets, the switching rate of SU packets and the average latency of SU packets, respectively. Then, we present numerical results to evaluate the performance of the system using the mini-slotted spectrum allocation strategy proposed in this chapter.

12.4.1 Performance Measures

The interruption rate of PU packets is defined as the number of PU packets being collided with SU packets per mini slot. When a PU packet arrives at the slot boundary, but the SU packet being transmitted misjudges this arrival, the tagged spectrum will be in the "disorder" condition, namely, the arriving PU packet is disrupted. Therefore, we give the interruption rate β_p of PU packets as follows:

$$\beta_p = \pi_{i,1,1}. \tag{12.18}$$

The throughput θ of SU packets is defined as the number of SU packets transmitted successfully per mini slot. An SU packet will be transmitted successfully except for being disrupted due to mistake detections. Therefore, we give the throughput θ of SU packets as follows:

$$\theta = \lambda - \beta_p. \tag{12.19}$$

The switching rate ζ_s of SU packets is defined as the average number of spectrum switching per mini slot. When the transmission of the SU packet occupying the spectrum is preempted, the SU will switch to another spectrum. Therefore, we give the switching rate ζ_s of SU packets as follows:

$$\zeta_s = \frac{\alpha \bar{p}_{\mathrm{md}} + \bar{\alpha} p_{\mathrm{fa}}}{\alpha \bar{p}_{\mathrm{md}} + \bar{\alpha} p_{\mathrm{fa}} + \bar{\alpha} \bar{p}_{\mathrm{fa}}} \sum_{i=1}^{\infty} \pi_{i,1,0} \tag{12.20}$$

where $\bar{p}_{fa} = 1 - p_{fa}$ and p_{fa} is the false alarm ratio given by Eq. (12.1), $\bar{p}_{md} = 1 - p_{md}$ and p_{md} is the mistake detection ratios given by Eq. (12.1), too.

The latency Y_s of an SU packet is defined as the duration in mini slots from the instant that an SU packet joins the system to the instant that SU packet leaves the system. Based on the analysis presented in Sect. 12.3, we can obtain the average latency $E[Y_s]$ of SU packets as follows:

$$E[Y_s] = \sum_{i=0}^{\infty} \sum_{j=1}^{m} \frac{i \times (\pi_{i,j,0} + \pi_{i,j,1})}{\lambda}. \tag{12.21}$$

12.4.2   Numerical Results

In order to evaluate quantitatively the influence of the system parameters on the system performance for the mini-slotted spectrum allocation strategy, we present numerical results with analysis and simulation. The system parameters are fixed as follows: $\lambda = 0.25$, $\mu = 0.4$, $\alpha = 0.6$ as an example for all the numerical results. From numerical results shown in the following figures, good agreements between the analysis results and the simulation results are observed.

Figure 12.2 illustrates the influence of the slot size m on the interruption rate β_p of PU packets with different mistake detection ratios p_{md}.

As can be seen in Fig. 12.2, for the same mistake detection ratio p_{md}, the interruption rate β_p of PU packets decreases with the enlargement of the slot size m. The reason is that the larger the slot size is, the less likely is that the transmission of the SU packet occupying the spectrum crosses the slot boundary, the interruption rate of PU packets will be lower. For the same slot size m, the interruption rate β_p of PU packets increases when the mistake detection ratio p_{md} increases. This

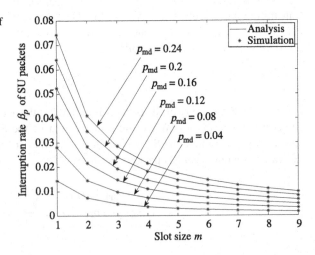

Fig. 12.2 Interruption rate of PU packets versus slot size

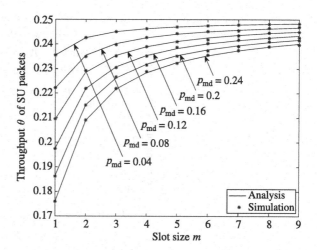

Fig. 12.3 Throughput of SU packets versus slot size

is because that the greater the mistake detection ratio is, the more likely is that the arriving PU packet and the SU packet will occupy the spectrum simultaneously, namely, the higher the possibility is that the arriving PU packet will be disrupted, so the interruption rate of PU packets will be higher.

Taking the false alarm ratio $p_{fa} = 0.08$ as an example, we show how the throughput θ of the SU packets changes with respect to the slot size m for different mistake detection ratios p_{md} in Fig. 12.3.

As can be seen in Fig. 12.3, for the same mistake detection ratio p_{md}, the throughput θ of SU packets increases with the enlargement of the slot size m. The reason is that as the slot size increases, there are more mini slots, during which the transmission of SU packets will not be disrupted, in a slot, then the higher the possibility is that an SU packet is transmitted successfully, so the throughput θ of SU packets will be greater. For the same slot size m, the throughput θ of SU packets increases when the mistake detection ratio p_{md} increases. This is because that the bigger the mistake detection ratio is, the more likely is that a PU packet and an SU packet will be collided, namely, more SU packets will be disrupted, this will certainly result in a lower throughput of SU packets.

Figure 12.4 describes the influence of the slot size m on the switching rate ζ_s of SU packets with different mistake detection ratios p_{md} and false alarm ratio p_{fa}.

From Fig. 12.4, we observe that for the same mistake detection ratio p_{md} and the same false alarm ratio p_{fa}, the switching rate ζ_s of SU packets decreases as the slot size m increases. The reason is that for a larger slot size, there are less potential preemption instants within a certain time period, which will reduce the switching rate of SU packets frequency, namely, the switching rate of SU packets will be lower. For the same slot size m and the same mistake detection ratio p_{md}, the switching rate ζ_s of SU packets increases as the false alarm ratio p_{fa} increases. This is because that the greater the false alarm ratio is, the more likely is that the SU will switch to another spectrum due to false alarms, so the switching rate of SU packets will be higher. For the same slot size m and the same false alarm ratio p_{fa},

Fig. 12.4 Switching rate of SU packets versus slot size

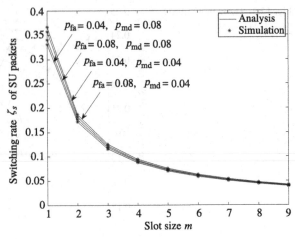

Fig. 12.5 Average latency of SU packets versus slot size

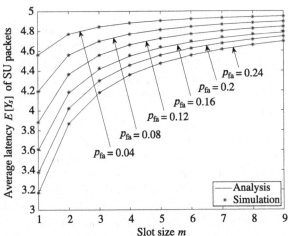

the switching rate ζ_s of SU packets increases when the mistake detection ratio p_{md} decreases. The reason is that the smaller the mistake detection ratio is, the more likely is that an SU packet will be preempted due to mistake detections, then the SU will more likely switch to another spectrum, so the switching rate of SU packets will be higher.

We examine the influence of the slot size m on the average latency $E[Y_s]$ of SU packets for different mistake detection ratios p_{md} in Fig. 12.5.

From Fig. 12.5, we find that for the same mistake detection ratio p_{md}, the average latency $E[Y_s]$ of SU packets increases as the slot size m increases. The reason is that the larger the slot size is, the more likely is that an SU packet will be transmitted successfully without interruption. The more the SU packets are transmitted successfully, the longer the average transmission time and the higher the average latency will be, this will therefore result in a higher average latency of SU packets. For the same slot size m, the average latency $E[Y_s]$ of SU packets decreases as the mistake detection ratio p_{md} increases. This is because that the greater the mistake detection ratio is, the more possible is that the spectrum being at the disorder state. Since the actual transmission time of the disrupted SU packet is shorter, the average latency of SU packets will decrease.

Summarizing the numerical results shown in Figs. 12.2, 12.3, 12.4, 12.5, we find that from the view point of the throughput of SU packets and the switching rate of SU packets, the proposal mini-slotted spectrum allocation strategy performs better than the conventional spectrum allocation strategy with a homogeneous structure. On the other hand, from the perspective of the average latency of SU packets, we find that the system performance of the mini-slotted spectrum allocation strategy is degraded a bit. That is to say, the mini-slotted spectrum allocation strategy proposed in this chapter is more appropriated to the delay tolerance application networks. Moreover, we conclude that there is a trade-off to be considered when setting the slot size in the mini-slotted spectrum allocation strategy.

12.5 Performance Optimization

To do performance optimization of the system, we construct a system profit function $F(m)$ to balance different performance measures presented in Sect. 12.4.1 as follows:

$$F(m) = f_1\theta - f_2E[Y_s] - f_3\zeta_s \tag{12.22}$$

where f_1, f_2 and f_3 are supposed to be the reward for transmitting an SU packet successfully, the cost for a mini slot due to the latency of an SU packet and the expense for one spectrum switching, respectively.

Setting $f_1 = 60$, $f_2 = 2.75$ and $f_3 = 0.3$ as an example, we plot how the system profit function $F(m)$ changes with respect to the slot size m for different arrival rates α of PU packets in Fig. 12.6.

It can be seen in Fig. 12.6, all values of the system profit function $F(m)$ experience two stages. In the first stage, the system profit function $F(m)$ increases with the enlargement of the slot size m. During this stage, the main influence factors are the switching rate of SU packets and the throughput of SU packets, the larger the slot size is, the less the switching rate of SU packets is, and the higher the throughput of SU packets is, so the greater the system profit function will be. In the second stage, the system profit function $F(m)$ decreases as the slot size m increases. During this period, the average latency of SU packets is the dominant measure, the larger the slot size is, the higher the average latency of SU packets is, so the lower the system profit function will be.

Fig. 12.6 System profit
function versus slot size

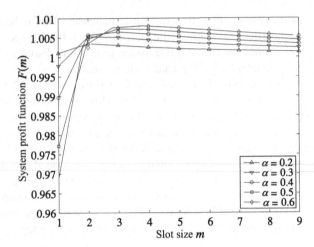

Slot size m

Table 12.1 Optimum slot size in proposed strategy

Arrival rates α of PU packets	Optimal slot sizes m^*	Maximum profits $F(m^*)$
0.2	2	1.0035
0.3	2	1.0053
0.4	3	1.0059
0.5	3	1.0073
0.6	4	1.0079

Conclusively, there is a maximum system profit function when the slot size is set to the optimal value m^*. That is

$$m^* = \underset{m \geq 1}{\mathrm{argmax}}\{F(m)\}$$

The optimal values m^* of slot size and the maximum values $F(m^*)$ of the system profit functions with different arrival rates α of PU packets are illustrated in Table 12.1.

12.6 Conclusion

In this chapter, we proposed a mini-slotted spectrum allocation strategy in CRNs, where the transmission of an SU packet is based on the mini-slot, the transmission of a PU packet is based on the slot. Based on working principle of the proposed spectrum allocation strategy, we built a discrete-time queueing model with possible collisions. We constructed a three-dimensional Markov chain and derived the formulas for some important performance measures. From the numerical results, we observed that the proposed mini-slotted spectrum allocation strategy can effectively improve the throughput of SU packets and decrease the spectrum switching frequency. Finally, considering the trade-off between different performance measures, we optimized the slot size by maximizing the system profit function.

Chapter 13
Channel Reservation Strategy with Imperfect Sensing Results

Channel reservation strategy in Cognitive Radio Networks (CRNs) is an effective technology for conserving communication resources. Considering the imperfect sensing of Secondary User (SU) packets, and the possible patience of SU packets, in this chapter, we propose a channel reservation strategy in a CRN. Aligned with the proposed channel reservation strategy, we establish a Continuous-Time Markov Chain (CTMC) model to capture the stochastic behavior of the two types of user packets, the Primary User (PU) packets and the SU packets. Then, in order to obtain the steady-state distribution of the system model, we present a new algorithm for solving the quasi-birth-and-death (QBD) process. Moreover, we derive numerical results for the system in terms of the throughput of SU packets, the average latency of SU packets, the switching rate of SU packets and the channel utilization. Finally, we present numerical results to evaluate the performance of the system using the proposed channel reservation strategy.

13.1 Introduction

With the rapid development of WCNs, the demand for spectrum resource grows daily. However, the current static spectrum resource allocation policy results in low spectrum utilization [Sala17a]. CRNs are proposed as a solution to enhance the spectrum utilization [Abed17, Xu17]. In CRNs, SUs are allowed to utilize the spectrum white spaces when the spectrum channels are unoccupied by PUs [Sala17b]. This means that the spectrum allocation can be realized dynamically in CRNs.

In dynamic spectrum allocation, many spectrum strategies have been proposed. In [Zhao15a], the authors proposed an adjustable channel bonding strategy, in which the number of channels used by an SU is variable. In [Zapp13], aiming to maximize the throughput of SU packets, the authors presented a resource discovery algorithm

© Springer Nature Singapore Pte Ltd. 2021
S. Jin, W. Yue, *Resource Management and Performance Analysis of Wireless Communication Networks*, https://doi.org/10.1007/978-981-15-7756-7_13

for channel aggregation strategy. In [Chak15], considering the channel reservation strategy, the authors presented numerical analysis for the optimal number of reserved channel.

These studies all have a common assumption that sensing results of SU packets were assumed to be perfect. In fact, spectrum sensing results of SU packets are usually imperfect in practice. Mistake detections and false alarms always occur in the processing of the dynamic spectrum access. In recent years, research into the imperfect sensing of SU packets in CRNs has proliferated. In [Muth13], considering one PU and multiple SUs in a distributed CRN, the authors investigated the influence of the mistake detections and false alarms on the throughput of SU packets. In [Rehm16], considering the face of imperfect sensing, the authors analyzed the delay and throughput of users in a CRN with a Go-Back-N hybrid automatic repeat request protocol. Other research has also shown that the sensing results of SU packets will affect the system performance to a variable extent [Behe15, Jin16c, Liu19, Wang13b, Xie12].

In addition, we note that QBD processes are always employed to evaluate the system performance in a CRN. However, the problem of solving the QBD process is often attributed to solve the associated linear equations. As known, there are two types of basic solutions, namely, direct methods and iterative methods. Direct methods are more suitable for solving systems with either unstructured matrices or a dense matrix, while iterative methods are appropriate for solving large sparse linear systems. The QBD process is perfectly applicable to large sparse linear systems, so a Successive Over Relaxation (SOR) [Hu08] method is one of the most effective iterative methods for deriving a solution. The SOR method has the advantages of easy programming, less computation and an unchanged coefficient matrix in the calculation process. However, the value of the relaxation factor of the SOR method substantially affects the convergence speed and precision.

In this chapter, with the imperfect sensing of SU packets and the degree of patience of the SU packets, we propose an imperfect sensing-based channel reservation strategy in CRNs, called "channel reservation strategy". Then, we present a two-dimensional CTMC model to capture the stochastic behavior of the two types of user packets, the PU packets and the SU packets. In order to obtain numerical solutions for the QBD process, we present a new algorithm for solving the QBD process that effectively fuses the Teaching-Learning-Based Optimization (TLBO) algorithm [Rao13] and the SOR method, namely TLBO-SOR algorithm. And then, we evaluate the system performance using numerical results.

The chapter is organized as follows. In Sect. 13.2, we describe the channel reservation strategy proposed in this chapter. Then, we present the system model in detail. In Sect. 13.3, we present a performance analysis by establishing a one-step transition rate matrix of the two-dimensional CTMC. We also present a TLBO-SOR algorithm to give the steady-state distribution of the QBD process. In Sect. 13.4, we derive performance measures and present numerical results to investigate the system performance. Finally, we draw our conclusions in Sect. 13.5.

13.2 Channel Reservation Strategy and System Model

In this section, we propose a channel reservation strategy in CRNs based on imperfect sensing. Then, we establish a CTMC model to capture the stochastic behavior of the two types of user packets, the PU packets and the SU packets, in CRNs.

13.2.1 Channel Reservation Strategy

We consider a centralized CRN consisting of M licensed channels and a central controller. The central controller allocates available channels to the users' packets. In order to enhance the throughput of SU packets, we set a buffer with a large capacity for SU packets, namely, all the newly arriving SU packets can queue at the end of the buffer waiting for future transmission. With the purpose of properly controlling the interference between the PU packets and the SU packets, we reserve N ($N \leq M$) licensed channels for SU packets. We note that if there are a large number of SU packets aggregated in the buffer, the QoS of SU packets will be undermined. Therefore, in order to make full use of the reserved N licensed channels and enhance the QoS of SU packets, we set an admission threshold H. If the number of SU packets aggregated in the buffer is greater than H, all the M licensed channels can be used opportunistically by SU packets, otherwise only N licensed channels can be used opportunistically by SU packets.

Considering that the transmission of packets is always influenced by the channel energy, we adopt an energy detection method to control the mistake detection ratio p_{md} and false alarm ratio p_{fa}. By using this method, the mistake detection ratio p_{md} and the false alarm ratio p_{fa} are adjusted by the energy detection threshold τ given as follows:

$$
\begin{cases}
p_{md} = 1 - Q_0\left(\left(\dfrac{\tau}{s_0^2} - \xi - 1\right)\sqrt{\dfrac{N_0}{2\xi + 1}}\right) \\[4mm]
p_{fa} = Q_0\left(\dfrac{\tau\sqrt{N_0}}{s_0^2}\right)
\end{cases}
\tag{13.1}
$$

where $Q_0(x) = 1/\sqrt{2\pi}\int_x^\infty e^{-x^2/2}dx$, N_0 is the number of times that a channel is detected, ξ is the Signal-to-Noise Ratio (SNR), and s_0 is the noise variance.

With this model, if a newly arriving PU packet attempts to preempt the channel being occupied by an SU packet, but the SU packet detects that the channel energy is less than the energy threshold τ, then a mistake detection occurs with the probability p_{md} of the SU packet being transmitted on this channel. Conversely, if there are no arriving PU packet attempts to preempt the channel being occupied by an SU packet, but the SU packet detects that the channel energy is greater than the energy threshold τ, a false alarm with the probability p_{fa} occurs.

We assume that if a mistake detection occurs for an SU packet being transmitted on one channel, this SU packet and the newly arriving packet will appear on the same channel. This induces a collision between two packets. In this case, the transmission of this SU packet is non-normally interrupted, and then the SU packet being transmitted on this channel and the newly arriving PU packet are dropped out of the system; if a mistake detection does not occur for an SU packet being transmitted on one channel, then this SU packet will return the channel to the newly arriving PU packet, and is normally interrupted.

Due to the cognitive ability of SU packets in CRNs, retrial feedback of normally interrupted SU packets is also considered. Considering that the degree of patience of SU packets usually and significantly depends on the system traffic load, we assume that the normally interrupted SU packets will go back to the buffer or depart the system.

In addition, we also assume that SU packets and PU packets will be transmitted under a First-Come First-Served (FCFS) discipline.

With the channel reservation strategy proposed in this chapter, the transmission processes of SU packets and PU packets are shown in Fig. 13.1. We call both of the SU packets and the PU packets the user packets.

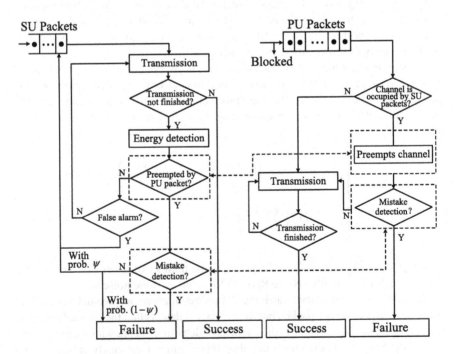

Fig. 13.1 Transmission process of user packets in system

In Fig. 13.1, the transmission process of PU packets shows that:

(1) If all M channels are being occupied by the other PU packets, the central controller will not allocate any a channel to a newly arriving PU packet, and the new PU packet will be blocked.

(2) If there are idle channels, the central controller will randomly allocate one idle channel to a newly arriving PU packet until its transmission finishes.

(3) If there are no idle channels, but there is at least one channel which is being occupied by an SU packet, the central controller will randomly allocate one channel which is being occupied by an SU packet to any newly arriving PU packet. In this case, if this SU packet is mistakenly detected, the newly arriving PU packet will be dropped out of the system. Otherwise, the newly arriving PU packet preempts this channel successfully until its transmission finishes.

In Fig. 13.1, the transmission process of SU packets shows that:

(1) If there are no idly available channels, the newly arriving SU packet queues at the end of the buffer. Otherwise, the central controller allocates an available channel to the newly arriving SU packet.

(2) If the number of SU packets waiting in the buffer is no more than H, and the number of SU packets being transmitted on the channels is less than the number N of reserved channels, and there is at least one available channel, then the central controller allocates one available channel to the SU packet queueing at the head of the buffer.

(3) If the number of SU packets in the system is greater than H, and there is at least one idle channel, then the central controller allocates one idle channel to the SU packet queueing at the head of the buffer.

(4) When a newly arriving PU packet attempts to preempt the channel being occupied by an SU packet, if a mistake detection occurs for this SU packet, then the preempted SU packet is dropped out of the system. Otherwise, the preempted SU packet returns its current channel to the newly arriving PU packet, and goes back to the head of the buffer with probability $\psi = N/M$ or departs the system with probability $\bar{\psi} = 1 - \psi$. In this chapter, we call probability ψ the retrial feedback probability.

(5) When no arriving PU packet claims the channel being occupied by an SU packet, and if a false alarm occurs for this SU packet, then it goes back to the head of the buffer, and is switched to an available channel by the central controller. Otherwise, the SU packet will be transmitted continuously on its current channel.

The interwind relationship between the transmission processes of SU packets and PU packets mentioned above is indicated by the dotted lines in Fig. 13.1.

13.2.2 System Model

In this system, the PU packets having a finite size's buffer are with preemptive priority and the SU packets having an infinite size's buffer are with low priority, to be transmitted on channels. Considering the continuous-time structure, we assume that the inter-arrival times and transmission times for both the SU packets and the PU packets are i.i.d. random variables following exponential distributions. Specifically, the inter-arrival times for the SU packets and the PU packets follow the exponential distributions with means $1/\lambda_1$ and $1/\lambda_2$, respectively, where λ_1 and λ_2 are the arrival rates of the SU packets and the PU packets, $\lambda_1 > 0$ and $\lambda_2 > 0$. We call λ_1 and λ_2 the arrival rates of the SU packets and the PU packets, respectively. The transmission times of an SU packet and a PU packet on one channel follow the exponential distributions with means $1/\mu_1$ seconds and $1/\mu_2$ seconds, respectively, where $\mu_1 > 0$ and $\mu_2 > 0$, called the service rates of the SU packets and the PU packets.

At the instant t, let $X(t) = i$ ($i = 0, 1, 2, \ldots$) indicate the number of SU packets in the system, and let $Y(t) = j$ ($j = 0, 1, 2, \ldots, M$) indicate the number of PU packets in the system. Then, $\{(X(t), Y(t)), t \geq 0\}$ constitutes a two-dimensional CTMC with the state space Ω given as follows:

$$\Omega = \{(i, j) : i \geq 0, 0 \leq j \leq M\}. \tag{13.2}$$

Let $\pi_{i,j}$ be the probability that the number of SU packets in the system is i and the number of PU packets in the system is j in the steady state. $\pi_{i,j}$ is then defined as follows:

$$\pi_{i,j} = \lim_{t \to \infty} \Pr\{X(t) = i, Y(t) = j\}, \quad i \geq 0, \ 0 \leq j \leq M. \tag{13.3}$$

Let π_i be probability of the system being at level i in the steady state. π_i can be given as follows:

$$\pi_i = (\pi_{i,0}, \pi_{i,1}, \pi_{i,2}, \ldots, \pi_{i,M}), \quad i \geq 0. \tag{13.4}$$

The steady-state distribution Π of the system is given as follows:

$$\Pi = (\pi_0, \pi_1, \pi_2, \ldots). \tag{13.5}$$

13.3 Performance Analysis and TLBO-SOR Algorithm

In this section, we first establish a one-step transition rate matrix of the Markov chain presented in this chapter. Then, we present the TLBO-SOR algorithm to obtain the steady-state distribution of the QBD process.

13.3.1   Performance Analysis

Let Q be a one-step transition rate matrix of the Markov chain $\{(X(t), Y(t)), t \geq 0\}$. The number of SU packets is defined as the system level, and the number of PU packets is defined as the system stage. Let $Q_{u,v}$ $(u, v \geq 0)$ be the one-step transition rate sub-matrix from the system level u to the system level v. Because there is no possibility for more than one SU packet arrival, $Q_{u,v} = 0$ $(|u - v| > 1)$. Therefore, according to the changes of the system levels u, the transition rate matrix Q of the Markov chain can be given as a block-structure form as follows:

$$Q = \begin{pmatrix} Q_{0,0} & Q_{0,1} & & & & \\ Q_{1,0} & Q_{1,1} & Q_{1,2} & & & \\ & Q_{2,1} & Q_{2,2} & Q_{2,3} & & \\ & & \ddots & \ddots & \ddots & \\ & & Q_{K,K-1} & Q_{K,K} & Q_{K,K+1} & \\ & & & Q_{K,K-1} & Q_{K,K} & Q_{K,K+1} \\ & & & & \ddots & \ddots & \ddots \end{pmatrix} \tag{13.6}$$

where $K = H + N + 1$.

The one-step transition rate matrix Q of the Markov chain $\{(X(t), Y(t)), t \geq 0\}$ can be discussed according to different system levels as follows:

(1) System level u changes to level $u + 1$ via a one-step transition. This means that an SU packet arrives at the system. Since the buffer prepared for SU packets is infinite, the one-step transition rate sub-matrixes $Q_{u,u+1}$ $(u \geq 0)$ are written as:

$$Q_{u,u+1} = \begin{pmatrix} \lambda_1 & & & & & \\ & \lambda_1 & & & & \\ & & \lambda_1 & & & \\ & & & \ddots & & \\ & & & & \lambda_1 & \\ & & & & & \lambda_1 \end{pmatrix}. \tag{13.7}$$

(2) System level u is fixed via a one-step transition. This means that there are no SU packet arrivals. We examine three different system circumstances to illustrate the one-step transition rate sub-matrix.

When $u = 0$ and $v = 0$, there are no packets in the system. This means that the PU packets cannot be interfered with SU packets, so a newly arriving PU packet will occupy an idle channel until its transmission finishes. This induces a change in the system stage from j to $j+1$ or $j-1$. In addition, if all M channels are being occupied by the other PU packets, a newly arriving PU packet will be blocked. Therefore, the one-step transition rate sub-matrix $\boldsymbol{Q}_{0,0}$ which is fixed to level 0 via a one-step transition can be given as follows:

$$
\boldsymbol{Q}_{0,0} = \begin{pmatrix}
g_0 & \lambda_2 & & & & \\
\mu_2 & g_1 & \lambda_2 & & & \\
& 2\mu_2 & g_2 & \lambda_2 & & \\
& & \ddots & & \ddots & \ddots \\
& & & (M-1)\mu_2 & g_{M-1} & \lambda_2 \\
& & & & M\mu_2 & g_M
\end{pmatrix}
\tag{13.8}
$$

where

$$
g_j = \begin{cases}
-\lambda_2 - \lambda_1 - j\mu_2, & j = 0, 1, 2, \ldots, M-1 \\
-\lambda_1 - M\mu_2, & j = M.
\end{cases}
$$

When $1 \le u \le N + H$ and $v = u$, there is one or more SU packets in the system, but the SU packets only occupy N channels at most. When the system stage $j = M$, the newly arriving PU packets will be blocked. When the system $0 \le j < M$, if the number of PU packets in the system is greater than $M - \min\{v, N\}$, and there is at least one channel that being occupied by an SU packet, the newly arriving PU packet will randomly preempt one channel that is being occupied by an SU packet. In this case, if there is no mistake detection of the SU packet, the preempted SU packets may return to the system with the probability ψ. This induces a change in the system stage from j to $j+1$ $(j \neq M)$, and the system level u becomes fixed. If a PU packet finishes transmission, the system stage will change from j to $j-1$ $(j \neq 0)$. Therefore, the one-step transition rate sub-matrixes $\boldsymbol{Q}_{u,u}$ which are fixed to level u via a one-step transition are given as follows:

$$
\boldsymbol{Q}_{u,u} = \begin{pmatrix}
z_0 & \lambda_2 & & & & & & & \\
\mu_2 & z_1 & \lambda_2 & & & & & & \\
& 2\mu_2 & z_2 & \lambda_2 & & & & & \\
& & \ddots & \ddots & & \ddots & & & \\
& & & (M-u)\mu_2 & z_{M-u} & \lambda_2\bar{P}_{\mathrm{md}}\psi & & & \\
& & & & (M-u+1)\mu_2 & z_{M-u+1} & \lambda_2\bar{P}_{\mathrm{md}}\psi & & \\
& & & & & \ddots & \ddots & \ddots & \\
& & & & & & (M-1)\mu_2 & z_{M-1} & \lambda_2\bar{P}_{\mathrm{md}}\psi \\
& & & & & & & M\mu_2 & z_M
\end{pmatrix}
\tag{13.9}
$$

where

$$
z_j = \begin{cases}
-\lambda_1 - \lambda_2 - \min\{u, N\}\mu_1 - j\mu_2, & j = 0, 1, 2, \ldots, M - \min\{u, N\} \\
-\lambda_1 - \lambda_2 - j\mu_2 - (M - j)\mu_1, \\
\qquad j = M - \min\{u, N\} + 1, M - \min\{u, N\} + 2, \ldots, M - 1 \\
-\lambda_1 - M\mu_2, \quad j = M.
\end{cases}
$$

When $u > N + H$ and $v = u$, there are more than $N + H$ SU packets in the system, and the SU packets can occupy all M channels opportunistically. When the system stage $j = M$, the newly arriving PU packets will be blocked. If system stage $0 \le j < M$, a newly arriving PU packet will randomly preempt one channel which is being occupied by an SU packet. In this case, if there is no mistake detection of the SU packet, the preempted SU packet may go back to the system with the probability ψ. This induces a change in the system stage from j to $j + 1$ $(j \neq M)$. If a PU packet finishes transmission normally, the system stage changes from j to $j - 1$ $(j \neq 0)$.

Therefore, the one-step transition rate sub-matrixes $\boldsymbol{Q}_{u,u}$ which are fixed to level u via a one-step transition are given as follows:

$$
\boldsymbol{Q}_{u,u} = \begin{pmatrix}
d_0 & \lambda_2 \bar{p}_{\mathrm{md}} \psi \\
\mu_2 & d_1 & \lambda_2 \bar{p}_{\mathrm{md}} \psi \\
& 2\mu_2 & d_2 & \lambda_2 \bar{p}_{\mathrm{md}} \psi \\
& & \ddots & \ddots & \ddots \\
& & & (M - 1)\mu_2 & d_{M-1} & \lambda_2 \bar{p}_{\mathrm{md}} \psi \\
& & & & M\mu_2 & d_M
\end{pmatrix}
\tag{13.10}
$$

where

$$
d_j = \begin{cases}
-\lambda_1 - \lambda_2 - (M - j)\mu_1 - j\mu_2, & j = 0, 1, 2, \ldots, M - 1 \\
-\lambda_1 - M\mu_2, & j = M.
\end{cases}
$$

(3) System level u changes to level $u - 1$ via a one-step transition. This means that an SU packet departs the system. If the transmission of an SU packet successfully finishes, the system stage becomes fixed in one-step transition rate sub-matrixes. When a newly arriving PU packet preempts one channel which is being occupied by an SU packet, if this induces a collision, then the preempted SU packet is non-normally interrupted, and departs the system. In this case, the system stage also becomes fixed in one-step transition rate sub-matrixes. Otherwise, the preempted SU packet is normally interrupted, and departs the system with probability $\bar{\psi}$. Then, the system stage j changes to $j + 1$ in one-step transition rate sub-matrixes.

We consider that only when the number of SU packets in the system is greater than $N + H$, all M licensed channels will be occupied by SU packets opportunistically. Otherwise, SU packets occupy N channels at most. Therefore, we examine two different system circumstances to illustrate one-step transition rate sub-matrixes.

When $u \leq N + H$, the one-step transition rate sub-matrixes $\boldsymbol{Q}_{u,u-1}$ change from level u to $u - 1$ via a one-step transition and are given as follows:

$$\boldsymbol{Q}_{u,u-1} = \begin{pmatrix} h_0 & 0 & & & & & & \\ & h_1 & 0 & & & & & \\ & & \ddots & \ddots & & & & \\ & & & h_{M-u} & \lambda_2 \bar{p}_{\mathrm{md}} \bar{\psi} & & & \\ & & & & h_{M-u+1} & \lambda_2 \bar{p}_{\mathrm{md}} \bar{\psi} & & \\ & & & & & \ddots & \ddots & \\ & & & & & & h_{M-1} & \lambda_2 \bar{p}_{\mathrm{md}} \bar{\psi} \\ & & & & & & & h_M \end{pmatrix} \quad (13.11)$$

where

$$h_j = \begin{cases} \min\{u, N\}\mu_1, & j = 0, 1, 2, \ldots, M - \min\{u, N\} - 1 \\ (M - j)\mu_1 + \lambda_2 p_{\mathrm{md}}, & j = M - \min\{u, N\}, M - \min\{u, N\} + 1, \\ & \qquad\qquad M - \min\{u, N\} + 2, \ldots, M - 1 \\ 0, & j = M. \end{cases}$$

When $u > N + H$, the one-step transition rate sub-matrixes $\boldsymbol{Q}_{u,u-1}$ change from level u to $u - 1$ via a one-step transition and are given as follows:

$$\boldsymbol{Q}_{u,u-1} = \begin{pmatrix} l_0 & \lambda_2 \bar{p}_{\mathrm{md}} \bar{\psi} & & & \\ & l_1 & \lambda_2 \bar{p}_{\mathrm{md}} \bar{\psi} & & \\ & & \ddots & \ddots & \\ & & & l_{M-1} & \lambda_2 \bar{p}_{\mathrm{md}} \bar{\psi} \\ & & & & l_M \end{pmatrix} \quad (13.12)$$

where

$$l_j = \begin{cases} (M - j)\mu_1 + \lambda_2 P_{\mathrm{md}}, & j = 0, 1, 2, \ldots, M - 1 \\ 0, & j = M. \end{cases}$$

From the structure of the one-step transition rate matrix Q, we find that Q is a blocked three-diagonal matrix and the system state transition occurs only in adjacent levels. Therefore, the stochastic process $\{(X(t), Y(t)),\ t \geq 0\}$ is a QBD process.

13.3.2   TLBO-SOR Algorithm

We find that the sub-matrix of the one-step transition rate Q starts to repeat after the level K. In order to employ a matrix-geometric solution method, we construct matrix $B[R_0]$ as follows:

$$B[R_0] = \begin{pmatrix} Q_{0,0} & Q_{0,1} \\ Q_{1,0} & Q_{1,1} & Q_{1,2} \\ & Q_{2,1} & Q_{2,2} & Q_{2,3} \\ & & \ddots & \ddots & \ddots \\ & & & Q_{N+H,N+H-1} & Q_{N+H,N+H} & Q_{N+H,N+H+1} \\ & & & & Q_{K,K-1} & Q_{K,K} + R_0 Q_{K,K-1} \end{pmatrix}$$

(13.13)

where matrix R_0, namely the rate matrix, is the minimum non-negative solution of the matrix equation as follows:

$$R_0^2 Q_{M+1,M} + R_0 Q_{M+1,M+1} + Q_{M+1,M+2} = 0.$$

(13.14)

Then, $\pi_0, \pi_1, \pi_2, \ldots, \pi_{K+1}$ satisfy the following set of linear equations:

$$\begin{cases} (\pi_0, \pi_1, \pi_2, \ldots, \pi_{K+1}) B[R_0] = (0, 0, 0, \ldots, 0, 1) \\ (\pi_0, \pi_1, \pi_2, \ldots, \pi_K) e_1 + \pi_{K+1} (I - R_0)^{-1} e_2 = 1 \end{cases}$$

(13.15)

where e_1 is a column vector with $(K + 1) \times (M + 1)$ elements and e_2 is a column vector with $M + 1$ elements, respectively. All elements of these vectors are equal to 1. And the number of zeros in parentheses above is $(K + 1) \times (M + 1)$.

Letting $X = (\pi_0, \pi_1, \pi_2, \ldots, \pi_{K+1})$, $b = (0, 0, 0, \ldots, 0, 1)$ and

$$A = \begin{pmatrix} B[R_0] & e_1 \\ & (I - R_0)^{-1} e_2 \end{pmatrix},$$

(13.16)

we can rewrite Eq. (13.15) as follows:

$$XA = b.$$

(13.17)

In most of the literature, iterative method has been employed to obtain solutions for Eq. (13.14), Gauss-Seidel iterative method has been used to obtain solutions for

Eq. (13.15). However, using this method may give rise to greater calculation errors when computing the steady-state distribution of the QBD process. This necessitates improving the computational precision for the steady-state distribution of the QBD process.

Letting $f(R) = ||R + (R^2 Q_{M+1,M} + Q_{M+1,M+2}) Q_{M+1,M+1}^{-1}||_2$, then the rate matrix R_0 can be obtained as follows:

$$R_0 = \underset{Sp(R)<1}{\operatorname{argmin}} \{||f(R)||_2\}. \tag{13.18}$$

We note that the TLBO algorithm is a very effective method for solving Eq. (13.18). In addition, the structure of the one-step transition rate matrix indicates that Eq. (13.17) is a large sparse linear system. The SOR method is one of the most effective iterative methods for solving large sparse linear system. Therefore, based on the TLBO algorithm and SOR method, a new hybrid algorithm is proposed for obtaining the steady-state distribution of the QBD process. In this chapter, we name this new algorithm the TLBO-SOR algorithm.

Different from the other nature-inspired algorithms, the TLBO algorithm is a parameter free algorithm and an efficient meta-heuristic optimization method based on the philosophy of teaching and learning. This algorithm has the advantages of being simpler, having fewer parameters, being easy to understand, and having a high degree of precision.

To the best of our knowledge, no research literature exists that examines solving the optimal relaxation factor for SOR based on TLBO algorithm. Therefore, in order to solve the optimal relaxation factor to achieve a good quality solution for the steady-state distribution of the QBD process, we develop a method to determine the optimal relaxation factor of SOR by using a TLBO algorithm, namely the TLBO-SOR algorithm.

Like other nature-inspired algorithms, the TLBO algorithm is a population-based intelligent algorithm, which uses a population of solutions to proceed to the global solutions. It comes from the research into the teaching-learning principle. In this principle, group of learners is considered as the population (R), and every learner is considered as an individual ($R_n, n = 1 : S$), where S is the population size. The learning result of a learner is analogous to his (or her) "fitness" (function value $f(R_n), n = 1 : S$). The teacher ($R_{teacher}$) is considered as the most knowledgeable person in a class who shares his/her knowledge with the students to improve the marks of class. A teacher tries to increase the mean value (R_{mean}) of the class up to his/her level. There are two parts in the TLBO: the "Teacher Phase" and the "Learner Phase". "Teacher Phase" means learning from the teacher and "Learner Phase" means learning through the interaction between learners.

For convenience, we define an indicator function $I(x)$ as follows:

$$I(x) = \begin{cases} 1, & x \geq 0 \\ 0, & x < 0. \end{cases} \tag{13.19}$$

Then, the main steps for the developed TLBO-SOR algorithm to obtain the steady-state distribution of the QBD process are given as follows:

Step 1: Set the population size S_1, the maximum number N_0 of generations and the current iteration $N_1 = 0$ for TLBO method. Initialize learners $\boldsymbol{R}_n$ ($n = 1 : S$).

Step 2: Compute $f(\boldsymbol{R}_n)$ ($n = 1 : S$) as fitness, $\boldsymbol{R}_{\text{mean}} = 1/S \sum_{n=1}^{S} \boldsymbol{R}_n$ as the mean value and $\boldsymbol{R}_{\text{teacher}} = \underset{n \in \{1,2,3,\dots,S\}}{argmin} \{\|f(\boldsymbol{R}_n)\|_2\}$ as a teacher.

    $N_1 = N_1 + 1$
    if $N_1 = N_0$
        go to **Step 6**
    endif

Step 3: Update learners $\boldsymbol{R}_n$ ($n = 1 : S$) to learners $\boldsymbol{D}_n$ ($n = 1 : S$) through teaching phase.

    $T_F = \text{round}$
    for $n = 1 : S$
        $\boldsymbol{C}_n = \boldsymbol{R}_n + \text{rand} \times (\boldsymbol{R}_{\text{teacher}} - T_F \boldsymbol{R}_{\text{mean}})$
        $\boldsymbol{D}_n = I(f(\boldsymbol{C}_n) - f(\boldsymbol{R}_n))\boldsymbol{R}_n + I(f(\boldsymbol{R}_n) - f(\boldsymbol{C}_n))\boldsymbol{C}_n$
    endfor

Step 4: Update learners $\boldsymbol{D}_n$ ($n = 1 : S$) to learners $\boldsymbol{W}_n$ ($n = 1 : S$) through learning phase.

    for $n = 1 : S$
        randomly select another $n' \neq n$
        $\boldsymbol{W}_n = \boldsymbol{D}_n + \text{rand} \times ((I(f(\boldsymbol{D}_{n'}) - f(\boldsymbol{D}_n))(\boldsymbol{D}_n - \boldsymbol{D}_{n'})$
        $+ I(f(\boldsymbol{D}_n) - f(\boldsymbol{D}_{n'}))(\boldsymbol{D}_{n'} - \boldsymbol{D}_n))$
    endfor

Step 5: Finish the update for learners $\boldsymbol{R}_n$ ($n = 1 : S$).

    $\boldsymbol{R}_n = \boldsymbol{W}_n$, go to **Step 2**

Step 6: Obtain the rate matrix $\boldsymbol{R}_0 = \boldsymbol{R}_{\text{teacher}}$, and the coefficient matrix $\boldsymbol{A}_{h'}$.

Step 7: Decompose the coefficient matrix $\boldsymbol{A}_{h'}$.

$$\boldsymbol{A}_{h'} = \boldsymbol{L} + \boldsymbol{D} + \boldsymbol{U}$$

$$\text{where } \boldsymbol{L} = \begin{pmatrix} 0 & 0 & \cdots & \cdots & 0 \\ a_{21} & 0 & \cdots & \cdots & 0 \\ a_{31} & a_{32} & 0 & \cdots & 0 \\ \vdots & \vdots & \ddots & \vdots & \vdots \\ a_{h'1} & a_{h'2} & \cdots & a_{h'h'-1} & 0 \end{pmatrix}$$

$$\boldsymbol{D} = \text{diag}(a_{11}, a_{22}, \dots, a_{h'h'})$$

$$\boldsymbol{U} = \begin{pmatrix} 0 & a_{12} & \cdots & \cdots & 0 \\ 0 & 0 & a_{23} & \cdots & 0 \\ \vdots & \vdots & \vdots & \ddots & 0 \\ 0 & 0 & \cdots & \cdots & a_{h'-1h'} \\ 0 & 0 & \cdots & 0 & 0 \end{pmatrix}$$

Step 8: Set the population size S_1, the maximum number N_3 of generations and the current iteration as $N_4 = 0$ for TLBO method. Set the maximum number N_2 of generations for SOR method. Initialize learners ξ_m $(m = 1 : S_1)$.

Step 9: Obtain approximate solutions Y_m $(m = 1 : S_1)$ with learners ξ_m $(m = 1 : S_1)$ with SOR method.

    for $m = 1 : S_1$
        $X_0 = 0$
        for $m_0 = 1 : N_2$
            $T_m = (D + \xi_m L)^{-1}(((1 - \xi_m)D - \xi_m U)X_0 + \xi_m b)$
            $X_0 = T_m$
        endfor
        $Y_m = T_m$
    endfor

Step 10: $Y^* = \underset{m \in \{1,2,3,\dots,S_1\}}{\text{argmin}} \{\|Y_m A - b\|_2\}$, record corresponding ξ as teacher ξ_{tch}.

$$\xi_{me} = \frac{1}{S_1} \sum_{m=1}^{S_1} \xi_m, \quad N_4 = N_4 + 1$$

    if $N_4 = N_3$
        go to **Step 17**
    endif

Step 11: Update learners ξ_m $(m = 1 : S_1)$ to learners ξ'_m $(m = 1 : S_1)$ through teaching phase.

    $T_F = \text{round}(1 + \text{rand})$
    for $m = 1 : S_1$
        $\xi'_m = \xi_m + \text{rand} \times (\xi_{tch} - T_F \xi_{me})$
    endfor

Step 12: Obtain approximate solution Y'_m $(m = 1 : S_1)$ with learner ξ'_m $(m = 1 : S_1)$ by using SOR method.

    for $m = 1 : S_1$
        $X_0 = 0$
        for $m_0 = 1 : N_2$
            $T_m = (D + \xi'_m L)^{-1}(((1 - \xi'_m)D - \xi_m U)X_0 + \xi'_m b)$
            $X_0 = T_m$
        endfor
        $Y'_m = T_m$
    endfor

Step 13: Update learners ξ'_m $(m = 1 : S_1)$ to learners ξ''_m $(m = 1 : S_1)$ through teaching phase.

    for $m = 1 : S_1$
        $x' = \|Y'_m A - b\|_2 - \|Y_m A - b\|_2$
        $\xi''_m = I(x')\xi_m + I(-x')\xi'_m$
    endfor

Step 14: Obtain approximate solution Y''_m ($m = 1 : S_1$) with learner ξ''_m ($m = 1 : S_1$) by using SOR method.

> **for** $m = 1 : S_1$
> $\quad X_0 = 0$
> $\quad$ **for** $m_0 = 1 : N_2$
> $\quad\quad T_m = (D + \xi''_m L)^{-1}(((1 - \xi''_m)D - \xi_m U)X_0 + \xi''_m b)$
> $\quad\quad X_0 = T_m$
> $\quad$ **endfor**
> $\quad Y'' = T_m$
> **endfor**

Step 15: Update learners ξ''_m ($m = 1 : S_1$) to learners ξ'''_m ($m = 1 : S_1$) through learning phase.

> **for** $m = 1 : S_1$
> $\quad$ randomly select another $m' \neq m$
> $\quad x'' = ||Y''_m A - b||_2 - ||Y''_{m'} A - b||_2$
> $\quad \xi'''_m = \xi''_m + \text{rand} \times (I(x'')(\xi''_m - \xi''_{m'}) + I(-x'')(\xi''_{m'} - \xi''_m))$
> **endfor**

Step 16: Finish the update for learners ξ_m ($m = 1 : S_1$).

> $\xi_m = \xi'''_m$

Step 17:

> **if** $N_4 < N_3$
> $\quad$ go to **Step 9**
> **else** $X = Y^*$
> **endif**

Step 18: Output $\pi_0, \pi_1, \pi_2, \ldots, \pi_{K+1}$.

From the structure of the transition rate matrix Q given in Eq. (13.6), we know π_i satisfies the matrix-geometric solution form as follows:

$$\pi_i = \pi_{K+1} R_0^{i-K-1}, \quad i = K + 2, K + 3, K + 4, \ldots. \tag{13.20}$$

13.4 Performance Measures and Numerical Results

In this section, we first derive performance measures of the system in terms of the throughput of SU packets, the average latency of SU packets, the channel utilization and the switching rate of SU packets, respectively. Then, we present numerical results to evaluate the performance of the system using the channel reservation strategy proposed in this chapter.

13.4.1 Performance Measures

The throughput θ of SU packets is defined as the number of SU packets that are transmitted successfully per second across the whole spectrum. When a newly arriving PU packet attempts to preempt one of the channels which is being occupied by SU packets, if there is a mistake detection of an SU packet, then the SU packet will be dropped from the system; if a mistake detection does not occur, the SU packet will depart the system with the probability $\bar{\psi}$. Therefore, we give the throughput θ of SU packets as follows

$$\theta = \lambda_1 - \lambda_2 \left(\sum_{i=1}^{K} \sum_{j=M-\min\{i,N\}}^{M-1} \pi_{i,j} + \sum_{i=K+1}^{\infty} \sum_{j=0}^{M-1} \pi_{i,j} \right) (p_{\text{md}} + \bar{p}_{\text{md}} \bar{\psi}).$$

$$(13.21)$$

The latency Y_s of an SU packet is defined as the duration from the instant an SU packet arrives at the system to the instant that the SU packet departs the system successfully. By using the total probability formula, we obtain the average value $E[N_s]$ for the number N_s of SU packets in the system as follows:

$$E[N_s] = \sum_{i=0}^{\infty} \sum_{j=0}^{M} i \pi_{i,j}. \qquad (13.22)$$

By using Eqs. (13.21) and (13.22), we can obtain the average latency $E[Y_s]$ of SU packets as follows:

$$E[Y_s] = \frac{E[N_s]}{\theta} = \frac{\displaystyle\sum_{i=0}^{\infty} \sum_{j=0}^{M} i \pi_{i,j}}{\theta}. \qquad (13.23)$$

The switching rate ζ_s of SU packets is defined as the number of times that the SU packets switch to the buffer from the channels due to a false alarm. When there is no PU packet which attempts to preempt one of channels being occupied by an SU packet and a false alarm occurs, then the false alarms induce a time switch between the current channel and the buffer. Therefore, we can give the switching rate ζ_s of SU packets as follows:

$$\zeta_s = \sum_{i=1}^{K} \sum_{j=M-\min\{i,N\}}^{M-1} (M-j)\pi_{i,j} + \sum_{i=K+1}^{\infty} \sum_{j=0}^{M-1} (M-j)\pi_{i,j}. \qquad (13.24)$$

The channel utilization U_c is defined as the probability that one channel is being occupied by a user packet (a PU packet or an SU packet). The channel utilization U_c

can be given by calculating the proportion of the average number of channels which are being occupied by user packets in relation to the total number of the channels. Therefore, we can also give the channel utilization U_c as follows:

$$U_c = \frac{\sum_{i=0}^{\infty} \sum_{k=0}^{M} \min\{i + j, M\}\pi_{i,j}}{M}. \tag{13.25}$$

13.4.2 Numerical Results

In order to investigate the influence of the number N of reserved channels and the energy detection threshold τ on the system performance, we carry our numerical results. We set the parameters shown in Eq. (13.1) to be $N_0 = 1, \xi = 100, s_0 = 0.5$ and $\tau \in [46, 50]$. The other parameters in relation to this model are set as follows: $M = 5, \lambda_1 = 2.4, 2.7, 3, \lambda_2 = 0.3, 0.4, 0.5, \mu_1 = \mu_2 = 0.8$. Unless otherwise specified, we use the same parameters as set out above in all numerical results.

Taking the number $N = 2$ of reserved channels as an example, we show the change trends of the throughput θ of SU packets, the average latency $E[Y_s]$ of SU packets, the switching rate ζ_s of SU packets and the channel utilization U_c with respect to the energy detection threshold τ in Figs. 13.2, 13.3, 13.2, 13.5, respectively.

From Fig. 13.2, we observe that if the arrival rates (λ_1 and λ_2) of user packets and the number N of reserved channels are given, the throughput θ of SU packets decreases as the energy detection threshold τ increases. This is because as the energy detection threshold increases, the mistake detection ratio increases. This will induce

Fig. 13.2 Throughput of SU packets versus energy detection threshold

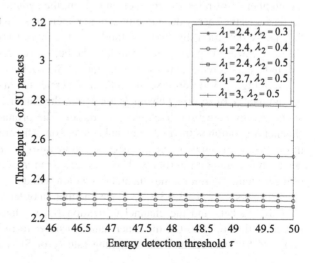

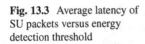

Fig. 13.3 Average latency of
SU packets versus energy
detection threshold

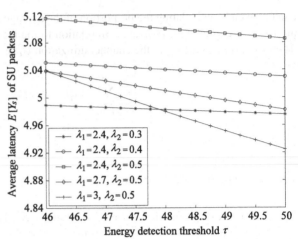

more SU packets being dropped out of the system. Therefore, the throughput of SU packets will decrease accordingly.

From Fig. 13.3, we observe that if the arrival rates (λ_1 and λ_2) of user packets and the number N of reserved channels are given, the average latency $E[Y_s]$ of SU packets decreases as the energy detection threshold τ increases. This is because the larger the energy detection threshold is, the higher the mistake detection ratio is, and the higher the probability is of a collision induced by a mistake detection. This leads more newly arriving PU packets to depart the system non-normally. Then, when the channels are allocated to the SU packets by the central controller, the waiting time of the SU packets becomes smaller. Therefore, the average latency of SU packets will decrease.

From Fig. 13.4, we observe that if the arrival rates (λ_1 and λ_2) of user packets and the number N of reserved channels are given, the switching rate ζ_s of SU packets decreases as the energy detection threshold τ increases. The reason is that the higher the energy detection threshold is, the lower the rate of false alarms is. Therefore, there are fewer SU packets that return to the buffer as a result of false alarms. This will induce a decrease in the switching rate of SU packets.

From Fig. 13.5, we observe that if the arrival rates (λ_1 and λ_2) of user packets and the number N of reserved channels are given, the channel utilization U_c decreases as the energy detection threshold τ increases. The primary factor influencing the channel utilization is the rate of mistake detection. The higher the energy detection threshold is, the higher the mistake detection ratio is. Therefore, many packets depart the system non-normally. This will lead to an increase in the idle probability of the system. Therefore, the channel utilization will decrease accordingly.

From Figs. 13.2, 13.3, 13.2, 13.5, we find that, in order to enhance the throughput θ of SU packets and the channel utilization U_c, we need to set a lower energy detection threshold τ. On the other hand, in order to reduce the average latency $E[Y_s]$ of SU packets and the switching rate ζ_s of SU packets, we need to set a

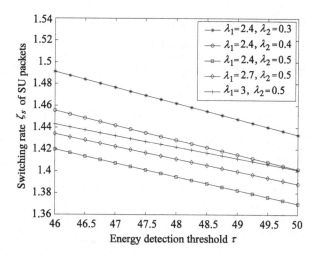

Fig. 13.4 Switching rate of SU packets versus energy detection threshold

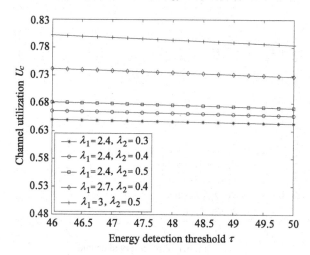

Fig. 13.5 Channel utilization versus energy detection threshold

higher energy detection threshold τ. The aim is to set a reasonable energy detection threshold τ for the purpose of balancing the system performance.

By setting an example as the energy detection threshold $\tau = 48$, we show the change trends of the throughput θ of SU packets, the average latency $E[Y_s]$ of SU packets, the switching rate ζ_s of SU packets and the channel utilization U_c in relation to the number N of reserved channel in Figs. 13.6, 13.7, 13.8, 13.9, respectively.

Looking at Fig. 13.6, we find that if the arrival rates (λ_1 and λ_2) of user packets and the energy detection threshold τ are given, the throughput θ of SU packets first decline, then increase, and then decline as the number N of reserved channels increases.

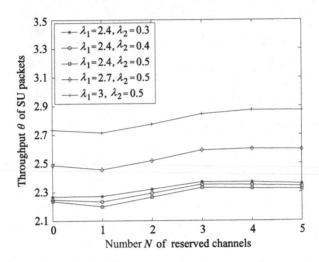

Fig. 13.6 Throughput of SU packets versus number of reserved channels

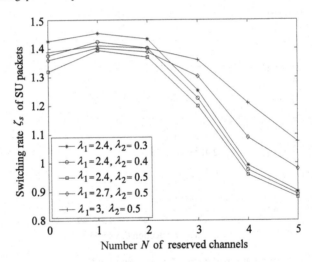

Fig. 13.7 Average latency of SU packets versus number of reserved channels

During the first declining stage, the throughput θ of SU packets decreases as the number N of reserved channels increases. The fewer reserved channels there are, the more the SU packets there will be transmitted on the non-reserved channels. In addition, a fewer number N of reserved channels induces a lower retrial probability. This means that when a newly arriving PU packet attempts to preempt the channels which are being occupied by SU packets, even though no mistake detection occurs, the SU packets which are preempted by PU packets depart the system with a higher probability. This induces a decrease in the throughput of SU packets.

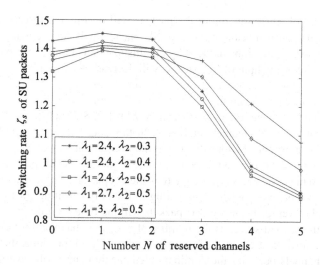

Fig. 13.8 Switching rate of SU packets versus number of reserved channels

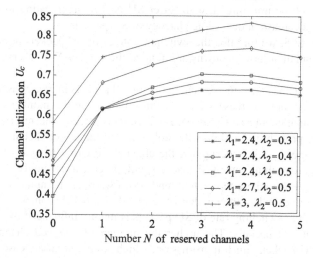

Fig. 13.9 Channel utilization versus number of reserved channels

During the increasing stage, the throughput θ of SU packets increases as the number N of reserved channels increases. Different from the first declining stage, in this stage, the SU packets which are preempted by PU packets have a higher retrial probability. The higher the number of reserved channels is, the more likely it is that the SU packets which are preempted by PU packets return to the buffer. This induces an increase in the throughput of SU packets.

During the last declining stage, the throughput θ of SU packets decreases as the number N of reserved channels increases. In this stage, the mistake detection ratio is a primary factor influencing the throughput of SU packets. As the number of reserved channels increases, more PU packets will occupy the reserved channels. This means that an increasing number of SU packets will non-normally depart the system because of collisions induced by the mistake detections of SU packets. The result being the throughput of SU packets will decrease accordingly.

Looking at Fig. 13.7, we find that the average latency $E[Y_s]$ of SU packets exhibits two stages as the number N of reserved channels increases.

During the first stage, the average latency $E[Y_s]$ of SU packets increases as the number N of reserved channels increases. In this stage, the retrial probability of SU packets is a primary factor influencing the average latency of SU packets. The higher the number of reserved channels is, the higher the retrial probability of SU packets is, so most SU packets preempted by the PU packets will return to the buffer. This leads to more SU packets aggregating in the buffer waiting for transmission. Therefore, the average latency of SU packets will increase accordingly.

During the second stage, as the number of reserved channels further increases, when the number of reserved channels is greater than a certain value, the number of available channels becomes the dominant element influencing the average latency of SU packets. The higher the number of reserved channels is, the sooner the SU packets will be transmitted, with fewer SU packets aggregating in the buffer, resulting in a decrease in the average latency of SU packets.

From Fig. 13.8, we find that the switching rate ζ_s of SU packets firstly shows a rising trend and then shows a downward trend as the number N of reserved channels increases.

During the rising stage, the switching rate of SU packets increases as the number of reserved channels increases. The reason is that when the number of reserved channels is smaller, the newly arriving PU packets occupy the reserved channel with a higher probability, and the idle probability of the reserved channel is lower. This leads the SU packets occupying the channels to easily trigger a false alarm. Therefore, the switching rate of SU packets will increase accordingly.

During the stage where the system tends downward, as the number of reserved channels increases, when the number of reserved channels becomes greater than a certain value, the switching rate of SU packets is mainly influenced by the average latency of the SU packets. The higher the number of reserved channels is, the sooner the SU packets finish transmission, which results in an increase in the idle probability of the channels. This will induce a decrease in the switching rate of SU packets.

Figure 13.9 shows that the channel utilization U_c firstly shows a rising trend and then shows a downward trend as the number N of reserved channels increases.

During the rising stage, the channel utilization increases as the number of reserved channels increases. The reason is that when the number of reserved channels is smaller, as the number of reserved channels increases, the SU packets waiting in the buffer have to timely access the reserved channels. This means that the reserved channels are fully used. In addition, with a smaller number of reserved channels, the number of SU packets queueing in the buffer awaiting transmission easily exceeds the buffer threshold, so the non-reserved channels are easily occupied by SU packets. This induces an increase in the channel utilization.

During the downward stage, as the number of reserved channels increases, the channel utilization decreases. When the number of reserved channels is greater than a certain value, most SU packets have access the reserved channels. This leads to fewer SU packets queueing in the buffer, and results in an excess or wastage of the non-reserved channels. Therefore, the channel utilization will decrease accordingly.

From Figs. 13.6, 13.7, 13.8, 13.9, we find that, in order to reduce the average latency $E[Y_s]$ of SU packets and the switching rate ζ_s of SU packets, we need to set a higher number N of reserved channels. But in this case, the throughput θ of SU packets and the channel utilization U_c cannot reaches their maximum. Therefore, the setting of a reasonable value for the number N of reserved channels is a key determinant for improving system performance.

13.5 Conclusion

In this chapter, considering the mistake detection ratio and the false alarm ratio, we proposed an imperfect sensing-based channel reservation strategy in CRNs. We firstly established a CTMC model to capture the stochastic behavior of the two types of user packets, the PU packets and the SU packets, and then also developed the TLBO-SOR algorithm to obtain the steady-state distribution of the QBD process. Based on the steady-state distribution of the system model, we mathematically estimated the system performance. Numerical results showed that the proposed strategy is feasible. The research work has potential applications in improving spectrum effectiveness in CRNs.

Chapter 14
Energy Saving Strategy in CRNs Based on a Priority Queue with Single Vacation

In order to improve spectrum efficiency and achieve greener communication in wireless applications, in this chapter, we consider Cognitive Radio Networks (CRNs) with an LTE-Advanced (LTE-A) structure and propose an energy saving strategy with a single-sleep mode. By establishing a preemptive priority queueing model with a single vacation, we capture the stochastic behavior of the proposed strategy. Using the method of a matrix-geometric solution, we derive performance measures of the system in terms of the average latency of Secondary User (SU) packets and the energy saving degree. Furthermore, we present numerical results to demonstrate the influence of the sleep parameter on the performance of the system using the proposed energy saving strategy with a single-sleep mode. Finally, by establishing the individual benefit function and the social benefit function, we investigate the Nash equilibrium and socially optimal behaviors of SU packets and present a pricing policy for SU packets to socially optimize the system performance.

14.1 Introduction

With the recent development of mobile Internet and the popularization of smart terminals, efficiency of energy consumption in WCNs has become increasingly important [Chen11a]. At the same time, the requirement for wireless spectrum resources has increased dramatically [Li15b], the waking spectrum resource is an invaluable commodity in the communication field [Huan15b].

LTE-A in 4G networks has gained great support from many communication companies and many mobile manufacturers [Šten10]. Any development in science and technology is inseparable from the support of infrastructure. As the main hardware device for wireless communication, Base Stations (BSs) play an important role in providing a wide coverage area and quick transmission rate. However, to support the operation of BSs, we pay the heavy price of high energy consumption [Chen11b]. Science community has a responsibility for developing green initiatives,

© Springer Nature Singapore Pte Ltd. 2021
S. Jin, W. Yue, *Resource Management and Performance Analysis of Wireless Communication Networks*, https://doi.org/10.1007/978-981-15-7756-7_14

more and more researchers are committing to combining their research with environmental protection goals [Li13, Spag15, Ting13]. In LTE-A, the sleep mode is introduced to decrease the energy consumption. In [Yang13b], considering that parts of BSs would be switched into sleep mode and the associated users would be transferred to neighboring cells, the authors formulated the minimum number of active BSs under user rate-guaranteed as a NP-hard problem. They designed an iterative set cover algorithm to solve the NP-hard problem. In [Sama16], the authors proposed a distributed learning algorithm, developed an opportunistic on/off strategy for BSs, and allowed the BSs to decide on whether to switch to a sleep mode or to an active mode for the purpose of minimizing the system cost function. HetNets are more energy efficient than macro-only deployment with the same capacity because of the lower power consumption of the small cells [Dini13]. In [Peng14], the authors studied the energy saving problem through closing some macro BSs and switching the traffic to some active micro BSs in heterogeneous cellular networks. In all the above literatures, energy saving has been considered as a key point.

On the other hand, in the traditional fixed spectrum allocation strategy [Sale15], the development of mobile communication technology is restricted, and the spectrum resource is not fully utilized. The emergence of the cognitive radio [Vara15] has broken the hegemonic governance of the fixed spectrum allocation strategy and weakened the strict protection of authorized communication. As an intelligent spectrum sharing technology, cognitive radio has the ability to sense the condition of wireless communication and learn from the environment to adjust the transmission parameters, such as frequency band, modulation mode and transmission power [Syed14]. In CRNs, the SUs are allowed to access the unused parts of the spectrum opportunistically, while the Primary Users (PUs) enjoy preemptive priority during the spectrum usage. As a result, CRNs effectively alleviate spectrum scarcity [Josh13].

With the aim of alleviating the spectrum scarcity crisis and reducing the energy consumption, we propose an energy saving strategy in CRNs with an LTE-A structure. LTE-A protocol permits the terminal device to discontinuously monitor the downlink data and the related processes. We introduce a sleep mode to the BS in CRNs. Based on the data traffic in communication networks, BSs will be switched between sleep period and awake period to conserve energy and attain higher spectrum efficiency.

In this chapter, we extend the analysis of our previous work [Jin15a] not only by providing additional explanations regarding the transition probability matrix, but also presenting a new algorithm to calculate the numerical solution for the rate matrix. Moreover, in this chapter, we use the Gravitation-Gravitational Search Algorithm (GSA) [Yazd14] to effectively solve the nonlinear optimization problem for the sake of obtaining the socially optimal arrival rate of SU packets with the social benefit function. Furthermore, in this chapter, we compare the Nash equilibrium arrival rate and the socially optimal arrival rate to present a pricing policy for SU packets.

The chapter is organized as follows. In Sect. 14.2, we describe the energy saving strategy in CRNs with a single-sleep mode and an LTE-A structure proposed in this chapter. Then, we present the system model in detail. In Sect. 14.3, we present a performance analysis of the system model in the steady state. Then, we obtain performance measures and present numerical results to evaluate the system performance. In Sect. 14.4, we firstly investigate the Nash equilibrium behavior and socially optimal behavior of SU packets in the energy saving strategy proposed in this chapter. Then, we propose an appropriate pricing policy for SU packets to optimize the system performance socially. Finally, we draw our conclusions in Sect. 14.5.

14.2 Energy Saving Strategy and System Model

In this section, we propose an energy saving strategy in CRNs with a single-sleep mode and an LTE-A structure. We call this energy saving strategy an "energy saving strategy with a single-sleep mode". Then, we establish a preemptive priority queueing model with a single vacation accordingly.

14.2.1 Energy Saving Strategy

In conventional CRNs, huge amounts of power are wasted due to BSs are always being awake even though there are no PU or SU packets that need to be transmitted or received. In this chapter, we introduce an energy saving strategy with a single-sleep mode as a way of reducing the energy consumption. In the strategy proposed in this chapter, we consider the CRNs with an LTE-A structure. Based on the stochastic behavior of PU and SU packets, as well as the operational characteristic of a sleep timer, the BS will be switched among awake periods, sleep periods and listening periods. In this network system, a buffer is also available for the SU packets' waiting.

(1) *Awake Period to Sleep Period*: During an awake period, the PU and SU packets are transmitted continuously. PU packets are transmitted with preemptive priority, while SU packets can only access the unused spectrum opportunistically.

   If all the packets in the system are completely transmitted, namely, the spectrum is idle and the buffer is empty, then the BS will enter a sleep period. Since the idle power is lower than the transmit power, more energy will be saved in the energy saving strategy with a single-sleep mode proposed in this chapter.

   What is noteworthy is that the BS can be switched to an awake period or a listening period from a sleep period.

(2) *Sleep Period to Awake Period*: At the beginning instant of a sleep period, a sleep timer with a random time length will be activated in order to restrain the maximum length of a sleep period, improve the spectrum efficiency and reduce the response time of SU packets. The sleep timer length is a time interval from the instant of a sleep timer is activated to the instant that the sleep timer expires.

A newly arriving PU packet will terminate a sleep period, and the BS will be awakened immediately. If there is no PU packet arrival at the system within the sleep timer length, the arriving SU packets (if any) will queue in the buffer. Once the sleep timer expires, the BS will enter into an awake period, and the SU packets queueing in the buffer will be transmitted following a First-Come First-Served (FCFS) discipline.

(3) *Sleep Period to Listening Period*: With the sleep mode, energy consumption can be reduced effectively. However, the average latency of SU packets becomes higher than before. From this point of view, we introduce a listening period as a transition phase from a sleep period to an awake period. In this way, the system performance with regard to response time will be improved.

If neither PU nor SU packet arrives at the system before the sleep timer expires, the BS will enter a listening period at the instant when the sleep time expires.

(4) *Listening Period to Awake Period*: Although no packet is transmitted during a listening period, the BS, which is more like a soldier on standby, waits for the arrival of a PU or an SU packet at any moment. Most of the transmission devices, such as the air interfaces which are shut down during sleep period will be activated in listening period. The power consumption in a listening period is lower than that in an awake period, but higher than that in a sleep period.

In order to make up the disadvantage that the power consumption in a listening period is higher than that in a sleep period, from the perspective of reducing the response time, we specify the listening period is a one-way transition period. During the listening period, either a newly arriving PU or SU packet will wake up the BS immediately, and the BS will return to an awake period with few delay.

The state transition of the energy saving strategy with a single-sleep mode proposed in this chapter is illustrated in Fig. 14.1.

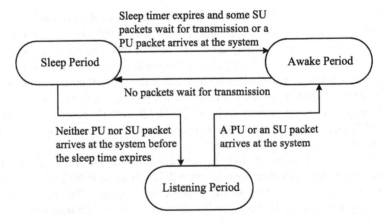

Fig. 14.1 State transition of proposed energy saving strategy

14.2.2 System Model

During an awake period, PU packets are transmitted with preemptive priority, while SU packets can only be opportunistically transmitted. We establish a preemptive priority queueing model with a single vacation to capture the working principle of the energy saving strategy with a single-sleep mode in CRNs with an LTE-A structure. The buffer to queue the SU packets is supposed to have an infinite capacity.

The time axis is segmented into a series of equal intervals, called slots, and the slot is marked as n ($n = 1, 2, 3, \ldots$). Following an Early Arrival System (EAS), we suppose that the arrival of data packets occurs at the beginning instant of a slot, marked as (n, n^+) ($n = 1, 2, 3, \ldots$), while the departure of data packets occurs at the end instant of a slot, marked as (n^-, n) ($n = 2, 3, 4, \ldots$), as shown in Fig. 14.2.

In Fig. 14.2, the instant that a data packet possibly arrives is marked as a solid circle and the instant that a data packet possibly departs is marked as a solid rectangle.

We assume that the arriving intervals of the PU packets and the SU packets follow geometric distributions with arrival rate λ_1 ($0 < \lambda_1 < 1$, $\bar{\lambda}_1 = 1 - \lambda_1$) and λ_2 ($0 < \lambda_2 < 1$, $\bar{\lambda}_2 = 1 - \lambda_2$), respectively. During an awake period, each transmission of a packet will take up the whole bandwidth of the spectrum, but the length in bits of a packet is variable, so the transmission time of a packet is also variable. In this chapter, we assume the transmission times of an SU and a PU packet follow geometric distributions with service rates μ_1 ($0 < \mu_1 < 1$, $\bar{\mu}_1 = 1 - \mu_1$) and μ_2 ($0 < \mu_2 < 1$, $\bar{\mu}_2 = 1 - \mu_2$), respectively. In addition, we assume the sleep timer length follows a geometric distribution with parameter δ ($\delta > 0$). We call δ the sleep parameter.

We describe the system model in an infinite state. Let $X_n = i$ ($i = 0, 1, 2, \ldots$) and $Y_n = j$ ($j = 0, 1, 2, 3$) be the number of SU packets in the system and the BS stage, respectively, at the instant n^+. $j = 0$ means the BS is in a sleep period; $j = 1$ means the BS is awake with the transmission of a PU packet; $j = 2$ means the BS is awake with the transmission of an SU packet; $j = 3$ means the BS is in a listening period. We also called the number of SU packets in the system as the system level. $\{(X_n, Y_n), n \geq 1\}$ constitutes a two-dimensional DTMC. The state space of the DTMC is given as follows:

$$\Omega = \{(i, j) : i \geq 0, j = 0, 1, 2, 3\}. \tag{14.1}$$

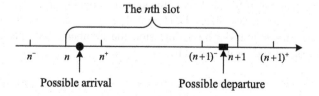

Fig. 14.2 EAS with possible arrival and departure instants

Let $\pi_{i,j}$ be the probability that the number of SU packets in the system is i and the number of PU packets in the system is j in the steady state. $\pi_{i,j}$ is then defined as follows:

$$\pi_{i,j} = \lim_{n \to \infty} \Pr\{X_n = i, \ Y_n = j\}, \quad i \geq 0, j = 0, 1, 2, 3. \tag{14.2}$$

By π_i ($i \geq 0$) we denote the steady-state probability vector $(\pi_{i,0}, \pi_{i,1}, \pi_{i,2}, \pi_{i,3})$. The steady-state distribution Π of the system. Π is given as follows:

$$\Pi = (\pi_0, \pi_1, \pi_2, \ldots). \tag{14.3}$$

Combining the balance equation and the normalization condition for the DTMC, we have

$$\begin{cases} \Pi P = \Pi \\ \Pi e = 1 \end{cases} \tag{14.4}$$

where e is a column vector with infinite elements and all elements of the vector are equal to 1. P is the state transition probability matrix of the DTMC $\{(X_n, Y_n), n \geq 1\}$.

14.3 Performance Analysis and Numerical Results

In this section, by using a matrix-geometric solution method, we give the steady-state analysis of the system model to obtain some performance measures. Then, we present numerical results to evaluate the performance of the system using the energy saving strategy with a single-sleep mode proposed in this chapter.

14.3.1 Performance Analysis

Considering there are four stages, namely, $j = 0, 1, 2, 3$, for a BS, we divide P into some sub-matrixes which are all of 4×4 structure. Each row or column of the sub-matrix represents a BS stage. Let $B_{i,k}$ be the transition probability sub-matrix for the number of SU packets in the system changing from the system level i ($i = 0, 1, 2, \ldots$) to the system level k ($k = 0, 1, 2, \ldots$). According to the principle of the one-step transition, during a time slot, the number of SU packets can be decreased by one, fixed, or increased by one.

We will deal with each sub-matrix in the transition probability P in detail.

(1) *Decreased Number of SU Packets*: At the instant $t = n^+$, there are i SU packets in the system, and after a time slot, the number of SU packets decreases to $k = i - 1$, where $i \geq 1$.

If $i = 1$, the transmission of the only SU packet in the system is completed, and there is no SU packet arrival at the system.

When the initial BS stage is $j = 2$: if no PU packet arrives at the system, then the BS stage will change to $j = 0$; if a PU packet arrives at the system, then the BS stage will change to $j = 1$.

Therefore, the transition probability sub-matrix $B_{1,0}$ is given as follows:

$$B_{1,0} = \bar{\lambda}_1 \mu_1 \begin{pmatrix} 0 & 0 & 0 & 0 \\ 0 & 0 & 0 & 0 \\ \bar{\lambda}_2 & \lambda_2 & 0 & 0 \\ 0 & 0 & 0 & 0 \end{pmatrix}. \tag{14.5}$$

If $i > 1$, one of SU packets in the system is completely transmitted, and there is no SU packet arrival at the system.

When the initial BS stage is $j = 2$: if a PU packet arrives at the system, then the BS stage will change to $j = 1$; if no PU packet arrives at the system, then the BS stage will be fixed at $j = 2$.

Therefore, the transition probability sub-matrix $B_{i,i-1}$ is given as follows:

$$B_{i,i-1} = \bar{\lambda}_1 \mu_1 \begin{pmatrix} 0 & 0 & 0 & 0 \\ 0 & 0 & 0 & 0 \\ 0 & \lambda_2 & \bar{\lambda}_2 & 0 \\ 0 & 0 & 0 & 0 \end{pmatrix}. \tag{14.6}$$

(2) *Fixed Number of SU Packets*: At the instant $t = n^+$, there are i SU packets in the system, and after a time slot, the number of SU packets remains at $k = i$, where $i \geq 0$.

If $i = 0$, there is no SU packet arrival at the system during the $(n + 1)$th slot.

When the initial BS stage is $j = 0$: if no PU packet arrives at the system, and the sleep timer is not over, then the BS stage will be fixed at $j = 0$; if a PU packet arrives at the system, then the BS stage will change to $j = 1$; if no PU packet arrives at the system, and the sleep timer is over, then the BS stage will change to $j = 3$.

When the initial BS stage is $j = 1$: if the transmission of the PU packet in the system is completed and no new PU packet arrives at the system, then the BS stage will change to $j = 0$; if the transmission of the PU packet in the system is completed and another PU packet arrives at the system, or if the transmission of the PU packet in the system is not completed, then the BS stage will be fixed at $j = 1$.

When the initial BS stage is $j = 3$: if a PU packet arrives at the system, then the BS stage will change to $j = 1$; if no PU packet arrives at the system, then the BS stage will be fixed at $j = 3$.

Therefore, the transition probability sub-matrix $\boldsymbol{B}_{0,0}$ is given as follows:

$$\boldsymbol{B}_{0,0} = \bar{\lambda}_1 \begin{pmatrix} \bar{\lambda}_2\bar{\delta} & \lambda_2 & 0 & \bar{\lambda}_2\bar{\delta} \\ \bar{\lambda}_2\mu_2 & \lambda_2\mu_2 + \bar{\mu}_2 & 0 & 0 \\ 0 & 0 & 0 & 0 \\ 0 & \lambda_2 & 0 & \bar{\lambda}_2 \end{pmatrix} \tag{14.7}$$

where δ is the sleep parameter defined in Sect. 14.2.2.

If $i \geq 1$, one case is that the transmission of the SU packet occupying the spectrum is not completed, but there is no newly arriving SU packet at the system during the $(n + 1)$th slot. The other is that the transmission of the SU packet occupying the spectrum is completed, but there is a newly arriving SU packet at the system during the $(n + 1)$th slot.

When the initial BS stage is $j = 0$: if no PU packet arrives at the system, and the sleep timer is not over, then the BS stage will be fixed at $j = 0$; if a PU packet arrives at the system, then the BS stage will change to $j = 1$; if no PU packet arrives at the system, and the sleep timer is over, then the BS stage will change to $j = 2$.

When the initial BS stage is $j = 1$: if the transmission of the PU packet in the system is completed and another PU packet arrives at the system, or if the transmission of the PU packet in the system is not completed, then the BS stage will be fixed at $j = 1$; if the transmission of the PU packet in the system is completed and no other PU packet arrives at the system, then the BS stage will change to $j = 2$.

When the initial BS stage is $j = 2$: if a PU packet arrives at the system, then the BS stage will change to $j = 1$; if no PU packet arrives at the system, then the BS stage will be fixed at $j = 2$.

Therefore, the transition probability sub-matrix $\boldsymbol{B}_{i,i}$ is given as follows:

$$\boldsymbol{B}_{i,i} = \begin{pmatrix} \bar{\lambda}_1 & 0 & 0 & 0 \\ 0 & \bar{\lambda}_1 & 0 & 0 \\ 0 & 0 & (\lambda_1\mu_1 + \bar{\lambda}_1\bar{\mu}_1) & 0 \\ 0 & 0 & 0 & 0 \end{pmatrix} \times \begin{pmatrix} \bar{\lambda}_2\bar{\delta} & \lambda_2 & \bar{\lambda}_2\delta & 0 \\ 0 & \lambda_2\mu_2 + \bar{\mu}_2 & \bar{\lambda}_2\mu_2 & 0 \\ 0 & \lambda_2 & \bar{\lambda}_2 & 0 \\ 0 & 0 & 0 & 0 \end{pmatrix}.$$

$$\tag{14.8}$$

(3) *Increased Number of SU Packets*: At the instant $t = n^+$, there are i SU packets in the system, and after a time slot, the number of SU packets increases to $k = i + 1$, where $i \geq 0$.

If $i = 0$, there is a newly arriving SU packet at the system during the $(n+1)$th slot.

When the initial BS stage is $j = 0$ or $j = 1$: the stage transitions are similar to those for the case of $i \geq 1$ in Item (2).

When the initial BS stage is $j = 3$: if a PU packet arrives at the system, then the BS stage will change to $j = 1$; if no PU packet arrives at the system, then the BS stage will change to $j = 2$.

Therefore, the transition probability sub-matrix $\boldsymbol{B}_{0,1}$ is given as follows:

$$
\boldsymbol{B}_{0,1} = \lambda_1 \begin{pmatrix}
\bar{\lambda}_2\bar{\delta} & \lambda_2 & \bar{\lambda}_2\delta & 0 \\
0 & \lambda_2\mu_2 + \bar{\mu}_2 & \bar{\lambda}_2\mu_2 & 0 \\
0 & 0 & 0 & 0 \\
0 & \lambda_2 & \bar{\lambda}_2 & 0
\end{pmatrix}. \tag{14.9}
$$

If $i \geq 1$, the transmission of an SU packet is not completed. Moreover, there is a newly arriving SU packet at the system during the $(n + 1)$th slot.

When the initial BS stage is $j = 0$, $j = 1$ or $j = 2$: the stage transitions are also similar to those for case of $i \geq 1$ in Item (2).

Therefore, the transition probability sub-matrix $\boldsymbol{B}_{i,i+1}$ is given as follows:

$$
\boldsymbol{B}_{i,i+1} = \begin{pmatrix}
\lambda_1 & 0 & 0 & 0 \\
0 & \lambda_1 & 0 & 0 \\
0 & 0 & \lambda_1\bar{\mu}_1 & 0 \\
0 & 0 & 0 & 0
\end{pmatrix} \times \begin{pmatrix}
\bar{\lambda}_2\bar{\delta} & \lambda_2 & \bar{\lambda}_2\delta & 0 \\
0 & \lambda_2\mu_2 + \bar{\mu}_2 & \bar{\lambda}_2\mu_2 & 0 \\
0 & \lambda_2 & \bar{\lambda}_2 & 0 \\
0 & 0 & 0 & 0
\end{pmatrix}. \tag{14.10}
$$

Up to now, all the sub-matrixes in $\boldsymbol{P}$ have been addressed. Starting from system level 3, all the sub-matrixes of the state transition probability matrix are repeated forever. By $\boldsymbol{A}_0$, $\boldsymbol{A}_1$ and $\boldsymbol{A}_2$, we denote $\boldsymbol{B}_{i,i-1}$ ($i \geq 2$), $\boldsymbol{B}_{i,i}$ ($i \geq 1$) and $\boldsymbol{B}_{i,i+1}$ ($i \geq 1$), respectively. $\boldsymbol{P}$ is a block tridiagonal matrix given as follows:

$$
\boldsymbol{P} = \begin{pmatrix}
\boldsymbol{B}_{0,0} & \boldsymbol{B}_{0,1} & & \\
\boldsymbol{B}_{1,0} & \boldsymbol{A}_1 & \boldsymbol{A}_2 & \\
& \boldsymbol{A}_0 & \boldsymbol{A}_1 & \boldsymbol{A}_2 \\
& & \ddots & \ddots & \ddots
\end{pmatrix}. \tag{14.11}
$$

The structure of $\boldsymbol{P}$ shows that the system transition occurs only in adjacent levels. Therefore, the two-dimensional DTMC $\{(X_n, Y_n), n \geq 1\}$ can be seen as a Quasi Birth-Death (QBD) process. By using a matrix-geometric solution method, we can derive the steady-state distribution $\boldsymbol{\Pi}$ of the system.

For the DTMC $\{(X_n, Y_n), n \geq 1\}$ with the transition probability matrix $\boldsymbol{P}$, the necessary and sufficient condition of positive recurrence is that the matrix quadratic equation

$$
\boldsymbol{R}^2\boldsymbol{A}_0 + \boldsymbol{R}\boldsymbol{A}_1 + \boldsymbol{A}_2 = \boldsymbol{R} \tag{14.12}
$$

has a minimal non-negative solution R, and the spectral radius $Sp(R) < 1$.

It is a difficult work to give the mathematical expression of the rate matrix R in close-form with the higher order matrix equation. We present an iteration algorithm to obtain the rate matrix R numerically. The main steps of the iteration algorithm are given as follows:

Step 1: Initialize a small constant ε (for example, $\varepsilon = 10^{-6}$) related to calculation accuracy and the rate matrix $R = 0$.

Step 2: Input A_0, A_1 and A_2.

Step 3: Calculate R^*.

$$R^* = (R^2 \times A_2 + A_0) \times (I - A_1)^{-1}$$

% I is an identity matrix.

Step 4:

if $\{||R - R^*||_\infty > \varepsilon\}$

% $||R - R^*||_\infty = \max\limits_{i \in \{1,2,3,4\}} \left\{ \sum\limits_{j=1}^{4} |r_{i,j} - r_{i,j}^*| \right\}$.

$R = R^*$

$R^* = (R^2 \times A_2 + A_0) \times (I - A_1)^{-1}$

go to **Step 4**

else

$R = R^*$

endif

Step 5: Output R.

Using the rate matrix R obtained in the above algorithm, we construct a stochastic matrix as follows:

$$B[R] = \begin{pmatrix} B_{00} & B_{01} \\ B_{10} & A_1 + RA_2 \end{pmatrix}. \tag{14.13}$$

Then, π_0 and π_1 satisfy the following set of linear equations:

$$\begin{cases} (\pi_0, \pi_1)B[R] = (\pi_0, \pi_1) \\ \pi_0 e + \pi_1(I - R)^{-1}e = 1 \end{cases} \tag{14.14}$$

where e is a column vector with 4 elements and all elements of the vector are equal to 1.

Based on Eq. (14.14), we further construct an augmented matrix as follows:

$$(\pi_0, \pi_1) \left(I - B[R] \Big|_{(I-R)^{-1}e}^{e} \right) = \underbrace{(0, 0, 0, \ldots, 0, 1)}_{8}. \tag{14.15}$$

By using the Gauss-Seidel method to solve Eq. (14.15), we can obtain π_0 and π_1.

From the structure of state transition probability matrix P, we know π_i ($i = 2, 3, 4, \ldots$) satisfies the matrix-geometric solution form as follows:

$$\pi_i = \pi_1 R^{i-1}, \quad i \geq 1. \tag{14.16}$$

By substituting π_1 obtained in Eq. (14.15) into Eq. (14.16), we can obtain π_i ($i = 2, 3, 4, \ldots$). Then, the steady-state distribution $\Pi = (\pi_0, \pi_1, \pi_2, \ldots)$ of the system can be given numerically.

14.3.2 Performance Measures

By using the total probability formula, the average number of SU packets in the system in the steady state is equal to $\sum_{i=0}^{\infty} i(\pi_{i,0} + \pi_{i,1} + \pi_{i,2} + \pi_{i,3})$. We define the latency Y_s of an SU packet as the duration from the instant an SU packet joins the system to the instant that SU packet is transmitted successfully. Based on the analysis presented in Sect. 14.3.1, we can obtain the average latency $E[Y_s]$ of SU packets as follows:

$$E[Y_s] = \frac{1}{\lambda_1} \sum_{i=0}^{\infty} i(\pi_{i,0} + \pi_{i,1} + \pi_{i,2} + \pi_{i,3}). \tag{14.17}$$

We define the energy saving degree γ_d as the overall level of energy conservation in the proposed energy saving strategy with a single-sleep mode. During an awake period, PU and SU packets are transmitted continuously, energy will be consumed normally. During a listening period, there are no packets to be transmitted or received, but the BS always waits for the arrival of PU or SU packets at any moment, hence, some energy will be consumed. During a sleep period, most of the transmission devices are shut down, the energy consumption will be minimized. Considering listening power is lower than transmit power but higher than idle power, we denote the energy saving level when the BS is in sleep period, listening period and awake period as 1, ρ ($0 < \rho < 1$) and 0, respectively. We give the energy saving degree γ_d as follows:

$$\gamma_d = \sum_{i=0}^{\infty} (1 \times \pi_{i,0} + \rho \times \pi_{i,3}). \tag{14.18}$$

14.3.3 Numerical Results

In order to estimate the influence of the sleep parameter δ on the energy saving strategy with a single-sleep mode proposed in this chapter, we present numerical

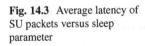

Fig. 14.3 Average latency of SU packets versus sleep parameter

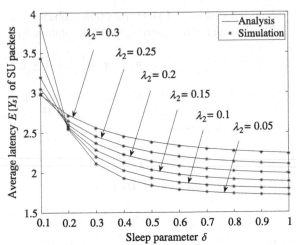

results with analysis and simulation to show the performance measures of the system. The system parameters are fixed as follows: $\mu_1 = 0.8$ and $\mu_2 = 0.7$ as an example for all the numerical results. From numerical results shown in the following figures, good agreements between the analysis results and the simulation results are observed.

By setting the arrival rate of SU packets as $\lambda_1 = 0.1$, we examine the influence of the sleep parameter δ on the average latency $E[Y_s]$ of SU packets for different arrival rates λ_2 of PU packets in Fig. 14.3.

In Fig. 14.3, we find that for the same arrival rate λ_2 of PU packets, the larger the sleep parameter δ is, the more likely it is that the BS is in an awake period, and the more timely the SU packets are transmitted. This results in a decrease in the average latency $E[Y_s]$ of SU packets.

We also observe that for a lesser sleep parameter δ, such as $\delta < 0.2$, the time length of a sleep period is greatly influenced by the arrival rate λ_2 of PU packets. That is to say, the longer the arriving interval of PU packets is, the more difficult it is for the sleep period to be terminated. Therefore, the waiting time for SU packets at the system will be longer, it leads to an increase in the average latency $E[Y_s]$ of SU packets. On the other hand, a larger sleep parameter δ, such as δ ($\delta \geq 0.2$), makes the time length of a sleep period shorter. As the arrival rate of PU packets increases, the BS is more likely to be awake with the transmission of PU packets. Therefore, the waiting time for SU packets at the system becomes longer, and that results in an increase in the average latency $E[Y_s]$ of SU packets.

By setting $\rho = 0.4$ as an example in numerical results, in Fig. 14.4, we illustrate the energy saving degree γ_d versus the sleep parameter δ for the different arrival rates λ_1 of SU packets and λ_2 of PU packets.

In Fig. 14.4, we find that when the arrival rate λ_1 of SU packets and the arrival rate λ_2 of PU packets are given, the larger the sleep parameter δ is, the more likely

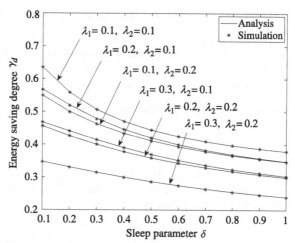

Fig. 14.4 Energy saving degree versus sleep parameter

the BS is in an awake period, thus the energy consumption increases and the energy saving degree γ_d decreases.

From Fig. 14.4, we also find that for the same sleep parameter δ and the same arrival rate λ_1 of SU packets (resp. the arrival rate λ_2 of PU packets), the higher the arrival rate λ_2 of PU packets (resp. the arrival rate λ_1 of SU packets) is, the easier it is for the sleep period to be terminated, and the more likely the BS will be in an awake period. This results in a higher energy consumption, so the energy saving degree decreases.

14.4   Analysis of Admission Fee

In this section, we first investigate the Nash equilibrium behavior and socially optimal behavior of SU packets in the energy saving strategy with a single-sleep mode proposed in this chapter. Then, we propose a pricing policy for SU packets to optimize the system performance socially. This issue can be addressed by imposing an appropriate admission fee for SU packets.

14.4.1   Behaviors of Nash Equilibrium and Social Optimization

A successful transmission means a reward for an SU packet. Every SU packet wants to access the system to be transmitted successfully to get reward. However, a higher arrival rate of SU packets will lead to a higher average latency of SU packets. For this, we study the Nash equilibrium behavior and the socially optimal behavior of

SU packets in this subsection. Based on the system model built in Sect. 14.2, we give some hypotheses as follows:

(1) The reward for an SU packet transmitted successfully is R_g.
(2) The cost of an SU packet staying in the system is C_g per slot.
(3) The benefit for all the SU packets can be added together.

Then, we give the individual benefit function $G_{ind}(\lambda_1)$ as follows:

$$G_{ind}(\lambda_1) = R_g - C_g E[Y_s]. \tag{14.19}$$

Under the condition that there is no pricing policy for an SU packet, by aggregating the individual benefits of all the SU packets, we give the social benefit function $G_{soc}(\lambda_1)$ as follows:

$$G_{soc}(\lambda_1) = \lambda_1(R_g - C_g E[Y_s]). \tag{14.20}$$

In order to explore the monotonic property of the individual benefit function $G_{ind}(\lambda_1)$ and the social benefit function $G_{soc}(\lambda_1)$, we present numerical results to illustrate the change trends of $G_{ind}(\lambda_1)$ and $G_{soc}(\lambda_1)$ in Figs. 14.5 and 14.6, respectively. Besides the system parameters given in Sect. 14.3.3, we set $R_g = 4.5$ and $C_g = 0.8$ in the numerical results.

Every SU is individually selfish and tries to access the system to get benefit. In Fig. 14.5, we find that with an increase in the arrival rate of SU packets, the individual benefit function shows a decreasing trend. The reason is that the average latency of SU packets will increase normally as the arrival rate of SU packets increases. We also observe that for each curve in Fig. 14.5, a unique value of λ_1

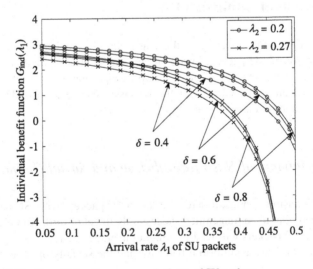

Fig. 14.5 Individual benefit function versus arrival rate of SU packets

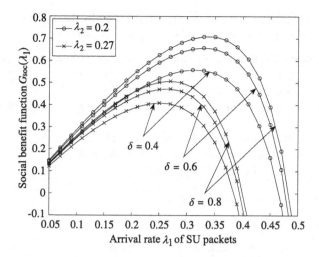

Fig. 14.6 Social benefit function versus arrival rate of SU packets

exists subject to $G_{ind}(\lambda_1) = 0$, and this is the Nash equilibrium arrival rate λ_1^e of SU packets.

In Fig. 14.6, we find that all the curves exhibit a property of concave on the whole. Loosely speaking, as the arrival rate of SU packets increases, the social benefit function firstly increases and then decreases. The reason is that when the arrival rate of SU packet is smaller, as the arrival rate of SU packet increases, the average latency of SU packets has no significant changes, however, more SUs will earn reward, so the value of the social benefit function will increase. When the arrival rate of SU packet is larger, the dominant factor influencing the value of the social benefit function is the average latency of SU packets, with the increase in the arrival rate of SU packet, the average latency of SU packets becomes greater. Therefore, the value of the social benefit function will decrease.

However, the mathematical expressions for the average latency of SU packets and the social benefit function are difficult to be given in close-forms, then the strict monotonicity of the social benefit function is difficult to be discussed. Therefore, neither the simple numerical algorithms nor the analytical approaches are inappropriate to be used to resolve the optimization problem involved in this chapter. Therefore, in order to obtain the socially optimal arrival rate of SU packets, we fall back on intelligent optimization algorithms with powerful global search ability. For this, we use the GSA to obtain the socially optimal arrival rate λ_1^* of SU packets and the maximum value of the social benefit function $G_{soc}(\lambda_1^*)$. We map the different arrival rates λ_1 of SU packets to the position of the agents, and the value of the social benefit function to the mass of an agent. The gravity force, the acceleration and the position charge of the agents are all produced by the existence of the mass of the agents. The main steps of the GSA are given as follows:

Step 1: Initialize a small constant ε (for example, $\varepsilon = 10^{-6}$) related to calculation accuracy, the number N of agents, the number L of necessary position updates and the number K of best solutions.

Step 2: Within the constraint condition $[0, 1)$, randomly set position (λ_1) of N agents.

% λ_1^h is the position of the hth agent, $h \in \{1, 2, 3, \ldots, N\}$.

Step 3: Calculate the inertial mass $M(h)$ for the hth agent.

$$G_{\text{soc}}(\lambda_1^t) = \lambda_1^t(R_g - C_g E[Y_s^t])$$
$$t \in \{1,2,3,\ldots,N\}$$

$$best = \max_{t \in \{1,2,3,\ldots,N\}} \{G_{\text{soc}}(\lambda_1^t)\}$$

$$worst = \min_{t \in \{1,2,3,\ldots,N\}} \{G_{\text{soc}}(\lambda_1^t)\}$$

$$m(h) = \frac{G_{\text{soc}}(\lambda_1^h) - worst}{best - worst}$$

$$M(h) = \frac{m(h)}{\sum\limits_{t=1}^{N} m(t)}, \quad h \in \{1, 2, 3, \ldots, N\}$$

% $E[Y_s^t]$ is the average latency of SU packets with the arrival rate λ_1^t.

Step 4: Calculate the gravity force $F(h)$ for the hth agent.

$$F_{ht} = G \frac{M(t)M(h)}{|\lambda_1^h - \lambda_1^t| + \varepsilon}$$

$$F(h) = \sum_{t=1, t \neq h}^{N} \text{rand} \times F_{ht}, \quad h, t \in \{1, 2, 3, \ldots, N\}$$

% rand is a random number selected in the interval $(0, 1)$.

Step 5: Calculate the acceleration $a(h)$ for the hth agent.

$$a(h) = \frac{F(h)}{M(h)}, \quad h \in \{1, 2, 3, \ldots, N\}$$

Step 6: Update velocity $V(h)$ and position λ_1^h for the hth agent.

$$V(h) = \text{rand}_h \times V(h) + a(h)$$
$$\lambda_1^h = \lambda_1^h + V(h), \quad h \in \{1, 2, 3, \ldots, N\}$$

Step 7:

    if the number of position updates does not reach the necessary limit L

      go to **Step 3**

    elseif the number of best solutions does not reach the upper limit K

$$G_{\text{soc}}(\lambda_1^t) = \lambda_1^t(R_g - C_g E[Y_s^t]), \quad t \in \{1, 2, 3, \ldots, N\}$$

$$\lambda[x] = \underset{t \in \{1,2,3,\ldots,N\}}{\text{argmax}} \{G_{\text{soc}}(\lambda_1^t)\}$$

$$x = x + 1$$

    go to **Step 2**

    % $\lambda[x]$ is an array that help to record K best solutions, $x \in \{1, 2, 3, \ldots, K\}$.

else
$$\lambda_1^* = \text{average}\{\lambda[x]\}$$
$$G_{\text{soc}}^* = \lambda_1^*(R_g - C_g E[Y_s^*])$$
endif
endif

Step 8: Output λ_1^* and $G_{\text{soc}}(\lambda_1^*)$.

In the algorithm above, the best solution is found through the position movement of agents. Agents are considered as objects and their performance is measured by their mass. All these objects are attracted to each other due to gravity. This force causes a movement of all objects globally towards the objects with heavier masses. The heavy masses correspond to good solutions of the problem. Inspired by physics, each agent has three specifications: position, inertial mass and gravitational force.

Substituting several groups of the given arrival rate λ_2 of PU packets and the sleep parameter δ into the algorithm, we obtain the socially optimal behavior of SU packets. The numerical results of the social optimal behavior of SU packets are shown in Table 14.1.

In Table 14.1, the estimates of λ_1^* and $G_{\text{soc}}(\lambda_1^*)$ are both accurate to four decimal places. Comparing the numerical results in Table 14.1 and Fig. 14.6, we find that the Nash equilibrium arrival rate λ_1^e is always greater than the corresponding socially optimal arrival rate λ_1^* when the arrival rate λ_2 of PU packets and the sleep parameter δ are given. The gap between λ_1^e and λ_1^* can be filled by charging the SU packets a reasonable fee.

14.4.2   Pricing Policy

We have the consensus that social optimization allows the system obtain its maximum benefit. Therefore, it is imperative to encourage a decrease in the Nash equilibrium arrival rate λ_1^e of SU packets. By introducing a pricing policy for SU packets, we set a spectrum admission fee f. We modify the individual benefit function $G_{\text{ind}}'(\lambda_1)$ as follows:

$$G_{\text{ind}}'(\lambda_1) = R_g - C_g E[Y_s] - f. \tag{14.21}$$

Table 14.1 Socially optimal arrival rate of SU packets

Arrival rates λ_2 of PU packets	Sleep parameters δ	Socially optimal arrival rates λ_1^* of SU packets	Maximum benefits $G_{\text{soc}}(\lambda_1^*)$
0.2	0.4	0.3061	0.5577
0.2	0.6	0.3347	0.6578
0.2	0.8	0.3365	0.7097
0.27	0.4	0.2459	0.4084
0.27	0.6	0.2571	0.4717
0.27	0.8	0.2691	0.507

Table 14.2 Numerical results for spectrum admission fee

Arrival rates λ_2 of PU packets	Sleep parameters δ	Optimal arrival rates λ_1^* of SU packets	Admission fees f
0.2	0.4	0.3061	1.8217
0.2	0.6	0.3347	1.9654
0.2	0.8	0.3365	2.1092
0.27	0.4	0.2459	1.6609
0.27	0.6	0.2571	1.8347
0.27	0.8	0.2691	1.8841

Letting $G'_{\text{ind}}(\lambda_1) = 0$, we calculate the admission fee f as follows:

$$f = R_g - C_g E[Y_s]. \tag{14.22}$$

Substituting the socially optimal arrival rate λ_1^* of SU packets given in Table 14.1 into Eq. (14.17), we calculate the average latency $E[Y_s]$ of SU packets. Afterwards, we obtain the spectrum admission fee f by Eq. (14.22).

For different arrival rates λ_2 of PU packets and sleep parameters δ, we present numerical results of the spectrum admission fee f in Table 14.2.

With the spectrum admission fee f, we modify the social benefit function $G'_{\text{soc}}(\lambda_1)$ as follows:

$$G'_{\text{soc}}(\lambda_1) = \lambda_1(R_g - C_g E[Y_s] - f) + \lambda_1 f$$
$$= \lambda_1(R_g - C_g E[Y_s]). \tag{14.23}$$

Comparing Eqs. (14.20) and (14.23), we find that the final expression of $G_{\text{soc}}(\lambda_1)$ and $G'_{\text{soc}}(\lambda_1)$ are the same. This is because that in spite of there being a pricing policy for SU packets, the aggregation of the spectrum admission fee f is still in the system. In another words, the money is just transferred from the SU packets side to the BS side, and this is the reason why the social benefit function does not change.

14.5   Conclusion

Taking into account the practical significance of improving the spectrum efficiency and achieving greener communication in wireless applications, in this chapter, we proposed an energy saving strategy with a single-sleep mode and an LTE-A structure in CRNs. Accordingly, we established a preemptive priority queueing model with a single vacation. We estimated the system performance in terms of the average latency of SU packets and the energy saving degree. Numerical results showed that there is a trade-off between the average latency of SU packets and the energy saving degree when setting the sleep parameter. Analytical results were compared with simulation results and good agreements were observed. Moreover, by establishing the individual benefit function and the social benefit function, we investigated the Nash equilibrium and socially optimal behaviors of SU packets and presented a pricing policy for SU packets to oblige the SU packets to optimize the system performance socially.

Chapter 15
Energy Saving Strategy in CRNs Based on a Priority Queue with Multiple Vacations

In order to meet the demand for more sustainable green communication, in this chapter, we propose a multiple-sleep mode for licensed channels in Cognitive Radio Networks (CRNs). Based on a dynamic spectrum access strategy with a multiple-sleep mode, we establish a Continuous-Time Markov chain (CTMC) model to capture the stochastic behavior of Primary User (PU) and Secondary User (SU) packets. By using the matrix-geometric solution method, we obtain the steady-state distribution of the system model. We derive performance measures of the system in terms of the throughput of SU packets, the average latency of SU packets, the energy saving rate of the system and the channel utilization. We also present numerical results to evaluate the influences of the service rate of one channel and the sleep parameter on the system performance measures. Finally, we construct a system cost function, and improve a Jaya algorithm employing an insect-population model to optimize the proposed energy saving strategy with a multiple-sleep mode.

15.1 Introduction

Radio spectrum is one of the most precious and limited resources in wireless communication systems. In the traditional framework of the communication resource allocation, a user has to obtain spectrum usage license from the respective government before transmitting and receiving data in the desired band [Mont01]. So that the remaining wireless spectrum suitable for WCNs is being exhausted. For this, CRN is proposed as a solution to solve the problem of spectrum scarcity [Mari12].

In a CRN, the licensed users are called PUs, while unlicensed users are called SUs [Yang13a]. The SUs with cognitive capability can dynamically access the licensed spectrum in an opportunistic way, so the available spectrum resource for the SUs is referred to be as a spectrum hole [Park19, Sult16]. In such network, users' data is always divided into several segments called packets. The users send request messages to mobile service switching center, namely, Base Station (BS), for

© Springer Nature Singapore Pte Ltd. 2021
S. Jin, W. Yue, *Resource Management and Performance Analysis of Wireless Communication Networks*, https://doi.org/10.1007/978-981-15-7756-7_15

transmitting data, and then the BS allocates available radio spectrum to the arriving user packets to complete data transmission.

With the rapid development of mobile terminals, such as mobile phones, laptop computers, tablet computers, POS peripherals and so on, more BSs are needed to be constructed. This results in more energy consumption and air pollution. To solve this problem, green communication in CRNs, called green CRNs, is proposed to reduce emission pollution, minimize operation cost and decrease energy consumption. Realizing green spectrum management is a strong challenge in the green CRNs. Certainly, in a green CRN, the study on how to enhance energy efficiency and spectrum efficiency is necessary and meaningful.

One of most efficient methods to enhance energy efficiency is to consider applying sleep mode and on/off mode that have been studied widely to the green spectrum management [Jin16a, Jin16b, Liu18]. In [Oh10], the authors studied a dynamic switching BS to reduce the energy consumption by considering the time varying characteristic of the traffic profile. In [Qiao12], the authors introduced a centralized sleep mode based on heterogeneous hierarchical cognitive radio sensor networks. Instead of adopting a unified and consistent BS-off scheme all over the network, in [Zhan13a], the authors proposed a clustering BS-off scheme and optimized BS on/off mechanism in each cluster separately. In [Xiao13], the authors proposed sleeping actions to find an optimal schedule in order to improve the energy efficiency in CRNs. In [Chen14], the authors developed a sleep mode for sensor nodes based on correlations among sensor data within sub-clusters in random cluster heads and sub-cluster heads. In [Chen15a], the authors proposed a distributed BS sleep scheduling scheme to maximize the energy efficiency with the constraint of spectral efficiency in relay-assisted cellular networks. In [Choi15], the authors proposed an adaptive cell zooming method to reduce the energy consumption in cellular networks by using an on/off technique. In [Wu15], the authors proposed a sleeping mode controller for pico BS in heterogeneous networks to save energy costs by adapting to a time-varying traffic load.

However, these researches are all to use a single-sleep mode and a single on/off mode to save energy consumption and to reduce emission pollution in a wireless communication system. In such wireless communication system with a single-sleep mode, the channels will switch to the awake state after a sleep period, even though the system buffer is empty. Actually, in the wireless communication system, the channels will enter another sleep period as long as there are no packets to be transmitted, according to a given transmission policy. Therefore, in such wireless communication system, obviously, with a multiple-sleep mode, more energy will be saved than that with a single-sleep mode.

On the other hand, in green CRNs, given a channel access protocol and a set of source-to-destination paths, the performance evaluations such as end-to-end throughput and packet delay are widely used. Also, in order to get the utmost out of the spectrum resource and meet the demands for the QoS of SUs, in the performance analysis, we should also consider some factors such the system throughput, the average latency and the channel utilization to optimize the system performance.

Nowadays, some scholars concentrate their research on the trade-off analysis for system performance in a communication network. In [Yang12], the authors proposed a general cooperative game-theoretical scheme to achieve the optimal performance trade-off between individual fairness and network energy efficiency. In [Teng13], the authors proposed an energy efficiency heuristic algorithm for cross-layer optimization based on the combination of QoS-aware flow control, routing selection, channel and power allocation. In [Li14a], the authors proved that the energy efficiency is a unimodal function, and then obtained the optimal sensing time and the maximal energy efficiency. In [Wu14], the authors presented an energy utility function and achieved the maximum data rate along with the energy saving rate of the system. In [Qu14], the authors designed a spectrum scheduling scheme, and then obtained a global solution to the joint problem of channel allocation and power control based on a Particle Swarm Optimization (PSO) algorithm. In [Wang15b], the authors investigated the robust energy efficiency maximization problem on the premise of ensuring the QoS of PUs. In [Liu16], the authors presented an opportunistic power control strategy with the purpose of minimizing the transmission power of SU packets by considering the uncertain channel gains in underlaying CRNs.

These researches pointed out that complicated nonlinear equations and nonlinear optimization problems will be involved to optimize the system parameters. It is of great significance to investigating an appropriate intelligent optimization algorithm to realize trade-off analysis for the system performance.

Motivated and inspired by the research work mentioned above, in this chapter, we present performance evaluation and optimization on green CRNs with a multiple-sleep mode by aiming to further improve the energy efficiency and enhance the spectrum utility. In addition, we will present an appropriate intelligent optimization algorithm to optimize the system parameters for the energy saving strategy with a multiple-sleep mode in green CRNs. In order to get the utmost out of the spectrum resource and meet the demands for QoS requirements of SUs, we also construct a system cost function, and apply a Jaya algorithm employing an insect-population model to optimize the energy saving strategy with a multiple-sleep mode proposed in this chapter.

The main contributions of this chapter can be listed as follows:

(1) For the purpose of conserving the energy consumption of BSs in green CRNs, we propose an energy saving strategy with a multiple-sleep mode.
(2) We establish a CTMC model to capture the stochastic behavior of user packets and present analyses to evaluate numerically the proposed energy saving strategy with a multiple-sleep mode.
(3) In order to get the utmost out of the spectrum resource and meet the demands for QoS requirements of SUs, we develop an improved Jaya algorithm to jointly optimize the service rate of a channel as well as the sleep parameter.

The chapter is organized as follows. In Sect. 15.2, we describe the energy saving strategy with a multiple-sleep mode proposed in this chapter. Then, we present the system model in detail. In Sect. 15.3, we present a performance analysis of the system model in the steady state. In Sect. 15.4, we derive performance measures and present numerical results to investigate the system performance. In Sect. 15.5, by analyzing the system cost and developing an improved Jaya algorithm, we optimize the system parameters in terms of the service rate of one channel and the sleep parameter. Our conclusions are drawn in Sect. 15.6.

15.2 Energy Saving Strategy and System Model

In this section, we propose an energy saving strategy with a multiple-sleep mode in green CRNs to reduce the energy consumption in BSs. Then, we establish a CTMC model to capture the stochastic behavior of the two types of user packets, the PU packets and the SU packets. We call both of the PU packets and the SU packets the user packets.

15.2.1 Energy Saving Strategy

Considering that more energy will be saved with a multiple-sleep mode than that with a single-sleep mode, in this chapter, we propose an energy saving strategy with a multiple-sleep mode. We suppose that a licensed spectrum with M channels is controlled by M ports in a BS, and there is a one-to-one relation among the M ports in a BS and the M channels. The ports in the BS will be switched between two states: the sleep state and the awake state.

When there are no packets to be transmitted on one channel, the corresponding port of the BS will be in the sleep state, namely, the hardware of the port and the port's application process will be closed. There may be multiple-sleep periods within one sleep state. At the beginning of a sleep period, a sleep timer is started.

When a PU packet arrives at the system during a sleep period, the sleep timer will be immediately terminated, one port in the BS will be switched to an awake state; Or when there are SU packets waiting for transmission in the system, once the sleep timer on one port in the BS expires, the corresponding port in the BS will be also switched to an awake state. On the contrary, if there are not any packet arrivals at the system when the current sleep timer expires, the system will enter another sleep period.

In summary, the state transition on one port in BS is illustrated in Fig. 15.1.

Based on the energy saving strategy with a multiple-sleep mode mentioned above, we discuss the activity of PU packets and SU packets, respectively. We will discuss the PU packet activity following three cases where PU packets arrive at the system.

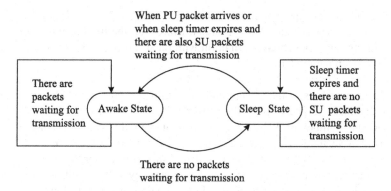

Fig. 15.1 State transition on one port in BS

(1) If all M channels are being occupied by the other PU packets, the newly arriving PU packet will be blocked. That is to say, we do not set a buffer for PU packets. This idea comes out of the consideration that the traffic of PU packets is light.

(2) If there are idle channels, the new arriving PU packet will randomly occupy one idle channel to finish the transmission.

(3) If there are no idle channels, but there is at least one channel which is being occupied by an SU packet, the newly arriving PU packet will randomly occupy the SU packet's channel preemptively.

The SU packet activity is assumed as follows: The newly arriving SU packet will enter the buffer prepared for SU packets. The SU packet queueing at the head of the buffer will occupy the channel opportunistically when the transmission of one user packet (one PU packet or one SU packet) in the system is completed or one port in the BS switches to an awake state normally. In addition, the SU packets interrupted by PU packets will be discarded by the system.

15.2.2 System Model

Based on the proposed energy saving strategy with a multiple-sleep mode mentioned in Sect. 15.2, we build a three-dimensional CTMC model. In this system model, there are two types of the data packets. One is the PU packets having a preemptive priority to be transmitted without a buffer. The other is the SU packets with low priority. An unlimited buffer called the system buffer is prepared for the SU packets and channels are used to transmit both PU packets and SU packets.

With the purpose of discussing the problem clearly and easily, we make the following assumptions: The inter-arrival times for both of the SU packets and the PU packets are i.i.d. random variables following exponential distributions with means $1/\lambda_1$ and $1/\lambda_2$, respectively, where λ_1 and λ_2 are the arrival rates of the SU packets and the PU packets, $\lambda_1 > 0$ and $\lambda_2 > 0$. The transmission time of a packet on one channel follows an exponential distribution with mean $1/\mu$ seconds, where $\mu > 0$. We call μ the service rate of one channel. Moreover, the time length of the sleep timer follows an exponential distribution with mean $1/\delta$ seconds, where $\delta > 0$. We call δ the sleep parameter.

Let $X(t) = i$ ($i = 0, 1, 2, \ldots$) and $Y(t) = j$ ($j = 0, 1, 2, \ldots, M$) indicate the total number of SUs in the system and the number of SU packets being serviced on the channels, respectively, at the instant t. Let $Z(t) = k$ ($k = 0, 1, 2, \ldots, M$) indicate the number of PU packets at instant t in the system. Using a three-dimensional vector $\{(X(t), Y(t), Z(t)),\ t \geq 0\}$ to record the stochastic behavior of PU packets and SU packets, we establish a three-dimensional CTMC model to capture our proposed spectrum energy saving strategy with a multiple-sleep mode in green CRNs. The state space of the Markov chain is given as follows:

$$\Omega = \{(i, j, k) : i \geq 0, j \geq 0, 0 \leq k \leq M - j\}. \tag{15.1}$$

Let $\pi_{i,j,k}$ be the probability that the total number of SUs in the system is i, the number of SU packets being serviced on the channels is j and the number of PU packets in the system is k in the steady state. $\pi_{i,j,k}$ is then given as follows:

$$\pi_{i,j,k} = \lim_{t \to \infty} \Pr\{X(t) = i, Y(t) = j, Z(t) = k\}, \quad i \geq 0, j \geq 0, 0 \leq k \leq M - j. \tag{15.2}$$

15.3   Performance Analysis

Let Q be a one-step transition rate matrix of the Markov chain $\{(X(t), Y(t), Z(t)),\ t \geq 0\}$, and $q_{(i,j,k),(l,m,n)}$ be the one-step transition rate from state (i, j, k) to state (l, m, n), where $(i, j, k), (l, m, n) \in \Omega$. The total number of SU packets is called the system level. According to the changes of the system levels, all the one-step transition rates from the original state (i, j, k) to the other possible state are discussed as follows.

(1) System level i changes to level $i + 1$ via a one-step transition. This means that there is an SU packet arrival. Since the buffer prepared for SU packets is infinite, the one-step transition rate $q_{(i,j,k),(i+1,j,k)}$ is written as:

$$q_{(i,j,k),(i+1,j,k)} = \lambda_1, \quad i, j, k \geq 0, \ j + k \leq M. \tag{15.3}$$

(2) System level i is fixed via a one-step transition. This means that there are no SU packet arrivals. If there is a PU packet arrival, the newly arriving PU packet will occupy an idle channel. This induces a change in the number of PU packets from k to $k + 1$. If the sleep timer for one port of the BS expires, then the SU packet queueing at the head of the buffer will immediately occupy the awake channel. In this case, there is one more SU packet being transmitted, namely, j

changes to $j + 1$. If the transmission of a PU packet occupying the channel is completed, the SU packet queueing at the head of the buffer will immediately occupy this channel. This means that k changes to $k - 1$ and j changes to $j + 1$.

Therefore, we can obtain the formula as follows:

$$
\begin{cases}
q_{(i,j,k),(i,j,k+1)} = \lambda_2, \quad i, j, k \geq 0, \ j + k < M \\
q_{(i,j,k),(i,j+1,k)} = (M - j - k)\delta, \quad i > j \geq 0, \ k \geq 0, \ j + k < M \\
q_{(i,j,k),(i,j+1,k-1)} = k\mu, \quad i > j > 0, \ k \geq 1, \ j + k < M \\
q_{(i,j,k),(i,j,k-1)} = k\mu, \quad i = j, \ j \geq 0, \ k \geq 1, \ j + k \leq M \\
q_{(i,j,k),(i,j,k)} = -\lambda_1 - \lambda_2 - (M - j - k)\delta - k\mu - j\mu, \\
\qquad\qquad\qquad\qquad i, j \geq 1, \ k \geq 0, \ j + k \leq M \\
q_{(i,j,k),(i,j,k)} = -\lambda_1 - \lambda_2 - k\mu, \quad i = 0, \ j = 0, \ 0 \leq k < M \\
q_{(i,j,k),(i,j,k)} = -\lambda_1 - M\mu, \quad i = 0, \ j = 0, \ k = M.
\end{cases}
\tag{15.4}
$$

(3) System level i changes to level $i - 1$ via one-step transition. This means that an SU packet departs the system. If there are no idle channels when a PU packet arrives at the system, then the newly arriving PU packet will randomly occupy one of channels on which an SU packet is being transmitted. This means that j changes to $j - 1$ and k changes to $k + 1$. If the transmission of an SU packet occupying the channel is completed normally, then j changes to $j - 1$ or it does not change.

Therefore, we can obtain a set of equations as follows:

$$
\begin{cases}
q_{(i,j,k),(i-1,j-1,k+1)} = \lambda_2, \quad i \geq j \geq 1, \ k \geq 0, \ j + k = M \\
q_{(i,j,k),(i-1,j-1,k)} = j\mu, \quad i = j, \ 1 \leq j \leq M, \ k \geq 0, \ j + k \leq M \\
q_{(i,j,k),(i-1,j,k)} = j\mu, \quad i > j \geq 1, \ k \geq 0, \ j + k \leq M.
\end{cases}
\tag{15.5}
$$

Since the PU packets have preemptive priority and there are M channels in the system, we can obtain the number S_i of all the states with the system level i as follows:

$$
S_i = \sum_{b=0}^{i} (M + 1 - b).
\tag{15.6}
$$

Let $Q_{u,v}$ be the one-step transition rate sub-matrix from the system level u to the system level v. Based on Eqs. (15.3)–(15.6), $Q_{u,v}$ can be figured out. The structure of the one-step transition rate Q is given as follows:

$$
Q = \begin{pmatrix}
Q_{0,0} & Q_{0,1} & Q_{0,2} & & & & \\
Q_{1,0} & Q_{1,1} & Q_{1,2} & & & & \\
& Q_{2,1} & Q_{2,2} & Q_{2,3} & & & \\
& & \ddots & \ddots & \ddots & & \\
& & & Q_{M,M-1} & Q_{M,M} & Q_{M,M+1} & \\
& & & & Q_{M+1,M} & Q_{M+1,M+1} & Q_{M+1,M+2} \\
& & & & & Q_{M+1,M} & Q_{M+1,M+1} & Q_{M+1,M+2} \\
& & & & & & \ddots & \ddots & \ddots
\end{pmatrix}.
$$

(15.7)

From the structure of the one-step transition rate matrix Q, we find that Q is a blocked three-diagonal matrix and the system state transition occurs only in adjacent levels. Therefore, the stochastic process $\{(X(t), Y(t), Z(t)), \ t \geq 0\}$ is a Quasi Birth-Death (QBD) process.

Let π_i be the steady-state distribution of the system being at level i. π_i can be given as follows:

$$
\begin{cases}
\pi_0 = (\pi_{0,0,0}, \pi_{0,0,1}, \ldots, \pi_{0,0,M}) \\
\pi_i = (\pi_{i,0,0}, \pi_{i,0,1}, \ldots, \pi_{i,0,M}, \pi_{i,1,0}, \pi_{i,1,1}, \ldots, \pi_{i,i,M-i}), \quad 0 < i \leq M \\
\pi_i = (\pi_{i,0,0}, \pi_{i,0,1}, \ldots, \pi_{i,0,M}, \pi_{i,1,0}, \pi_{i,1,1}, \ldots, \pi_{i,j,M-j}, \ldots, \pi_{i,M,0}), \\
\qquad\qquad\qquad\qquad i \geq M + 1.
\end{cases}
$$

(15.8)

The steady-state distribution Π of the system is given as follows:

$$
\Pi = (\pi_0, \pi_1, \pi_2, \ldots). \tag{15.9}
$$

We find that the rows of the one-step transition rate Q start to repeat after the $\sum_{i=0}^{M} S_i$ row. In order to employ a matrix-geometric solution method, we construct the new matrix $B[R]$ as follows:

$$
B[R] = \begin{pmatrix}
Q_{0,0} & Q_{0,1} & Q_{0,2} & & & \\
Q_{1,0} & Q_{1,1} & Q_{1,2} & & & \\
& Q_{2,1} & Q_{2,2} & Q_{2,3} & & \\
& & \ddots & \ddots & \ddots & \\
& & & Q_{M+1,M} & Q_{M+1,M+1} & Q_{M+1,M+2} \\
& & & & Q_{M+1,M} & Q_{M+1,M+1} + R Q_{M+1,M}
\end{pmatrix}
$$

(15.10)

where matrix R is the minimum non-negative solution of the matrix equation $R^2 Q_{M+1,M} + R Q_{M+1,M+1} + Q_{M+1,M+2} = 0$.

The steady-state distribution Π satisfies the following set of equations:

$$
\begin{cases}
(\pi_0, \pi_1, \pi_2, \ldots, \pi_{M+1}) B[R] = 0 \\
\pi_0 e + \pi_1 e + \pi_2 e + \ldots + \pi_M e + \pi_{M+1}(I - R)^{-1} e = 1 \\
\pi_i = \pi_{M+1} R^{i-M-1}, \quad i \geq M + 1
\end{cases}
\tag{15.11}
$$

where e is a three-dimensional column vector and all elements of the vector are equal to 1.

The steady-state distribution Π can be obtained based on Eqs. (15.10) and (15.11).

15.4 Performance Measures and Numerical Results

In this section, we first derive performance measures of the system in terms of the throughput of SU packets, the average latency of SU packets, the energy saving rate of the system and the channel utilization, respectively. Then, we present numerical results to evaluate the performance of the system using the energy saving strategy with a multiple-sleep mode proposed in this chapter.

15.4.1 Performance Measures

The throughput θ of SU packets is defined as the number of SU packets transmitted successfully per second across the whole spectrum. An arriving SU packet can be successfully transmitted only when the SU packet is interrupted by a PU packet. If there are no idle channels, the BS will randomly assign one of the channels being occupied by SU packets to a newly arriving PU packet. This induces an interruption in the transmission of SU packets. Therefore, we give the throughput θ of SU packets as follows:

$$
\theta = \lambda_1(1 - \beta_s)
\tag{15.12}
$$

where β_s is the interruption rate of SU packets given by

$$
\beta_s = \lambda_2 \sum_{k=0}^{M-1} \sum_{i=M-k}^{\infty} \pi_{i,M-k,k}.
\tag{15.13}
$$

The latency Y_s of an SU packet is defined as the duration from the instant an SU packet arrives at the system to the instant this SU packet departs the system successfully. By using the total probability formula, the average value $E[N_s]$ for the number N_s of SU packets in the system is given as follows:

$$E[N_s] = \sum_{i=0}^{\infty} \sum_{j=0}^{\min\{i,M\}} \sum_{k=0}^{M-j} i\pi_{i,j,k}. \qquad (15.14)$$

By using Eqs. (15.12) and (15.14), we can obtain the average latency $E[Y_s]$ of SU packets as follows:

$$E[Y_s] = \frac{E[N_s]}{\theta} = \frac{\sum_{i=0}^{\infty} \sum_{j=0}^{\min\{i,M\}} \sum_{k=0}^{M-j} i\pi_{i,j,k}}{\theta}. \qquad (15.15)$$

The energy saving rate γ of the system is defined as the energy conservation in the BS per second. In our proposed energy saving strategy with a multiple-sleep mode, for one port in the BS, there are two states, namely, the sleep state and the awake state. Since some ports in the BS will be turned off when their corresponding channels are sleeping, energy can be conserved during the sleep state. However, when the sleeping channels return to the awake state from the sleep state, additional energy will be consumed to activate the closed ports. Let g_1 be the energy conservation per second when one port in the BS is in the sleep state, and let g_2 be the energy consumption for each switching procedure of one port in the BS from the sleep state to the awake state. We give the energy saving rate γ of the system as follows:

$$\gamma = \sum_{i=0}^{\infty} \sum_{j=0}^{\min\{i,M\}} \sum_{k=0}^{M-j} ((g_1 - g_2\delta)(M - j - k) - g_2\lambda_2)\pi_{i,j,k}. \qquad (15.16)$$

The channel utilization U_c is defined as the probability that one channel is being occupied by a user packet (a PU packet or an SU packet). The channel utilization U_c can be given by calculating the proportion of the average number of channels which are being occupied by user packets in relation to the total number of the channels. Therefore, we give the channel utilization U_c as follows:

$$U_c = \frac{1}{M} \sum_{i=0}^{\infty} \sum_{k=0}^{M} \sum_{j=0}^{\min\{i,M-k\}} \pi_{i,j,k}(k + j). \qquad (15.17)$$

15.4.2   Numerical Results

With the purpose of verifying the feasibility of the proposed strategy with a multiple-sleep mode and verifying the accuracy of the theoretical analysis with the system model, we use Matlab 2010a to carry out analysis experiments, MyEclipse2014 to carry out simulation experiments. We set 2^{16} as the capacity for SU packets to approximate infinity in numerical results. Considering a WiFi network with bandwidth of 7 Mbps to 56 Mbps, average packet length of 1760 Byte and unit time of 1 ms, we carry out numerical results by using these system parameters. All the numerical results are provided using the Inter (R) Core (TM) i7-4790 CPU @3.6 GHz, 6GB RAM.

The system parameters are fixed as follows: $M = 4$, $g_1 = 2$, $g_2 = 0.2$, $\lambda_1 = 0.2, 0.4, 0.6$ and $\lambda_2 = 0.2, 0.25, 0.3$ as an example for all the numerical results. From numerical results shown in the following figures, good agreements between the analysis results and the simulation results are observed.

In order to conspicuously reflect the deviation for the throughput θ of SU packets caused by the sleep parameter δ and the service rate μ of one channel, we introduce the normalized throughput θ' of SU packets as follows:

$$\theta' = \frac{\theta - \min\{\theta\}}{\max\{\theta\} - \min\{\theta\}}. \tag{15.18}$$

By setting the service rate of one channel as $\mu = 0.3$ as an example, we respectively show the change trend for the normalized throughput θ' of SU packets, the average latency $E[Y_s]$ of SU packets, the energy saving rate γ of the system and the channel utilization U_c in relation to the sleep parameter δ in Figs. 15.2, 15.3, 15.4, 15.5.

Fig. 15.2 Normalized throughput of SU packets versus sleep parameter

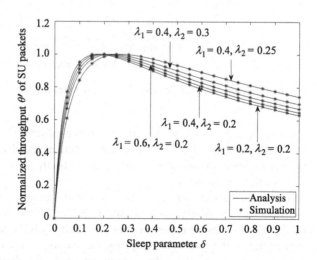

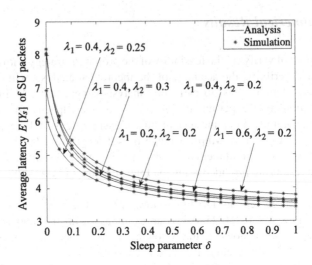

Fig. 15.3 Average latency of SU packets versus sleep parameter

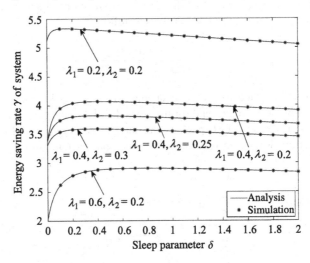

Fig. 15.4 Energy saving rate of system versus sleep parameter

Looking at Fig. 15.2, we find that if the arrival rates (λ_1 and λ_2) of user packets and the service rate μ of one channel are given, as the sleep parameter δ increases, the normalized throughput θ' of SU packets firstly shows a rising trend and then shows a downward trend.

During the rising stage, the normalized throughput of SU packets is mainly influenced by the sleep parameter. Recall that the SU packet queueing at the head of the buffer occupies the channel opportunistically only when the transmission of one user packet in the system is completed or one port of the BS switches to an awake state normally. The greater the sleep parameter is, the shorter the average length of

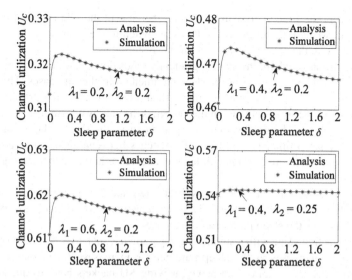

Fig. 15.5 Channel utilization versus sleep parameter

the sleep timer is and the more SU packets can be transmitted per second, which results in an increase in the normalized throughput of SU packets.

During the downward stage, when the sleep parameter is greater than a certain value, the normalized throughput of SU packets is mainly influenced by the arrivals of PU packets. A greater sleep parameter means the BS can assign channels to SU packets at the right time. This will lead more SU packets to be interrupted by PU packets. That is to say, the interrupted rate of SU packets is greater, and the fewer SU packet transmissions there will be that are finished successfully. Therefore, the normalized throughput of SU packets will decrease.

From Fig. 15.3, we observe that if the arrival rates (λ_1 and λ_2) of user packets and the service rate μ of one channel are given, the average latency $E[Y_s]$ of SU packets decreases as the sleep parameter δ increases. The reason is that as the sleep parameter increases, the average time length of the sleep timer decreases, the sooner the ports in the BS switch to an awake state from a sleep state, and the waiting time of the SU packets in the buffer becomes shorter. Then, the average latency of SU packets will decrease accordingly.

Looking at Fig. 15.4, we find that if the arrival rates (λ_1 and λ_2) of user packets and the service rate μ of one channel are given, the energy saving rate γ of the system shows a gradual downward stage after an initial fast rising stage as the sleep parameter δ increases.

During the fast rising stage, the energy saving rate γ of the system increases as the sleep parameter δ increases. The reason is that when the sleep parameter is lesser, the primary factor influencing the energy saving rate of the system is the probability of the system being idle. The greater the sleep parameter is, the shorter the average time length of the sleep timer is, so the SU packets will be transmitted at the appropriate time. This will lead to an increase in the idle probability of the system, and so the ports in the BS will easily switch to the sleep state. This results in an increase in the energy saving rate of the system.

During the stage where the system gradually tends downward, the energy saving rate γ of the system decreases as the sleep parameter δ further increases. The reason is that when the sleep parameter is greater than a certain value, the primary factor influencing the energy saving rate of the system is the energy consumption generated by state switching. The greater the sleep parameter is, the shorter the average time length of the sleep timer is, and the state switches between awake state and sleep state will be more frequent. This will lead to a decrease in the energy saving rate of the system.

In Fig. 15.5, we find that if the arrival rates (λ_1 and λ_2) of user packets and the service rate μ of one channel are given, the channel utilization U_c exhibits two obvious stages as the sleep parameter δ increases.

During the first stage with a smaller sleep parameter, such as $\delta \leq 0.2$, the channel utilization U_c increases as the sleep parameter δ increases. Recall that when the ports in the BS are sleeping, the newly arriving SU packets have to queue at the tail of buffer waiting for future transmission. When the sleep parameter is smaller, there will be more SU packets queueing in the buffer. For this case, as the sleep parameter increases, the ports in the BS will easily switch to the awake state. This leads to more channels being occupied by SU packets, and the channel utilization will increase accordingly.

During the second stage with a larger sleep parameter, such as $\delta \geq 0.2$, the channel utilization U_c decreases as the sleep parameter δ increases. Different from the first stage, when the sleep parameter is greater than a certain value, there will be fewer SU packets queueing at the buffer. For this case, the primary factor influencing the channel utilization is the traffic. The larger the sleep parameter is, the less SU packets will aggregate before the ports in the BS switch to awake state, and more ports in the BS will stay asleep. This will lead to a decrease in the channel utilization.

With the purpose of investigating the change trend for the normalized throughput θ' of SU packets in relation to the service rate μ of one channel, we set system parameters $\lambda_1 = 0.2, 0.8, 1.2$ and $\delta = 0.25$ as examples in Fig. 15.6.

Figure 15.6 shows that if the arrival rates (λ_1 and λ_2) of user packets and the sleep parameter δ are given, as the service rate μ of one channel increases, the normalized throughput θ' of SU packets firstly increases quickly, then increases more slowly and tends to 1.

At the beginning stage, the interruption rate of SU packets is a primary factor influencing the normalized throughput θ' of SU packets. The higher the service rate of one channel is, the fewer SU packets are interrupted by PU packets, and the more SU packets there are that can finish transmission successfully. This leads to an increase in the normalized throughput θ' of SU packets. However, as the service rate of one channel further increases, the influence of the interruption rate of SU packets on the normalized throughput θ' of SU packets decreases. The level of traffic is a primary factor influencing the normalized throughput θ' of SU packets. The greater the service rate of one channel is, the smaller the system load is, so the normalized

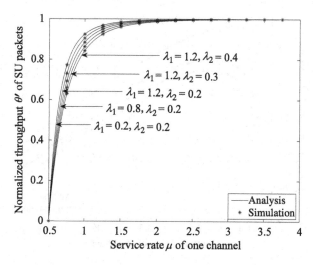

Fig. 15.6 Normalized throughput of SU packets versus service rate of one channel

throughput θ' of SU packets increases slowly. Obviously, when the service rate of one channel is high enough, most SU packets are transmitted without interruption, so the normalized throughput θ' of SU packets tends to 1.

By setting sleep parameter $\delta = 0.25$ as an example, we respectively show the change trend for the average latency $E[Y_s]$ of SU packets and the energy saving rate γ of the system in relation to the service rate μ of one channel in Figs. 15.7 and 15.8.

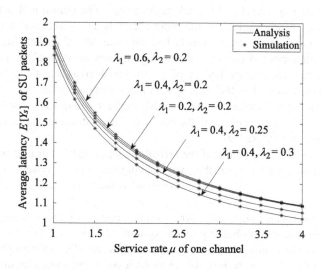

Fig. 15.7 Average latency of SU packets versus service rate of one channel

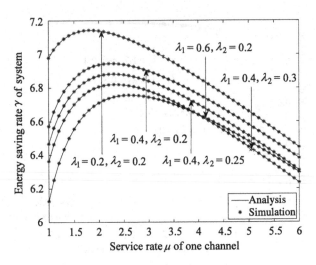

Fig. 15.8 Energy saving rate of system versus service rate of one channel

In Fig. 15.7, we find that if the arrival rates (λ_1 and λ_2) of user packets and the sleep parameter δ are given, the average latency $E[Y_s]$ of SU packets decreases as the service rate μ of one channel increases. The reason is that the higher the service rate of one channel is, the fewer SU packets there will be queueing in the buffer. This leads to a decrease in the waiting time of SU packets queueing in the buffer. Hence, the average latency $E[Y_s]$ of SU packets decreases.

In addition, from Figs. 15.3 and 15.7, we find that for the same arrival rate λ_1 of SU packets and the sleep parameter δ, as the arrival rate λ_2 of PU packets increases, the average latency $E[Y_s]$ of SU packets decreases. The reason is that the higher the arrival rate of PU packets is, the more channels are awake, and the more SU packets access the channels in a timely fashion once the PU packets' transmission is over. This induces a decrease in the average latency of SU packets. Conversely, we also find that the average latency of SU packets increases as the arrival rate of SU packets increases when the other system parameters are given. Since the higher the arrival rate of SU packets is, the more SU packets will queue at the buffer, and the longer the SU packets queue in the buffer is. This will lead to an increase in the average latency of SU packets.

In Fig. 15.8, we observe that if the arrival rates (λ_1 and λ_2) of user packets and the sleep parameter δ are given, the energy saving rate γ of the system firstly shows a rising trend and then shows a downward trend as the service rate μ of one channel increases.

During the rising stage, the energy saving rate γ of the system is mainly influenced by the idle probability of the system. It is obvious that the higher the service rate of one channel is, the more the channels are idle, and the higher the idle probability of the system is, resulting in an increase in the energy saving rate γ of the system.

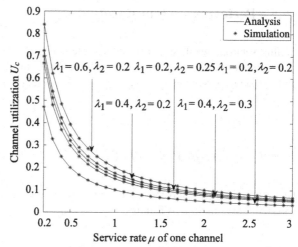

Fig. 15.9 Channel utilization versus service rate of one channel

During the stage where the system tends downward, as the service rate increases, the energy saving rate γ of the system is mainly influenced by the energy consumption generated by a state switch. The higher the service rate of one channel is, the more the energy consumption generated by a one-time state switch is. This induces a decrease in the energy saving rate γ of the system.

From Figs. 15.4 and 15.8, we also note that when the arrival rate λ_2 of PU packets, the service rate μ of one channel and the sleep parameter δ are given, as the arrival rate λ_1 of SU packets increases, the energy saving rate γ of the system decreases. The reason is that the higher the arrival rate of SU packets is, the smaller the probability of ports in the BS being in a sleep state is, so the energy saving rate γ of the system will decrease. This observation is in line with our intuition.

Taking the sleep parameter $\delta = 0.6$ as an example, we show the change trend for the channel utilization U_c in relation to the service rate μ of one channel in Fig. 15.9.

In Fig. 15.9, we observe that if the arrival rates (λ_1 and λ_2) of user packets and the sleep parameter δ are given, the channel utilization U_c decreases as the service rate μ of one channel increases. The reason is that the higher the service rate of one channel is, the fewer SU packets will queue at the buffer waiting for transmission, and the more channels where will be that are asleep. Therefore, the channel utilization will decrease.

The experiments show that the system performance measures are mainly influenced by the sleep parameter and the service rate of one channel. In order to get the utmost out of the spectrum resource and meet the demands for QoS requirements of SUs, we optimize the system performance with reasonable parameters in terms of the service rate of one channel and the sleep parameter.

15.5 Performance Optimization

In this section, we first construct a system cost function to trade off different performance measures. Then, we jointly optimize the service rate of one channel and the sleep parameter in the energy saving strategy with a multiple-sleep mode proposed in this chapter for improving the system performance.

15.5.1 Analysis of System Cost

By trading off different performance measures obtained in Sect. 15.4.1, we construct a system cost function as follows:

$$F(\mu, \delta) = f_1 E[Y_s] + \frac{f_2}{\gamma} + \frac{f_3}{\theta} + \frac{f_4}{U_c} \qquad (15.19)$$

where f_1, f_2, f_3 and f_4 are the impact factors of the average latency $E[Y_s]$ of SU packets, the energy saving rate γ of the system, the throughput θ of SU packets and the channel utilization U_c to the system cost function, respectively.

By minimizing the system cost function, the optimal combination (μ^*, δ^*) is given as follows:

$$(\mu^*, \delta^*) = \underset{\mu>0, \delta>0}{\mathrm{argmin}} \{F(\mu, \delta)\}. \qquad (15.20)$$

In order to obviously demonstrate the change trend of the system cost function, we set system parameters $\lambda_1 = 0.3, 0.4, 0.5$, $\lambda_2 = 0.2, 0.3, 0.35$, $f_1 = 1$, $f_2 = 50$, $f_3 = 2$, $f_4 = 2$ and $M = 4$ in the numerical results.

Taking the service rate of one channel as $\mu = 0.6$, we investigate the change trend for the system cost function $F(\mu, \delta)$ in relation to the sleep parameter δ in Fig. 15.10.

From Fig. 15.10, we find that for all the combinations of the arrival rate λ_1 of SU packets and the arrival rate λ_2 of PU packets, if the service rate μ of one channel is given, the system cost function $F(\mu, \delta)$ firstly shows a downward trend and then shows a rising trend as the sleep parameter δ increases.

As shown in Figs. 15.2, 15.3, 15.4, 15.5, when the sleep parameter is lower, as the sleep parameter increases, all the throughput of SU packets, the energy saving rate of the system and the channel utilization will increase, while the average latency of SU packets will decrease. Those results reduce the system cost, so the system cost function shows a downward trend. As the sleep parameter further increases, the throughput of SU packets, the energy saving rate of the system and the channel utilization decrease. This causes an increase in the system cost. The average latency of SU packets decreases as the sleep parameter increases and this leads a decrease in the system cost. However, the average latency of SU packets is not a primary factor

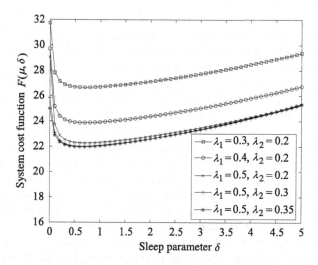

Fig. 15.10 System cost function versus sleep parameter

influencing the system cost in this case. Therefore, the system cost function shows a rising trend after an initial downward trend.

Taking the sleep parameter $\delta = 0.8$ as an example, we investigate the change trend for the system cost function $F(\mu, \delta)$ in relation to the service rate μ of one channel in Fig. 15.11.

From Fig. 15.11, we find that for all the combinations of the arrival rate λ_1 of SU packets and the arrival rate λ_2 of PU packets, if the sleep parameter δ is given, the

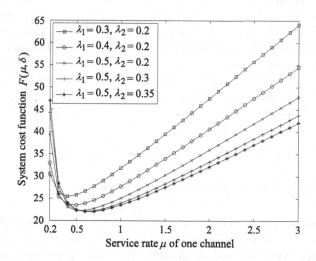

Fig. 15.11 System cost function versus service rate of one channel

system cost function $F(\mu, \delta)$ also firstly shows a downward trend and then shows a rising trend as the service rate μ of one channel increases.

Figures 15.6, 15.7, 15.8 show that when the service rate of one channel is lower, as the service rate of one channel increases, the average latency of SU packets will decrease, while the energy saving rate of the system and the throughput of SU packets will increase. This causes a decrease in the system cost. Although Fig. 15.9 shows that the decrease in channel utilization induces an increase in the system cost, the channel utilization is not a primary factor influencing the system cost in this case. Therefore, the system cost function shows a downward trend initially. However, when the service rate of one channel is greater than a certain value, the influences of the average latency of SU packets and throughput of SU packets on the system cost become weaker. At the same time, the influences of the energy saving rate of the system and channel utilization on the system cost become stronger. This results in the system cost function showing a rising trend.

Looking at Figs. 15.10 and 15.11, we conclude that there is an optimal sleep parameter δ and an optimal service rate μ of one channel which allows the system cost function to reach local minimums. If we determine the optimal service rate of one channel according to a fixed sleep parameter, in order to decrease the average latency of SU packets and increase the throughput of SU packets, the value of service rate of one channel should be set higher. This leads to a lower channel utilization which results in a higher system cost. Conversely, if we determine the optimal sleep parameter according to a fixed service rate of one channel, in order to increase the energy saving rate of the system, the channel utilization and the throughput of SU packets, the value of the sleep parameter should be set lower. This leads to a greater average latency of SU packets which also results in a higher system cost. Therefore, the local minimum of the system cost function may often not be the global minimum of the system cost function. Therefore, it is necessary to avoid some local minima and find a global minimum for the system cost function. In addition, it is difficult to give an analytical expression for the system cost function $F(\mu, \delta)$ in a closed form. By using conventional optimization methods, such as the steepest descent method or Newton's method, we cannot quickly obtain the global minimum for the system cost function. Therefore, we turn to an intelligent optimization algorithm with a strong global convergence ability to minimize the system cost function, and obtain the optimal combination (μ^*, δ^*) of the service rate of one channel and the sleep parameter.

15.5.2   Optimization of System Parameters

Based on the concept that the solution obtained for a given problem should move towards the best solution and should avoid the worst solution, a Jaya optimization algorithm has been proposed in [Rao16]. The Jaya optimization algorithm is a simple but powerful searching algorithm for solving optimization problems. However, we note that the initializing population with a uniform distribution will influence the development ability of the algorithm. In order to enhance the searching ability of the algorithm, we develop an improved Jaya algorithm by adopting an insect-population model to generate a chaotic population as a more diverse initial population. The main steps for the improved Jaya algorithm to obtain an optimal combination (μ^*, δ^*) with a global minimum for the system cost are as follows:

Step 1: Set the maximum iteration D, the maximum sleep parameter $\delta = \omega$, the number of channels M, the arrival rate λ_1 of SU packets, the population size N and the maximum service rate $\mu = \varepsilon$ of one channel. Initialize the current iteration as $d = 0$.

Step 2: Within the constraint condition $\mu \in [\lambda_1/M, \varepsilon]$ and $\delta \in [0, \omega]$, initialize populations $(\mu, \delta)_a$, $(a = 1, 2, 3, \ldots, N)$.

$\mu_{a+1} = 3.85 \times \mu_a(1 - \mu_a)$, $\mu_1 = \text{rand}$

$\delta_{a+1} = 3.85 \times \delta_a(1 - \delta_b)$, $\delta_1 = \text{rand}$

$$(\mu, \delta)_a = \left(\frac{\mu_a - \min\{\mu_a\}}{\max\{\mu_a\} - \min\{\mu_a\}} \times \left(\varepsilon - \frac{\lambda_1}{M} \right), \frac{\delta_a}{\max\{\delta_a\} - \min\{\delta_a\}} \times \omega \right)$$

% rand is a random number selected in the interval $(0, 1)$.

Step 3: Calculate the best candidate $(\mu, \delta)_{\text{best}}$ and the worst candidate $(\mu, \delta)_{\text{worst}}$.

$$(\mu, \delta)_{\text{best}} = \underset{a \in \{1,2,3,\ldots,N\}}{\arg\min} \{F((\mu, \delta)_a)\}$$

$$(\mu, \delta)_{\text{worst}} = \underset{a \in \{1,2,3,\ldots,N\}}{\arg\max} \{F((\mu, \delta)_a)\}$$

Step 4:

    for $a = 1 : N$

        $(\mu, \delta)_a^* = (\mu, \delta)_a + \text{rand} \times ((\mu, \delta)_{\text{best}} - (\mu, \delta)_a) - \text{rand} \times ((\mu, \delta)_{\text{worst}} - (\mu, \delta)_a)$

        if $F((\mu, \delta)_a) > F((\mu, \delta)_a^*)$

            $(\mu, \delta)_a = (\mu, \delta)_a^*$

        endif

    endfor

Step 5:

    if $d < D$

        $d = d + 1$

        go to **Step 3**

    else $(\mu^*, \delta^*) = \underset{a \in \{1,2,3,\ldots,N\}}{\arg\min} \{F((\mu, \delta)_a)\}$

    endif

Step 6: Output (μ^*, δ^*) as the optimal combination.

By setting the same parameters as used in Figs. 15.10 and 15.11, we obtain the optimal combination (μ^*, δ^*) of the service rate of one channel and the sleep parameter with global minimum of the system cost function in Table 15.1.

Table 15.1 Optimum combination of parameters in proposed strategy

Arrival rates λ_1 of SU packets	Arrival rates λ_2 of PU packets	Optimum combinations (μ^*, δ^*)	Minimum costs $F(\mu^*, \delta^*)$
0.3	0.2	(0.4,0.87)	25.44
0.3	0.3	(0.47,0.74)	25.08
0.3	0.35	(0.5,0.69)	24.95
0.4	0.2	(0.47,0.79)	23.43
0.4	0.3	(0.54,0.68)	23.17
0.4	0.35	(0.57,0.64)	23.08
0.5	0.2	(0.54,0.74)	22.19
0.5	0.3	(0.61,0.64)	22.01
0.5	0.35	(0.64,0.6)	21.93

15.6   Conclusion

In this chapter, considering that more energy will be saved with a multiple-sleep mode than that with a single-sleep mode, we proposed an energy saving strategy with a multiple-sleep mode in green CRNs. We established a CTMC model to capture the stochastic behavior of the two types of user packets, the PU packets and the SU packets, and then mathematically estimated the system performance by using the steady-state distribution of the system. In addition, the feasibility of the proposed energy saving strategy with a multiple-sleep mode is validated by numerical results with analysis and simulation. Trading off different performance measures, we constructed a system cost function. Finally, by using an insect-population model with chaotic characteristics to initialize population, we developed an improved Jaya searching algorithm and jointly optimized the service rate of one channel and the sleep parameter with a global minimum for the system cost. Numerical results showed that by setting reasonable parameter combinations in terms of the service rate of one channel and the sleep parameter, the global minimum of the system cost can be obtained. Therefore, the proposed energy saving strategy with a multiple-sleep mode for energy saving in green CRNs is effective. The research work has potential applications to improve energy saving strategies in green CRNs.

Part III
Resource Management and Performance Analysis on Cloud Computing

Part III discusses the Virtual Machine (VM) allocation and sleep mode in cloud computing systems aiming to realize green cloud computing. From the perspective of multiple servers, we have an insight into queueing models with task migrations, wake-up thresholds, variable service rates, partial vacations, and second optional services.

There are six chapters in Part III, beginning with Chap. 16.

In Chap. 16, we propose a VM scheduling strategy with a speed switch and a multiple-sleep mode to improve the energy efficiency of Cloud Data Center (CDC). Commensurate with our proposal, we develop a continuous-time queueing model with an adaptive service rate and a partial synchronous vacation. In Chap. 17, aiming to achieve greener, more efficient computing in CDC, we propose an energy-efficient VM allocation strategy with an asynchronous multiple-sleep mode and an adaptive task-migration scheme. The VMs hosted in a virtual cluster are divided into two modules, namely, Module I and Module II. In Chap. 18, we propose a clustered VM allocation strategy based on a sleep mode with a wake-up threshold. Under the proposed strategy, all the VMs are dominated by a control server, where several sleep timers, a task counter, and a VM scheduler are deployed. In Chap. 19, considering the high energy consumption and the establishment of a loyal client base in cloud computing systems, we propose a sleep mode-based cloud architecture with a free service and a registration service. In Chap. 20, we present a task scheduling strategy with a sleep-delay timer and a wake-up threshold aiming to satisfy the response performance of cloud users while reducing the energy consumption in a cloud computing system. In Chap. 21, we propose an energy-saving VM allocation scheme with the constraint of response performance to aim a green cloud computing system. We establish a queueing model with multiple servers to capture the stochastic behavior of tasks in the CDC with the proposed scheme.

Part III
Resource Management and Performance Analysis on Cloud Computing

Chapter 16
Speed Switch and Multiple-Sleep Mode

In this chapter, we propose a Virtual Machine (VM) scheduling strategy with a speed switch and a multiple-sleep mode to improve the energy efficiency of cloud data centers. In accordance with the current traffic loads, a proportion of VMs operate at a low speed or a high speed, while the remaining VMs either sleep or operate at a high speed. Commensurate with our proposal, we develop a continuous-time queueing model with an adaptive service rate and a partial synchronous vacation. We derive performance measures of the system in terms of the energy saving level of the system and the average latency of tasks, respectively. We present numerical results to evaluate the performance of the system using the proposed VM scheduling strategy. We also establish a system profit function to achieve a trade-off among different performance measures and develop an improved Firefly algorithm to obtain the optimal sleep parameter.

16.1 Introduction

The rapid development of information technology (IT) and the explosive growth in global data have generated enormous demand for cloud computing. Consequently, Cloud Data Centers (CDCs) are growing exponentially, both in number and in size, to provide universal service. International Data Corporation (IDC) predicts that the total number of CDCs deployed worldwide will peak at 8.6 million in 2017 [Hint16]. Currently, high energy consumption and serious environmental pollution in WCNs are significant factors restricting the development of CDCs. One of the key challenges in constructing green CDCs is reducing energy consumption without seriously degrading the QoS.

In CDCs, besides the necessary energy consumption produced by providing service for cloud users, a large amount of energy is wasted maintaining excess service capacity [Gao12, Hame16, Sali12, Zhao19]. All the VMs in CDCs remain

© Springer Nature Singapore Pte Ltd. 2021
S. Jin, W. Yue, *Resource Management and Performance Analysis of Wireless Communication Networks*, https://doi.org/10.1007/978-981-15-7756-7_16

open waiting for the arrivals of cloud users, even in the night and early morning. During those hours, the utilization of VMs is merely 5 to 10%. However, the energy consumption of an idle VM is 60 to 80% of that of a busy VM [Duan15]. In addition, inappropriate VM scheduling can also result in superfluous energy consumption. Researchers have therefore directed their focus on improving energy efficiency by reducing the amount of wasted energy in CDCs.

The energy consumption of a VM is approximately in line with the CPU utilization, so the most direct method of conserving energy is to operate all the VMs at lower voltage and frequency [Fara15, Qava14]. One of the common techniques for optimizing energy consumption in CDCs is to engage Dynamic Power Management (DPM). DPM refers to dynamic CPU energy consumption and CPU processing speed adjustment according to the current traffic load. In [Li16], the author proved that if the application environment and average energy consumption are given, there is an optimal speed scheme that minimizes the average response time of tasks. In [Wang11b], the authors presented a workload predictor based on online Bayes classifier and a DPM technique based on an adaptive reinforcement learning algorithm to reduce the energy consumption in stochastic dynamic systems.

In [Chen16], the authors proposed a Dynamic Voltage and Frequency Scaling (DVFS) scheme based on DPM technique, by which the best fitting voltage and frequency for a multi-core embedded system are dynamically predicted. All the methods based on DPM technique mentioned above can improve energy efficiency from the perspective of reducing the energy consumption of each VM in CDCs. However, all the VMs in the CDCs remain open all the time, even though there are no tasks in CDCs. Even when operating at low-speed and in low-voltage mode, the accumulated energy consumption by thousands of VMs in CDCs is significant.

In respect to the low utilization of VMs in CDCs, inducing some VMs to enter a sleep state or a power-off state during lower workload hours can also save energy [Dabb15a, Shen17]. In [Chou16], the authors proposed a DynSleep scheme. DynSleep dynamically postpones the processing of some tasks, creating longer idle periods. It says that the use of a deep sleep mode can save more energy. In [Dabb15b], the authors proposed an integrated energy-aware resource provisioning framework for CDCs. This framework first predicts the number of cloud users that will arrive at CDCs in the near future, then estimates the number of VMs that are needed to serve those cloud users. In [Liao15b], the authors proposed an energy-efficient strategy, which dynamically switches two backup groups of servers on and off according to different thresholds. By using the methods above, energy can be conserved by decreasing the number of VMs running in the system. However, few methods can accurately estimate the behavior of tasks. Their arrivals and departures are stochastic. Pushing VMs to enter a sleep state or a power-off state based on only the predicted behavior of tasks is very risky, and might lead to a significant sacrifice of the response performance.

In this chapter, by applying a DPM technique and introducing a sleep mode, we propose a VM scheduling strategy with a speed switch and a multiple-sleep mode. Typically, if the traffic load is very heavy, all the VMs in CDCs will operate at a high speed so that cloud users can be served faster and the average latency can be reduced. On the other hand, if the traffic load is very light, some VMs will operate at a low speed while the remaining VMs go to sleep so that energy consumption can be greatly reduced without significant response performance degradation. Accordingly, we establish a continuous-time queueing model with an adaptive service rate and a partial synchronous vacation to investigate the behavior of cloud users and all the VMs in CDCs with the proposed VM scheduling strategy. From the perspective of the total number of cloud users in the CDC and the state of all the VMs, we construct a two-dimensional Markov chain to analyze the queueing model. Moreover, we mathematically and numerically evaluate the energy saving level of the system and the average latency of tasks. In order to achieve a reasonable balance between different performance measures, we establish a system profit function. Finally, we develop an improved Firefly algorithm to search the optimal sleep parameter and the maximum system profit function.

The chapter is organized as follows. In Sect. 16.2, we describe the VM scheduling strategy with a speed switch and a multi-sleep mode proposed in this chapter. Then, we present the system model in detail. In Sect. 16.3, we present a performance analysis of the system model, through the analysis of the transition rate matrix and the steady-state distribution. In Sect. 16.4, we obtain performance measures and present numerical results to evaluate the system performance. In Sect. 16.5, we establish a system profit function and develop an improved Firefly algorithm to optimize the sleep parameter. Finally, we draw our conclusions in Sect. 16.6.

16.2   Virtual Machine Scheduling Strategy and System Model

In this section, we first propose a VM scheduling strategy for improving the energy efficiency in CDCs by using a speed switch and a multi-sleep mode. Then, we construct a continuous-time queueing model with an adaptive service rate and a partial synchronous vacation.

16.2.1   Virtual Machine Scheduling Strategy

In conventional CDCs, all the VMs remain open regardless of traffic load. This results in a large amount of energy being wasted, which is referred to as idle energy consumption. Furthermore, inappropriate VM scheduling also generates additional energy consumption, referred to as luxury energy consumption. In order to improve the energy efficiency of CDCs, we propose a VM scheduling strategy with an adaptive service rate and a partial synchronous vacation to capture the stochastic behavior of the system.

In the VM scheduling strategy proposed in this chapter, all the VMs in the CDC are divided into one of two modules, namely, a base-line module or a reserve module. The VMs in the base-line module are always active, and their processing speed can be switched between a low speed and a high speed in accordance with the traffic load. The VMs in the reserve module can be awakened from multiple sleeps.

Based on the stochastic behavior of cloud users, as well as the operational characteristics of sleep timers, the CDC will be converted in the following three cases:

Case I: The VMs in the base-line module operate at a low speed while the VMsin the reserve module are asleep. The rate of energy-conservation in the CDC is the most significant in this case.

Case II: The VMs in the base-line module operate at a high speed while the VMs in the reserve module are asleep. The rate of energy-conservation in the CDC is relatively obvious in this case.

Case III: The VMs in the base-line module operate at a high speed while the VMs in the reserve module are awake and operate at a high speed. The response performance in the CDC is most ideal in this case.

To avoid frequently switching the processing speed of VMs in the base-line module, we use a dual-threshold, marked as ω_1 ($\omega_1 = 0, 1, 2, \ldots$) and ω_2 ($\omega_2 = 0, 1, 2, \ldots$), to jointly control the VMs processing speed in the base-line module, in which we set $0 < \omega_2 < \omega_1$. When the number of cloud users in the CDC exceeds the threshold ω_1, all the VMs in the base-line module will operate at a high speed. When the number of cloud users in the CDC is less than the threshold ω_2, all the VMs in the base-line module will operate at a low speed. To guarantee the QoS in the CDC even when the traffic load is heavy, we use another threshold, called the activation threshold ω_3, to wake up the VMs in the reserve module. If the number of cloud users waiting in the CDC buffer exceeds the threshold ω_3, all the VMs in the reserve module will be awakened and operate at a high speed after the sleep timer expires. Otherwise, the sleep timer will be restarted with a random duration, and all the VMs in the reserve module will go to sleep again.

For convenience of presentation, we denote the number of VMs in the base-line module as n, and the number of VMs in the reserve module as m. To avoid the appearance that all the VMs in the reserve module are awake while the VMs in the base-line module operate at a low speed, we set $(n - \omega_2) \geq m$. To ensure all the cloud users in the CDC buffer can be served once the VMs in the reserve module are awakened, we set $0 < \omega_3 < m$.

According to the VM scheduling strategy proposed in this chapter, the transition among the three CDC cases is illustrated in Fig. 16.1.

In Case I, each cloud user is served immediately on a VM available in the base-line module at a low speed. However, with the arrivals of the cloud users, more VMs in the base-line module will be occupied. We call the VMs being occupied by cloud users as busy VMs. When the number of busy VMs in the base-line module exceeds the threshold ω_1, all the VMs in the base-line module will be switched to a high speed, namely, the CDC will be converted to the Case II state. The cloud users that have not received service yet will be served continuously on the same VM, but at a high speed. In this CDC case, there are no cloud users waiting in the CDC buffer.

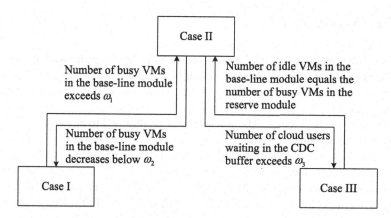

Fig. 16.1 Transition among three CDC cases in proposed strategy

Therefore, when the sleep timer expires, this sleep timer will be restarted with a random duration, and all the VMs in the reserve module will go to sleep again.

In Case II, if there are idle VMs in the base-line module, the incoming cloud users will be served immediately in the base-line module at a high speed. Otherwise, the cloud users have to wait in the CDC buffer. On the one hand, with the arrivals of cloud users, more cloud users will queue in the CDC buffer. When the sleep timer expires, if the number of cloud users waiting in the CDC buffer exceeds the activation threshold ω_3, all the VMs in the reserve module will be awakened and operate at a high speed directly, namely, the CDC will be converted to the Case III state. Then, all the cloud users in the CDC buffer will be served immediately in the reserve module at a high speed. Otherwise, the CDC will remain in the Case II state. As service continues, cloud users that have finished being served depart, so fewer VMs in the base-line module will be busy. When the number of busy VMs in the base-line module decreases below the threshold ω_2, all the VMs in the base-line module will be switched to a low speed, namely, the CDC will be converted to the Case I state. The cloud users queueing in the CDC buffer will be served continuously on the same VM, but at a low speed.

In Case III, if there are idle VMs in either the base-line module or the reserve module, the incoming cloud users will be served immediately at a high speed. Otherwise, the cloud users will queue in the CDC buffer. However, as cloud users that have finished being served depart, fewer VMs in both the base-line module and the reserve module will be busy. When the number of idle VMs in the base-line module is equal to the number of busy VMs in the reserve module, the cloud users queueing in the CDC buffer will be migrated to the idle VMs in the base-line module and served at a high speed. Then, the sleep timer will be restarted with a random duration, and all the VMs in the reserve module will go to sleep again, namely, the CDC will be converted to the Case II state.

16.2.2   System Model

In this subsection, we establish a continuous-time queueing model with an adaptive service rate and a partial synchronous vacation to capture the stochastic behavior of tasks from cloud users in the CDC by using the proposed VM scheduling strategy on the VMs. In this system model, there are several independent VMs. A VM can only serve one task at a time. The system buffer is supposed to be infinite.

We assume that the inter-arrival time of tasks follows an exponential distribution with mean $1/\lambda$, where $\lambda > 0$, called the arrival rate of tasks. We assume that the service time of a task when the system is in the Case I state follows an exponential distribution with mean $1/\mu_l$ seconds, where $\mu_l > 0$. The service time of a task when the system is in either the Case II state or the Case III state follows an exponential distribution with mean $1/\mu_h$ seconds, where $\mu_h > 0$. We call μ_l the service rate in the Case I, μ_h the service rate in the Case II or the Case III. Furthermore, we assume that the energy consumption of a VM during the sleep state is J_v ($J_v > 0$), the energy consumption of an idle VM is J_o ($J_o > J_v$), the energy consumption of a busy VM operating at the low speed and the high speed are J_l and J_h ($J_h > J_l$), respectively. And, the additional energy consumption of a VM switching to a high speed from a low speed is J_a ($J_a > 0$), and that of a VM being woken up from a sleep state is J_b ($J_b > 0$). In addition, we assume that the time length of a sleep timer follows an exponential distribution with mean $1/\delta$ seconds, where $\delta > 0$. Here, we refer to δ as the sleep parameter.

Let random variable $N(t) = i$ ($i \in \{0, 1, 2, \ldots\}$) be the total number of tasks in the system at instant t, which is called the system level. Let random variable $C(t) = j$ ($j \in \{1, 2, 3\}$) be the system case at instant t. $j = 1, 2, 3$ represents the system being in the states of Case I, Case II and Case III, respectively. $\{(N(t), C(t)), t \geq 0\}$ constitutes a two-dimensional Continuous-Time Markov Chain (CTMC). The state space Ω of the CTMC is given as follows:

$$\Omega = \{(i, j) : i \geq 0, j = 1, 2, 3\}. \tag{16.1}$$

For the two-dimensional CTMC, we define $\pi_{i,j}$ as the probability when the system level is i and the system case is j in the steady state. $\pi_{i,j}$ is given as follows:

$$\pi_{i,j} = \lim_{t \to \infty} \Pr\{N(t) = i, C(t) = j\}, \quad i \geq 0, j = 1, 2, 3. \tag{16.2}$$

We define π_i as the probability vector when the system level is i in the steady state. π_i can be given as follows:

$$\pi_i = \begin{cases} (\pi_{i,1}, \pi_{i,2}), & 0 \leq i \leq n \\ (\pi_{i,2}, \pi_{i,3}), & i \geq n+1. \end{cases} \tag{16.3}$$

The steady-state distribution $\mathbf{\Pi}$ of the CTMC is composed of π_i ($i \geq 0$). $\mathbf{\Pi}$ is given as follows:

$$\mathbf{\Pi} = (\pi_0, \pi_1, \pi_2, \ldots). \tag{16.4}$$

16.3 Performance Analysis

In this section, we first discuss the transition rate matrix of the two-dimensional CTMC. Then, we derive the steady-state distribution of the system model.

16.3.1 Transition Rate Matrix

According to the VM scheduling strategy proposed in this chapter, the system case is related to the system level. The relation between the system level and the system case is illustrated in Table 16.1.

Based on Table 16.1, we illustrate the state transition of the system model in Fig. 16.2.

Let Q be the one-step state transition rate matrix of the two-dimensional CTMC $\{(N(t), C(t)), t \geq 0\}$. As shown in Table 16.1, each system level has at most two corresponding system cases, so we separate Q into sub-matrices of 2×2 structure. Let $Q_{u,v}$ be the one-step state transition rate sub-matrix for the system

Table 16.1 Relation between system level and system case

System levels	Initial system cases before one-step transition	Possible system cases after one-step transition
$[0, \omega_2 - 1]$	Case I	Case I
ω_2	Case I	Case I
	Case II	Case I or Case II
$[\omega_2 + 1, \omega_1 - 1]$	Case I	Case I
	Case II	Case II
ω_1	Case I	Case I or Case II
	Case II	Case II
$[\omega_1 + 1, n]$	Case II	Case II
$n + 1$	Case II	Case II
	Case III	Case II or Case III
$[n + 2, n + \omega_3]$	Case II	Case II
	Case III	Case III
$[n + \omega_3 + 1, \infty)$	Case II	Case II or Case III
	Case III	Case III

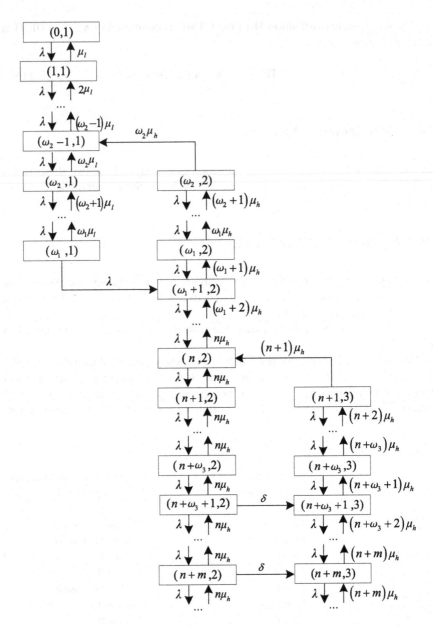

Fig. 16.2 State transition of system model

level changing from u ($u = 0, 1, 2, \ldots$) to v ($v = 0, 1, 2, \ldots$). For clarity, $\boldsymbol{Q}_{u,u}$, $\boldsymbol{Q}_{u,u-1}$ and $\boldsymbol{Q}_{u,u+1}$ are abbreviated as $\boldsymbol{A}_u$, $\boldsymbol{B}_u$ and $\boldsymbol{C}_u$, respectively.

For the initial system level $u = 0$, $\boldsymbol{A}_0$ indicates the state transition rates from $(0, 1)$ to $(0, 1)$, $(0, 1)$ to $(0, 2)$, $(0, 2)$ to $(0, 1)$ and $(0, 2)$ to $(0, 2)$. $\boldsymbol{C}_0$ indicates the state transition rates from $(0, 1)$ to $(1, 1)$, $(0, 1)$ to $(1, 2)$, $(0, 2)$ to $(1, 1)$ and $(0, 2)$ to $(1, 2)$.

For the initial system level $1 \leq u \leq n - 1$, $\boldsymbol{A}_u$ indicates the state transition rates from $(u, 1)$ to $(u, 1)$, $(u, 1)$ to $(u, 2)$, $(u, 2)$ to $(u, 1)$ and $(u, 2)$ to $(u, 2)$. $\boldsymbol{B}_u$ indicates the state transition rates from $(u, 1)$ to $(u-1, 1)$, $(u, 1)$ to $(u-1, 2)$, $(u, 2)$ to $(u-1, 1)$ and $(u, 2)$ to $(u-1, 2)$. $\boldsymbol{C}_u$ indicates the state transition rates from $(u, 1)$ to $(u + 1, 1)$, $(u, 1)$ to $(u + 1, 2)$, $(u, 2)$ to $(u + 1, 1)$ and $(u, 2)$ to $(u + 1, 2)$.

For the initial system level $u = n$, $\boldsymbol{A}_n$ indicates the state transition rates from $(n, 1)$ to $(n, 1)$, $(n, 1)$ to $(n, 2)$, $(n, 2)$ to $(n, 1)$ and $(n, 2)$ to $(n, 2)$. $\boldsymbol{B}_n$ indicates the state transition rates from $(n, 1)$ to $(n - 1, 1)$, $(n, 1)$ to $(n - 1, 2)$, $(n, 2)$ to $(n - 1, 1)$ and $(n, 2)$ to $(n - 1, 2)$. $\boldsymbol{C}_n$ indicates the state transition rates from $(n, 1)$ to $(n + 1, 2)$, $(n, 1)$ to $(n + 1, 3)$, $(n, 2)$ to $(n + 1, 2)$ and $(n, 2)$ to $(n + 1, 3)$.

For the initial system level $u = n + 1$, $\boldsymbol{A}_{n+1}$ indicates the state transition rates from $(n + 1, 2)$ to $(n + 1, 2)$, $(n + 1, 2)$ to $(n + 1, 3)$, $(n + 1, 3)$ to $(n + 1, 2)$ and $(n + 1, 3)$ to $(n + 1, 3)$. $\boldsymbol{B}_{n+1}$ indicates the state transition rates from $(n + 1, 2)$ to $(n, 1)$, $(n+1, 2)$ to $(n, 2)$, $(n+1, 3)$ to $(n, 1)$ and $(n+1, 3)$ to $(n, 2)$. $\boldsymbol{C}_{n+1}$ indicates the state transition rates from $(n+1, 2)$ to $(n+2, 2)$, $(n+1, 2)$ to $(n+2, 3)$, $(n+1, 3)$ to $(n + 2, 2)$ and $(n + 1, 3)$ to $(n + 1, 3)$.

For the initial system level $n + 2 \leq u < \infty$, $\boldsymbol{A}_u$ indicates the state transition rates from $(u, 2)$ to $(u, 2)$, $(u, 2)$ to $(u, 3)$, $(u, 3)$ to $(u, 2)$ and $(u, 3)$ to $(u, 3)$. $\boldsymbol{B}_u$ indicates the state transition rates from $(u, 2)$ to $(u - 1, 2)$, $(u, 2)$ to $(u - 1, 3)$, $(u, 3)$ to $(u-1, 2)$ and $(u, 3)$ to $(u-1, 3)$. $\boldsymbol{C}_u$ indicates the state transition rates from $(u, 2)$ to $(u + 1, 2)$, $(u, 2)$ to $(u + 1, 3)$, $(u, 3)$ to $(u + 1, 2)$ and $(u, 3)$ to $(u + 1, 3)$.

All the sub-matrices in $\boldsymbol{Q}$ can be addressed according to the state transition rates shown in Fig. 16.2. We find that starting from the system level $(n + m + 1)$, all the sub-matrices in $\boldsymbol{Q}$ are repeated forever. Then, $\boldsymbol{Q}$ is given as follows:

$$
\boldsymbol{Q} = \begin{pmatrix}
\boldsymbol{A}_0 & \boldsymbol{C}_0 & & & & & & & \\
\boldsymbol{B}_1 & \boldsymbol{A}_1 & \boldsymbol{C}_1 & & & & & & \\
& \ddots & \ddots & \ddots & & & & & \\
& & \boldsymbol{B}_{n-1} & \boldsymbol{A}_{n-1} & \boldsymbol{C}_{n-1} & & & & \\
& & & \boldsymbol{B}_n & \boldsymbol{A}_n & \boldsymbol{C}_n & & & \\
& & & & \boldsymbol{B}_{n+1} & \boldsymbol{A}_{n+1} & \boldsymbol{C}_{n+1} & & \\
& & & & & \boldsymbol{B}_{n+2} & \boldsymbol{A}_{n+2} & \boldsymbol{C}_{n+2} & \\
& & & & & & \ddots & \ddots & \ddots \\
& & & & & & & \boldsymbol{B}_{n+m} & \boldsymbol{A}_{n+m} & \boldsymbol{C}_{n+m} \\
& & & & & & & & \ddots & \ddots & \ddots
\end{pmatrix}. \tag{16.5}
$$

The block-tridiagonal structure of Q shows that the state transitions occur only between adjacent system levels. Hence, the two-dimensional CTMC $\{(N(t), C(t)),\ t \geq 0\}$ can be seen as a Quasi Birth-Death (QBD) process.

16.3.2 Steady-State Distribution

For the CTMC $\{(N(t), C(t)),\ t \geq 0\}$ with the one-step state transition rate matrix Q, the necessary and sufficient conditions for positive recurrence are that the matrix quadratic equation:

$$R^2 B_{n+m} + R A_{n+m} + C_{n+m} = 0 \tag{16.6}$$

has a minimal non-negative solution R and that the spectral radius $Sp(R) < 1$, where 0 is a zero matrix of order 2×2.

We assume the rate matrix $R = \begin{pmatrix} r_{11} & r_{12} \\ 0 & r_{22} \end{pmatrix}$, then substitute R, A_{n+m}, B_{n+m} and C_{n+m} into Eq. (16.6), so we have

$$\begin{pmatrix} n\mu_h r_{11}^2 & (n+m)\mu_h(r_{11}+r_{22})r_{12} \\ 0 & (n+m)\mu_h r_{22}^2 \end{pmatrix}$$

$$+ \begin{pmatrix} -(\lambda + n\mu_h + \delta)r_{11} & r_{11}\delta - (\lambda + (n+m)\mu_h)r_{12} \\ 0 & -(\lambda + (n+m)\mu_h)r_{22} \end{pmatrix} + \begin{pmatrix} \lambda & 0 \\ 0 & \lambda \end{pmatrix} = \begin{pmatrix} 0 & 0 \\ 0 & 0 \end{pmatrix}. \tag{16.7}$$

By solving Eq. (16.7), we can derive r_{11}, r_{22} and r_{12} as follows:

$$\begin{cases} r_{11} = \dfrac{(\lambda + n\mu_h + \delta) - \sqrt{(\lambda + n\mu_h + \delta)^2 - 4n\lambda\mu_h}}{2n\mu_h} \\[3mm] r_{22} = \dfrac{\lambda}{(n+m)\mu_h} \\[3mm] r_{12} = \dfrac{r_{11}\delta}{\lambda + (n+m)(1 - r_{11} - r_{22})\mu_h}. \end{cases} \tag{16.8}$$

The rate matrix R has been given in closed-form. Note that $Sp(R) = \max\{r_{11}, r_{22}\}$ and r_{11} can be proved mathematically to be less than 1. Therefore, the necessary and sufficient condition for positive recurrence of the CTMC $\{(N(t), C(t)),\ t \geq 0\}$ is equivalent to $r_{22} < 1$, that is, $\lambda < (n+m)\mu_h$.

With the rate matrix R obtained, we construct a square matrix $B[R]$ as follows:

$$
B[R] = \begin{pmatrix}
A_0 & C_0 & & & & & & & \\
B_1 & A_1 & C_1 & & & & & & \\
& \ddots & \ddots & \ddots & & & & & \\
& & B_{n-1} & A_{n-1} & C_{n-1} & & & & \\
& & & B_n & A_n & C_n & & & \\
& & & & B_{n+1} & A_{n+1} & C_{n+1} & & \\
& & & & & B_{n+2} & A_{n+2} & C_{n+2} & \\
& & & & & & \ddots & \ddots & \ddots \\
& & & & & & B_{n+m-1} & A_{n+m-1} & C_{n+m-1} \\
& & & & & & & B_{n+m} & R \times B_{n+m} + A_{n+m}
\end{pmatrix}.
$$

$$(16.9)$$

By using a matrix-geometric solution method, we can give an equation set as follows:

$$
\begin{cases}
(\pi_0, \pi_1, \pi_2, \ldots, \pi_{n+m})B[R] = \underbrace{(0, 0, 0, \ldots, 0)}_{2(n+m+1)} \\[4mm]
(\pi_0, \pi_1, \pi_2, \ldots, \pi_{n+m-1})e_1 + \pi_{n+m}(I - R)^{-1}e_2 = 1
\end{cases}
$$

$$(16.10)$$

where e_1 is a column vector with $2 \times (n + m)$ elements and e_2 is a column vector with 2 elements, respectively. All elements of these vectors are equal to 1.

We further construct an augmented matrix as follows:

$$
(\pi_0, \pi_1, \pi_2, \ldots, \pi_{n+m}) \left(B[R] \begin{array}{|c} e_1 \\ \hline (I - R)^{-1}e_2 \end{array} \right) = \underbrace{(0, 0, 0, \ldots, 0, 1)}_{2(n+m+1)}.
$$

$$(16.11)$$

Applying the Gauss-Seidel method to solve Eq. (16.11), we can obtain $\pi_0, \pi_1, \pi_2, \ldots, \pi_{n+m}$. From the structure of the transition rate matrix Q, we know π_i ($i = n+m+1, n+m+2, n+m+3, \ldots$) satisfies the matrix-geometric solution form as follows:

$$
\pi_i = \pi_{n+m} R^{i-(n+m)}, \quad i \geq n+m. \tag{16.12}
$$

Substituting π_{n+m} obtained in Eq. (16.11) into Eq. (16.12), we can obtain π_i ($i = n+m+1, n+m+2, n+m+3, \ldots$). Then the steady-state distribution $\Pi = (\pi_0, \pi_1, \pi_2, \ldots)$ of the system can be given mathematically.

16.4 Performance Measures and Numerical Results

In this section, by using the performance analysis presented in Sect. 16.3, we derive performance measures of the system in terms of the energy saving level of the system and the average latency of tasks, respectively. Then, we present numerical results to evaluate the performance of the system using the VM scheduling strategy proposed in this chapter.

16.4.1 Performance Measures

The energy saving level γ_l of the system is defined as the ratio of the difference between the energy consumption of the conventional CDC and that of the CDC with the proposed VM scheduling strategy over the energy consumption of the conventional CDC.

The energy consumption g_1 of the CDC with the proposed VM scheduling strategy is given as follows:

$$g_1 = g_2 + g_3 + g_4 + g_5 \tag{16.13}$$

where g_2, g_3 and g_4 are the average energy consumption when the system is in the states of Case I, Case II and Case III, respectively, g_5 is the additional energy consumption caused by speed switching and VM activation. g_2, g_3, g_4 and g_5 are given as follows:

$$g_2 = \sum_{i=0}^{\omega_1} \pi_{i,1}(iJ_l + (n-i)J_o + mJ_v), \tag{16.14}$$

$$g_3 = \sum_{i=\omega_2}^{n} \pi_{i,2}(iJ_h + (n-i)J_o + mJ_v) + \sum_{i=n+1}^{\infty} \pi_{i,2}(nJ_h + mJ_v), \tag{16.15}$$

$$g_4 = \sum_{i=n+1}^{n+m} \pi_{i,3}(iJ_h + (n+m-i)J_o) + \sum_{i=n++m+1}^{\infty} \pi_{i,3}((n+m)J_h), \tag{16.16}$$

$$g_5 = \sum_{i=n+\omega_3+1}^{\infty} \pi_{i,2}\delta m J_b + \pi_{\omega_1,1}\lambda n J_a. \tag{16.17}$$

In the conventional CDC, the energy consumption g_1' is given as follows:

$$g_1' = (n + m) J_h \left(\frac{\lambda}{(n+m)\mu_h} \right) + (n + m) J_o \left(1 - \frac{\lambda}{(n+m)\mu_h} \right)$$

$$= \frac{\lambda J_h}{\mu_h} + J_o \left(n + m - \frac{\lambda}{\mu_h} \right). \tag{16.18}$$

Combining Eqs. (16.13) and (16.18), we give the energy saving level γ_l of the system in the CDC with the proposed VM scheduling strategy as follows:

$$\gamma_l = \frac{g_1' - g_1}{g_1'}. \tag{16.19}$$

Based on the steady-state distribution obtained in Sect. 16.3.2, we give the average value $E[L_t]$ for the number L_t of tasks waiting in the system buffer as follows:

$$E[L_t] = \sum_{i=n+1}^{\infty} (i - n)\pi_{i,2} + \sum_{i=n+m+1}^{\infty} (i - (n + m)) \pi_{i,3}. \tag{16.20}$$

We define the latency Y_t of a task as the duration from the instant a task arrives at the system to the instant this task is served. By using Eq. (16.20), we can obtain the average latency $E[Y_t]$ of tasks as follows:

$$E[Y_t] = \frac{E[L_t]}{\lambda}$$

$$= \frac{1}{\lambda} \left(\sum_{i=n+1}^{\infty} (i - n)\pi_{i,2} + \sum_{i=n+m+1}^{\infty} (i - (n + m)) \pi_{i,3} \right). \tag{16.21}$$

16.4.2 Numerical Results

In order to evaluate the response performance and the energy saving effect of the CDC with the VM scheduling strategy proposed in this chapter, we present numerical results with analysis and simulation. The analysis results are obtained based on Eq. (16.10) using Matlab 2011a. The simulation results are obtained by averaging over 10 independent runs using MyEclipse 2014. Good agreements between the analysis results and the simulation results are observed. With our proposed strategy, all the system parameters satisfy $0 < \omega_2 < \omega_1$, $0 < \omega_3 < m \leq n - \omega_2$, $0 < J_v < J_o < J_l < J_h$, $0 < J_a < J_b$ and $0 < \lambda < (n + m)\mu_h$.

We list the system parameters settings in Table 16.2 as an example for all the numerical results, where J_v is the energy consumption level of a sleeping VM, J_o is

Table 16.2 Parameter settings in numerical results

Parameters	Values
Total number $(n + m)$ of VMs in the system	50
Service rate μ_l when VM operates at the low speed	0.01 ms^{-1}
Service rate μ_h when VM operates at the high speed	0.02 ms^{-1}
Dual-threshold ω_1, ω_2	20, 10
Activation threshold ω_3	15
J_v of a sleeping VM	0.2 mJ
J_o of an idle VM	0.4 mJ
J_l of a busy VM operating at the low speed	0.45 mJ
J_h of a busy VM operating at the high speed	0.5 mJ
J_a of a VM caused by speed switch	1mJ
J_b of a VM caused by activation	2 mJ

the energy consumption level of an idle VM, J_l is the energy consumption level of a busy VM operating at the low speed, J_h is the energy consumption level of a busy VM operating at the high speed, J_a is the additional energy consumption level of a VM caused by speed switch, and J_b is the additional energy consumption level of a VM caused by activation defined in Sect. 16.2.2, respectively.

To elucidate the better energy saving effect of the proposed VM scheduling strategy, we carry out a numerical comparison between the proposed VM scheduling strategy and the conventional DPM strategy. In conventional DPM strategy, all the VMs are open all the time, but their processing speed can be switched between a low speed and a high speed according to the traffic load of the system at that time.

By setting the number of VMs in the reserve module $m = 20$ as an example, we examine the influence of the arrival rate λ of tasks on the energy saving level γ_l of the system for different sleep parameters δ in Fig. 16.3a. By setting the sleep parameter $\delta = 0.05$ as an example, we examine the influence of the arrival rate λ of tasks on the energy saving level γ_l of the system for different numbers m of VMs in the reserve module in Fig. 16.3b. In Fig. 16.3, the solid line represents the analysis results for the energy saving level of the system with the proposed VM scheduling strategy, and the dotted line represents the analysis results for the energy saving level of the system when using the conventional DPM strategy.

In Fig. 16.3, we observe that for the same sleep parameter δ and the same number m of VMs in the reserve module, the energy saving level γ_l of the system initially decreases gradually then decrease sharply as the arrival rate λ of tasks increases. When λ is smaller (such as $\lambda < 0.5$ for $\delta = 0.05$ and $m = 20$), as λ increases, it becomes more possible that the number of tasks in the system will exceed the threshold ω_1. That is, all the VMs in the base-line module will be switched to the high speed from the low speed. Since the energy consumption of a VM operating at the high speed is greater than that operating at the low speed, the energy consumption will increase, and the energy saving level of the system will decrease as the arrival rate of tasks increases. When λ is larger (such as $\lambda > 0.5$ for $\delta = 0.05$

Fig. 16.3 Energy saving
level of system versus arrival
rate of tasks

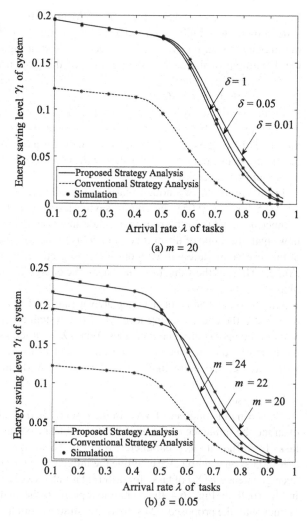

(a) $m = 20$

(b) $\delta = 0.05$

and $m = 20$), all the VMs in the base-line module will be busy, and the incoming
tasks will wait in the system buffer. As λ increases, the number of tasks waiting in
the system buffer is more likely to exceed the activation threshold ω_3, so the VMs in
the reserve module will be awakened after the sleep timer expires. Since the energy
consumption of a VM operating at the high speed is greater than that of a VM being
asleep, the energy saving due to the sleep mode is greater than that due to switching
to the low speed from the high speed, the energy saving level of the system will
decrease as the arrival rate of tasks increases.

With our proposed strategy, the energy consumption from the highest to the
lowest in sequence are: high speed mode, low speed mode and sleep mode. We note
that, for a VM, the energy consumption difference between sleep mode and high
speed mode is greater than that between low speed mode and high speed mode.

Moreover, the energy consumption of a VM in the reserve module switching from a sleep mode to a high speed mode is far greater than that of a VM in the base-line module switching from a low speed mode to a high speed mode. Therefore, the downtrend of the energy saving level of the system at the preceding stage is weaker than that at the following one, and the energy saving level of the system drops abruptly from $\lambda = 0.5$.

From Fig. 16.3a, we notice that for the same arrival rate λ of tasks, the energy saving level γ_l of the system decreases as the sleep parameter δ increases. The larger the value of δ is, the more likely the VMs in the reserve module will be awake. Thus, the energy consumption of the system will increase, and the energy saving level of the system will decrease.

From Fig. 16.3b, we notice that for a smaller arrival rate λ of tasks (such as $\lambda < 0.5$ for $\delta = 0.05$), the energy saving level γ_l of the system increases as the number m of VMs in the reserve module increases. When λ is smaller, no matter how small the value of the number n of VMs in the base-line module is, all the VMs in the reserve module will be more likely to go to sleep again after the sleep timer expires. Thus, as the value of m increases, the energy saving level of the system will increase.

On the other hand, for a larger arrival rate λ of tasks (such as $\lambda > 0.55$ for $\delta = 0.05$), the energy saving level γ_l of the system decreases as the number m of VMs in the reserve module increases. When λ is larger, as the number n of VMs in the base-line module decreases, the number of tasks waiting in the system buffer will increase, and more likely it is that all the VMs in the reserve module will awaken. Thus, as the value of m increases, the energy saving level of the system will decrease.

By setting the number of VMs in the reserve module $m = 20$ as an example, we examine the influence of the arrival rate λ of tasks on the average latency $E[Y_t]$ of tasks for different sleep parameters δ in Fig. 16.4a. By setting the sleep parameter $\delta = 0.05$ as an example, we examine the influence of the arrival rate λ of tasks on the average latency $E[Y_t]$ of tasks for different numbers m of VMs in the reserve module in Fig. 16.4b. In Fig. 16.4, the solid line represents the analysis results of the average latency with the proposed VM scheduling strategy, and the dotted line represents the analysis results of the average latency of tasks when using the conventional DPM strategy.

From Fig. 16.4, we observe that for the same sleep parameter δ and the same number m of VMs in the reserve module, the average latency $E[Y_t]$ of tasks initially increases from 0, then decreases slightly before finally increasing sharply as the arrival rate λ of tasks increases. When λ is relatively small (such as $0 < \lambda < 0.65$ for $\delta = 0.05$ and $m = 20$), the VMs in the reserve module are more likely to be asleep, only the VMs in the base-line module will be available. If all the VMs in the base-line module are busy, the incoming tasks have to wait in the system buffer. Thus, the average latency of tasks will increase gradually. When λ is moderate (such as $0.65 < \lambda < 0.85$ for $\delta = 0.05$ and $m = 20$), the VMs in the reserve module are more likely to be awakened after the sleep timer expires, so all the tasks waiting in the system buffer can be served in the reserve module. Furthermore, the larger the

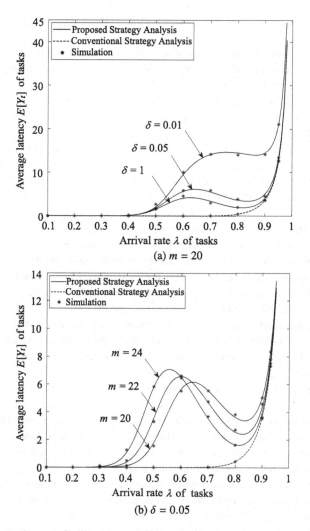

Fig. 16.4 Average latency of tasks versus arrival rate of tasks

value of λ is, the earlier the VMs in the reserve module will be awakened. Thus, the average latency of tasks will decrease. When λ further increases (such as $\lambda > 0.85$ for $\delta = 0.05$ and $m = 20$), the number of tasks waiting in the system buffer increases rapidly, even though all the VMs in both the base-line module and the reserve module are busy. In this case, the system tends to be unsteady, so the average latency of tasks will increase sharply.

From Fig. 16.4a, we notice that for the same arrival rate λ of tasks, the average latency $E[Y_t]$ of tasks decreases as the sleep parameter δ increases. A larger δ will shorten the average sleep time of VMs in the reserve module. The tasks waiting in the system buffer can be served earlier in the reserve module. That is, the tasks will be waiting a shorter time in the system buffer, so the average latency of tasks will decrease.

From Fig. 16.4b, we notice that for a smaller arrival rate λ of tasks (such as $\lambda < 0.6$), the average latency $E[Y_t]$ of tasks increases as the number m of VMs in the reserved module increases. Note that the total number $(n + m)$ of VMs in the system is fixed. Hence, the increase in the number m of VMs in the reserve module means a decrease in the number n of VMs in the base-line module. When λ is smaller, there are fewer tasks in the system, and the number of tasks waiting in the system buffer is more likely to be below the activation threshold ω_3 no matter how small the value of n is. That is, the tasks can only be served in the base-line module, and the average waiting time of tasks will increase as the value of n decreases. Thus, in this situation, the average latency of tasks will increase as the value of m increases.

On the other hand, for a larger arrival rate λ of tasks (such as $\lambda > 0.65$), the average latency $E[Y_t]$ of tasks decreases as the number m of VMs in the reserve module increases. When λ is large, there are many tasks in the system, and the number of tasks waiting in the system buffer is more likely to be higher than the activation threshold ω_3 no matter how great the value of n is. Furthermore, the larger the value of m is, the smaller the value of n is, and the earlier the VMs in the reserve module be awakened synchronously. That is, not only will the VMs in the base-line module be involved in the service, but also the VMs in the reserve module will be involved in the service. Thus, the average latency of tasks will decrease as the value of m increases.

Concluding with the numerical results shown in Figs. 16.3, 16.4, we find that, compared with the conventional CDCs, the energy saving level of the system in CDCs using the proposed VM scheduling strategy increases significantly, while the average latency of tasks increases slightly. What's more, the larger the value of the sleep parameter is, the greater the difference is between the performance measures of the conventional CDCs and those of the CDCs using the strategy proposed in this chapter. Therefore, a trade-off among the average latency of tasks and the energy saving rate of the system should be considered when setting the sleep parameter in our proposed VM scheduling strategy. We also find that as the sleep parameter increases, the average latency of tasks decreases, while both the energy saving level of the system decrease too.

16.5   Performance Optimization

To do performance optimization of the system, we establish a system profit function $F(\delta)$ to improve the trade-off among different performance measures presented in Sect. 16.4.1 as follows:

$$F(\delta) = f_1 \times \gamma_l - f_2 \times E[Y_t] \tag{16.22}$$

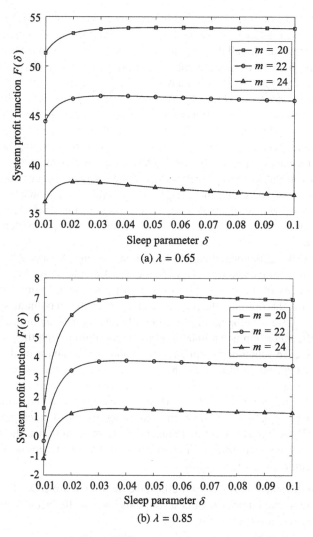

Fig. 16.5 System profit function versus sleep parameter

where f_1 and f_2 are the factors impacting on the system profit function, that being the energy saving level of the system and the average latency of tasks, respectively, on the system profit function. γ_l and $E[Y_t]$ have been obtained in Eqs. (16.19)–(16.21).

Using the system parameters given in Table 16.2, and setting $f_1 = 500$ and $f_2 = 1$ as an example in Fig. 16.5, we illustrate how the system profit function $F(\delta)$ changes along with the sleep parameter δ for different arrival rates λ of tasks and numbers m of VMs in the reserve module.

In Fig. 16.5, we show the system profit function $F(\delta)$ versus the sleep parameter δ for the cases of arrival rates of tasks being $\lambda = 0.65$ and $\lambda = 0.85$ as examples. From Fig. 16.5, we find that for all the combinations of arrival rates λ of tasks and numbers m of VMs in the reserve module, the system profit function $F(\delta)$ firstly increases and then decreases as the sleep parameter δ increases. We recall that both of the average latency of tasks and the energy saving level of the system will decrease as the sleep parameter increases. When the sleep parameter is smaller, the downward trend of the average latency of tasks is bigger than that of the energy saving level of the system, and the average latency of tasks is a dominant impacting factor. Hence, there is an increasing stage in Fig. 16.5. When the sleep parameter is greater, the downward trend of the energy saving level of the system is bigger than that of the average latency of tasks, and the energy saving level of the system is a dominant impacting factor. Hence, there is a decreasing stage in Fig. 16.5. Therefore, there is a maximum system profit function $F(\delta^*)$ when the sleep parameter is set to the optimal value δ^*.

We note that the mathematical expression of the energy saving level γ_l of the system and the average latency $E[Y_t]$ of tasks are difficult to give in a closed-form. The monotonicity of the system profit function is uncertain. In order to obtain the exact value for the optimal sleep parameter δ^* with the maximum system profit function $F(\delta^*)$, we develop an improved Firefly intelligent searching algorithm.

The Firefly algorithm is a population-based algorithm to find the global optimal value of objective function by imitating the collective behavior of fireflies. In the Firefly algorithm, all fireflies are randomly distributed in the search space, then the less bright ones will move towards the brighter ones because the attractiveness of fireflies is proportional to their brightness. We note that the initial positions of fireflies have great influence on the searching ability of intelligent searching algorithms. By using a chaotic mapping mechanism to initialize the positions of fireflies, we develop an improved Firefly algorithm. The main steps for the improved Firefly algorithm are given as follows:

Step 1: Initialize the number N of fireflies, the iteration number X of each firefly's position, the number K of best solutions, the chaotic factor ξ, the maximum attractiveness β_0, the step size α.
Set the sequence number of the best solutions as $x = 1$.

Step 2: Initialize the position of each firefly in the interval $[0.01, 0.10]$ using chaotic equations. δ_h is the position of the hth firefly, $h \in \{1, 2, 3, \ldots, N\}$.
$\delta_1 = 0.09 \times \text{rand} + 0.01$
% rand is a random number selected in the interval $(0, 1)$.
for $h = 2 : N$
$\delta_h = \xi \times (\delta_{h-1} - 0.01) \times (1 - (\delta_{h-1} - 0.01)/0.9) + 0.01$
endfor

Step 3: Calculate the self brightness $I_0(h)$ of each firefly, $h \in \{1, 2, 3, \ldots, N\}$.
$I_0(h) = F(\delta_h) = f_1 \gamma_l - f_2 E[Y_t]$
% γ_l and $E[Y_t]$ are the energy saving level of the system and the average latency of tasks with the sleep parameter δ_h in this chapter, respectively.

Step 4: For each firefly, update the position and the self brightness. Initialize the
current iteration as $t = 1$.

while $t \leq X$

for $i = 1 : N$

for $j = 1 : N$

$I(j, i) = I_0(j) \times e^{-E[Y_t] \times |\delta_j - \delta_i|}$

% Calculate the relative brightness $I(j, i)$ for the jth firefly to the
ith firefly.

if $I(j, i) > I_0(i)$

$\beta(j, i) = \beta_0 \times e^{-E[Y_t] \times |\delta_j - \delta_i|}$

% Calculate the attractiveness $\beta(j, i)$ of the jth firefly to the ith
firefly.

$\delta_i = \beta(j, i) \times (\delta_j - \delta_i) + \alpha \times (\text{rand} - 0.5)$

% Calculate the move distance δ_i of the ith firefly to the jth
firefly $\alpha \times (\text{rand} - 0.5)$ is the disturbing factor to break away
from the local maximum.

$\delta_i = \delta_i + \delta_i$

$I_0(i) = F(\delta_i) = f_1 \gamma_l - f_2 E[Y_t]$

% Update the position δ_i and the self brightness $I_0(i)$ of the ith
firefly.

$t = t + 1$

endif

endfor

endfor

endwhile

Step 5: Select the maximum self brightness and the best position of N fireflies as
one of the best solutions.

$\delta[x] = \underset{h \in \{1,2,3,...,N\}}{\text{argmax}} \{I_0(h)\}$

$x = x + 1$

% $\delta[x]$ is an array that help to record K best solutions.

Step 6:

if

$x \leq K$

go to **Step** 4

else

$\delta^* = \text{average}\{\delta[x]\}$

$F(\delta^*) = f_1 \gamma_l - f_2 E[Y_t]$

% γ_l and $E[Y_t]$ are the energy saving level of the system and the average
latency of tasks with the sleep parameter δ^* in this chapter, respectively.

endif

Step 7: Output δ^* and $F(\delta^*)$.

Table 16.3 Optimum sleep parameter in proposed strategy

Arrival rates λ of tasks	Numbers m of VMs in the reserve module	Optimum sleep parameters δ^*	Maximum profits $F(\delta^*)$
0.65	20	0.055	53.921
	22	0.034	46.9805
	24	0.023	38.2877
0.75	20	0.039	26.3022
	22	0.028	19.8894
	24	0.023	13.5192
0.85	20	0.049	7.0671
	22	0.039	3.8076
	24	0.035	1.3757

In the improved Firefly algorithm, we set $N = 10^4$, $X = 10^{12}$, $K = 100$, $\beta_0 = 0.01$ and $\alpha = 0.2$. Then, we obtain the optimal sleep parameter δ^* and the corresponding maximum system profit function $F(\delta^*)$ of the system in Table 16.3.

From Table 16.3, we observe that for the same arrival rate λ of tasks, both of the optimal sleep parameter δ^* and corresponding maximum system profit function $F(\delta^*)$ decreases as the number m of VMs in the reserve module increases. For the same number m of VMs in the reserve module, as the arrival rate λ of tasks increases, the optimal sleep parameter δ^* firstly decreases and then increases, while the corresponding maximum system profit function $F(\delta^*)$ continuously decreases.

16.6   Conclusion

In this chapter, in order to improve the energy efficiency, we proposed a VM scheduling strategy with a speed switch and a multi-sleep mode in CDCs. By applying DPM technology and introducing a sleep mode, our proposed strategy is shown to improve energy efficiency significantly by reducing both the luxury energy consumption and the idle energy consumption. We established a continuous-time queueing model with an adaptive service rate and a partial synchronous vacation and obtained performance measures in terms of the energy saving level of the system and the average latency of tasks. Numerical results with analysis and simulation showed that on the premise of guaranteeing the QoS of CDCs, the energy saving effect of CDCs with our proposed VM scheduling strategy is remarkably more efficient when compared with conventional CDCs. We also established a system profit function to achieve a trade-off among different performance measures and developed an improved Firefly algorithm to obtain the optimal sleep parameter.

Chapter 17
Virtual Machine Allocation Strategy

In this chapter, we propose an energy-efficient Virtual Machine (VM) allocation strategy with an asynchronous multi-sleep mode and an adaptive task-migration scheme to achieve greener, more efficient computing in cloud data centers. The VMs hosted in a virtual cluster are divided into two modules, namely, Module I and Module II. The VMs in Module I are always awake, whereas the VMs in Module II will go to sleep independently, if possible. Accordingly, we build a queueing model with partial asynchronous multiple vacations to capture the working principle of the proposed strategy. We derive performance measures of the system in terms of the average latency of tasks and the energy saving rate of the system. We present numerical results to validate the proposed VM allocation strategy and to show the influence of the system parameters on performance measures. Finally, we construct a system cost function to trade off different performance measures and develop an intelligent searching algorithm to jointly optimize the number of VMs in Module II and the sleeping parameters.

17.1 Introduction

Cloud data centers are growing exponentially in number and size to accommodate an escalating number of users and an expansion in applications. In the current "Cisco Global Cloud Index", IT manufacturer Cisco predicts that by 2019, more than four-fifths of the workload in data centers will be handled in Cloud Data Centers (CDCs) [Hint16]. As a result, the tremendous amount of energy consumption and carbon dioxide emissions from CDCs in WCNs are becoming a great concern worldwide. According to a report from Natural Resources Defense Council (NRDC), CDC energy consumption is estimated to reach 140 billion kilowatt hours by 2020, which will be responsible for the emission of nearly 150 million tons of carbon pollution [Jin16d]. Therefore, producing energy-efficient systems has become a focus for the development and operation of CDCs.

© Springer Nature Singapore Pte Ltd. 2021
S. Jin, W. Yue, *Resource Management and Performance Analysis of Wireless Communication Networks*, https://doi.org/10.1007/978-981-15-7756-7_17

In CDCs, an enormous amount of energy can be wasted due to excessive provisioning [Hadd17, Jin19a, Jin19b, Sing16a], while Service Level Agreement (SLA) violations can be risked by insufficient provisioning [Hasa17, Naka17]. In [Aria17], by introducing DVFS methods as part of a consolidation approach, the authors proposed a fuzzy multi-criteria and multi-objective resource management solution to reduce energy consumption and alleviate SLA violation. In [Son17], the authors proposed a dynamic overbooking strategy, allocating a more precise amount of resources to VMs and traffic with a dynamically changing workload. In this strategy, both of the energy consumption and the SLA violations were considered. In [Hoss15], for the purpose of minimizing the energy consumption, the authors introduced an optimal utilization level of a host to execute a certain number of instructions. Furthermore, they proposed a VM scheduling algorithm based on unsurpassed utilization level in order to derive the optimal energy consumption while satisfying a given QoS requirement. The literature mentioned above has contributed to reducing energy consumption while guaranteeing response performance in CDCs. However, the energy consumption generated by idle hosts in CDCs has been ignored.

The use of a sleep mode is an efficient approach for reducing the energy consumption in data centers [Luo17]. In [Duan15], the authors proposed a dynamic idle interval prediction scheme that can estimate the future idle interval length of a CPU and thereby choose the most cost-effective sleep state to minimize the power consumption during runtime. In [Sarj11], the authors proposed two energy models based on the statistical analysis of a server's operational behavior. With these models, the Energy Savings Engine (ESE) in the cloud provider decided either to migrate the VMs from a lightly-loaded server and then put the machine into a sleep mode, or to keep the current server running and ready for receiving any new tasks. In [Liu12], the authors proposed a sleep state management model to balance the system's energy consumption and the response performance. In this model, idle nodes were classified into different groups according to their sleep states. In the resource allocation process, nodes with the highest level of readiness were preferentially provided to the application. This research emphasized applying a sleep mode to a Physical Machine (PM).

To improve the energy efficiency of CDCs, Jin et al. proposed an energy-efficient strategy with a speed switch on PMs and a synchronous multi-sleep mode on partial VMs [Jin17a]. In [Jin17b], by applying DPM technology to PMs and introducing synchronous semi-sleep modes to partial VMs, the authors proposed a VM scheduling strategy for reducing energy consumption in CDCs. Both of the studies mentioned above applied a synchronous sleep mode to the VMs. However, there has so far been no research into the effect of asynchronous sleep modes on the level of VMs in CDCs.

In 1995, the Particle Swarm Optimization (PSO) algorithm was developed as an effective tool for function optimization. Since then, numerous research studies on improving the searching ability of PSO algorithms have appeared. In [Cao16], to enhance the performance of PSO algorithms, the authors improved PSO algorithms by introducing a nonlinear dynamic inertia weight and two dynamic learning factors.

In [Zhan16], the authors proposed a PSO algorithm based on an adaptive inertia weight and chaos optimization, which enhanced the local optimization ability of the PSO algorithm and helped objective functions easily jump out of local optimum. In [Tian18], the author presented a new PSO algorithm by introducing chaotic maps (Tent and Logistic), a Gaussian mutation mechanism, and a local re-initialization strategy into the standard PSO algorithm. The chaotic map is utilized to generate uniformly distributed particles for the purpose of improving the quality of the initial population. From the research mentioned above, we note that the searching ability of PSO algorithms are greatly influenced by the inertia weight and the initial positions of particles.

Inspired by the literature mentioned above, in this chapter, we propose an energy-efficient strategy for VM allocation over CDCs. We note that letting all the VMs in a virtual cluster go to sleep may degrade the quality of cloud service. Taking both the response performance and the energy conservation level into consideration, we divide the VMs in a virtual cluster into two parts: Module I and Module II. The VMs in Module I stay awake all the time to provide an instant cloud service for accomplishing tasks, while the VMs in Module II may go to sleep whenever possible to reduce energy consumption. The energy consumption of a VM is related to the processing speed of the VM. Generally speaking, the higher the processing speed is, the more energy will be consumed. In the strategy proposed in this chapter, the VMs in Module I process tasks at a higher speed to guarantee the response performance, while the VMs in Module II process tasks at a lower speed to save more energy. In order to further enhance the energy efficiency of the proposed strategy, we introduce an adaptive task-migration scheme which shifts an unfinished task in Module II to an idle VM in Module II. When an idle VM appears in Module I, a task being processed on a VM in Module II will migrate to the idle VM in Module I, and then the just evacuated VM in Module II will go to sleep independently. To analyze the proposed strategy, we build a queueing model with partial asynchronous multiple vacations by using a matrix-geometric solution method, and investigate the system performance through theoretical analysis and simulation experiments. Finally, in order to optimize the proposed strategy, we construct a system cost function to balance different system performance levels, and improve the PSO algorithm to obtain reasonable system parameter settings.

The main contributions of this chapter are summarized as follows:

(1) For reducing energy consumption and achieving greener cloud computing, we propose an energy-efficient VM allocation strategy with an asynchronous multi-sleep mode and an adaptive task-migration scheme.
(2) We present a method to model the proposed VM allocation strategy and to evaluate the system performance in terms of the average latency of tasks and the energy saving rate of the system.
(3) By improving an intelligent searching algorithm with a chaotic mapping mechanism and a nonlinear decreasing inertia weight, we develop an improved PSO algorithm and optimize the proposed VM allocation strategy to trade off different performance measures.

The chapter is organized as follows. In Sect. 17.2, we describe the energy-efficient VM allocation strategy with an asynchronous multi-sleep mode and an adaptive task-migration scheme proposed in this chapter. Then, we present the system model in detail. In Sect. 17.3, we present a performance analysis of the system model, through the analysis of the transition rate matrix and the steady-state distribution. In Sect. 17.4, we obtain performance measures and present numerical results to evaluate the system performance. In Sect. 17.5, we build a system cost function and develop an improved PSO algorithm to optimize the system performance. Our conclusions are drawn in Sect. 17.6.

17.2 Virtual Machine Allocation Strategy and System Model

In this section, we propose an energy-efficient VM allocation strategy with an asynchronous multi-sleep mode and an adaptive task-migration scheme. Then, we establish a continuous-time multiple-server queueing model with partial asynchronous multiple vacations.

17.2.1 Virtual Machine Allocation Strategy

In conventional CDCs, all the VMs remain open waiting for the arrival of tasks regardless of current traffic. This may result in a great deal of energy wastage. To get around this problem, a VM allocation strategy with an asynchronous multi-sleep mode and an adaptive task-migration scheme is proposed in this chapter. It should be emphasized that the asynchronous multi-sleep mode considered in this chapter is at the level of VMs rather than that of PMs.

Given the processing capability and the energy conservation level, all the VMs hosted in a virtual cluster are divided into two modules, namely, Module I and Module II. The VMs in Module I stay awake all the time and run at a high speed when tasks arrive. Whereas, the VMs in Module II switch between the sleep state and the busy state.

For a busy VM in Module II, state transition only happens at the instant when a task is completely processed. Given that a task is completely processed in Module II, if the system buffer is empty, the evacuated VM in Module II will go to sleep. Once a VM in Module II switches to the sleep state, a sleep timer will be started, the data in the memory will be saved to a hibernation file on the hard disk, and then the power of the other accessories, except for the memory, will be cut off, so that the VM will no longer be available for processing tasks in the system. Given that a task is completely processed in Module I, if the system buffer is empty and there is at least one task being processed in Module II, one of the tasks being processed in Module II will be migrated to Module I, and then the evacuated VM in Module II will go to sleep. We note that the task-migration considered in this chapter is an online VM-migration between different modules within a virtual cluster.

For a sleeping VM in Module II, state transition only happens at the instant when a sleep timer expires. At the moment that a sleep timer expires, the sleeping VM in Module II will listen to the system and decide whether to keep sleeping or to wake up. If the system buffer is empty, another sleep timer will be started and the sleeping VM in Module II will begin another sleep period, so that multiple sleep periods are formed. Otherwise, the sleeping VM in Module II will wake up to process the first task waiting in the system buffer at a lower speed. Once a VM in Module II switches to the awake state, the corresponding sleep timer will be turned off, the data of the hibernation file on the hard disk will be read into the memory, and then the power of all accessories will be turned on, so that the VM will be available for processing tasks in the system.

With the sleep mode proposed in this chapter, energy consumption could be saved, but the incoming tasks may not receive timely service. We speculate that the average latency of tasks is lower with a smaller number of VMs in Module II, while the energy saving rate of the system is higher with a suitable number of VMs in Module II. We also speculate that the average latency of tasks is lower with a shorter sleep period, while the energy saving rate of the system is higher with a longer sleep period. Given this, we should optimize the strategy proposed in this chapter by trading off the average latency and the energy saving rate of the system.

We show the state transition of a virtual cluster in the CDC with the VM allocation strategy proposed in this chapter in Fig. 17.1.

As shown in Fig. 17.1, in the proposed strategy, the numbers of the VMs in Module I and Module II are denoted as c and d, respectively. All the VMs hosted in one virtual cluster are dominated by a control server, in which several sleep timers and a VM scheduler are deployed. Each sleep timer is responsible for controlling

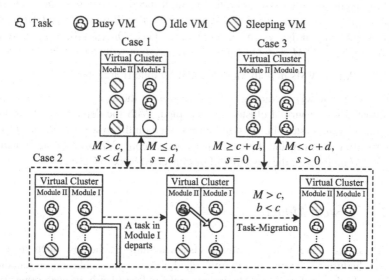

Fig. 17.1 State transition in CDC with proposed VM allocation strategy

the sleep time of a VM in Module II. The numbers of tasks in the system, busy VMs in Module I and sleeping VMs in Module II are denoted as M, b and s, respectively. Given these parameters, the VM scheduler adjusts the VM state.

According to the state of VMs both in Module I and in Module II, we consider three cases as follows:

Case 1:   There is at least one idle VM in Module I, and all the VMs in Module II are sleeping.

In Case 1, each arriving task could be processed immediately at a high speed in Module I. However, as more tasks arrive at the system, more VMs in Module I will be occupied. If there are no VMs available, a newly incoming task has to wait in the system buffer. Once a sleep timer expires, the corresponding VM in Module II will wake up to process the first task queueing in the system buffer at a low speed, and then the system will be converted to Case 2.

Case 2:   All the VMs in Module I are busy, and there is at least one sleeping VM in Module II.

In Case 2, with the departures of the tasks, more VMs in Module II will go to sleep. At the moment a task process is completed in Module I and there are no tasks waiting in the system buffer, namely, $M > c$ and $b < c$, one of the tasks being processed in Module II will be migrated to Module I, then the just evacuated VM in Module II will go to sleep. When all the VMs in Module II are asleep, namely, $M \leq c$ and $s = d$, the system will be converted back to Case 1.

We note that for Case 2, there are no VMs available in the system, so a newly incoming task will queue in the system buffer. When a task is completely processed on one of the VMs in Module I, the just evacuated VM in Module I will process the first task queueing in the system buffer at a high speed. Also, when one of the sleep timers expires, the corresponding VM in Module II will wake up and process the first task queueing in the system buffer at a low speed. Once all the VMs in Module II wake up, namely, $M \geq c + d$ and $s = 0$, the system will be converted to Case 3.

Case 3:   All the VMs in both Module I and Module II are busy.

In Case 3, a newly incoming task has to wait in the system buffer since all the VMs hosted in the virtual cluster are occupied. With the departures of the tasks, more tasks in the system buffer will be processed on the evacuated VMs. Once the system buffer is empty and there is at least one sleeping VM in Module II, namely, $M < c + d$ and $s > 0$, the system will be converted back to Case 2.

17.2.2   System Model

In CDCs, there are many available task scheduling schemes, such as event-driven scheduling schemes, preemptive scheduling schemes and random scheduling schemes. In this system, the VMs process tasks independently of each other. When no tasks are processed at a VM, the VM will go into a sleep period as a vacation, or will go into multiple sleep periods as multiple vacations. We model this system as a queueing model with partial asynchronous multiple vacations by using the proposed VM allocation strategy to quantify the effects of the VMs in Module II and the sleep parameter. We assume that an available VM can be assigned to the first task queueing in the unlimited system buffer.

The system model is described to be in an infinite state. Let random variable $N(t) = i$ ($i \geq 0$) be the total number of tasks in system at instant t. $N(t)$ is also called the system level. Let random variable $J(t) = j$ ($0 \leq j \leq d$) be the number of busy VMs in Module II at instant t. $J(t)$ is also called the system stage. $\{(N(t), J(t)), t \geq 0\}$ constitutes a two-dimensional continuous-time stochastic process with the state-space $\boldsymbol{\Omega}$ as follows:

$$\boldsymbol{\Omega} = \{(i,0) : 0 \leq i \leq c\} \cup \{(i,j) : c < i \leq c+d, \ 0 \leq j \leq i-c\}$$

$$\cup \{(i,j) : i > c+d, \ 0 \leq j \leq d\}. \tag{17.1}$$

In our research, we focus on user initiated tasks, and we make the following assumptions. We suppose that the inter-arrival time of tasks, the service time of a task processed in Module I and in Module II, and the time length of a sleep timer are i.i.d. random variables. We assume that the inter-arrival time of tasks follows an exponential distribution with mean $1/\lambda$, where $\lambda > 0$, called the arrival rate of tasks, the service times of a task processed in Module I and in Module II are assumed to follow exponential distributions with means $1/\mu_1$ seconds and $1/\mu_2$ seconds, respectively, where $\mu_1 > 0$ and $\mu_2 > 0$, called the service rate of tasks on the VM in Module I and the service rate of tasks on the VM in Module II. In addition, the time length of a sleep timer is assumed to follow an exponential distribution with mean $1/\delta$ seconds, where $\delta > 0$, called the sleep parameter. It should be noted that in the system model, we assume that no time is taken for a task to migrate or for a sleeping VM to wake up.

Based on the assumptions above, $\{(N(t), J(t)), t \geq 0\}$ can be regarded as a two-dimensional Continuous-Time Markov Chain (CTMC).

We define $\pi_{i,j}$ as the probability of the system level being equal to i and the system stage being equal to j. $\pi_{i,j}$ is then given as follows:

$$\pi_{i,j} = \lim_{t \to \infty} \Pr\{N(t) = i, J(t) = j\}, \quad (i,j) \in \boldsymbol{\Omega}. \tag{17.2}$$

We define $\boldsymbol{\pi}_i$ as the probability of the system level being equal to i in the steady state. $\boldsymbol{\pi}_i$ can be given as follows:

$$\boldsymbol{\pi}_i = \begin{cases} \pi_{i,0}, & 0 \leq i \leq c \\ (\pi_{i,0}, \pi_{i,1}, \pi_{i,2}, \ldots, \pi_{i,i-c}), & c < i \leq c+d \\ (\pi_{i,0}, \pi_{i,1}, \pi_{i,2}, \ldots, \pi_{i,d}), & i > c+d. \end{cases} \tag{17.3}$$

The steady-state distribution $\boldsymbol{\Pi}$ of the two-dimensional CTMC is composed of π_i $(i \geq 0)$. $\boldsymbol{\Pi}$ is given as follows:

$$\boldsymbol{\Pi} = (\pi_0, \pi_1, \pi_2, \ldots). \tag{17.4}$$

17.3 Performance Analysis

In this section, we present a performance analysis of the system model, through the analysis of the transition rate matrix and the steady-state distribution.

17.3.1 Transition Rate Matrix

Let $\boldsymbol{Q}$ be the one-step state transition rate matrix of the two-dimensional CTMC $\{(N(t), J(t)), \ t \geq 0\}$. Based on the system level, $\boldsymbol{Q}$ is separated into several sub-matrices. Let $\boldsymbol{Q}_{k,l}$ be the one-step state transition rate sub-matrix for the system level changing from k $(k = 0, 1, 2, \ldots)$ to l $(l = 0, 1, 2, \ldots)$. For convenience of presentation, we denote $\boldsymbol{Q}_{k,k}$, $\boldsymbol{Q}_{k,k-1}$ and $\boldsymbol{Q}_{k,k+1}$ as $\boldsymbol{A}_k$, $\boldsymbol{B}_k$ and $\boldsymbol{C}_k$, respectively. $\boldsymbol{A}_k$, $\boldsymbol{B}_k$ and $\boldsymbol{C}_k$ are discussed in the following cases.

(1) When the initial system level k ranges from 0 to c, k VMs in Module I are busy and all the VMs in Module II are sleeping.

For the case of $k = 0$, there are no tasks at all in the system. This means that the possible state transitions are from $(0, 0)$ to $(1, 0)$ and from $(0, 0)$ to $(0, 0)$. If a task arrives at the system, the system level will increase by one but the system stage will remain unchanged, namely, the system state will transform to $(1, 0)$ from $(0, 0)$ with the transition rate λ. Otherwise, the system state will remain fixed at $(0, 0)$ with the transition rate $-\lambda$. Thus, $\boldsymbol{A}_0$ and $\boldsymbol{C}_0$ are given as follows:

$$\boldsymbol{A}_0 = -\lambda, \quad \boldsymbol{C}_0 = \lambda. \tag{17.5}$$

For the case of $0 < k \leq c$, all the tasks in the system are being processed on the VMs in Module I. If a task is completely processed, the system level will decrease by one but the system stage will remain unchanged, namely, the system state will transform to $(k - 1, 0)$ from $(k, 0)$ with the transition rate $k\mu_1$. If a task arrives at the system, the system level will increase by one but the system stage will remain unchanged, namely, the system state will transfer to $(k + 1, 0)$ from $(k, 0)$ with the transition rate λ. Otherwise, the system state will remain fixed at $(k, 0)$ with the transition rate $-(\lambda + k\mu_1)$.

Thus, $\boldsymbol{A}_k$, $\boldsymbol{B}_k$ and $\boldsymbol{C}_k$ are given as follows:

$$\boldsymbol{A}_k = -(\lambda + k\mu_1), \quad \boldsymbol{B}_k = k\mu_1, \quad \boldsymbol{C}_k = \lambda. \tag{17.6}$$

(2) When the initial system level k ranges from $(c + 1)$ to $(c + d)$, all the VMs in Module I are busy, while at most $(k - c)$ VMs in Module II are busy.

For the case of $k = c + x$ $(x = 1, 2, 3, \ldots, d - 1)$, the number of busy VMs in Module I is c, while in Module II, there are at most x busy VMs.

If one of the sleep timers expires, the corresponding VM in Module II will wake up and process the first task queueing in the buffer. Consequently, the system level k will remain fixed but the system stage n will increase by one, namely, the system state will transform to $(k, n + 1)$ from (k, n) with the transition rate $(d - n)\delta$. Otherwise, the system state will remain fixed: when the system buffer is not empty, the transition rate is $-h_n$, where $h_n = \lambda + c\mu_1 + n\mu_2 + (d - n)\delta$; when the system buffer is empty, the transition rate is $-(\lambda + c\mu_1 + x\mu_2)$.

Thus, A_k is a rectangular $(x + 1) \times (x + 1)$ matrix and is given as follows:

$$
A_k = \begin{pmatrix}
-h_0 & d\delta & & & \\
& -h_1 & (d-1)\delta & & \\
& & \ddots & \ddots & \\
& & & -h_{x-1} & (d-x+1)\delta \\
& & & & -(\lambda + c\mu_1 + x\mu_2)
\end{pmatrix}. \tag{17.7}
$$

If a task is completely processed and there is at least one task in the system buffer, the first task queueing in the system buffer will occupy the evacuated VM to receive service. Consequently, the system level will decrease by one, but the system stage will remain fixed, namely, the system state will transform to $(k - 1, n)$ from (k, n) with the transition rate $(c\mu_1 + n\mu_2)$, where n $(0 \le n \le x)$ is the number of busy VMs in Module II.

If a task is completely processed on the VM in Module I and there are no tasks in the system buffer, one of the tasks being processed in Module II will migrate to the evacuated VM in Module I and the just-evacuated VM in Module II will start sleeping.

If a task is completely processed on the VM in Module II and there are no tasks in the system buffer, the evacuated VM in Module II will start sleeping directly. Consequently, both the system level and the system stage will decrease by one, namely, the system state will transform to $(k - 1, x - 1)$ from (k, x) with the transition rate $(c\mu_1 + x\mu_2)$.

Thus, B_k is a rectangular $(x + 1) \times x$ matrix and is given as follows:

$$
B_k = \begin{pmatrix}
c\mu_1 & & & \\
& c\mu_1 + \mu_2 & & \\
& & \ddots & \\
& & & c\mu_1 + (x-1)\mu_2 \\
& & & c\mu_1 + x\mu_2
\end{pmatrix}. \tag{17.8}
$$

None of VMs in Module II will wake up before their corresponding sleep timers expire, even though the system buffer is not empty. If a task arrives at the system before one of the sleep timers expires, the system level will increase by one but the system stage will remain fixed, namely, the system state will transform to $(k + 1, n)$ from (k, n) with the transition rate λ.

Thus, C_k is a rectangular $(x + 1) \times (x + 2)$ matrix and is given as follows:

$$C_k = \begin{pmatrix} \lambda & & & & 0 \\ & \lambda & & & 0 \\ & & \ddots & & \vdots \\ & & & \lambda & 0 \\ & & & & \lambda & 0 \end{pmatrix}. \tag{17.9}$$

For the case of $k = c + d$, the number of tasks in the system is equal to the total number of VMs. This is really just a specialized case discussed previously. A_k and C_k are square matrices of the order $(d+1) \times (d+1)$, B_k is a rectangular $(d + 1) \times d$ matrix. A_k, B_k and C_k are given as follows:

$$A_k = \begin{pmatrix} -h_0 & d\delta & & & \\ & -h_1 & (d-1)\delta & & \\ & & \ddots & \ddots & \\ & & & -h_{d-1} & \delta \\ & & & & -h_d \end{pmatrix}. \tag{17.10}$$

$$B_k = \begin{pmatrix} c\mu_1 & & & \\ & c\mu_1 + \mu_2 & & \\ & & \ddots & \\ & & & c\mu_1 + (d-1)\mu_2 \\ & & & c\mu_1 + d\mu_2 \end{pmatrix}, \tag{17.11}$$

$$C_k = \begin{pmatrix} \lambda & & & \\ & \lambda & & \\ & & \ddots & \\ & & & \lambda \\ & & & & \lambda \end{pmatrix}, \tag{17.12}$$

(3) When the initial system level is greater than the total number of VMs, namely, $k > c + d$, all the VMs in Module I are busy, while the VMs in Module II are either busy or sleeping. A_k, B_k and C_k are square matrices of the order

$(d + 1) \times (d + 1)$. Similar to the discussion in Item (2), the sub-matrices A_k, B_k and C_k are given as follows:

$$A_k = \begin{pmatrix} -h_0 & d\delta & & & \\ & -h_1 & (d-1)\delta & & \\ & & \ddots & \ddots & \\ & & & -h_{d-1} & \delta \\ & & & & -h_d \end{pmatrix}. \tag{17.13}$$

$$B_k = \begin{pmatrix} c\mu_1 & & & & \\ & c\mu_1 + \mu_2 & & & \\ & & \ddots & & \\ & & & c\mu_1 + (d-1)\mu_2 & \\ & & & & c\mu_1 + d\mu_2 \end{pmatrix}, \tag{17.14}$$

$$C_k = \begin{pmatrix} \lambda & & & & \\ & \lambda & & & \\ & & \ddots & & \\ & & & \lambda & \\ & & & & \lambda \end{pmatrix}, \tag{17.15}$$

Now, all the sub-matrices in the one-step state transition rate matrix Q have been addressed. Starting from the system level $(c + d)$, the sub-matrices A_k and C_k in Q are repeated forever. Starting from the system level $(c + d + 1)$, the sub-matrices B_k in Q are repeated forever. The repetitive sub-matrices A_k, B_k and C_k are represented by A, B and C, respectively. For this, Q is written as follows:

$$Q = \begin{pmatrix} A_0 & C_0 & & & & & & \\ B_1 & A_1 & C_1 & & & & & \\ & \ddots & \ddots & \ddots & & & & \\ & & B_c & A_c & C_c & & & \\ & & & B_{c+1} & A_{c+1} & C_{c+1} & & \\ & & & & \ddots & \ddots & \ddots & \\ & & & & & B_{c+d} & A & C \\ & & & & & & B & A & C \\ & & & & & & & \ddots & \ddots & \ddots \end{pmatrix}. \tag{17.16}$$

The block-tridiagonal structure of the one-step state transition rate matrix Q shows that the state transitions occur only between adjacent system levels. We know that the two-dimensional CTMC $\{(N(t), J(t)), \ t \geq 0\}$ can be seen as a Quasi Birth-Death (QBD) process.

17.3.2   Steady-State Distribution

For the QBD process $\{(N(t), J(t)), \ t \geq 0\}$ with the one-step state transition rate matrix $\boldsymbol{Q}$, the necessary and sufficient condition for positive recurrence is that the matrix quadratic equation

$$R^2 B + RA + C = 0 \tag{17.17}$$

has the minimal non-negative solution $\boldsymbol{R}$ with the spectral radius $\mathrm{Sp}(\boldsymbol{R}) < 1$. This solution, called the rate matrix and denoted by $\boldsymbol{R}$, can be explicitly determined.

From Sect. 17.3.1, we find that the sub-matrices $\boldsymbol{A}$, $\boldsymbol{B}$ and $\boldsymbol{C}$ are upper-triangular matrices. Therefore, the rate matrix $\boldsymbol{R}$ must be an upper-triangular matrix and can be expressed as follows:

$$R = \begin{pmatrix} r_0 & r_{01} & r_{02} & \cdots & r_{0d-1} & r_{0d} \\ & r_1 & r_{12} & \cdots & r_{1d-1} & r_{1d} \\ & & r_2 & \cdots & r_{2d-1} & r_{2d} \\ & & & \ddots & \vdots & \vdots \\ & & & & r_{d-1} & r_{d-1d} \\ & & & & & r_d \end{pmatrix}. \tag{17.18}$$

Then, the elements of $\boldsymbol{R}^2$ are

$$\begin{cases} (R^2)_{kk} = r_k^2, & 0 \leq k \leq d, \\ (R^2)_{jk} = \sum_{i=j}^{k} r_{ji} r_{ik}, & 0 \leq j \leq d-1, j+1 \leq k \leq d. \end{cases} \tag{17.19}$$

By substituting $\boldsymbol{R}^2$, $\boldsymbol{R}$, $\boldsymbol{A}$, $\boldsymbol{B}$ and $\boldsymbol{C}$ into Eq. (17.17), the following set of equations can be generated:

$$\begin{cases} (c\mu_1 + k\mu_2)r_k^2 - h_k r_k + \lambda = 0, & 0 \leq k \leq d \\ (c\mu_1 + k\mu_2)\sum_{i=j}^{k} r_{ji} r_{ik} - h_k r_{jk} + (d-k+1)\delta r_{j,k-1} = 0, & \\ \qquad 0 \leq j \leq d-1, \ j+1 \leq k \leq d. \end{cases} \tag{17.20}$$

The necessary and sufficient condition for the system being stable is $\rho = \lambda(c\mu_1 + d\mu_2)^{-1} < 1$. We analyze the system model and evaluate the system performance under the condition that $\rho < 1$. If the traffic load $\rho < 1$, it can be proven that the first equation of Eq. (17.20) has two real roots $0 < r_k < 1$ and $r_k^* \geq 1$. Note that the diagonal elements of R are r_k ($0 \leq k \leq d$) and the spectral radius of R satisfies the following equation:

$$\text{Sp}(R) = \max\{r_0, r_1, r_2, \ldots, r_d\} < 1. \tag{17.21}$$

The off-diagonal elements of R satisfy the last equation of Eq. (17.20). It is an arduous task to give a general expression for r_{jk} ($0 \leq j \leq d-1, j+1 \leq k \leq d$) in closed-form, so we recursively compute the off-diagonal elements based on the diagonal elements.

Since the QBD process with the one-step state transition rate matrix Q is positive recurrent, the stationary distribution is easily expressed in the matrix-geometric solution form with the rate matrix R as follows:

$$\pi_i = \pi_{c+d} R^{i-(c+d)}, \quad i \geq c+d. \tag{17.22}$$

In order to obtain the unknown stationary distribution $\pi_0, \pi_1, \pi_2, \ldots, \pi_{c+d}$, we construct a square matrix $B[R]$ of the order $(c+1/2 \times (d+1)(d+2)) \times (c+1/2 \times (d+1)(d+2))$ as follows:

$$B[R] = \begin{pmatrix} A_0 & C_0 & & & & & & \\ B_1 & A_1 & C_1 & & & & & \\ & \ddots & \ddots & \ddots & & & & \\ & & B_c & A_c & C_c & & & \\ & & & B_{c+1} & A_{c+1} & C_{c+1} & & \\ & & & & \ddots & \ddots & \ddots & \\ & & & & & B_{c+d-1} & A_{c+d-1} & C_{c+d-1} \\ & & & & & & B_{c+d} & RB+A \end{pmatrix}. \tag{17.23}$$

By using a matrix-geometric solution method, we can construct an augmented matrix equation as follows:

$$(\pi_0, \pi_1, \pi_2, \ldots, \pi_{c+d})\left(B[R] \begin{matrix} e_1 \\ (I-R)^{-1}e_2 \end{matrix} \right) = (0, 0, 0, \ldots, 0, 1) \tag{17.24}$$

where e_1 is a column vector with $c + 1/2 \times d(d+1)$ elements and e_2 is a column vector with $d+1$ elements, respectively. All elements of these vectors are equal to 1. The number of zeros in parentheses to the right of Eq. (17.24) is $c + 1/2 \times (d+1)(d+2)$.

By using the Gauss-Seidel method to solve Eq. (17.24), we can obtain $\pi_0, \pi_1, \pi_2, \ldots, \pi_{c+d}$. Substituting π_{c+d} obtained in Eq. (17.24) into Eq. (17.22), we can obtain π_i ($i = c+d+1, c+d+2, c+d+3, \ldots$). Then, the steady-state distribution $\Pi = (\pi_0, \pi_1, \pi_2, \ldots)$ of the system can be given mathematically.

17.4 Performance Measures and Numerical Results

In this section, we first derive performance measures of the system in terms of the average latency of tasks and the energy saving rate of the system, respectively. Then, we present numerical results to evaluate the performance of the system using the VM allocation strategy proposed in this chapter.

17.4.1 Performance Measures

We define the latency Y_t of a task as the duration from the instant a task arrives at the system to the instant this task is completely processed.

Based on the steady-state distribution of the system model given in Sect. 17.3.2, we obtain the average latency $E[Y_t]$ of tasks as follows:

$$E[Y_t] = \frac{1}{\lambda}\left(\sum_{i=0}^{c} i\pi_{i,0} + \sum_{i=c+1}^{c+d}\sum_{j=0}^{i-c} i\pi_{i,j} + \sum_{i=c+d+1}^{\infty}\sum_{j=0}^{d} i\pi_{i,j}\right). \tag{17.25}$$

In our proposed VM allocation strategy, energy consumption can be reduced during the sleep period. We let g_1 ($g_1 > 0$) be the energy consumption per second for a busy VM in Module II, and g_2 ($g_2 > 0$) be the energy consumption per second for a sleeping VM in Module II. Obviously, $g_1 > g_2$. We note that additional energy will be consumed when a task migrates from Module II to Module I, when a VM in Module II listens to the system buffer, as well as when a VM in Module II wakes up from a sleep state. Let g_3 ($g_3 > 0$), g_4 ($g_4 > 0$) and g_5 ($g_5 > 0$) be the energy consumption for each migration, listening and wake-up, respectively.

We define the energy saving rate γ of the system as the energy conservation per second with our proposed strategy. Based on the discussions above and the steady-state distribution of the system model given in Sect. 17.3.2, we give the energy saving rate γ of the system as follows:

$$\gamma = (g_1 - g_2)\sum_{i=0}^{\infty}\sum_{j=0}^{d}(d-j)\pi_{i,j} - \left(g_3 \sum_{i=c+1}^{c+d}\sum_{j=1}^{d} c\mu_1\pi_{i,j}\right.$$

$$\left. + g_4 \sum_{i=0}^{\infty}\sum_{j=0}^{d}\delta(d-j)\pi_{i,j} + g_5 \sum_{i=c+j+1}^{\infty}\sum_{j=0}^{d-1}\delta(d-j)\pi_{i,j}\right). \tag{17.26}$$

17.4.2 Numerical Results

In order to evaluate the average latency of tasks and the energy saving rate of the system with the VM allocation strategy proposed in this chapter, we present numerical results with analysis and simulation. The analysis results are obtained based on Eqs. (17.25) and (17.26) using Matlab 2011a. The simulation results are obtained by using MyEclipse 2014. We create a JOB class with attributes in terms of UNARRIVE, WAIT, RUNHIGH, RUNLOW and FINISH to record the task state. We also create a SERVER class with attributes in terms of SLEEP, IDLE, BUSYLOW and BUSYHIGH to record the state of a VM. Good agreements between the analysis results and the simulation results are observed.

We list the system parameters settings in Table 17.1 as an example for all the numerical results, where g_1 is the energy consumption per second of a busy VM in Module II, g_2 is the energy consumption per second of a sleeping VM in Module II, g_3 is the additional energy consumption for each migration, g_4 is the additional energy consumption for each listening, and g_5 is the additional energy consumption for each wake-up defined in Sect. 17.4.1, respectively.

We note that, with different parameter settings, as long as the system is stable, trends for all the performance measures will show only slight variations.

Figure 17.2 examines the influence of the sleep parameter δ on the average latency $E[Y_t]$ of the tasks for different numbers d of VMs in Module II.

In Fig. 17.2, we show the average latency $E[Y_t]$ versus the sleep parameter δ for the service rate $\mu_2 = 0.1$ in Module II as an example. From Fig. 17.2, we observe that if there are less VMs in Module II (such as $d < 24$), the average latency $E[Y_t]$ of tasks remains nearly constant across all the values of the sleep parameter δ. For this case, the capability of the VMs in Module I is strong enough to process all the arriving tasks, and there are no tasks waiting in the system buffer. As a result, it is likely that the VMs in Module II keep sleeping. Therefore, the average latency of tasks is approximately the average service time μ_1^{-1} of tasks processed in Module I.

From Fig. 17.2, we also observe that if there are more VMs in Module II (such as $d = 24, 41, 44, 50$), the average latency $E[Y_t]$ of tasks initially decreases sharply from a high value, then decreases slightly before finally converging to a certain value as the sleep parameter δ increases. For this case, the processing capability of

Table 17.1 Parameter settings in numerical results

Parameters	Values
Total number $c + d$ of VMs in the system	50
Arrival rate λ of tasks	$4.5 \, \text{ms}^{-1}$
Service rate μ_1 of a task on the VM in Module I	$0.2 \, \text{ms}^{-1}$
g_1 of a busy VM in Module II	0.5 mJ
g_2 of a sleeping VM in Module II	0.1 mJ
g_3 of each migration	0.2 mJ
g_4 of each listening	0.15 mJ
g_5 of each wake-up	0.2 mJ

Fig. 17.2 Average latency of
tasks versus sleep parameter

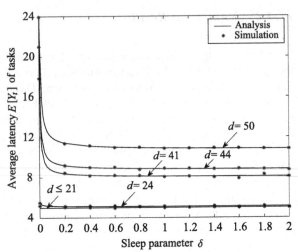

the VMs in Module I is insufficient to cope with the existing traffic load, so some arriving tasks have to wait in the system buffer. As a result, the VMs in Module II are more likely to be awake after a sleep period and process the tasks waiting in the system buffer. The influence of the sleep parameter on the average latency of tasks is explained as follows.

When the sleep parameter δ is relatively small (such as $0 < \delta < 0.4$ for $d = 41$), the tasks arriving in the sleep period will have to wait longer in the system buffer. This results in a higher average latency of tasks. For this case, the influence on the average latency of tasks exerted by the sleep parameter is greater than that exerted by the arrival rate of tasks and the service rate of tasks. Consequently, the average latency of tasks will decrease sharply as the sleep parameter increases.

When the sleep parameter δ becomes larger (such as $0.4 < \delta < 2$ for $d = 41$), the tasks arriving during a sleep period will be processed earlier. This results in a lower average latency of tasks. For this case, the arrival rate of tasks and the service rate of tasks are the dominate factors influencing the average latency of tasks. Consequently, there is only a slight decreasing trend in the average latency of tasks in respect to the sleep parameter.

From Fig. 17.2, we also notice that for the same sleep parameter δ, the average latency $E[Y_t]$ of tasks increases as the number d of VMs in Module II increases. As the number of VMs in Module II increases, and the system capability becomes weaker, tasks will sojourn longer in the system. This will inevitably increase the average latency of tasks.

Figure 17.3 examines the influence of the sleep parameter δ on the energy saving rate γ of the system for different numbers d of VMs in Module II.

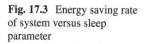

Fig. 17.3 Energy saving rate
of system versus sleep
parameter

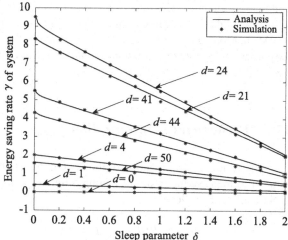

In Fig. 17.3, we show the energy saving rate γ of the system versus the sleep parameter δ for the service rate of a task on the VM in Module II, $\mu_2 = 0.1$ as an example. From Fig. 17.3, we observe that for the same number d of VMs in Module II, the energy saving rate γ of the system decreases as the sleep parameter δ increases. The larger the sleep parameter is, the more frequently the VM in Module II listens to the system buffer and consumes additional energy. Therefore, the energy saving rate of the system will decrease.

From Fig. 17.3, we also notice that for the same sleep parameter δ, either too few or too many VMs being deployed in Module II leads to a lower energy saving rate γ of the system. When the number of VMs in Module II is very small (such as $d = 0, 1, 4$), less energy can be saved even though all the VMs in Module II are sleeping. This results in a lower energy saving rate of the system. When the number of VMs in Module II is very large (such as $d = 41, 42, 50$), the system capability gets weaker. There is hardly any chance for the VMs in Module II to go to sleep. This results in a lower energy saving rate of the system.

In Figs. 17.2 and 17.3, the experiment results with $d = 0$ are for the conventional strategy where all the VMs always stay awake. The experiment results with $d = 50$ are for the conventional strategy where all the VMs are under an asynchronous multi-sleep mode. Compared to the conventional strategy where all the VMs always stay awake, our proposed strategy results in greater energy consumption without significantly affecting the response performance. Compared to the conventional strategy where all the VMs are under an asynchronous multi-sleep mode, our proposed strategy performs better in guaranteeing the response performance at the cost of occasional degradation in the energy saving effect.

Comparing the results shown in Figs. 17.2 and 17.3, we find that a larger sleep parameter leads to not only a shorter average latency of tasks but also a lower energy saving rate of the system, while a smaller sleep parameter leads to not only a higher energy saving rate of the system but also a higher average latency of tasks. We also find that the energy saving rate of the system is higher with a moderate number of VMs in Module II, while the average latency of tasks is lower with a smaller number of VMs in Module II. Therefore, a trade-off between the average latency of tasks and the energy saving rate of the system should be aimed for when setting the number of VMs in Module II and the sleep parameter in our proposed VM allocation strategy.

17.5 Performance Optimization

To do performance optimization of the system, we establish a system cost function $F(d, \delta)$ to improve the trade-off among different performance measures presented in Sect. 17.4.1 as follows:

$$F(d, \delta) = f_1 \times E[Y_t] - f_2 \times \gamma \tag{17.27}$$

where f_1 and f_2 are treated as the impact factors for the average latency $E[Y_t]$ of tasks and the energy saving rate γ of the system on the system cost function.

We note that the mathematical expressions for the average latency $E[Y_t]$ of tasks and the energy saving rate γ of the system are difficult to express in closed-forms. The monotonicity of the system cost function is uncertain. In order to jointly optimize the number of the VMs in Module II and the sleep parameter with the minimum system cost function, we turn to the PSO intelligent searching algorithm.

Compared with other intelligent optimization algorithms, the PSO algorithm is simple to implement, and there are few parameters that require adjusting. However, the PSO algorithm has the disadvantage of premature convergence and easily falling into a local extreme. In order to efficiently control the global and local search of the PSO algorithm, we develop an improved PSO algorithm by introducing a chaotic mapping mechanism and a nonlinear decreasing inertia weight. The main steps for the improved PSO algorithm are given as follows:

Step 1: Initialize the number N of particles, the maximum number of iterations $iter_{max}$ for each particle's position, the cognitive acceleration coefficients c_1, the social acceleration coefficients c_2, the maximal inertia weight w_{max}, the minimal inertia weight w_{min}, the upper search boundary U_b, the lower search boundary L_b, the number X of total VMs, and set the initial number of VMs in Module II as $d = 0$.

Step 2: Set the optimal combination (d^*, δ^*) for the number of VMs in Module II and the sleep parameter, and calculate the corresponding fitness F^*.
$(d^*, \delta^*) = (0, U_b)$, $F^* = F(d^*, \delta^*) = f_1 E[Y_t] - f_2 \gamma$

Step 3: Initialize position δ_i for the ith ($i = 1, 2, 3, \ldots, N$) particle by using a chaotic equation.

$\delta_1 = \text{rand}$

% rand is a random number selected in the interval (0, 1).

for $i = 2 : N$

    if $\delta_{i-1} < 0.5$

        $\delta_i = 1.5 \times \delta_{i-1} + 0.25$

    else

        $\delta_i = 0.5 \times \delta_{i-1} - 0.25$

    endif

endfor

for $i = 1 : N$

    $\delta_i = L_b + ((\ln(\delta_i + 0.5) + \ln 2)/\ln 3) \times (U_b - L_b)$

endfor

Step 4: For each particle ($i = 1, 2, 3, \ldots, N$), initialize the personal best position pb_i and the velocity v_i, and calculate the fitness F_i.

$pb_i = \delta_i$, $v_i = \text{randn}$, $F_i = F(d, pb_i) = f_1 E[Y_t] - f_2 \gamma$

% randn returns a sample from the standard normal distribution.

Step 5: Select the global best position gb among all the personal best positions with the number d of VMs in module II.

$$gb = \underset{i \in \{1,2,3,\ldots,N\}}{\text{argmin}} \{F_i\}$$

Step 6: Set the initial number of iterations as $iter = 1$.

Step 7: Update the inertia weight with the strategy of nonlinear decreasing inertia weight.

$$w = (w_{\max} - w_{\min})\left(\frac{iter}{iter_{\max}}\right)^2 + (w_{\min} - w_{\max})\left(2 \times i \frac{iter}{iter_{\max}}\right) + w_{\max}$$

Step 8: For each particle ($i = 1, 2, 3, \ldots, N$), update the position δ_i, the velocity v_i, the fitness F_i and the personal best position pb_i.

$v_i = w \times v_i + c_1 \times \text{rand} \times (pb_i - \delta_i) + c_2 \times \text{rand} \times (gb - \delta_i)$, $\delta_i = \delta_i + v_i$

$F_i' = F(d, \delta_i) = f_1 E[Y_t] - f_2 \gamma$

if $F_i' < F_i$

    $F_i = F_i'$

    $pb_i = \delta_i$

endif

Step 9: Select the global best position gb among all the personal best positions with the number d of VMs in module II.

$$gb = \underset{i \in \{1,2,3,\ldots,N\}}{\text{argmin}} \{F_i\}$$

Step 10: Check the number of iterations.

    if $iter < iter_{\max}$

        $iter = iter + 1$

        go to **Step 7**

    endif

Step 11: Select the optimal combination (d^*, δ^*) for the number of VMs in Module II and the sleep parameter.

Table 17.2 Optimum combination of parameters in proposed strategy

Service rates μ_2	Optimum combinations (d^*, δ^*)	Minimum costs $F(d^*, \delta^*)$
0.05	(22,0.0001)	11.9046
0.1	(23,0.033)	11.8698
0.15	(25,0.0816)	11.6391
0.2	(49,0.1593)	11.041

$$F' = F(d, gb) = f_1 E[Y_t] - f_2 \gamma$$
$$\textbf{if } F' < F^*$$
$$\quad F^* = F'$$
$$\quad (d^*, \delta^*) = (d, gb)$$
$$\textbf{endif}$$

Step 12: Check the number of VMs in Module II.

$$\quad \textbf{if } d < X$$
$$\quad\quad d = d + 1$$
$$\quad\quad \text{go to } \textbf{Step 3}$$
$$\quad \textbf{endif}$$

Step 13: Output the optimal combination (d^*, δ^*) and the minimum system costs F^*.

In the improved PSO algorithm, we use the system parameters given in Table 17.1, and set $f_1 = 4$, $f_2 = 1$, $N = 100$, $iter_{max} = 200$, $c_1 = c_2 = 1.4962$, $w_{max} = 0.95$, $w_{min} = 0.4$, $U_b = 2$, $L_b = 0$ and $X = 50$. For different service rates μ_2, we obtain the optimal combination (d^*, δ^*) for the number of VMs in Module II and the sleep parameter with the minimum system cost function $F(d^*, \delta^*)$ in Table 17.2.

The optimization results in Table 17.2 depend on the arrival intensity of tasks, the serving capability of VMs and the cloud capacity. By substituting the arrival rate λ, the service rate μ_2 of a task on the VM in Module II, and the total number $(c + d)$ of VMs in a virtual cluster, etc. into the algorithm in Table 17.2, the optimal parameter combination (d^*, δ^*) for the number of VMs in Module II and the sleep parameter can be obtained for the proposed strategy.

17.6   Conclusion

In this chapter, with the aim of reducing energy consumption and achieving greener computing, we proposed an energy-efficient VM allocation strategy with an asynchronous multi-sleep mode and an adaptive task-migration scheme, we established a queueing model with partial asynchronous multiple vacations and derived the steady-state distribution of the system model. The queueing model quantified the effects of the number of VMs in Module II and the sleep parameter. These effects were evaluated by two performance measures: the average latency of tasks and the energy saving rate of the system. Numerical results showed that the energy saving rate of the system is higher with a moderate number of VMs in Module II, while the average latency of tasks is lower with a smaller number of VMs in Module II. Accordingly, we built a system cost function to investigate a trade-off between different performance measures. By introducing a chaotic mapping mechanism and a nonlinear decreasing inertia weight, we developed an improved PSO algorithm and jointly optimized the number of VMs in Module II and the sleep parameter with the minimum value of the system cost function.

Chapter 18
Clustered Virtual Machine Allocation Strategy

In this chapter, we propose a clustered Virtual Machine (VM) allocation strategy based on a sleep-mode with wake-up threshold to achieve green computing. The VMs in a cloud data center are clustered into two modules, namely, Module I and Module II. The VMs in Module I remain awake at all times, while the VMs in Module II go to sleep under a light workload. We build a queue with an N-policy and asynchronous vacations for partial servers to capture the stochastic behavior of tasks with the proposed strategy. We derive performance measures of the system in terms of the average latency of tasks and the energy saving rate of the system, respectively. Furthermore, we present numerical results to demonstrate the impact of the system parameters on the system performance. Finally, we construct a system cost function to trade off different performance measures and develop an intelligent searching algorithm to jointly optimize the number of the VMs in Module II, the wake-up threshold and the sleep parameter.

18.1 Introduction

The energy consumption in Cloud Data Centers (CDCs) in WCNs has been steadily increasing over the last few years [Qie19, Zhou18]. As a result, the minimization of power and energy consumption in a cloud computing system has become a challenging problem and has received significant attention recently [Jin19c, Mare18].

Applying sleep mode is an effective method for reducing energy consumption in CDCs [Khoj18]. In [Duan15], for the purpose to minimize the energy consumption during runtime, the authors presented a dynamic idle interval prediction scheme to estimate the future idle interval length of a CPU and thereby choose the most cost-effective sleep state. In [Chou16], the authors proposed a fine-grain power management scheme for data center workloads. This proposed scheme could dynamically postpone the processing of some tasks, create longer idle periods and

© Springer Nature Singapore Pte Ltd. 2021
S. Jin, W. Yue, *Resource Management and Performance Analysis of Wireless Communication Networks*, https://doi.org/10.1007/978-981-15-7756-7_18

promote the use of a deeper sleep mode. In [Luo17], based on flow preemption and power-aware routing, the authors proposed a dynamic adaptive scheduling algorithm to reduce energy consumption by decreasing the ratio of low utilization devices and putting more devices into sleep mode. In the literature mentioned above, we note that a sleep mode was only applied to a Physical Machine (PM) rather than a VM. This weakens the application flexibility of sleep modes.

The queueing models with vacation are used as an important way to evaluate the performance measures of the CDCs with sleep modes [Do18]. Multiple-server queueing models with various types of vacations have been studied by many researchers. In [Jin18], the authors proposed a M/M/c queue with N-policy and synchronous vacations. In [Tian01], the authors analyzed the equilibrium theory for M/M/c queue with an N-policy and asynchronous vacations. In [Jin17a], the authors proposed a queueing system with synchronous vacations for partial servers. The research mentioned above has contribution to enrich the queueing system with an N-policy and synchronous vacations, an N-policy and asynchronous vacations, as well as asynchronous vacations for partial servers. However, there has so far been no research on the queueing system with an N-policy and asynchronous vacations for partial servers. Teaching-Learning-Based Optimization (TLBO) algorithm is an effective tool for function optimization [Budd19]. Since the appearance of TLBO algorithm, many researchers have taken their effort to improve the searching capability of TLBO algorithms. In [Yu16], as an effort to enhance the performance of TLBO algorithms, the authors introduced the mutation crossover operation to increase population diversity, and applied the chaotic perturbation mechanism to make the TLBO algorithm avoid falling into the local optimum. In [Ji17], inspired by the concept of historical population, the authors added two new process, namely the self-feedback learning process as well as the mutation and crossover process to the TLBO algorithm. The added processes improved the exploration ability compared to the original TLBO algorithm. In [Li17], in order to enhance the diversification of population, the authors increased the number of teachers, introduced new students and performed local search around the potentially optimal solutions. Enhancing the population diversity as well as improving the global and local searching ability are potential methods to extend the traditional TLBO algorithms.

Inspired by the research mentioned above, in this chapter, we firstly propose a clustered VM allocation strategy based on a sleep mode with wake-up threshold, and build a queue with an N-policy and asynchronous vacations for partial servers. And then, we mathematically and numerically evaluate the system performance in terms of the average latency of tasks and the energy saving rate of the system. Finally, we develop an improved TLBO algorithm by introducing a cube chaotic mapping mechanism for the grade initialization and an exponentially decreasing strategy for the teaching process to optimize the strategy proposed in this chapter with the minimum value of the system cost.

The chapter is organized as follows. In Sect. 18.2, we describe the clustered VM allocation strategy based on a sleep mode proposed in this chapter. Then, we present the system model in detail. In Sect. 18.3, we present a performance analysis of the system model, through the analysis of the transition rate matrix and the steady-state distribution. In Sect. 18.4, we obtain performance measures and present numerical results to evaluate the system performance. In Sect. 18.5, we develop an improved TLBO algorithm to jointly optimize the number of the VMs in Module II, the wake-up threshold and the sleep parameter. Our conclusions are drawn in Sect. 18.6.

18.2   Clustered Virtual Machine Allocation Strategy and System Model

In this section, in order to enhance the energy efficiency in a cloud computing system, we first propose a clustered VM allocation strategy based on a sleep mode with wake-up threshold. Then, we establish a queue with an N-policy and asynchronous vacations for partial servers to capture the strategy proposed in this chapter.

18.2.1   Clustered Virtual Machine Allocation Strategy

It should be noted that additional energy will be consumed when a VM frequently switches from the sleep state to the awake state, while the system performance will be degraded when all the VMs are put in a sleep mode. To get around these problems, based on a sleep mode with wake-up threshold, we propose a clustered VM allocation strategy.

The VMs in a CDC are clustered into two modules, namely, Module I and Module II. The VMs in Module I remain awake all the time and operate at a higher speed. The VMs in Module II will go to sleep independently when there are no tasks in the system buffer. For the purpose of improving the energy efficiency of CDCs, we introduce a wake-up threshold to a periodic sleep mode. At the end epoch of a sleep period, if the number of the tasks gathered in the system buffer reaches or exceeds a certain value, namely wake-up threshold, the corresponding VM in Module II will wake up independently and operate at a lower speed. Otherwise, the VM in Module II will begin another sleep period with a new sleep timer.

With the strategy proposed in this chapter, all the VMs are dominated by a control server, where several sleep timers, a task counter, and a VM scheduler are deployed. When a VM in Module II is scheduled to go to sleep from an awake state or begin another sleep period, a sleep timer with a random value will be started to control the time length of a sleep period. The task counter keeps working and counts the tasks waiting in the system buffer. At the moment that a sleep timer expires, the VM scheduler adjusts the state of the corresponding VM according to the value in the task counter.

Figure 18.1 illustrates the working flow of a VM with the clustered VM allocation strategy based on a sleep mode with wake-up threshold proposed in this chapter.

Fig. 18.1 Working flow of
VM with proposed strategy

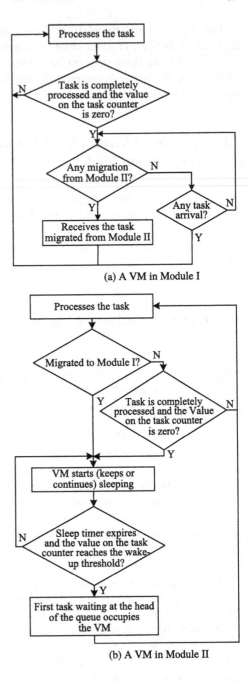

(a) A VM in Module I

(b) A VM in Module II

As shown in Fig. 18.1, the VMs in Module I keep awake all the time, while the VMs in Module II switches between sleep state and awake state. Next, we discuss the state transition for the VMs in Module II.

(1) *Awake State to Sleep State*: For a busy VM in Module II, the state transition from an awake state to a sleep state occurs only at the instant when a task either in Module I or Module II is completely processed. When a task is completely processed in Module I, if the VM scheduler monitors that the value in the task counter is zero and there is at least one task being processed in Module II, one of the tasks being processed in Module II will be migrated to Module I, and then the evacuated VM in Module II will go to sleep. When a task is completely processed in Module II, if the VM scheduler monitors that the value in the task counter is zero, the evacuated VM in Module II will go to sleep directly. We note that the task-migration considered in this chapter is an online VM-migration between different modules within a CDC, and this task-migration is to make the VMs in Module II go to sleep earlier.

(2) *Sleep State to Awake State*: For a sleeping VM in Module II, the state transition from a sleep state to an awake state occurs only at the end epoch of a sleep period. When a sleep timer expires, if the VM scheduler monitors that the value in the task counter is equal to or greater than the wake-up threshold, the corresponding VM in Module II will wake up to process the first task waiting in the system buffer at a lower speed. Otherwise, a new sleep timer will be started and the VM in Module II will begin another sleep period.

To summarize, at the moment that a sleep timer expires, the VMs in Module II will wake up only if the number of the tasks waiting in the system buffer reaches the wake-up threshold. The wake-up mechanism in our proposed strategy effectively improves the energy conservation via extending sleep time by delaying the instant for a VM to wake up. This is exactly the starting point for the N-policy sleep mode with our proposed strategy.

18.2.2   System Model

We model the system as a queueing system with an N-policy and asynchronous vacations for partial independent servers in the VMs to process the tasks according to the proposed clustered VM allocation strategy based on a sleep mode and a wake-up threshold.

In this system model, the buffer capacity is supposed to be infinite. Moreover, the numbers of the VMs in Module I and Module II are denoted as c and d, respectively. Let $I(t)$ be the total number of tasks in the system at instant t. $I(t)$ is called the system level, $I(t) \geq 0$. Let $J(t)$ be the number of busy VMs in Module II at instant t. $J(t)$ is called the system stage, $J(t) = 0, 1, 2, \ldots, d$.

Thus, the behavior of the system under consideration can be described in terms of the two-dimensional continuous-time stochastic process $\{(I(t), J(t)), \ t \geq 0\}$.

The state-space Ω of the two-dimensional continuous-time stochastic process $\{(I(t), J(t)), \ t \geq 0\}$ is given as follows:

$$
\begin{aligned}
\Omega =\{&(i, 0) : i = 0, 1, 2, \ldots, c\} \\
&\cup \{(i, j) : i = c + 1, c + 2, c + 3, \ldots, c + d, \ j = 0, 1, 2, \ldots, i - c\} \\
&\cup \{(i, j) : i \geq c + d + 1, \ j = 0, 1, 2, \ldots, d\}.
\end{aligned} \tag{18.1}
$$

We assume that all the tasks in the system are homogenous and all the VMs in a module are identical, where each task is for one VM. We assume that the inter-arrival time of tasks follows an exponential distribution with mean $1/\lambda$, where $\lambda > 0$, called the arrival rate of tasks. In addition, the service times of tasks processed in Module I and in Module II are assumed to be exponentially distributed with mean $1/\mu_1$ seconds and $1/\mu_2$ seconds, respectively, where $\mu_1 > 0$ and $0 < \mu_2 < \mu_1$. We call μ_1 the service rate in Module I and μ_2 the service rate in Module II. Furthermore, the time length of a sleep timer is assumed to follow an exponential distribution with mean of $1/\delta$ seconds, where $\delta > 0$. Here, δ is called the sleep parameter.

Based on the assumptions above, the two-dimensional continuous-time stochastic process $\{(I(t), J(t)), \ t \geq 0\}$ can be seen as a two-dimensional Continuous-Time Markov Chain (CTMC).

We define $\pi_{i,j}$ as the probability of the system level being equal to i and the system stage being equal to j in the steady state. $\pi_{i,j}$ is then given as follows:

$$
\pi_{i,j} = \lim_{t \to \infty} \Pr\{I(t) = i, J(t) = j\}, \quad (i, j) \in \Omega. \tag{18.2}
$$

And then, we form the probability vectors $\boldsymbol{\pi}_i$ in the steady state as follows:

$$
\boldsymbol{\pi}_i = \begin{cases}
\pi_{i,0}, & i = 0, 1, 2, \ldots, c \\
(\pi_{i,0}, \pi_{i,1}, \pi_{i,2}, \ldots, \pi_{i,i-c}), & i = c + 1, c + 2, c + 3, \ldots, c + d \\
(\pi_{i,0}, \pi_{i,1}, \pi_{i,2}, \ldots, \pi_{i,d}), & i \geq c + d + 1.
\end{cases} \tag{18.3}
$$

The steady-state distribution $\boldsymbol{\Pi}$ of the two-dimensional CTMC is composed of $\boldsymbol{\pi}_i$ ($i \geq 0$). $\boldsymbol{\Pi}$ is given as follows:

$$
\boldsymbol{\Pi} = (\boldsymbol{\pi}_0, \boldsymbol{\pi}_1, \boldsymbol{\pi}_2, \ldots). \tag{18.4}
$$

18.3 Performance Analysis

In this section, we present a performance analysis of the system model, through the analysis of the transition rate matrix and the steady-state distribution.

18.3.1 Transition Rate Matrix

We denote Q as the one-step state transition rate matrix of the two-dimensional CTMC $\{(I(t), J(t)), t \geq 0\}$. According to the system level, we separate the one-step state transition rate matrix Q of the two-dimensional CTMC $\{(I(t), J(t)), t \geq 0\}$ into some sub-matrices with different orders. To clearly represent the sub-matrices, we denote $Q_{k,k'}$ as the one-step state transition rate sub-matrix for the system level changing from k ($k \geq 0$) to k' ($k' \geq 0$). Based on the initial system level k, we discuss the one-step state transition rate sub-matrix $Q_{k,k'}$ by the following three cases.

(1) *Initial System Level $k = 0, 1, 2, \ldots, c$*: The number of busy VMs in Module I is k, while the number of busy VMs in Module II is 0.

$k = 0$ means no tasks exist in the system at all. Therefore, the possible state transitions are from state $(0, 0)$ to state $(1, 0)$ and from state $(0, 0)$ to state $(0, 0)$. If there is a task arrival at the system, the system level will be increased by one, while the system stage will not change, the transition rate from state $(0, 0)$ to state $(1, 0)$ is λ. If no task arrival occurs in the system, the system state will be fixed at state $(0, 0)$, the transition rate is $-\lambda$.

Consequently, the sub-matrices $Q_{0, 1}$ and $Q_{0, 0}$ are actually quantities given as follows:

$$Q_{0, 1} = \lambda, \quad Q_{0, 0} = -\lambda. \tag{18.5}$$

$k = 1, 2, 3, \ldots, c$ means all the tasks in the system are receiving service from the VMs in Module I. If one of the tasks is completely processed and departs the system, the system level will be decreased by one, while the system stage will not change, the transition rate from state $(k, 0)$ to state $(k - 1, 0)$ is $k\mu_1$. If there is a task arrival at the system, the system level will be increased by one, while the system stage will not change, the transition rate from state $(k, 0)$ to state $(k + 1, 0)$ is λ. If neither an arrival nor a departure occurs, the system state will be fixed at state $(k, 0)$, the transition rate is $-(\lambda + k\mu_1)$.

Consequently, the sub-matrices $Q_{k,k-1}$, $Q_{k,k+1}$ and $Q_{k,k}$ are also actually quantities given as follows:

$$Q_{k,k-1} = k\mu_1, \quad Q_{k,k+1} = \lambda, \quad Q_{k,k} = -(\lambda + k\mu_1), \quad k = 1, 2, 3, \ldots, c. \tag{18.6}$$

(2) *Initial System Level $k = c + x$ ($x = 1, 2, 3, \ldots, d - 1$)*: The number of busy VMs in Module I is c, while the number of busy VMs in Module II is less than or equal to x.

   If there is at least one task queueing in the system buffer when a task is completely processed on a VM, whether the VM belongs to Module I or Module II, the task waiting at the head of the queue will be scheduled to the evacuated VM. For this case, the system level will be decreased by one, while the system stage will not change, the transition rate from state (k, n) to state $(k - 1, n)$ is $(c\mu_1 + n\mu_2)$, where n ($n = 0, 1, 2, \ldots, x$) is the number of busy VMs in Module II.

   If there are no tasks queueing in the system buffer when a task is completely processed on a VM, whether the VM belongs to Module I or Module II, the number of sleeping VMs will be increased by one. If the evacuated VM belongs to Module I, one of the tasks receiving service in Module II will be migrated to the evacuated VM in Module I, and the just-evacuated VM in Module II will go to sleep. If the evacuated VM belongs to Module II, the evacuated VM in Module II will start sleeping directly. For both of these two cases, both the system level and the system stage will be decreased by one, the transition rate from state (k, x) to state $(k - 1, x - 1)$ is $(c\mu_1 + x\mu_2)$.

   Consequently, the sub-matrix $\boldsymbol{Q}_{k,k-1}$ is a rectangular $(x + 1) \times x$ matrix given as follows:

$$
\boldsymbol{Q}_{k,k-1} = \begin{pmatrix} c\mu_1 & & & \\ & c\mu_1 + \mu_2 & & \\ & & \ddots & \\ & & & c\mu_1 + (x - 1)\mu_2 \\ & & & c\mu_1 + x\mu_2 \end{pmatrix},
$$

$$
k = c + x, x = 1, 2, 3, \ldots, d - 1. \tag{18.7}
$$

   Before any of the sleep timers in Module II expires, if there is a task arrival at the system, the newly arriving task has to queue in the system buffer. For this case, the system level will be increased by one, while the system stage will not change, the transition rate from state (k, n) to state $(k+1, n)$ is λ. Consequently, the sub-matrix $\boldsymbol{Q}_{k,k+1}$ is a rectangular $(x+1) \times (x+2)$ matrix given as follows:

$$
\boldsymbol{Q}_{k,k+1} = \begin{pmatrix} \lambda & & & & 0 \\ & \lambda & & & 0 \\ & & \ddots & & \vdots \\ & & & \lambda & 0 \\ & & & \lambda & 0 \end{pmatrix},
$$

$$
k = c + x, \ x = 1, 2, 3, \ldots, d - 1. \tag{18.8}
$$

At the moment that one of the sleep timers expires, the corresponding VM in Module II will decide whether to wake up or continue sleeping according to the number of tasks queueing in the system buffer. If this number is equal to or greater than the wake-up threshold N, the corresponding VM in Module II will wake up and prepare to process the task waiting at the head of the queue. For this case, the system level will not change, while the system stage will be increased by one, the transition rate from state (k, n) to state $(k, n + 1)$ is $(d - n)\delta$, where n $(n = 0, 1, 2, \ldots, x)$ is the number of busy VMs in Module II. Otherwise, the corresponding VM in Module II will continue sleeping and there will be no state transitions at all.

Before any of the sleep timers in Module II expires, if neither an arrival nor a departure occurs, the system state will not change. When the number of tasks queueing in the system buffer is equal to or greater than the wake-up threshold N, the transition rate is $(-h_n - (d - n)\delta)$, where $h_n = \lambda + c\mu_1 + n\mu_2$. When the number of tasks queueing in the system buffer is less than the wake-up threshold N, the transition rate is $-h_n$.

Consequently, the sub-matrix $\boldsymbol{Q}_{k,k}$ is a rectangular $(x + 1) \times (x + 1)$ matrix given as follows:

$$\boldsymbol{Q}_{k,k} = \text{diag}\,(-h_0, -h_1, \ldots, -h_x),$$

$$k = c + x, \ x = 1, 2, 3, \ldots, \min\{N, d\} - 1, \tag{18.9}$$

$$\boldsymbol{Q}_{k,k} = \begin{pmatrix} -h_0 - d\delta & & d\delta & & & & \\ & \ddots & & \ddots & & & \\ & & -h_{x-N} - (d - (x - N))\delta & (d - (x - N))\delta & & & \\ & & & -h_{x-N+1} & 0 & & \\ & & & & \ddots & \ddots & \\ & & & & & -h_{x-1} & 0 \\ & & & & & & -h_x \end{pmatrix},$$

$$k = c + x, \ x = N, N + 1, N + 2, \ldots, d - 1, \ N \le d. \tag{18.10}$$

(3) *Initial System Level* $k = c + x, \ x \ge d$: The number of busy VMs in Module I is c, while the number of busy VMs in Module II is less than or equal to d.

For the case of $k = c + d$, $\boldsymbol{Q}_{k,k-1}$ is a rectangular matrix of order $(d+1) \times d$. For the case of $k = c + x, \ x \ge d + 1$, $\boldsymbol{Q}_{k,k-1}$ is square matrices of the order $(d + 1) \times (d + 1)$. For the case of $k = c + x, \ x \ge d$, $\boldsymbol{Q}_{k,k+1}$ and $\boldsymbol{Q}_{k,k}$ are all square matrices of the order $(d + 1) \times (d + 1)$.

Referencing to the discussions in Item (2), the sub-matrices $Q_{k,k-1}$, $Q_{k,k+1}$ and $Q_{k,k}$ are given as follows:

$$Q_{k,k-1} = \begin{pmatrix} c\mu_1 & & & & \\ & c\mu_1 + \mu_2 & & & \\ & & \ddots & & \\ & & & c\mu_1 + (d-1)\mu_2 & \\ & & & & c\mu_1 + d\mu_2 \end{pmatrix}, \quad k = c+d,$$

(18.11)

$$Q_{k,k-1} = \mathrm{diag}\,(c\mu_1, c\mu_1 + \mu_2, \ldots, c\mu_1 + d\mu_2),$$
$$k = c+x, \ x \geq d+1, \qquad (18.12)$$

$$Q_{k,k+1} = \mathrm{diag}\,(\lambda, \lambda, \lambda, \ldots, \lambda), \quad k = c+x, \ x \geq d, \qquad (18.13)$$

$$Q_{k,k} = \mathrm{diag}\,(-h_0, -h_1, -h_2, \ldots, -h_d),$$
$$k = c+x, \ d \leq x \leq N-1, \ N > d, \qquad (18.14)$$

$$Q_{k,k} = \begin{pmatrix} -h_0 - d\delta & & d\delta & & & \\ \ddots & & & \ddots & & \\ & -h_{x-N} - (d-(x-N))\delta & (d-(x-N))\delta & & & \\ & & -h_{x-N+1} & 0 & & \\ & & & \ddots & & \ddots \\ & & & & -h_{d-1} & 0 \\ & & & & & -h_d \end{pmatrix},$$

$$k = c+x, \ \max\{N, d\} \leq x \leq N+d-1, \qquad (18.15)$$

$$Q_{k,k} = \begin{pmatrix} -h_0 - d\delta & d\delta & & & \\ & -h_1 - (d-1)\delta & (d-1)\delta & & \\ & & \ddots & & \ddots \\ & & & -h_{d-1} - \delta & \delta \\ & & & & -h_d \end{pmatrix},$$

$$k = c+x, \ x \geq d+N. \qquad (18.16)$$

So far, all the sub-matrices in the one-step state transition rate matrix Q have been addressed. In the one-step state transition rate matrix Q, the sub-matrices $Q_{k,k-1}$ are repeated forever starting from the system level $(c + d + 1)$, the sub-matrices $Q_{k,k+1}$ are repeated forever starting from the system level $(c + d)$, and the sub-matrices $Q_{k,k}$ are repeated forever starting from the system level $(c + d + N)$.

For presentation purposes, the repetitive sub-matrices $Q_{k,k-1}$, $Q_{k,k+1}$ and Q_k are represented by B, C and A, respectively.

Then, the one-step state transition rate matrix Q is written as follows:

$$Q = \begin{pmatrix} Q_{0,0} Q_{0,1} \\ Q_{1,0} Q_{1,1} & Q_{1,2} \\ \ddots & \ddots & \ddots \\ & Q_{c+d,c+d-1} Q_{c+d,c+d} & C \\ & B & Q_{c+d+1,c+d+2} & C \\ & & \ddots & \ddots & \ddots \\ & & B & Q_{c+d+N-1,c+d+N-1} C \\ & & & B & A & C \\ & & & & \ddots & \ddots & \ddots \end{pmatrix}.$$

$$(18.17)$$

Obviously, the obtained form of the one-step state transition rate matrix Q is none other than a Quasi Birth-Death (QBD) matrix, so the CTMC $\{(I(t), J(t)), \ t \geq 0\}$ is also called a QBD CTMC.

Several methods can be used to resolve the steady-state transition probabilities. In our study, we consider a matrix-geometric solution method which is vastly used to analyze the QBD CTMC in the steady state.

18.3.2 Steady-State Distribution

To analyze the QBD CTMC $\{(I(t), J(t)), \ t \geq 0\}$ by using a matrix-geometric solution method, we need to solve for the minimal non-negative solution of the matrix quadratic equation:

$$R^2 B + RA + C = 0, \tag{18.18}$$

and this solution, called the rate matrix and denoted by R, can be explicitly determined.

In Sect. 18.3.1, the sub-matrices A, B and C are deduced to be upper-triangular matrices. Thus, the rate matrix R is an upper-triangular matrix. We denote the unknown element of the rate matrix R in line u ($u = 0, 1, 2, \ldots, d$) column

v ($v = 0, 1, 2, \ldots, d$) as $r_{u,v}$, and $r_u = r_{u,u}$ ($u = 0, 1, 2, \ldots, d$). The rate matrix R can be written as follows:

$$R = \begin{pmatrix} r_0 & r_{0,1} & r_{0,2} & \cdots & r_{0,d-1} & r_{0,d} \\ & r_1 & r_{1,2} & \cdots & r_{1,d-1} & r_{1,d} \\ & & r_2 & \cdots & r_{2,d-1} & r_{2,d} \\ & & & \ddots & \vdots & \vdots \\ & & & & r_{d-1} & r_{d-1,d} \\ & & & & & r_d \end{pmatrix}. \tag{18.19}$$

Using the elements in R, we calculate the elements of R^2 as follows:

$$(R^2)_{u,u} = r_u^2, \quad u = 0, 1, 2, \ldots, d. \tag{18.20}$$

$$(R^2)_{u,v} = \sum_{i=u}^{v} r_{u,i} r_{i,v}, \quad u = 0, 1, 2, \ldots, d-1, v = u+1, u+2, u+3, \ldots, d. \tag{18.21}$$

Next, by substituting the R^2, R, A, B and C into Eq. (18.18), we obtain a set of equations as follows:

$$\begin{cases} (c\mu_1 + u\mu_2) r_u^2 - (h_u + (d-u)\delta) r_u + \lambda = 0, \quad u = 0, 1, 2, \ldots, d \\ (c\mu_1 + v\mu_2) \sum_{i=u}^{v} r_{u,i} r_{i,v} - (h_v + (d-v)\delta) r_{u,v} + (d-v+1)\delta r_{u,v-1} = 0, \\ \qquad u = 0, 1, 2, \ldots, d-1, \ v = u+1, u+2, u+3, \ldots, d. \end{cases} \tag{18.22}$$

With the constraint that the traffic load $\rho = \lambda(c\mu_1 + d\mu_2)^{-1} < 1$, we prove that the first equation of Eq. (18.22) has two real roots r_u ($0 < r_u < 1$) and r_u^* ($r_u^* \geq 1$). It should be noted that the diagonal elements of R are r_u ($u = 0, 1, 2, \ldots, d$).

It is too arduous to deduce a general expression of $r_{u,v}$ ($u = 0, 1, 2, \ldots, d-1, v = u+1, u+2, u+3, \ldots, d$) in closed-form. Therefore, based on the last equation of Eq. (18.22), we recursively compute the off-diagonal elements starting from the diagonal elements obtained in the first equation of Eq. (18.22).

By using a matrix-geometric solution method, we have

$$\pi_i = \pi_{c+d+N} R^{i-(c+d+N)}, \quad i \geq c+d+N, \tag{18.23}$$

and the unknown steady-state probability vectors $\pi_0, \pi_1, \pi_2, \ldots, \pi_{c+d+N}$ can be obtained by solving the augmented matrix equation as follows:

$$(\pi_0, \pi_1, \pi_2, \ldots, \pi_{c+d+N}) \left(B[R] \begin{matrix} e_1 \\ (I - R)^{-1} e_2 \end{matrix} \right) = (0, 0, 0, \ldots, 0, 1)$$

$$(18.24)$$

where e_1 is a column vector with $c + (d/2 + N) \times (d + 1)$ elements and e_2 is a column vector with $d + 1$ elements, respectively. All elements of these vectors are equal to 1. And the number of zeros in parentheses to the right of Eq. (18.24) is $c + ((d + 2)/2 + N) \times (d + 1)$.

$$B[R] = \begin{pmatrix} Q_{0,0} Q_{0,1} & & & & & \\ Q_{1,0} Q_{1,1} & Q_{1,2} & & & & \\ & \ddots & \ddots & \ddots & & \\ & & Q_{c+d,c+d-1} Q_{c+d,c+d} & C & & \\ & & B & Q_{c+d+1,c+d+2} & C & \\ & & & \ddots & \ddots & \ddots \\ & & & & B & Q_{c+d+N-1,c+d+N-1} & C \\ & & & & & B & RB + A \end{pmatrix}.$$

$$(18.25)$$

Finally, we obtain π_i $(i = 0, 1, 2, \ldots, c + d + N)$ by using the Gauss-Seidel method to solve Eq. (18.24), and obtain π_i $(i \geq c + d + N + 1)$ by substituting π_{c+d+N} obtained in Eq. (18.24) into Eq. (18.23).

So far, the steady-state distribution $\Pi = (\pi_0, \pi_1, \pi_2, \ldots)$ of the system have been given mathematically.

18.4 Performance Measures and Numerical Results

In this section, by using the performance analysis presented in Sect. 18.3, we derive performance measures of the system in terms of the average latency of tasks and the energy saving rate of the system, respectively. Then, we present numerical results to evaluate the performance of the system using the clustered VM allocation strategy proposed in this chapter.

18.4.1 Performance Measures

We define the latency Y_t of a task as the duration from the instant a task arrives at the system to the instant this task is completely processed.

Based on the steady-state distribution of the system model given in Sect. 18.3.2, we obtain the average latency $E[Y_t]$ of tasks as follows:

$$E[Y_t] = \frac{1}{\lambda}\left(\sum_{i=0}^{c} i\pi_{i,0} + \sum_{i=c+1}^{c+d}\sum_{j=0}^{i-c} i\pi_{i,j} + \sum_{i=c+d+1}^{\infty}\sum_{j=0}^{d} i\pi_{i,j}\right). \qquad (18.26)$$

We define the energy saving rate γ of the system as the energy conservation per second.

With the strategy proposed in this chapter, the energy consumption of a VM in awake state is higher than that in sleep state. Indeed, additional energy will be consumed when a task is migrated from Module II to Module I, when a VM in Module II listens to the system buffer, as well as when a VM in Module II wakes up from a sleep state.

Based on the discussions above and the steady-state distribution of the system model given in Sect. 18.3.2, we give the energy saving rate γ of the system with our proposed strategy as follows:

$$\gamma = (g1 - g_2)\sum_{i=0}^{\infty}\sum_{j=0}^{d}(d-j)\pi_{i,j} - \left(g_3\sum_{i=c+1}^{c+d}\sum_{j=1}^{d}c\mu_1\pi_{i,j}\right.$$

$$\left. +g_4\sum_{i=0}^{\infty}\sum_{j=0}^{d}\delta(d-j)\pi_{i,j} + g_5\sum_{i=c+j+N}^{\infty}\sum_{j=0}^{d-1}\delta(d-j)\pi_{i,j}\right) \qquad (18.27)$$

where g_1 ($g_1 > 0$) is the energy consumption per second for a busy VM in Module II, g_2 ($g_2 > 0$) is the energy consumption per second for a sleeping VM in Module II, g_3 ($g_3 > 0$) is the energy consumption for each task-migration, g_4 ($g_4 > 0$) is the energy consumption for each listening, g_5 ($g_5 > 0$) be the energy consumption for each wake-up from a sleep state to an awake state.

18.4.2 Numerical Results

In this subsection, we present numerical results with analysis and simulation to evaluate the impacts of the system parameters on the system performance in terms of the average latency of tasks and the energy saving rate of the system.

Matlab has the capabilities of the efficient calculations and the powerful display. Analysis experiments are performed in the Matlab R2011a environment based on

Eqs. (18.26) and (18.27). Java language has the object-oriented features supporting the representation of multiple attributes through the definition of a class. Simulation experiments are implemented in the MyEclipse 2014 environment using the Java language. In the simulation experiment, a TASK class is defined to include the attributes of UNARRIVE, WAIT, RUNHIGH, RUNLOW and FINISH, a VM class is defined to include the attributes of SLEEP, IDLE, BUSYLOW and BUSYHIGH. Good agreements between the analysis results and the simulation results are observed.

The system parameters are fixed as follows: $c + d = 50$, $\lambda = 7$ s^{-1}, $\mu_1 = 0.2$ s^{-1}, $\mu_2 = 0.1$ s^{-1}, $g_1 = 0.5$ mW, $g_2 = 0.1$ mW, $g_3 = 0.2$ mJ, $g_4 = 0.15$ mJ, $g_5 = 0.2$ mJ as an example for all the numerical results. Good agreements between the analysis results and the simulation results are observed.

Figure 18.2 shows the average latency $E[Y_t]$ of tasks versus the sleep parameter δ for different numbers d of the VMs in Module II and different wake-up thresholds N.

Figure 18.2a indicates that the average latency $E[Y_t]$ of tasks increases with the increase of the number d of the VMs in Module II. For a given wake-up threshold and a given sleep parameter, the more VMs deployed in Module II is, the weaker the system capability becomes, and the longer the tasks will sojourn in the system. This results in a larger average latency of tasks.

Figure 18.2b indicates that the average latency $E[Y_t]$ of tasks increases with the increase of the value for threshold N. This is due to the fact that when the number of the VMs in Module II and the sleep parameter are given, as the value for the wake-up threshold increases, the more possible is that the VMs in Module II will continue sleeping even their sleep timers expire, that is to say, more tasks will accumulate in the system buffer before being processed. Consequently, the average latency $E[Y_t]$ of tasks will increase.

From Fig. 18.2, we also observe that for any number d of the VMs in Module II and any value for the wake-up threshold N, as the sleep parameter δ increases, the average latency $E[Y_t]$ of tasks initially decreases sharply and then decreases gradually. When the sleep parameter δ is smaller (such as $0 < \delta < 0.4$ for $N = 11$ and $d = 15$), the time length of a sleep period is relatively long, so the tasks arriving during the sleep period have to wait longer in the system buffer. This leads to a larger average latency of tasks. For this case, the sleep parameter has a greater impact on the average latency of tasks than any of the other factors, such as the arrival rate of tasks and the service rate of a task on a VM. Thus, the average latency of tasks will decrease sharply as the sleep parameter increases.

When the sleep parameter δ is larger (such as $0.4 < \delta < 1.6$ for $N = 11$ and $d = 15$), the time length of a sleep period is relatively short, so the tasks arriving during the sleep period will be processed earlier. This leads to a lower average latency of tasks. For this case, the arrival rate of tasks and the service rate of VMs play a dominate role in influencing the average latency of tasks. Thus, the average latency of tasks will decrease gradually as the sleep parameter increases.

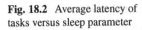

Fig. 18.2 Average latency of tasks versus sleep parameter

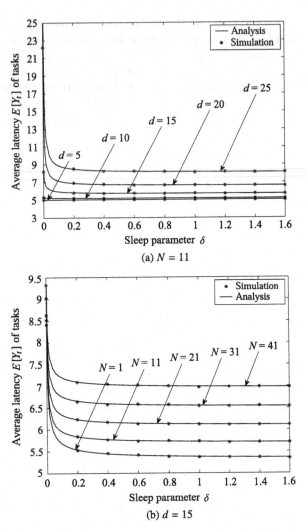

Figure 18.3 depicts the resulting energy saving rate γ of the system versus the sleep parameter δ for different numbers d of the VMs in Module II and different wake-up thresholds N.

Figure 18.3a indicates that deploying either relatively too few or too many VMs in Module II results in a lower energy saving rate γ of the system. When the number of VMs in Module II is smaller (such as $d = 5$ for $N = 11$), even though all the VMs in Module II are being asleep, less energy will be saved. This leads to a lower energy saving rate of the system. When the number of VMs in Module II is bigger (such as $d = 25$ for $N = 11$), the system capability is relatively weaker. This is to say, once a VM in Module II wakes up, this VM has little opportunity to go to sleep again. This leads to a lower energy saving rate of the system too.

Fig. 18.3 Energy saving rate of system versus sleep parameter

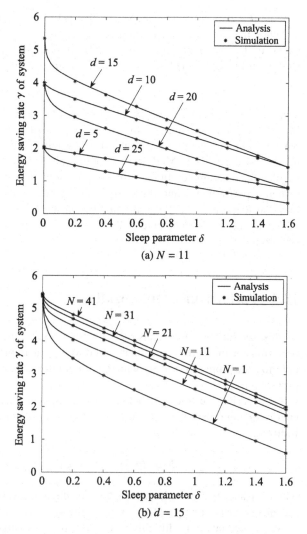

(a) $N = 11$

(b) $d = 15$

Figure 18.3b indicates that a larger value for the wake-up threshold N results in a higher energy saving rate γ of the system. This is due to the fact that when the number of the VMs in Module II and the sleep parameter are given, the higher the value for the wake-up threshold is, the later a VM in Module II will wake up from a sleep state, and the longer the VMs in Module II will remain asleep. This leads to a higher energy saving rate of the system.

From Fig. 18.3, we also notice that for any number d of the VMs in Module II and any value for the wake-up threshold N, the energy saving rate γ of the system decreases as the sleep parameter δ increases. For one thing, the larger the sleep parameter is, the shorter the time length of a sleep period is, and the earlier a VM in

Module II wakes up from a sleep state, that is to say, less energy will be saved. For the other, the larger the sleep parameter is, the more frequently the VM in Module II listens to the system buffer or wakes up from a sleep state, that is to say, additional energy will be consumed. Consequently, the energy saving rate of the system will decrease.

Combining the results shown in Figs. 18.2 and 18.3, we find that a lower average latency of tasks can be obtained with a smaller number of the VMs in Module II, a smaller value for the wake-up threshold and a larger sleep parameter, whereas a higher energy saving rate of the system can be obtained with a moderate number of the VMs in Module II, a larger value for the wake-up threshold and a smaller sleep parameter. In the actual application of the proposed strategy, both the response performance and the energy saving rate of the system should be taken into consideration. Therefore, the number of the VMs in Module II, the wake-up threshold and the sleep parameter should be jointly optimized to balance the system performance.

18.5 Performance Optimization

In this section, we first construct a system cost function to balance different performance measures. And then, we develop an improved TLBO algorithm to optimize the proposed strategy with the minimum system cost to improve the system performance.

Considering the trade-off between the average latency $E[Y_t]$ of tasks and the energy saving rate of the system, we establish a system cost function $F(d, N, \delta)$ as follows:

$$F(d, N, \delta) = f_1 \times E[Y_t] - f_2 \times \gamma \qquad (18.28)$$

where f_1 and f_2 are the impact factors for the average latency $E[Y_t]$ of tasks and the energy saving rate γ of the system to the system cost function.

We note that, it is difficult to express the mathematical expressions for the average latency $E[Y_t]$ of tasks or the energy saving rate γ of the system in a closed-form. Besides, we cannot determine the monotonicity of the system cost function. For the purpose of jointly optimizing the number of the VMs in Module II, the wake-up threshold and the sleep parameter with the minimum system cost function, we turn to an intelligent optimization algorithm.

TLBO algorithm is one of the most efficient algorithms to provide good quality solutions in a reasonable computation time. However, there are still some drawbacks with the TLBO algorithm, such as low population diversity and poor searching ability. By introducing a cube chaotic mapping mechanism to disperse the initial grades of students and using an exponentially decreasing strategy for the teaching process to enhance the searching ability, we develop an improved TLBO algorithm. The main steps for the improved TLBO algorithm are given as follows:

Step 1: Initialize the population size Num, the maximum number $iter_{max}$ of the iterations for the teaching-learning process, the maximal weight w_{max} for teacher's teaching process, the minimal weight w_{min} for teacher's teaching process, the variation sections of the sleep parameter $[0, \delta_{max}]$, the number X of total VMs.

Step 2: Set the initial number d of VMs in Module II, the initial value N for the wake-up threshold, the current optimal combination (d^*, N^*, δ^*), and calculate the corresponding fitness F^*.

$$d = 0, \ N = 1, \ (d^*, N^*, \delta^*) = (d, N, \delta_{max}), \ F^* = F(d^*, N^*, \delta^*)$$

Step 3: Initialize grade δ_i for the ith ($i = 1, 2, 3, \ldots, Num$) student by using cube chaotic equation.

$\delta_1 = rand_1$

% $rand_1$ returns a sample in the interval $(-1, 1)$.

for $i = 2 : Num$

$\quad \delta_i = 4 \times (\delta_{i-1})^3 - 3 \times \delta_{i-1}$

endfor

for $i = 1 : Num$

$\quad \delta_i = (\delta_i + 1) \times \delta_{max}/2$

endfor

Step 4: Calculate the average grade δ_{mean} of current students, select the best grade $\delta_{teacher}$ as the teacher's grade, and calculate the best fitness F_{best}.

$$\delta_{mean} = \underset{i \in \{1,2,3,\ldots,Num\}}{mean} \{\delta_i\}$$

$$\delta_{teacher} = \underset{i \in \{1,2,3,\ldots,Num\}}{argmin} \{F(d, N, \delta_i)\}$$

$$F_{best} = \underset{i \in \{1,2,3,\ldots,Num\}}{min} \{F(d, N, \delta_i)\}$$

Step 5: Set the initial number of the iterations as $iter = 1$.

Step 6: Update each student's grade δ_i ($i = 1, 2, 3, \ldots, Num$) via the teacher's teaching process.

$$G = round(1 + rand), \ w = w_{max} - (w_{max} - w_{min})(iter/iter_{max})^{1/iter}$$

% round represents the rounding operation.

% rand is a random number selected in the interval $(0, 1)$.

$\delta_i' = w \times \delta_i + rand \times (\delta_{teacher} - G \times \delta_{mean})$

if $F(d, N, \delta_i') < F(d, N, \delta_i)$

$\quad \delta_i = \delta_i'$

endif

Step 7: Update each student's grade δ_i ($i = 1, 2, 3, \ldots, Num$) via the students' learning process among each others.

Randomly select the jth ($j \neq i$) student

if $F(d, N, \delta_j) < F(d, N, \delta_i)$

$\quad \delta_i = \delta_i + rand \times (\delta_j - \delta_i)$

else

$\quad \delta_i = \delta_i + rand \times (\delta_i - \delta_j)$

endif

Step 8: For the current iteration, select the best fitness F_{gbest} and the best grade δ_{gbest}.

$$F' = \min_{i \in \{1,2,3,\ldots,Num\}} \{F(d, N, \delta_i)\}$$

if $F' < F_{best}$

$\quad F_{gbest} = F'$, $\delta_{gbest} = \underset{i \in \{1,2,3,\ldots,Num\}}{argmin} \{F(d, N, \delta_i)\}$

else

$\quad F_{gbest} = F'$, $\delta_{gbest} = \underset{i \in \{1,2,3,\ldots,Num\}}{argmin} \{F(d, N, \delta_i)\}$

endif

Step 9: Check the number of iterations.

$\quad$ **if** $iter < iter_{max}$

$\quad\quad iter = iter + 1$

$\quad\quad$ go to **Step 4**

$\quad$ **endif**

Step 10: Update the current optimal combination (d^*, N^*, δ^*).

$\quad$ **if** $F_{gbest} < F^*$

$\quad\quad F^* = F_{gbest}$

$\quad\quad (d^*, N^*, \delta^*) = (d, N, \delta_{gbest})$

$\quad$ **endif**

Step 11: Check the number of VMs in Module II.

$\quad$ **if** $d < X$

$\quad\quad d = d + 1$

$\quad\quad$ go to **Step 3**

$\quad$ **endif**

Step 12: Check the value for the wake-up threshold.

$\quad$ **if** $N < X$

$\quad\quad N = N + 1$

$\quad\quad$ go to **Step 3**

$\quad$ **endif**

Step 13: Output the optimal combination (d^*, N^*, δ^*).

Using the system parameters given in Sect. 18.4.2, and setting $Num = 100$, $iter_{max} = 200$, $w_{max} = 0.8$, $w_{min} = 0.1$, $\delta_{max} = 2$, $X = 50$, $f_1 = 4$, $f_2 = 1$, we execute the improved TLBO algorithm. With different service rates μ_2 of a task on the VM in Module II, we obtain the optimal combination (d^*, N^*, δ^*) for the number of VMs in Module II, the wake-up threshold and the sleep parameter with the minimum system costs $F(d^*, N^*, \delta^*)$ in Table 18.1.

The cloud capacity, the arrival intensity of tasks and the serving capability of VMs greatly influence the optimal outcomes in Table 18.1. With the optimization outcomes shown in Table 18.1, we can trade off the average latency of $E[Y_t]$ tasks and the energy saving rate of the system for the proposed strategy.

Table 18.1 Optimum combination of parameters in proposed strategy

Service rates μ_2	Optimum combinations (d^*, N^*, δ^*)	Minimum costs $F(d^*, N^*, \delta^*)$
0.05	(9, 2, 0.0001)	17.2012
0.1	(10, 10, 0.0403)	17.1864
0.15	(11, 4, 0.0945)	17.0491
0.2	(49, 2, 0.2846)	16.1823

18.6 Conclusion

In this chapter, in order to enhance the energy efficiency in a cloud computing system, we proposed a clustered VM allocation strategy with a sleep mode. Considering the sleep mode with wake-up threshold in the proposed strategy, we established a queue with an N-policy and asynchronous vacations for partial servers. Based on the stochastic behavior of tasks with the proposed strategy, we built the QBD matrix and resolved the steady-state transition probabilities by using a matrix-geometric solution method. Moreover, we evaluated the system performance in terms of the average latency of tasks and the energy saving rate of the system. Numerical results with analysis and simulation showed that the average latency of tasks is lower with a smaller number of the VMs in Module II, a smaller value for the wake-up threshold and a larger sleep parameter, while the energy saving rate of the system is higher with a moderate number of the VMs in Module II, a larger value for the wake-up threshold and a smaller sleep parameter. For this, we constructed a system cost function to balance different performance measures. By introducing a cube chaotic mapping mechanism for the grade initialization and an exponentially decreasing strategy for the teaching process, we developed an improved TLBO algorithm and optimized the proposed strategy with the minimum value of the system cost function.

Chapter 19
Pricing Policy for Registration Service

Considering the energy efficiency and the registration service in the cloud computing systems, in this chapter, we propose a sleep-mode based cloud architecture, in which free service and optional registration service are provided on the same server. Regarding the free service as the first essential service, the registration service as the second optional service and the sleep state as the vacation, we build an asynchronous multiple-vacation queueing model with a second optional service. We derive performance measures of the system in terms of the energy saving rate of the system and the average latency of the anonymous users who select the registration service. Moreover, by constructing the individual benefit function and the social benefit function, and developing an improved Bat algorithm, we present a pricing policy with an appropriate registration fee to correspond the user behaviors of Nash equilibrium and social optimization.

19.1 Introduction

Cloud computing is offering utility-oriented IT services to users worldwide [Buyy10, Huss15]. There are a wide variety of cloud services, and each service has its own unique strengths and limitations. Therefore, selecting a proper cloud service becomes a challenge for potential cloud users [Gui13]. In order to appeal to more users, cloud vendors always provide free service to anonymous users. If an anonymous user is satisfied with the free service and likely to receive better service next time, the anonymous user may well register as a Very Important Person (VIP) user.

On the other hand, in cloud computing systems, the energy consumption of the under-utilized resources accounts for a substantial amount of the actual energy use [Dhan15]. Therefore, understanding how to provide a registration service with an energy efficient cloud architecture is an important issue confronting cloud vendors.

© Springer Nature Singapore Pte Ltd. 2021
S. Jin, W. Yue, *Resource Management and Performance Analysis of Wireless Communication Networks*, https://doi.org/10.1007/978-981-15-7756-7_19

In the face of fierce competition in the cloud environment, there is an increasing number studies on how to reasonably build a cloud architecture that attracts more cloud users. In [Shi17], the authors proposed a structure named Joint Cloud Corporation Environment (JCCE), which offers a mutual benefit and win-win Joint Cloud environment for global cloud service providers. In [Guo11], the authors described the Policy-Based Market-Oriented Cloud Service Management Architecture (PBMOCSMA), showing how to provide flexible, dynamic and extensible policy-based management capabilities to cloud vendors.

With growing interest in green cloud computing systems and carbon emissions reduction, the need to develop strategies for building energy efficient cloud architecture is becoming more pressing. In [Ye10], the authors presented a Virtual Machine (VM) based energy-efficient data center architecture for cloud computing. They investigated the potential performance overheads caused by server consolidation and live migration of VM technology. In [Hu13b], the authors proposed a globally collaborative mechanism of the Green Private Cloud Computing and built a Green Private Cloud Architecture model with virtualization technology.

However, in all of these studies, achieving a green cloud computing system and at the same time cultivating powerful groups of loyal cloud users had not been considered. We know that the design and development of competitive cloud computing systems require not only improving energy efficiency but also attracting more cloud users [Ahn18]. Therefore, researches need to consider improving energy efficiency and also attracting more cloud users at the same time when construct such systems.

On the other hand, to run a cloud computing system well, a numerical evaluation of the system performance is needed. For this, queueing theory with a second optional service was considered to be suitable for modeling the registration service in cloud computing systems. Queueing theory with a second optional service was first formulated in [Mada00]. Following this formulation, various papers on a queueing model with a second optional service were published. In [Sing11], the authors investigated a single server bulk queueing system with state-dependent rates and a second optional service. In this queueing system, the service time of the essential service followed the general distribution whereas that of an optional service followed the exponential distribution. In [Ghor16], the authors considered a two-phase tandem queueing model with a second optional service and random feedback. In [Wei16], the authors discussed a discrete-time Geom/G/1 retrial queue with balking customers and a second optional service, where the retrial time followed a geometrical distribution.

Also, queueing theory with a vacation mechanism is suitable for modeling the sleep mode in cloud computing systems. The vacation model terminology first appeared in the 1970s. In [Dosh86], the author wrote an excellent survey paper on vacation models. Numerous papers on vacation models have appeared since that time. In [Gary00], the authors analyzed a multiple-vacation queueing model, where the server is subject to breakdown while in operation. In [Jain17], the authors considered an asynchronous vacation policy for the multi-server repair problem with server breakdown and two types of spares, first type unit and second type unit.

However, we note that all the mentioned papers were studied in either the queueing model with only the second optional service or the queueing model with only the vacation mechanism. For this reason, these researches could not be used to model the stochastic behavior of the networks in an energy efficient cloud architecture with a free service and a registration service. A new queueing model should be constructed by considering both of the second optional service and the vacation mechanism to capture the stochastic behavior of the systems.

In this chapter, we firstly propose a sleep mode-based cloud architecture where a free service and a registration service are provided on the same server. The newly vacated server will enter the sleep state once there are no users waiting in the system buffer. We classify the users into two categories: anonymous users and VIP users. An anonymous user who demands for free cloud service is likely to register as a VIP user to receive better service next time. In order to investigate the system performance of the proposed cloud architecture, we build an asynchronous multiple-vacation queueing model with a second optional service. To analyze the queueing model, we also construct a three-dimensional Markov chain from the perspective of the total number of anonymous users, the number of servers running normally and the number of anonymous users applying for the registration service to analyze the queueing model. Moreover, we evaluate the system performance in terms of the average latency of the anonymous users who select the registration service and the energy saving rate of the system in a cloud environment. We note that if more anonymous users join the system, the average latency of anonymous users is higher, and the QoS for anonymous users is lower. If fewer anonymous users join the system, the energy saving rate of the system is higher, but the benefit for the cloud vendor is lower. Aiming to get a better trade-off between the QoS for anonymous users and the benefit for cloud vendor, we investigate the Nash equilibrium arrival rate of anonymous users, and then we develop an improved Bat algorithm to search the socially optimal arrival rate of anonymous users. We present a pricing policy to impose an appropriate registration fee on the anonymous users who select the registration service so that the Nash equilibrium arrival rate coincides with the socially optimal arrival rate.

The chapter is organized as follows. In Sect. 19.2, we describe the sleep mode-based cloud architecture with a second optional service proposed in this chapter. Then, we present the system model in detail. In Sect. 19.3, we present a performance analysis of the system model, through the analysis of the transition rate matrix and the steady-state distribution. In Sect. 19.4, we obtain performance measures and present numerical results to evaluate the system performance. In Sect. 19.5, by constructing the individual benefit function and the social benefit function, and developing an improved Bat algorithm, we present a pricing policy with an appropriate registration fee to correspond the user behaviors of Nash equilibrium and social optimization. Finally, we draw our conclusions in Sect. 19.6.

19.2   Cloud Architecture and System Model

In this section, we propose a sleep mode-based cloud architecture with free service and registration service. Then, we establish an asynchronous multiple-vacation queueing model with a second optional service.

19.2.1   Cloud Architecture

It is a common practice for cloud vendors to offer free service to attract new anonymous users. In conventional cloud computing systems, all the VMs always keep awake, even though there are no users to be serviced. This results in a large amount of wasted energy.

Considering the energy efficiency and the registration service in the cloud computing systems, we propose a sleep mode-based cloud architecture shown in Fig. 19.1.

In a cloud environment, the configuration of Physical Machines (PMs) is usually very high. Several VMs are deployed to a PM and each VM runs its own operating system independently. This makes it possible to implement a sleep mode at the VM level.

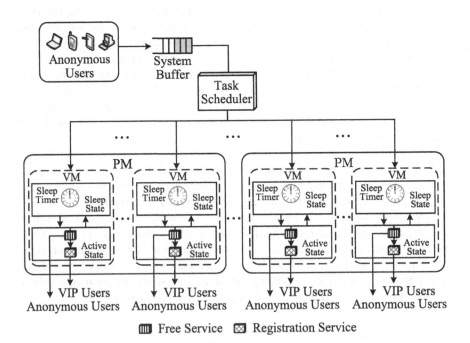

Fig. 19.1 Sleep mode-based cloud architecture

(1) When an anonymous user enters the system, the anonymous user will queue in the system buffer waiting for a free cloud service. Once there is at least one new vacated VM, or one VM on any PM that has just woke up, this VM will be allocated by the task scheduler to the first anonymous user queueing in the system buffer, so the anonymous user just getting the VM will receive the free cloud service.

(2) After the completion of the free cloud service, the anonymous user selects whether to receive the registration service according to their level of service satisfaction. If the anonymous user opts to register as a VIP user after completing a registration process, the user has to pay a reasonable fee, ensuring a higher QoS the next time. Otherwise, the anonymous user will leave the system directly and remain as an anonymous user.

(3) If there are no anonymous users waiting in the system buffer and a user departs a VM, namely, a VM is evacuated, the VM will enter the sleep mode. Once the VM enters the sleep state, a sleep timer with a random duration will be activated to control the time length of a sleep period. At the end of the sleep period, if there are no anonymous users in the system buffer, another sleep timer will be activated, and the VM will begin another sleep period. Otherwise, the VM will return to the active state and wake up in order to serve the anonymous users in the system buffer.

We next build a queueing model with the proposed architecture to mathematically evaluate and reasonably optimize the system performance.

19.2.2 System Model

We model the system with the proposed sleep mode-based cloud architecture shown in Fig. 19.1 as an asynchronous multiple-vacation queueing model with the free service as the first essential service and the registration service as the second optional service.

The buffer in the system is supposed to be infinite. Let c be the total number of VMs in the system. Let random variable $N(t) = i$ ($i = 0, 1, 2, \ldots$) be the total number of anonymous users in system at instant t. Let random variable $Y(t) = j$ ($j = 0, 1, 2, \ldots, \min\{i, c\}$) be the number of VMs running normally at instant t. Let random variable $S(t) = k$ ($k = 0, 1, 2, \ldots, j$) be the number of anonymous users experiencing registration service at instant t. We call $N(t)$ the system level, $Y(t)$ the system state and $S(t)$ the system phase. The behavior of the system model under consideration can be described in terms of the regular irreducible three-dimensional continuous-time stochastic process $\{(N(t), Y(t), S(t)), \ t \geq 0\}$ with state space $\boldsymbol{\Omega}$ as follows:

$$\boldsymbol{\Omega} = \{(i, j, k) : i \geq 0, 0 \leq j \leq \min\{i, c\}, 0 \leq k \leq j\}. \tag{19.1}$$

In this model, we assume that the inter-arrival time of anonymous users follows an exponential distribution with mean $1/\lambda$, where $\lambda > 0$, called the arrival rate of anonymous users. We assume that the free service time and the registration service time of an anonymous user follow exponential distributions with means $1/\mu_1$ seconds and $1/\mu_2$ seconds, respectively, where $\mu_1 > 0$ and $\mu_2 > 0$. We call μ_1 the free service rate and μ_2 the registration service rate. Moreover, we assume that an anonymous user either opts for the registration service with probability q, or opts not to select the registration service with probability $\bar{q}$ ($\bar{q} = 1 - q$). Furthermore, we assume that the time length of the sleep timer follows an exponential distribution with mean $1/\delta$, where $\delta > 0$, called the sleep parameter.

Based on the assumptions above, we conclude that the stochastic process $\{(N(t), Y(t), S(t)), \ t \geq 0\}$ is a three-dimensional Continuous-Time Markov Chain (CTMC).

We analyze the system model and evaluate the system performance under a condition of the system being in the steady state. The traffic load ρ of the system model can be given as follows:

$$\rho = \frac{\lambda(\mu_2 + q\mu_1)}{c\mu_1\mu_2}. \tag{19.2}$$

The necessary and sufficient condition for the system being stable is $\rho < 1$.

We define $\pi_{i,j,k}$ as the probability distribution of the system level being equal to i, the system state being equal to j and the system phase being equal to k in the steady state. $\pi_{i,j,k}$ is given as follows:

$$\pi_{i,j,k} = \lim_{t \to \infty} \Pr\{N(t) = i, Y(t) = j, S(t) = k\}, \quad (i, j, k) \in \mathbf{\Omega}. \tag{19.3}$$

We define the vector $\boldsymbol{\pi}_i$ as probability of the system level being equal to i in the state. The steady-state distribution $\mathbf{\Pi}$ of the three-dimensional CTMC is written as a partitioned vector. The partitioned vector can be given as follows:

$$\mathbf{\Pi} = (\boldsymbol{\pi}_0, \boldsymbol{\pi}_1, \boldsymbol{\pi}_2, \ldots). \tag{19.4}$$

19.3   Performance Analysis

In this section, we present a performance analysis of the system model, through the analysis of the transition rate matrix and the steady-state distribution.

19.3.1   Transition Rate Matrix

The necessary step in analyzing the steady-state distribution of the system model is to construct the transition rate matrix.

Let Q be the one-step state transition rate matrix of the three-dimensional CTMC $\{(N(t), Y(t), S(t)), t \geq 0\}$. Let $Q_{x,y}$ be the one-step state transition rate sub-matrix for the system level changing from x ($x = 0, 1, 2, \ldots$) to y ($y = 0, 1, 2, \ldots$). For convenience of presentation, we denote $Q_{x,x}$ as A_x, $Q_{x,x-1}$ as B_x and $Q_{x,x+1}$ as C_x. We discuss A_x, B_x and C_x by the following three specific cases.

(1) *System Level Remains Fixed*: If $x = 0$, it means that there are no anonymous users in the system and all the VMs are in the sleep state. If there are no anonymous users arriving at the system within the sleep timer, all of the system level, the system state and the system phase remain fixed, and the transition rate is $-\lambda$.

   Thus, the sub-matrix A_0 is given as follows:

$$A_0 = -\lambda. \tag{19.5}$$

   If $x \geq 1$, it means that there is at least one anonymous user in the system.

   In the case where the number of VMs working normally is less than $\min\{c, x\}$, there is at least one VM being in the sleep state. If there are no anonymous users arriving at the system within the sleep timer and no anonymous users selecting or finishing registration service, all of the system level, the system state and the system phase remain fixed, and the transition rate is $-(\lambda + (c - j)\delta + (j - k)q\mu_1 + k\mu_2)$, where j is the number of VMs working normally in the system, and k is the number of anonymous users experiencing registration service in the system.

   In the case where the number of VMs working normally is equal to $\min\{c, x\}$, if the number of anonymous users is greater than the number of VMs, namely, $\min\{c, x\} = c$, all the VMs are working normally without sleep. If the number of VMs is greater than the number of anonymous users, namely, $\min\{c, x\} = x$, there are no anonymous users waiting in the system buffer. Hence, the VMs in the sleep state will enter another sleep period at the end of a sleep period. If there are no anonymous users arriving at the system, and no anonymous selecting or finishing registration service, all of the system level, the system state and the system phase remain fixed, and the transition rate is $-(\lambda + (j - k)q\mu_1 + k\mu_2)$.

   For the two cases mentioned above, if there is one anonymous user opting for the registration service, both the system level and the system phase remain fixed, while the system phase increases by one, and the transition rate is $(j - k)q\mu_1$. If there is one VM changing to the active state, both the system level and the system phase remain fixed, whereas the system state increases by one, and the transition rate is $(c - j)\delta$. We denote $-(\lambda + (c - j)\delta + (j - k)q\mu_1 + k\mu_2)$ as ω_1, $(c - j)\delta$ as ω_2 and $-(\lambda + (j - k)q\mu_1 + k\mu_2)$ as ω_3.

Thus, the sub-matrix A_x with an order of $(1/2 \times (\min\{x, c\}+1)(\min\{x, c\}+2)) \times (1/2 \times (\min\{x, c\}+1)(\min\{x, c\}+2))$ is given as follows:

$$
A_x =
\begin{pmatrix}
\omega_1 & \omega_2 & & & & & & & & & & \\
\omega_1 & q\mu_1 & \omega_2 & & & & & & & & & \\
& \omega_1 & 0 & \omega_2 & & & & & & & & \\
& & \omega_1 & 2q\mu_1 & 0 & \omega_2 & & & & & & \\
& & & \omega_1 & q\mu_1 & 0 & \omega_2 & & & & & \\
& & & & \omega_1 & 0 & 0 & & \omega_2 & & & \\
& & & & & \ddots & \ddots & & \ddots & & \ddots & \\
& & & & & & \omega_1 & (x-1)q\mu_1 & 0 & 0 & \cdots & \omega_2 \\
& & & & & & & \ddots & \ddots\ddots & \ddots & & \ddots \\
& & & & & & & \omega_1 & 0 & 0 & 0 & \cdots & \omega_2 \\
& & & & & & & & \omega_3 & xq\mu_1 & 0 & \\
& & & & & & & & & \omega_3 & (x-1)q\mu_1 & 0 \\
& & & & & & & & & & \ddots & \ddots\ddots \\
& & & & & & & & & & & \omega_3 & 2q\mu_1 & 0 \\
& & & & & & & & & & & & \omega_3 & q\mu_1 \\
& & & & & & & & & & & & & \omega_3 \\
\end{pmatrix}
\tag{19.6}
$$

(2) *System Level Decreases*: If $1 \leq x \leq c$, it means that there are no more than c anonymous users.

When the number of anonymous users is equal to the number of VMs working normally in the system, if the system level decreases by one, the number of VMs working normally decreases by one. When an anonymous user finishes the free service and departs the system, both the system level and the system state decrease by one, whereas the system phase remains unchanged. Therefore, the transition rate is $(j - k)\bar{q}\mu_1$, where j is the number of VMs working normally in the system, and k is the number of anonymous users experiencing registration service in the system. When an anonymous user finishes the registration service and departs the system, all of the system level, the system state and the system phase decrease by one, and the transition rate is $k\mu_2$.

When the number of anonymous users is more than the number of VMs working normally in the system, if the system level decreases by one, the number of VMs working normally remains fixed. When an anonymous user finishes the free service and departs the system, the system level decreases by one, whereas the system state and the system case remain unchanged. Therefore, the transition rate is $(j - k)\bar{q}\mu_1$, where j is the number of VMs working normally in the system, and k is the number of anonymous users experiencing registration service in the system. When an anonymous user finishes the registration service and departs the system, both the system level and the system phase decrease by one, while the system state remains unchanged, and the transition rate is $k\mu_2$.

Thus, the sub-matrix $\boldsymbol{B}_x$ with an order of $(1/2 \times (\min\{x, c\}+1)(\min\{x, c\}+2)) \times (1/2 \times x(x + 1))$ is given as follows:

$$
\boldsymbol{B}_x=
\begin{pmatrix}
0 & & & & & & & & & & & \\
0 & \bar{q}\mu_1 & & & & & & & & & & \\
 & \mu_2 & 0 & & & & & & & & & \\
 & & 0 & 2\bar{q}\mu_1 & & & & & & & & \\
 & & & \mu_2 & \bar{q}\mu_1 & & & & & & & \\
 & & & & 2\mu_2 & 0 & & & & & & \\
 & & & & & & \ddots & \ddots & & & & \\
 & & & & & & 0 & (x-1)\bar{q}\mu_1 & & & & \\
 & & & & & & \mu_2 & (x-2)\bar{q}\mu_1 & & & & \\
 & & & & & & & & \ddots & \ddots & & \\
 & & & & & & & & (x-2)\mu_2 & \bar{q}\mu_1 & & \\
 & & & & & & & & & (x-1)\mu_2 & 0 & \\
 & & & & & & 0 & x\bar{q}\mu_1 & & & & \\
 & & & & & & \mu_2 & (x-1)\bar{q}\mu_1 & & & & \\
 & & & & & & & & \ddots & \ddots & & \\
 & & & & & & & & (x-2)\mu_2 & 2\bar{q}\mu_1 & & \\
 & & & & & & & & & (x-1)\mu_2 & \bar{q}\mu_1 & \\
 & & & & & & & & & & x\mu_2 &
\end{pmatrix}.
$$

$$\tag{19.7}$$

If $x > c$, it means that the number of anonymous users is more than the number of VMs in the system. Therefore, if the system level decreases by one, the number j of VMs working normally remains fixed.

When an anonymous user finishes the free service and departs the system, the system level decreases by one, while the system state and the system phase remain unchanged, the transition rate is $(j - k)\bar{q}\mu_1$, where j is the number of VMs working normally in the system, and k is the number of anonymous users experiencing registration service in the system. When an anonymous user finishes the registration service and departs the system, both the system level and the system phase decrease by one, while the system state remains unchanged, and the transition rate is $k\mu_2$.

Thus, the sub-matrix $\boldsymbol{B}_x$ with an order of $(1/2 \times (c+1)(c+2)) \times (1/2 \times (c+1)(c+2))$ is given as follows:

$$
\boldsymbol{B}_x =
\begin{pmatrix}
0 & & & & & & & & \\
\bar{q}\mu_1 & & & & & & & & \\
\mu_2 & 0 & & & & & & & \\
 & 0 & 2\bar{q}\mu_1 & & & & & & \\
 & & \mu_2 & \bar{q}\mu_1 & & & & & \\
 & & & 2\mu_2 & 0 & & & & \\
 & & & & & \ddots & \ddots & & \\
 & & & & & & 0 & c\bar{q}\mu_1 & \\
 & & & & & & \mu_2 & (c-1)\bar{q}\mu_1 & \\
 & & & & & & & \ddots & \ddots \\
 & & & & & & & & (c-1)\mu_2 & \bar{q}\mu_1 \\
 & & & & & & & & & c\mu_2 & 0
\end{pmatrix}
. \tag{19.8}
$$

(3) *System Level Increases*: No matter how many anonymous users are there in the system, how many VMs are working normally or how many anonymous users are applying for registration service, as long as there is one anonymous user arriving at the system, the system level increases by one, while the system state and the system phase remain fixed, and the transition rate is λ.

Thus, the sub-matrix $\boldsymbol{C}_x$ with an order of $(1/2 \times (\min\{x, c\}+1)(\min\{x, c\}+2)) \times (1/2 \times (\min\{x, c\}+1)(\min\{x, c\}+2))$ is given as follows:

$$
\boldsymbol{C}_x =
\begin{pmatrix}
\lambda & & & \\
 & \lambda & & \\
 & & \ddots & \\
 & & & \lambda
\end{pmatrix}
. \tag{19.9}
$$

Now, all the sub-matrices in the one-step state transition rate matrix $\boldsymbol{Q}$ have been addressed. Starting from system level c, the sub-matrices $\boldsymbol{A}_x$ and $\boldsymbol{C}_x$ are repeated forever. Starting from system level $(c+1)$, the sub-matrices $\boldsymbol{B}_x$ are repeated forever.

The repetitive sub-matrices A_x, B_x and C_x are represented by A, B and C. For this, we write Q as follows:

$$
Q = \begin{pmatrix}
A_0 & C_0 & & & & & & \\
B_1 & A_1 & C_1 & & & & & \\
& B_2 & A_2 & C_2 & & & & \\
& & \ddots & \ddots & \ddots & & & \\
& & & B_{c-1} & A_{c-1} & C_{c-1} & & \\
& & & & B_c & A & C & \\
& & & & & B & A & C \\
& & & & & & \ddots & \ddots & \ddots
\end{pmatrix}. \tag{19.10}
$$

The block-tridiagonal structure of Q shows that the state transitions occur only between adjacent levels. We know that the three-dimensional CTMC $\{(N(t), Y(t), S(t)), t \geq 0\}$ can be seen as a Quasi Birth-Death (QBD) process.

19.3.2 Steady-State Distribution

To analyze this QBD process, we need to solve the matrix quadratic equation $R^2 B + RA + C = 0$. The necessary and sufficient conditions for positive recurrence are that the matrix quadratic equation $R^2 B + RA + C = 0$ has a minimal non-negative solution R and a spectral radius $\text{Sp}(R) < 1$.

By using the consistency technique formula to tackle the one-step transition rate matrix Q and dividing Q by the absolute value X of the minimum element in matrix A, the one-step transition rate matrix Q can be tackled as follows:

$$
Q' = Q/X. \tag{19.11}
$$

The matrix quadratic equation can be modified as $R^2 B' + RA' + C' = 0$, where $A' = A/X$, $B' = B/X$ and $C' = C/X$. However, it is difficult to give the mathematical expression of the rate matrix R in a closed-form with a higher-order matrix equation. Here, we propose an iteration algorithm shown are given as follows:

Step 1: Initialize the error precision ε (for example, $\varepsilon = 10^{-8}$). Initialize c, λ, μ_1, μ_2, δ and q as needed. Initialize the rate matrix $R = 0$ with an order of $m \times m$, where $m = (1/2 \times (c+1)(c+2))$.

Step 2: Tackle Q by using the consistency technique formula and get A', B' and C'.
$Q' = Q/X$, $A' = A/X$, $B' = B/X$, $C' = C/X$
% X is absolute value of the minimum element in matrix A.

Step 3: Calculate R^*.

$$R^* = R^2 \times B' + R \times (I + A') + C'.$$

% I is an identity matrix.

Step 4:

    while $\|R - R^*\|_\infty > \varepsilon$

    % $\|R - R^*\|_\infty = \max\left\{ \sum_{i=1}^{m} \sum_{j=1}^{m} |r_{i,j} - r_{i,j}^*| \right\}$, where $r_{i,j}$ and $r_{i,j}^*$ are

    elements in R and R^*, respectively.

    $R = R^*$

    $R^* = R^2 \times B' + R \times (I + A') + C'$

    endwhile

Step 5: $R = R^*$.

Step 6: Output R.

Using the rate matrix R obtained from algorithm, we construct a square matrix $B[R]$ as follows:

$$B[R] = \begin{pmatrix} A_0 & C_0 & & & & \\ B_1 & A_1 & C_1 & & & \\ & B_2 & A_2 & C_2 & & \\ & & \ddots & \ddots & \ddots & \\ & & & B_{c-1} & A_{c-1} & C_{c-1} \\ & & & & B_c & RB + A \end{pmatrix}. \tag{19.12}$$

By using a matrix-geometric solution method, we can give a set of linear equations as follows:

$$\begin{cases} (\pi_0, \pi_1, \pi_2, \ldots, \pi_c) B[R] = 0 \\ (\pi_0, \pi_1, \pi_2, \ldots, \pi_{c-1}) e_1 + \pi_c (I - R)^{-1} e_2 = 1 \end{cases} \tag{19.13}$$

where e_1 is a column vector with $1/2 \times c \times (c+1) \times (c+2)$ elements and e_2 is a column vector with $1/2 \times (c+1) \times (c+2)$ elements, respectively. All elements of these vectors are equal to 1.

By using the Gauss-Seidel method to solve Eq. (19.13), we can obtain $\pi_0, \pi_1, \pi_2, \ldots, \pi_c$. From the structure of the one-step state transition rate matrix Q, we know that π_i ($i = c+1, c+2, c+3, \ldots$) satisfies the matrix-geometric solution form as follows:

$$\pi_i = \pi_c R^{i-c}, \quad i \geq c + 1. \tag{19.14}$$

By substituting π_c obtained from Eq. (19.13) into Eq. (19.14), we can obtain π_i ($i = c+1, c+2, c+3, \ldots$). The steady-state distribution $\Pi = (\pi_0, \pi_1, \pi_2, \ldots)$ of the system can be given numerically.

19.4 Performance Measures and Numerical Results

In this section, we first derive performance measures of the system in terms of the average latency of anonymous users and the energy saving rate of the system, respectively. Then, we present numerical results to evaluate the performance of the system using the sleep mode-based cloud architecture proposed in this chapter.

19.4.1 Performance Measures

We define the latency of a user as the duration from the instant a user arrives at the system to the instant this user completes service and departs the system. Let Y_{a1} denote the latency of an anonymous user who experiences only the free service, and Y_{a2} denote the latency of an anonymous user who selects the registration service. We note that the latency of a user includes the time period waiting in the system buffer and the time period getting service from the cloud computing system.

Based on the steady-state distribution of the system model given in Sect. 19.3.2, we give the average value $E[L_a]$ for the number L_a of anonymous users queueing in the system buffer as follows:

$$E[L_a] = \sum_{i=c}^{\infty} \sum_{j=0}^{c} \sum_{k=0}^{j} (i - c)\pi_{i,j,k}. \tag{19.15}$$

By using Eq. (19.15), we can obtain the average waiting time $E[W]$ of anonymous users as follows:

$$E[W] = \frac{E[L_a]}{\lambda} = \frac{1}{\lambda} \left(\sum_{i=c}^{\infty} \sum_{j=0}^{c} \sum_{k=0}^{j} (i - c)\pi_{i,j,k} \right). \tag{19.16}$$

The average value $E[X_1]$ of the service time X_1 of the anonymous user experiencing only the free service is given as follows:

$$E[X_1] = \frac{1}{\mu_1} \tag{19.17}$$

where μ_1 is the free service rate defined in Sect. 19.2.2.

We give the average latency $E[Y_{a1}]$ of the anonymous users who experience only the free service as follows:

$$E[Y_{a1}] = E[W] + E[X_1] = \frac{1}{\lambda} \left(\sum_{i=c}^{\infty} \sum_{j=0}^{c} \sum_{k=0}^{j} (i - c)\pi_{i,j,k} \right) + \frac{1}{\mu_1} \tag{19.18}$$

where $E[W]$ is given by Eq. (19.16) and $E[X_1]$ is given by Eq. (19.17).

The average value $E[X_2]$ of the service time X_2 of the anonymous user selecting the registration service is the sum of the average free service time $1/\mu_1$ and the average registration service time $1/\mu_2$. It follows that

$$E[X_2] = \frac{1}{\mu_1} + \frac{1}{\mu_2} \tag{19.19}$$

where μ_2 is the registration service rate defined in Sect. 19.2.2.

We give the average latency $E[Y_{a2}]$ of the anonymous users who select the registration service as follows:

$$E[Y_{a2}] = E[W] + E[X_2] = \frac{1}{\lambda}\left(\sum_{i=c}^{\infty}\sum_{j=0}^{c}\sum_{k=0}^{j}(i-c)\pi_{i,j,k}\right) + \frac{1}{\mu_1} + \frac{1}{\mu_2} \tag{19.20}$$

where $E[W]$ is given by Eq. (19.16) and $E[X_2]$ is given by Eq. (19.19).

We define the energy saving rate γ of the system as the energy conservation per second for system with the proposed sleep mode-based cloud architecture. During the active state of the VM, the energy will be consumed normally, while during the sleep state of the VM, the energy will be saved. However, additional energy will be consumed when a VM switches from the sleep state to the active state.

Let g_1 be the energy consumption per second when a VM provides the free service, g_2 be the energy consumption per second when a VM provides the registration service, g_3 be the energy consumption per second when a VM is in the sleep state, and g_4 be the energy consumption when a VM switches from the sleep state to the active state. We give the energy saving rate γ of the system as follows:

$$\gamma = (1-q)g_1 + q(g_1 + g_2 - g_3)\sum_{i=0}^{\infty}\sum_{j=0}^{c}\sum_{k=0}^{j}(c-j)\pi_{i,j,k}$$

$$- g_4\sum_{i=1}^{\infty}\sum_{j=0}^{c}\sum_{k=0}^{j}(c-j)\pi_{i,j,k} \times \delta. \tag{19.21}$$

19.4.2 Numerical Results

In order to evaluate the system performance of the sleep mode-based cloud architecture proposed in this chapter, we present numerical results with analysis and simulation in this subsection. The analysis results are carried out in Matlab 2010a on Intel (R) Core (TM) i7-4790 CPU @ 3.6 GHz, 6 GB RAM. The simulation results are obtained by averaging over 10 dependent runs using MyEclipse 2014. We create

Fig. 19.2 Average latency versus arrival rate of anonymous users

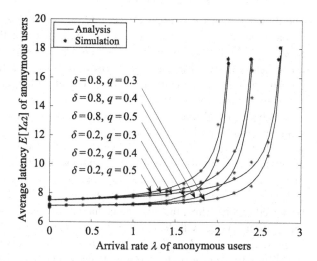

a JOB class with attributes in terms of UNARRIVE, WAIT, RUN, FINISH and TYPE to record the user state and the user type. We also create a SERVER class with attributes in terms of SLEEP and BUSY to record the state of a VM. Good agreements between the analysis results and the simulation results are observed.

We assume that the intensity of the free service is μ_1 anonymous users per second and the intensity of the registration service is μ_2 anonymous users per second. By setting $c = 10$, $\mu_1 = 0.5$ s^{-1} and $\mu_2 = 0.2$ s^{-1} as an example, we show how the average latency $E[Y_{a2}]$ of anonymous users who select the registration service changes versus the arrival rate λ of anonymous users for different registration probabilities q and different sleep parameters δ in Fig. 19.2.

As can be seen from Fig. 19.2, for all the registration probabilities q and all the sleep parameters δ, the average latency $E[Y_{a2}]$ of the anonymous users who select the registration service increases as the arrival rate λ of anonymous users increases. The reason is that the larger the arrival rate of anonymous users is, the more anonymous users will queue in the system buffer and the longer the newly arriving anonymous users, including the anonymous users who will select the registration service later, will wait before they gain access to the VMs. Thus, the average latency of the anonymous users who select the registration service will increase.

We also find that when the arrival rate λ of anonymous users is smaller (such as $\lambda < 1.2$), for the same sleep parameter δ, the influence of the registration probability q on the average latency $E[Y_{a2}]$ of the anonymous users who select the registration service is relatively small, since the VM is more likely to be asleep with a smaller arrival rate of anonymous users. In this case, the sleep parameter is the dominant factor in influencing the waiting time of anonymous users queueing in the system buffer.

When the arrival rate λ of anonymous users is larger (such as $\lambda > 1.2$), for the same sleep parameter δ, the average latency $E[Y_{a2}]$ of the anonymous users who select the registration service increases as the registration probability q increases.

The reason is that for a larger arrival rate of anonymous users, the more likely it is that a newly arriving anonymous user will queue in the system buffer before accessing a VM. The larger the registration probability is, the more anonymous users there will be who select the registration service. Therefore, the anonymous users queueing in system buffer will have to wait for longer, and the average latency of the anonymous users who select the registration service will increase.

When the arrival rate λ of anonymous users is smaller (such as $\lambda < 1.8$), for the same registration probability q, the average latency $E[Y_{a2}]$ of the anonymous users who select the registration service decreases as the sleep parameter δ increases. We note that with a smaller arrival rate of anonymous users, it is less possible that all of VMs are active. The larger the sleep parameter is, the shorter the time length of a sleep period is, namely, the anonymous users arriving during the sleep period can be served earlier. Thus, the average latency of the anonymous users who select the registration service will decrease. When the arrival rate λ of anonymous users is larger (such as $\lambda > 1.8$), for the same registration probability q, the influence of the sleep parameter δ on the average latency $E[Y_{a2}]$ of the anonymous users who select the registration service is relatively small, since all the VMs are less likely to be asleep with a larger value of λ.

By setting $g_1 = 5$ mW, $g_2 = 4$ mW, $g_3 = 0.5$ mW and $g_4 = 6$ mW as an example, we illustrate how the energy saving rate γ of the system changes versus the arrival rate λ of anonymous users for different registration probabilities q and different sleep parameters δ in Fig. 19.3.

As can be seen from Fig. 19.3, for all the registration probabilities q and all the sleep parameters δ, the energy saving rate γ of the system shows a downtrend decreasing to 0 as the arrival rate λ of anonymous users increases. As the arrival rate of anonymous users increases, the VMs are more likely to be active. Since there is

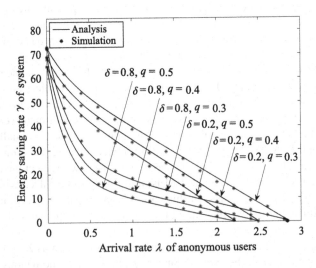

Fig. 19.3 Energy saving rate of system versus arrival rate of anonymous users

no energy conservation during active state, the energy saving rate of the system will decrease. When the arrival rate of anonymous users reaches a certain value (such as $\lambda \geq 2.5$ for $\delta = 0.8$ and $q = 0.4$), all the VMs will be more likely to remain awake without sleep. Therefore, the energy saving rate of the system tends to 0.

We find that for the same arrival rate λ of anonymous users and the same sleep parameter δ, the energy saving rate γ of the system decreases as the registration probability q increases. The greater the registration probability is, the more likely it is that an anonymous user will select the registration service, which forces the VMs to operate for a longer time, namely, sleep for a shorter time. Therefore, the energy saving rate of the system will decrease.

We also observe that for the same arrival rate λ of anonymous users and the same registration probability q, the energy saving rate γ of the system decreases as the sleep parameter δ increases. The greater the sleep parameter is, the shorter the time length of a sleep period is. Once there are anonymous users arriving during the sleep period, the VMs will be more likely to enter an active state earlier when the sleep parameter is larger rather than smaller. Therefore, the energy saving rate of the system will show a downtrend as the sleep parameter becomes larger.

We note that a larger arrival rate of anonymous users means a higher average latency for the anonymous users who select the registration service and a smaller energy saving benefit for the system, but a greater benefit for the cloud vendor. A smaller arrival rate of anonymous users means a shorter average latency for the anonymous users who select the registration service and a greater energy saving benefit for the system, but a smaller benefit for the cloud vendor. Regulating the arrival behavior of anonymous users is therefore the key component in optimizing the system performance.

19.5 Analysis of Registration Fee

In this section, we first investigate the Nash equilibrium behavior and the socially optimal behavior of anonymous users in the cloud architecture proposed in this chapter. Then, we impose an appropriate registration fee on the anonymous users who select the registration service with a pricing policy to maximize the value of the social benefit function.

19.5.1 Behaviors of Nash Equilibrium and Social Optimization

In the proposed cloud architecture, we note that all the anonymous users independently make decisions to access the system in order to maximize their benefits. In other words, they are selfish. However, the decision should be made at a social level. Therefore, we will investigate both the Nash equilibrium behavior and the socially optimal behavior of anonymous users. To this end, we present a few hypotheses as follows:

(1) An anonymous user's reward from completed free service is R_{g1}; an anonymous user's reward from completed registration service is R_{g2}.
(2) The cost of an anonymous user staying in the system is C_{g1} per second; the benefit of the system from the energy saving is R_{g3} for each milliwatt.
(3) The benefit for each anonymous user is identical, and the benefits for all the anonymous users are additive.
(4) When there is at least one available VM in the system, a newly arriving anonymous user's reward from completed service should be greater than its cost to the system, namely, $R_{g1}/C_{g1} > 1/\mu_1$ and $R_{g2}/C_{g1} > 1/\mu_2$.
(5) The maximal arrival rate denoted as λ_{max} is set to ensure the system is always stable.

We define the individual benefit function $G_{ind}(\lambda)$ as follows:

$$G_{ind}(\lambda) = R_{g1} + q R_{g2} - C_{g1}\big(q E[Y_{a1}] + (1 - q)E[Y_{a2}]\big) \qquad (19.22)$$

where $E[Y_{a1}]$ is the average latency of the anonymous users who experience only the free service, given in Eq. (19.18); $E[Y_{a2}]$ is the average latency of the anonymous users who select the registration service, given in Eq. (19.20).

Each anonymous user has two strategies: join the system, or balk the system. In the case of $G_{ind}(0) \geq 0$, even if all the anonymous users join the system, they all enjoy a non-negative benefit. Therefore, joining the system is an equilibrium strategy, namely, $\lambda^e = \lambda_{max}$ is the Nash equilibrium arrival rate of anonymous users. In the case of $G_{ind}(\lambda_{max}) \leq 0$, even if no other anonymous user joins the system, the net benefit of an anonymous user who joins is non-positive. Therefore, balking the system is an equilibrium strategy, namely, $\lambda^e = 0$ is the Nash equilibrium arrival rate of anonymous users. In the case of $G_{ind}(0) < G_{ind}(\lambda) < G_{ind}(\lambda_{max})$, a unique equilibrium strategy exists, the Nash equilibrium arrival rate λ^e of anonymous users is subject to $G_{ind}(\lambda^e) = 0$.

Under the condition that there is no pricing policy for an anonymous user, by aggregating the individual benefits of all the anonymous users and the energy saving rate of the system, the social benefit function $G_{soc}(\lambda)$ can be obtained as follows:

$$G_{soc}(\lambda) = \lambda\big(R_{g1} + q R_{g2} - C_{g1}\big(q E[Y_{a1}] + (1 - q)E[Y_{a2}]\big)\big) + R_{g3}\gamma. \qquad (19.23)$$

By maximizing the value of the social benefit function $G_{soc}(\lambda)$ in Eq. (19.23), we can write the socially optimal arrival rate λ^* of anonymous users as follows:

$$\lambda^* = \underset{0 \leq \lambda \leq \lambda_{max}}{\mathrm{argmax}} \{G_{soc}(\lambda)\} \qquad (19.24)$$

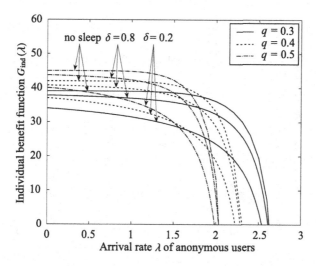

Fig. 19.4 Individual benefit function versus arrival rate of anonymous users

where "argmax" stands for the argument of the maximum. In other words, the set of points from "argmax" makes the social benefit function $G_{soc}(\lambda)$ attain its maximum value.

In order to explore the monotonic property of the individual benefit function $G_{ind}(\lambda)$ and the social benefit function $G_{soc}(\lambda)$, we present numerical results to illustrate the changing trends of $G_{ind}(\lambda)$ and $G_{soc}(\lambda)$. Besides the system parameters given in Sect. 19.4.2, we set $R_{g1} = 50$, $R_{g2} = 80$, $C_{g1} = 10$ and $R_{g3} = 0.1$ as examples in the numerical results.

Figure 19.4 demonstrates the individual benefit function $G_{ind}(\lambda)$ versus the arrival rate λ of anonymous users for different registration probabilities q and sleep parameters δ.

In Fig. 19.4, we find that for all the registration probabilities q and all the sleep parameters δ, as the arrival rate λ of anonymous users increases, the individual benefit function $G_{ind}(\lambda)$ continually decreases and tends to a negative value. As the arrival rate of anonymous users increases, the sojourn time of an anonymous user increases, hence the individual benefit function decreases.

When the arrival rate of anonymous users reaches a certain value (such as $\lambda \geq 2.3$ for $\delta = 0.8$ and $q = 0.4$), the system state tends towards becoming unstable and the sojourn time of an anonymous user increases sharply, hence the individual benefit function tends to a negative value. We also find that for each curve of the individual benefit function $G_{ind}(\lambda)$, a unique arrival rate exists subject to $G_{ind}(\lambda) = 0$, and this is the Nash equilibrium arrival rate λ^e of anonymous users.

Figure 19.5 demonstrates the social benefit function $G_{soc}(\lambda)$ versus the arrival rate λ of anonymous users for different registration probabilities q and sleep parameters δ.

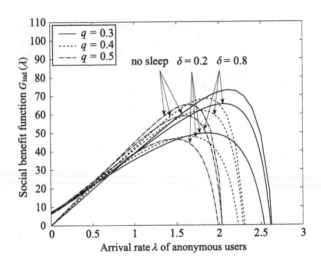

Fig. 19.5 Social benefit function versus arrival rate of anonymous users

In Fig. 19.5, we find that all the curves exhibit the property of being concave, namely, as the arrival rate of anonymous users increases, the social benefit function firstly increases, and then decreases. When the arrival rate of anonymous users is lower, the average latency of anonymous users does not change significantly as the arrival rate of anonymous users increases. In this case, the reward for anonymous users is the dominant factor influencing the social benefit. The higher the arrival rate of anonymous users is, the higher the reward anonymous users will earn, which results in a greater social benefit. When the arrival rate of anonymous users is higher, the dominant factor influencing the value of the social benefit function is the average latency of anonymous users. With an increase in the arrival rate of anonymous users, the average latency of anonymous users increases, resulting in a smaller social benefit. Therefore, all the curves exhibit the property of being concave. The maximum social benefit is the peak value for each curve, and the corresponding arrival rate is the socially optimal arrival rate λ^* of anonymous users.

Obviously, for all the registration probabilities q and all the sleep parameters δ, the Nash equilibrium arrival rate λ^e of anonymous users in Fig. 19.4 is always higher than the socially optimal arrival rate λ^* of anonymous users in Fig. 19.5. That is, more anonymous users will join the system under Nash equilibrium behavior. In order to restrain λ^e and make it equal to λ^*, we impose an appropriate registration fee on the anonymous users who select the registration service with a pricing policy.

19.5.2 Pricing Policy

In order to regulate the arrival behavior of anonymous users, we charge an appropriate registration fee f to the anonymous users who select the registration service. With the registration fee f, the individual benefit function $G'_{\text{ind}}(\lambda)$ of an anonymous user is modified as follows:

$$G'_{\text{ind}}(\lambda) = R_{g1} + q(R_{g2} - f) - C_{g1}\big(q E[Y_{a1}] + (1 - q)E[Y_{a2}]\big). \qquad (19.25)$$

Accordingly, the social benefit function $G'_{\text{soc}}(\lambda)$ is given as follows:

$$G'_{\text{soc}}(\lambda) = \lambda\big(R_{g1} + q(R_{g2} - f) - C_{g1}\big(q E[Y_{a1}] + (1 - q)E[Y_{a2}]\big)\big) + R_{g3}\gamma + \lambda q f$$

$$= \lambda\big(R_{g1} + q R_{g2} - C_{g1}\big(q E[Y_{a1}] + (1 - q)E[Y_{a2}]\big)\big) + R_{g3}\gamma. \qquad (19.26)$$

Comparing Eqs. (19.23) and (19.26), we find that the registration fee has no effect on the social benefit. This is because that the registration fee is just transferred from the anonymous users who select the registration service to the cloud vendor. We note that anonymous users and the cloud vendor combine to constitute a system, so the value of the social benefit function remains unchanged.

However, the mathematical expression for the social benefit $G'_{\text{soc}}(\lambda)$ are difficult to obtain in closed-forms, and the strict monotonicity of the social benefit function $G'_{\text{soc}}(\lambda)$ is difficult to explain. For this, we turn to an intelligent searching algorithm based on the echolocation behavior of bats, the Bat algorithm, to obtain the socially optimal arrival rate λ^* of anonymous users.

By introducing an adaptive step adjusted by the number of iterations, we develop an improved Bat algorithm. At the beginning of the algorithm, we set up a sizeable step to avoid local optimization. With an increase in the number of iterations, the step span decreases gradually, thus the algorithm converges rapidly and the searching result is more accurate. The main steps for the improved Bat algorithm proposed in this chapter are given as follows:

Step 1: Initialize the number N of bats, loudness A_0, pulse rate R_0, the maximum search frequency $f_{\max}$, the minimum search frequency $f_{\min}$, upper search bound U_b, lower search bound L_b, the minimum moving step $step_{\min}$, volume attenuation coefficient η, searching frequency enhancement factor ϕ. Set the initial number of iterations as $iter = 1$, the maximum iterations as $iter_{\max}$.

Step 2: Initialize the position, the loudness and the pulse rate for each bat.

    for $i = 1 : N$

    $\lambda_i = L_b + (U_b - L_b) \times \text{rand}$

    % rand is a random number selected in the interval $(0, 1)$.

    $A_i = A_0$

    $r_i = R_0$

endfor

Step 3: Calculate the fitness for each bat.

$$G_{soc}(\lambda_i) = \lambda_i\big(R_{g1} + qR_{g2} - C_{g1}\big(qE[Y_{a1}] + (1-q)E[Y_{a2}]\big)\big) + R_{g3}\gamma$$

$$\lambda^* = \underset{i\in\{1,2,3,\ldots,N\}}{argmax} \ \{G_{soc}(\lambda_i)\}$$

% λ^* is current optimal position.

Step 4: Update the position and the fitness for each bat.

for $i = 1 : N$

$$f_i = f_{min} + (f_{max} - f_{min}) \times rand$$
$$v_i = v_i + (\lambda_i - \lambda^*)f_i$$
$$\lambda_i = \lambda_i + v_i$$

if $r_i < rand$

$$\lambda_i = \lambda^* + (1/(2 \times iter) + step_{min}) \times randn$$

% randn returns a sample from the "standard normal" distribution.

endif

$$G'_{soc}(\lambda_i) = \lambda_i\big(R_{g1} + qR_{g2} - C_{g1}\big(qE[Y_{a1}] + (1-q)E[Y_{a2}]\big)\big) + R_{g3}\gamma$$

if $\big(G'_{soc}(\lambda_i) > G_{soc}(\lambda_i)\big)$ and $\big(A_i > rand\big)$

$$G_{soc}(\lambda_i) = G'_{soc}(\lambda_i)$$
$$A_i = \eta A_i$$
$$r_i = R_0(1 - \exp(-\phi \times iter))$$

endif

endfor

Step 5: Select the optimal position among all the bats.

$$\lambda^* = \underset{i\in\{1,2,3,\ldots,N\}}{argmax} \ \{G_{soc}(\lambda_i)\}.$$

Step 6: Check the number of iterations.

if $iter < iter_{max}$

$$iter = iter + 1$$

go to **Step4**

endif

Step 7: Output the optimal position λ^* and the maximum fitness $G_{soc}(\lambda^*)$.

Applying the system parameters in Figs. 19.4 and 19.5 to the improved Bat algorithm, we obtain the numerical results for the socially optimal arrival rate λ^* of anonymous users with different registration probabilities q and sleep parameters δ.

By substituting the socially optimal arrival rate λ^* of anonymous users into Eq. (19.25) and setting $G'_{ind}(\lambda) = 0$, we can obtain the registration fee f as follows:

$$f = \frac{1}{q} \times \big(R_{g1} + qR_{g2} - C_{g1}\big(qE[Y_{a1}] + (1-q)E[Y_{a2}]\big)\big)\big|_{\lambda=\lambda^*} \tag{19.27}$$

where $\big(qE[Y_{a1}] + (1-q)E[Y_{a2}]\big)\big|_{\lambda=\lambda^*}$ is the average latency with the socially optimal arrival rate λ^* of anonymous users.

For different registration probabilities q and sleep parameters δ, in Table 19.1, we present numerical results for the pricing policy in terms of the socially optimal

Table 19.1 Numerical results for registration fee

Sleep parameters δ	Registration probabilities q	Socially optimal arrival rates λ^*	Maximum social benefits $G_{soc}(\lambda^*)$	Registration fees f
No sleep	0.3	2.1256	73.0759	114.5963
No sleep	0.4	1.8489	68.6578	92.836
No sleep	0.5	1.6465	65.3641	79.3976
0.8	0.3	2.056	65.4861	105.0306
0.8	0.4	1.7981	61.9026	85.2437
0.8	0.5	1.602	59.2212	73.3388
0.2	0.3	1.8647	49.7374	78.455
0.2	0.4	1.642	47.9301	69.9957
0.2	0.5	1.4728	46.618	60.8772

arrival rate λ^* of anonymous users, the maximum social benefit $G_{soc}(\lambda^*)$ of anonymous users and the registration fee f charged to the anonymous users who select the registration service.

From Table 19.1, we find that for the same sleep parameter δ, the registration fee f decreases as the registration probability q increases. We note that as the registration probability increases, the average latency of anonymous users will increase accordingly, and then the sojourn cost of anonymous users will increase. In this case, the anonymous users are reluctant to access the system. In order to attract more anonymous users to access the system, a lower registration fee f should be set. We also find that for the same registration probability q, the registration fee f increases as the sleep parameter δ increases. We note that as the sleep parameter increases, the average latency of anonymous users will decrease accordingly, and then the sojourn cost of anonymous users will decrease, which will lead to more anonymous users joining the system. Therefore, the registration fee should be set higher. Compared with no sleep state for the VMs, for the same registration probability q, the registration fee f is lower. When there is a sleep state for the VMs, the average latency of anonymous users will increase, and then the sojourn cost of anonymous users will increase. Therefore, the registration fee should be set lower to attract more anonymous users to access the system.

19.6 Conclusion

In this chapter, considering the high energy consumption and the establishment of a loyal and stable client base in a cloud computing system, we proposed a sleep mode-based cloud architecture with a free service and a registration service. Accordingly, we presented a method to model and evaluate the proposed cloud architecture by establishing an asynchronous multiple-vacation queueing model with a second optional service. We provided numerical results to investigate the impacts of the sleep parameter and the registration probability on the average latency of the anonymous users who select the registration service and the energy saving rate of the system. From the perspective of economics, we constructed an individual benefit function to investigate the Nash equilibrium behavior of anonymous users. Furthermore, we developed an improved Bat algorithm to obtain the socially optimal arrival rate of anonymous users. With numerical results, we found that the individually optimal arrival rate is always higher than the socially optimal arrival rate of anonymous users. For this, we presented a method for motivating anonymous users to accept the socially optimal strategy by charging an appropriate registration fee on the anonymous users who select the registration service.

Chapter 20
Energy-Efficient Task Scheduling Strategy

For the purpose of satisfying the service level agreement of cloud users while improving the energy efficiency in cloud computing system, in this chapter, we propose an energy-efficient task scheduling strategy with a sleep-delay timer and a wake-up threshold. Accordingly, we build a synchronous vacation queueing model with vacation-delay and an N-policy. By using the matrix-geometric solution method, we analyze the queueing model in the steady state. We derive performance measures of the system in terms of the average latency of tasks and the energy saving rate of the system. We present numerical results to evaluate the performance of the system using the proposed energy-efficient task scheduling strategy. Finally, we construct a system cost function to trade off different performance measures and develop an improved Genetic algorithm to jointly optimize the system parameters with the proposed strategy.

20.1 Introduction

Cloud computing is a style of computing in which dynamically scalable and virtualized resources are provided as a service over the Internet [Fati19, Mond19, Olok17].

In a cloud computing system, there are two key actors: cloud providers and cloud users [Madn17a]. The cloud providers hold enormous computing resources in data centers [Zhan18]. They rent the resources out to the cloud users on a pay-per-use basis. Therefore, the cloud providers want to improve resource utilization and maximize their profit, while the cloud users, who have applications of various loads, attempt to receive service from different cloud providers at the lowest expense possible [Abdu16].

How to achieve a higher resource utilization in WCNs, while at the same time offer a lower cost to the cloud users is a key part of a cloud provider's management strategy. Some researchers have conducted analyses to minimize the

© Springer Nature Singapore Pte Ltd. 2021
S. Jin, W. Yue, *Resource Management and Performance Analysis of Wireless Communication Networks*, https://doi.org/10.1007/978-981-15-7756-7_20

construction period and increase the system utilization by scheduling several cloud tasks on different Virtual Machines (VMs) [Abdu14]. However, this scheduling method requires all servers to keep incoming tasks active, leading to dramatic levels of energy consumption and elevated carbon dioxide emissions [Duan18, Li14b, Wang19]. Electricity usage is estimated to continue rising to approximately 73 billion kWh by 2020 [Zaka17].

Therefore, one of the current challenges in cloud computing is to reduce energy consumption while guaranteeing the quality of user experience.

Cloud computing, with its unprecedented computing capability, has become a popular paradigm yet has raised concerns from enterprises [You16]. Recently, many scholars have carried out fruitful research on cloud management and cloud optimization. In [Xia15], the authors presented a queueing network-based performance framework with dynamic voltage scaling with the purpose of conserving power consumption. In [Chen15b], the authors presented a method of Minimum Expectation Execution Energy with Performance Constraints (ME^3PC), by which the energy consumption can be effectively conserved under certain performance constraints. In [Chen16], the authors proposed a DVFS scheme by which the most suitable voltage and frequency for the multi-core embedded system could be dynamically forecasted. Aiming to achieve adaptive regulations for different requirements in a cloud computing system, in [Shen17], the authors proposed a Genetic algorithm to achieve adaptive regulations for different requirements of energy and performance in cloud tasks.

All the aforementioned researches have sought to conserve energy consumption, but has ignored the fact that even though no tasks are being processed, all the VMs remain awake.

Putting idle VMs in sleep mode is a way of conserving power consumption when the traffic load is light [Khos17a]. In [Kemp15], the author investigated a vacation queueing system with a finite buffer in which the transmission is restarted if the number of packets in the buffer reaches a threshold at the epoch when a sleep period ends. For the purpose of reducing the carbon footprint of data centers, in [Mcba16], the authors presented a combined approach using an energy conservation method of dynamic voltage/frequency scaling and sleep mode. In order to efficiently control the traffic load on each VM, in [Lawa16], the authors introduced a vacation mechanism with threshold policy. With this mechanism both the energy consumption and the system cost could be cut down. In [Sing16b], the authors presented a deep-sleep mode in a cloud computing system to conserve energy consumption and improve the resource utility. Putting idle VMs into a sleep state can to some extent conserve energy consumption. However, continually switching VMs between asleep and awake states can cause response penalties.

Based on this research background, we consider an effective strategy to satisfy the response performance of cloud users while also trying to improve the energy efficiency in cloud computing systems.

In making a compromise between the performance degradation and the energy conservation, we need to evaluate the strategy performance and optimize the strategy parameters.

We note that a Genetic algorithm is a heuristic method used to search for near-optimal solution in a large solution space [Madn17b]. Genetic algorithm originated as an effective tool for function optimization in the 1960s. Since then, considerable research on improving the searching ability of Genetic algorithm has been carried out. In order to improve the convergence rate of the Genetic algorithm, in [Qiu15], the authors implemented the rank-based roulette wheel scheme in the selection mimics, where the better individuals have more chance in reproducing offspring. For the purpose of introducing population diversity, in [Huan15a], the authors randomly and uniformly selected the initial parent chromosome from among the top-tier chromosomes, whereupon the second parent selected from the lower-tier chromosomes. In [Jian12], the authors presented an adaptive Genetic algorithm, changing the crossover probability and the mutation probability according to the level of fitness. This method can prevent the Genetic algorithm from getting stuck at a local optimal solution.

As mentioned above, the searching ability of a Genetic algorithm is greatly influenced by the crossover probability and the mutation probability. This can conceptually be applied to optimizing the system parameters and to enhancing the overall performance for the greening of cloud computing.

This chapter is a substantial and appropriate extension of our previous work [Jin19d] appearing in the conference proceedings. In the paper of the proceedings [Jin19d], we proposed a task scheduling strategy with a sleep-delay timer and a wake-up threshold to satisfy the response performance of cloud users while reducing the energy consumption in a cloud computing system. In the numerical results, the performance of the system and the effects of the design parameters based on the average latency of tasks of tasks and the energy saving rate of the system were evaluated.

Therefore, in this chapter, with the advent of energy shortages and a rise in greenhouse gas emissions, we extend the work [Jin19d] to propose a more effective strategy for the greening of cloud computing. This is to further satisfy the response performance of cloud users on the premise of ensuring a higher energy efficiency in cloud computing by proposing a task scheduling strategy with a sleep-delay timer and a wake-up threshold. Moreover, to optimize the parameter settings for the cloud providers, we analyze the system performance, derive the system performance measures, present a system cost function and propose an enhanced intelligent searching algorithm using a compromise between different performance measures.

The following summarizes the main contributions of this chapter.

(1) We provide additional analyses for the task scheduling strategy proposed in this chapter, and present the two forms of the state transition based on the relationship between the number of VMs in the system and the wake-up threshold.
(2) We evaluate the system performance of the energy efficient task scheduling strategy with a two-dimensional Continuous-Time Markov Chain (CTMC).
(3) We present new numerical results with analysis and simulation to evaluate the proposed strategy. We also make a comparison between our proposed strategy and conventional strategies.

(4) By considering both the average latency of tasks and the energy saving rate of the system, we establish a system cost function to make a compromise between the performance degradation and the energy efficiency.

(5) By dynamically adjusting the crossover probability and the mutation probability, and initializing the individuals with chaotic equations, we develop an improved Genetic algorithm to jointly optimize the system parameters with the proposed strategy.

The chapter is organized as follows. In Sect. 20.2, we describe the energy-efficient task scheduling strategy proposed in this chapter. Then, we present the system model in detail. In Sect. 20.3, we present a performance analysis of the system model, through the analysis of the transition rate matrix and the steady-state distribution. In Sect. 20.4, we obtain performance measures and present numerical results to evaluate the system performance. In Sect. 20.5, we construct a system cost function and develop an improved Genetic algorithm to jointly optimize the system parameters with the proposed strategy. Finally, we draw our conclusions in Sect. 20.6.

20.2 Energy-Efficient Task Scheduling Strategy and System Model

In this section, by introducing the sleep-delay parameter and the wake-up threshold, we first propose a sleep mode-based energy-efficient task scheduling strategy. Then, we model the proposed task scheduling strategy as a vacation queueing system.

20.2.1 Energy-Efficient Task Scheduling Strategy

Cloud services are provided over the cloud computing environment via distributed software and hardware. On a Physical Machine (PM), several VMs can be deployed. Even if no tasks need processing, sets of VMs located on one or more PMs will remain awake. As a result, massive amounts of energy are wasted, so energy consumption has become a special concern for cloud providers.

Generally speaking, the PMs in Cloud Data Centers (CDCs) are highly configured. In order to ensure high availability and improve parallel processing capability, each VM hosted on a PM works within its own Operating System (OS). This arrangement offers the theoretical foundation for the feasible and practical implementation of a sleep mode on each VM.

Putting idle VMs to sleep is considered to be a useful method of reducing energy consumption in a cloud computing system. However, the sleep mode may degrade the response performance. Therefore, we propose an energy efficient task scheduling strategy for cloud computing by introducing a sleep-delay timer and a wake-up threshold. Considering the negative effect from the sleep mode, we set a sleep-delay timer on a PM to guarantee the quality of experience for cloud users. When the system becomes empty, namely, all the VMs are idle, the PM will not go to sleep immediately, but remain awake for a random time length under the control of a sleep-delay timer. Tasks arriving at the system during the sleep-delay period will receive immediate service. If and only if no tasks arrive at the system before the sleep-delay timer expires will the PM go into periodical sleep, where multiple sleep periods constitute a sleep state.

Frequent state switches will certainly cause additional energy consumption and lead to extra latency. Thus, we set a critical wake-up threshold N to improve the energy efficiency. At the epoch when a sleep period is completed, if there are fewer tasks waiting in the system buffer than the wake-up threshold N, the PM will move into an awake state and all the tasks queueing in the system buffer will receive service accordingly. If not, a new sleep period is initiated, namely, the PM remains asleep. As a result, the energy consumption of each PM can be efficiently conserved.

For the task scheduling strategy proposed in this chapter, a PM is in the awake state, the sleep state, or the sleep-delay state.

(1) *Awake State*: During the awake state, there is at least one VM busy with task processing. The tasks in the system receive service in accordance with a First-Come First-Served (FCFS) discipline.
(2) *Sleep-Delay State*: In order to extend the awake period and improve the response performance, once all the tasks in the system are completely executed, a sleep-delay timer with a random time length will be started, and all the VMs will remain active within the constraint of the sleep-delay timer. We call this state the sleep-delay state. A new task arriving at the system during the sleep-delay period will receive service promptly. At the epoch when the sleep-delay timer expires, if there are no tasks queueing in the system buffer, the PM will switch into sleep state.
(3) *Sleep State*: A sleep timer with a random time length will also be started as soon as the PM enters the sleep state. In the sleep state, the power of some accessories, with the exception of the memory, will be cut off, and the tasks in the system will not be served. All the tasks arriving at the system during the sleep state have to wait in the system buffer. When the sleep time expires, if there are fewer tasks waiting in the system buffer than the threshold N, another sleep period will be started. The time duration of this sleep period is controlled by a new sleep timer. Otherwise, the PM will switch to the awake state, and all the VMs in the PM will wake up and prepare to serve all the tasks queueing in the system buffer.

As a summary, the state transition of the PM with the task scheduling strategy proposed in this chapter is plotted in Fig. 20.1.

20.2.2 System Model

In this chapter, we establish a synchronous multiple vacation queueing system with a vacation-delay and an N-policy to model the proposed strategy. Note that the

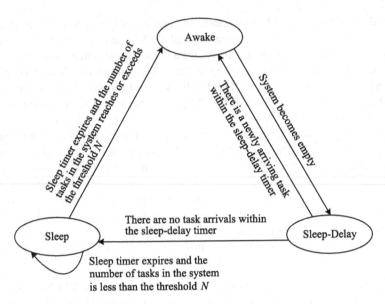

Fig. 20.1 Transition among three states in proposed strategy

behavior of all the PMs in a CDC is stochastically homogeneous. We consider a tagged PM to evaluate the task scheduling strategy.

We assume that the inter-arrival time of tasks follows an exponential distribution with mean $1/\lambda$, where $\lambda > 0$, called the arrival rate of tasks. The execution time of a task by a VM is assumed to follow an exponential distribution with mean $1/\mu$ seconds, where $\mu > 0$. We call μ the service rate. The time length for the sleep-delay timer is supposed to follow an exponential distribution with mean $1/\beta$ seconds, where $\beta > 0$. The time length for the sleep timer is supposed to follow an exponential distribution with mean $1/\delta$ seconds, where $\delta > 0$. We call β the sleep parameter and δ the sleep-delay parameter.

The system buffer is supposed to have an infinite capacity.

By $X(t) = i$ $(i = 0, 1, 2, \ldots)$ we denote the number of tasks in the system at the epoch t. By $Y(t) = j$ $(j = 0, 1, 2)$ we denote the PM state at the epoch t. $j = 0$ means the PM is asleep, $j = 1$ means the PM is awake, and $j = 2$ means the PM is in a sleep-delay state. $X(t)$ and $Y(t)$is are called the system level and the PM state, respectively. The stochastic behavior of the queueing system under consideration can be given in a two-dimensional CTMC $\{(X(t), Y(t)), \ t \geq 0\}$. The CTMC is regular irreducible and has an infinite state space as follows:

$$\Omega = \{(0, 0), (0, 2)\} \cup \{(i, j) : i \geq 1, j = 0, 1, 2\}. \tag{20.1}$$

Under the assumption that the CTMC is positive recurrent, the steady-state distribution $\pi_{i,j}$ is defined as follows:

$$\pi_{i,j} = \lim_{t \to \infty} \Pr\{X(t) = i, \ Y(t) = j\}, \quad (i, j) \in \mathbf{\Omega}. \tag{20.2}$$

Let us form the row vector π_i as

$$\pi_i = (\pi_{i,0}, \pi_{i,1}, \pi_{i,2}), \quad i = 0, 1, 2, \ldots. \tag{20.3}$$

The steady-state distribution $\mathbf{\Pi}$ of the CTMC can be partitioned as follows:

$$\mathbf{\Pi} = (\pi_0, \pi_1, \pi_2, \ldots). \tag{20.4}$$

20.3 Performance Analysis

In this section, we present a performance analysis of the system model, through the analysis of the transition rate matrix and the steady-state distribution.

20.3.1 Transition Rate Matrix

To analyze the CTMC in the steady state, one of the key steps is to construct a transition rate matrix $\mathbf{Q}$. Based on the relationship between the number k of VMs in the cloud computing system under consideration and the wake-up threshold N with our proposed strategy, we will give two forms of transition rate matrices.

Let $\mathbf{Q}_{x,y}$ be the sub-matrices of $\mathbf{Q}$ for the system level jumping from x ($x = 0, 1, 2, \ldots$) to y ($y = 0, 1, 2, \ldots$). For the sake of convenience in presentation, we denote $\mathbf{Q}_{x,x}$ as A_x, $\mathbf{Q}_{x,x-1}$ as B_x and $\mathbf{Q}_{x,x+1}$ as C_x. Considering the size relationship between the number k of VMs in the cloud computing system and the wake-up threshold N, we discuss in detail A_x, B_x and C_x for the following two cases.

Case I: $N \leq k$

(1) When the initial level x is 0, the sub-matrices A_0 and C_0 are given as follows:

$$A_0 = \begin{pmatrix} -\lambda & 0 & 0 \\ 0 & 0 & 0 \\ \beta & 0 & -(\lambda + \beta) \end{pmatrix}, \tag{20.5}$$

$$C_0 = \begin{pmatrix} \lambda & 0 & 0 \\ 0 & 0 & 0 \\ 0 & \lambda & 0 \end{pmatrix}. \tag{20.6}$$

(2) When the initial level x is 1, the sub-matrices A_1, B_1 and C_1 are given as follows:

$$A_1 = \begin{pmatrix} -\lambda & 0 & 0 \\ 0 & -(\lambda + \mu) & 0 \\ 0 & 0 & 0 \end{pmatrix},$$ (20.7)

$$B_1 = \begin{pmatrix} 0 & 0 & 0 \\ 0 & 0 & \mu \\ 0 & 0 & 0 \end{pmatrix},$$ (20.8)

$$C_1 = \begin{pmatrix} \lambda & 0 & 0 \\ 0 & \lambda & 0 \\ 0 & 0 & 0 \end{pmatrix}.$$ (20.9)

(3) When the initial level x ranges from 2 to $N - 1$, the sub-matrices A_x, B_x and C_x are given as follows:

$$A_x = \begin{pmatrix} -\lambda & 0 & 0 \\ 0 & -(\lambda + x\mu) & 0 \\ 0 & 0 & 0 \end{pmatrix},$$ (20.10)

$$B_x = \begin{pmatrix} 0 & 0 & 0 \\ 0 & x\mu & 0 \\ 0 & 0 & 0 \end{pmatrix},$$ (20.11)

$$C_x = \begin{pmatrix} \lambda & 0 & 0 \\ 0 & \lambda & 0 \\ 0 & 0 & 0 \end{pmatrix}.$$ (20.12)

(4) When the initial level x ranges from N to $+\infty$, the sub-matrices A_x, B_x and C_x are given as follows:

$$A_x = \begin{pmatrix} -(\lambda + \delta) & \delta & 0 \\ 0 & -(\lambda + \min\{x, k\}\mu) & 0 \\ 0 & 0 & 0 \end{pmatrix},$$ (20.13)

$$B_x = \begin{pmatrix} 0 & 0 & 0 \\ 0 & \min\{x, k\}\mu & 0 \\ 0 & 0 & 0 \end{pmatrix},$$ (20.14)

$$C_x = \begin{pmatrix} \lambda & 0 & 0 \\ 0 & \lambda & 0 \\ 0 & 0 & 0 \end{pmatrix}.$$ (20.15)

From the transition rate matrix Q in the case of $N \leq k$, we find that all the sub-matrices A_x and B_x are repeated for $x \geq k$, and all the sub-matrices C_x are repeated for $x \geq 1$. For the convenience of presentation, we represent the repetitive sub-matrices A_x ($x \geq k$), B_x ($x \geq k$) and C_x ($x \geq 1$) as A, B and C, respectively. Therefore, the transition rate matrix Q for Case I can be given as follows:

$$
Q = \begin{pmatrix}
A_0 & C_0 & & & & & & \\
B_1 & A_1 & C & & & & & \\
& \ddots & \ddots & \ddots & & & & \\
& & B_{N-1} & A_{N-1} & C & & & \\
& & & B_N & A_N & C & & \\
& & & & \ddots & \ddots & \ddots & \\
& & & & & B_{k-1} & A_{k-1} & C \\
& & & & & & B & A & C \\
& & & & & & & \ddots & \ddots & \ddots
\end{pmatrix}. \tag{20.16}
$$

Case II: $N > k$

(1) When the initial level x is 0, the sub-matrices A_0 and C_0 are the same as those given in Case I.
(2) When the initial level x is 1, the sub-matrices A_1, B_1 and C_1 are the same as those given in Case I, too.
(3) When the initial level x ranges from 2 to $N - 1$, the sub-matrices A_x, B_x and C_x can be given as follows:

$$
A_x = \begin{pmatrix}
-\lambda & 0 & 0 \\
0 & -\lambda - \min\{x, k\}\mu & 0 \\
0 & 0 & 0
\end{pmatrix}, \tag{20.17}
$$

$$
B_x = \begin{pmatrix}
0 & 0 & 0 \\
0 & \min\{x, k\}\mu & 0 \\
0 & 0 & 0
\end{pmatrix}, \tag{20.18}
$$

$$
C_x = \begin{pmatrix}
\lambda & 0 & 0 \\
0 & \lambda & 0 \\
0 & 0 & 0
\end{pmatrix}. \tag{20.19}
$$

(4) When the initial level x ranges from N to $+\infty$, the sub-matrices A_x, B_x and C_x are given as follows:

$$
A_x = \begin{pmatrix}
-(\lambda + \delta) & \delta & 0 \\
0 & -(\lambda + k\mu) & 0 \\
0 & 0 & 0
\end{pmatrix}, \tag{20.20}
$$

$$B_x = \begin{pmatrix} 0 & 0 & 0 \\ 0 & k\mu & 0 \\ 0 & 0 & 0 \end{pmatrix}, \tag{20.21}$$

$$C_x = \begin{pmatrix} \lambda & 0 & 0 \\ 0 & \lambda & 0 \\ 0 & 0 & 0 \end{pmatrix}. \tag{20.22}$$

From the transition rate matrix Q in the case of $N > k$, we find that all the sub-matrices A_x and B_x are repeated for $x \geq N$, and all the sub-matrices C_x are repeated for $x \geq 1$. We denote the repetitive sub-matrices A_x ($x \geq N$), B_x ($x \geq N$) and C_x ($x \geq 1$) as A, B and C, respectively. Transition rate matrix Q for Case II is given as follows:

$$Q = \begin{pmatrix} A_0 & C_0 & & & & & & \\ B_1 & A_1 & C & & & & & \\ & \ddots & \ddots & \ddots & & & & \\ & & B_{k-1} & A_{k-1} & C & & & \\ & & & B_k & A_k & C & & \\ & & & & \ddots & \ddots & \ddots & \\ & & & & & B_{N-1} & A_{N-1} & C \\ & & & & & & B & A & C \\ & & & & & & & \ddots & \ddots & \ddots \end{pmatrix}. \tag{20.23}$$

Now, all the sub-matrices in the transition rate matrix Q have been addressed. From the block-tridiagonal structure of the transition rate matrix Q given in Eqs. (20.16) and (20.23), we note that the two-dimensional CTMC $\{(X(t), Y(t)),\ t \geq 0\}$ is in fact a Quasi Birth-Death (QBD) process.

20.3.2 Steady-State Distribution

To analyze the two-dimensional stochastic process $\{(X(t), Y(t)),\ t \geq 0\}$, we need to obtain the minimal non-negative solution to the following matrix quadratic equation:

$$R^2 B + RA + C = 0. \tag{20.24}$$

We call the solution R of Eq. (20.24) the rate matrix.

Since the coefficient matrices in Eq. (20.24) are all upper triangular, and all the elements of the third row and the third column are zero, the rate matrix R can be written as follows:

$$R = \begin{pmatrix} r_{11} & r_{12} & 0 \\ 0 & r_{22} & 0 \\ 0 & 0 & 0 \end{pmatrix} \tag{20.25}$$

where r_{ij} is the element that needs to be solved.

Substituting Eq. (20.25) into Eq. (20.24) yields the following set of equations:

$$\begin{cases} k\mu(r_{11}r_{12} + r_{12}r_{22}) + \delta r_{11} - (\lambda + k\mu)r_{12} = 0 \\ k\mu r_{22}^2 - (\lambda + k\mu)r_{22} + \lambda = 0 \\ -(\lambda + \delta)r_{11} + \lambda = 0. \end{cases} \tag{20.26}$$

By solving Eq. (20.26) under the condition that $\rho = \lambda/(k\mu) < 1$, we can derive the minimal non-negative solution R of Eq. (20.24) as follows:

$$R = \begin{pmatrix} \dfrac{\lambda}{\lambda + \delta} & \rho & 0 \\ 0 & \rho & 0 \\ 0 & 0 & 0 \end{pmatrix}. \tag{20.27}$$

Based on the analytical result of R, we can easily obtain the spectral radius $\mathrm{Sp}(R) = \max\{\lambda/(\lambda + \delta), \rho\} < 1$. Therefore, the QBD is positive recurrent, and the steady-state distribution of the CTMC exists.

Let u be the minimum value of the number k of VMs and the wake-up threshold N, namely, $u = \min\{k, N\}$. Let v be the maximum value of the number k of VMs and the wake-up threshold N, namely, $v = \max\{k, N\}$. Then, we construct an augmented matrix as follows:

$$B[R] = \begin{pmatrix} A_0 & C_0 & & & & & \\ B_1 & A_1 & C & & & & \\ & \ddots & \ddots & \ddots & & & \\ & & B_u & A_u & C & & \\ & & & \ddots & \ddots & \ddots & \\ & & & & B_{v-1} & A_{v-1} & C \\ & & & & & B & A + RB \end{pmatrix}. \tag{20.28}$$

By using the equilibrium equation and the normalization condition, we give a set of linear equations as follows:

$$\begin{cases} (\pi_0, \pi_1, \pi_2, \ldots, \pi_v) B[R] = 0 \\ (\pi_0, \pi_1, \pi_2, \ldots, \pi_{v-1}) e_1 + \pi_v (I - R)^{-1} e_2 = 1 \end{cases} \qquad (20.29)$$

where e_1 is a column vector with $3 \times v$ elements and e_2 is a column vector with 3 elements, respectively. All elements of these vectors are equal to 1.

By using the Gauss-Seidel method, we obtain $\pi_0, \pi_1, \pi_2, \ldots, \pi_v$. Note that the CTMC $\{(X(t), Y(t)), \ t \geq 0\}$ is a QBD process, we can give the steady-state distribution of the CTMC as follows:

$$\pi_i = \pi_v R^{i-v}, \quad i \geq v + 1. \qquad (20.30)$$

By substituting π_v obtained in Eq. (20.29) into Eq. (20.30), we get π_i $(i = v + 1, v + 2, v + 3, \ldots)$. Therefore, the steady-state distribution $\Pi = (\pi_0, \pi_1, \pi_2, \ldots)$ of the CTMC can be presented numerically.

20.4   Performance Measures and Numerical Results

In this section, we first derive performance measures of the system in terms of the average latency of tasks and the energy saving rate of the system, respectively. Then, we present numerical results to evaluate the performance of the system using the energy-efficient task scheduling strategy proposed in this chapter.

20.4.1   Performance Measures

We define the latency Y_t of a task as the duration from the epoch a task arrives at the cloud computing system to the epoch that task successfully departs the cloud computing system. That is to say, by adding the average waiting time of a task in the system buffer and the service time of a task on the VM, we can obtain the latency of a task. Based on the model analysis in Sect. 20.3.2, we give the average value $E[N_t]$ for the number N_t of tasks in the system as follows:

$$E[N_t] = \sum_{i=0}^{\infty} i (\pi_{i,0} + \pi_{i,1} + \pi_{i,2}). \qquad (20.31)$$

By using the analysis presented in Sect. 20.3, we can obtain the average latency $E[Y_t]$ of tasks as follows:

$$E[Y_t] = \frac{1}{\lambda} \left(\sum_{i=0}^{\infty} i(\pi_{i,0} + \pi_{i,1} + \pi_{i,2}) \right). \tag{20.32}$$

Within our proposed strategy, we define the energy saving rate γ of the system as the decreased energy consumption per second. During both the awake state and the sleep-delay state, the PM consumes energy normally. During the sleep state, energy consumption is reduced. Moreover, at each sleep period completion instant, the listening process and the state transition will consume additional energy.

In deriving the energy saving rate of the system, by g_1 we denote the energy consumption per second when the VMs are awake or idle in the sleep-delay state, by g_2 we denote the energy consumption per second when the VMs are asleep, by g_3 we denote the additional energy consumption for the VMs to wake up from the sleep state, and by g_4 we denote the additional energy consumption when the VMs listen to the cloud computing system. Based on these values, we give the energy saving rate γ of the system as follows:

$$\gamma = (g_1 - g_2) \sum_{i=0}^{\infty} \pi_{i,0} - g_3 \sum_{i=1}^{\infty} \pi_{i,0}\delta - g_4 \sum_{i=0}^{\infty} \pi_{i,0}\delta. \tag{20.33}$$

20.4.2 Numerical Results

We carry out numerical results with analysis and simulation to investigate the strategy proposed in this chapter and validate the system model. All the experiments are performed on a personal computer configured with AMD Ryzen 3 2200, CPU @ 3.6 GHz, 16 GB RAM and 4T disk. Good agreements between the analysis results and the simulation results are observed.

We list the system parameters settings in Table 20.1 as an example for all the numerical results, where g_1 is the energy consumption level of a busy VM, g_2 is the energy consumption level of a sleeping VM, g_3 is the energy consumption level for each switching, and g_4 is the energy consumption level for each listening defined in Sect. 20.4.1, respectively.

In all of the following figures, the analytical results and the simulation results are illustrated as lines and markers, respectively.

Figure 20.2 illustrates how the average latency $E[Y_t]$ of tasks changes with the sleep parameter δ for the different wake-up thresholds N and the different sleep-delay parameters β. The larger the sleep-delay parameter is, the less likely it is that the tasks arriving at the system after the awake state are served immediately, so the average latency of tasks is greater. As the sleep parameter increases, the tasks arriving at the system within a sleep period need to wait a shorter time for

Table 20.1 Parameter
settings in numerical results

Parameters	Values
Total number k of VMs in the system	20
Arrival rate λ of tasks	0.4 tasks/second
Service rate μ of a task	0.2 tasks/second
g_1 of a busy VM	20 mW
g_2 of a sleeping VM	2 mW
g_3 of each switching	12 mJ
g_4 of each listening	4 mJ

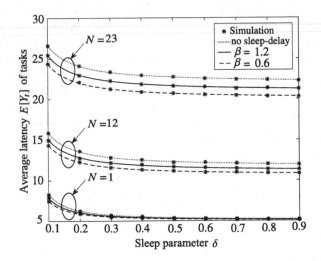

Fig. 20.2 Average latency of tasks versus sleep parameter

the completion of the sleep period, so the average latency of tasks decreases. As the
wake-up threshold increases, the tasks have to wait longer in the sleep state, hence
the average latency of tasks increases.

Figure 20.3 illustrates how the energy saving rate γ of the system changes with
the sleep parameter δ for the different wake-up thresholds N and the different sleep-
delay parameters β. The larger the sleep-delay parameter is, the more likely it is that
the PM will switch into the sleep state from the sleep-delay state, so more energy
will be conserved. On the other hand, with a larger sleep parameter, the sleep period
gets shorter, and the listening frequency increases. Additional energy consumption
resulting from listening increases, and the energy conservation decreases. As the
wake-up threshold increases, the PM stays in the sleep state longer accumulating
tasks, hence the energy saving rate of the system increases.

In Figs. 20.2 and 20.3, the statistical results with $\beta = +\infty$ are for a conventional
strategy without a sleep-delay scheme, and the statistical results with $N = 1$ are
for a conventional synchronous multi-sleep strategy without a wake-up threshold.
Compared to the conventional strategy without a sleep-delay scheme as results
with $\beta = +\infty$ in Fig. 20.2, we observe that our proposed strategy performs

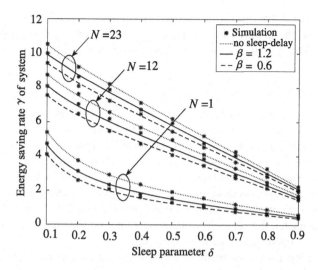

Fig. 20.3 Energy saving rate of system versus sleep parameter

better in guaranteeing the response performance. Compared to the conventional synchronous multi-sleep strategy without a wake-up threshold as results with $N = 1$ in Fig. 20.3, we observe that our proposed strategy is more effective at reducing energy consumption.

Based on the statistical results shown in Figs. 20.2 and 20.3, we conclude that the system performance of the proposed strategy is determined by the system parameters in terms of the sleep parameter, the sleep-delay parameter and the wake-up threshold. For the non-real-time applications where response performance is highly valued, the sleep parameter should be set larger, while the wake-up threshold and the sleep-delay parameter should be set smaller. Conversely, for the non-real-time applications where energy conservation is urgently required, the sleep parameter should be set smaller, and the wake-up threshold and the sleep-delay parameter should be set larger. Thus, a compromise between response performance and energy conservation should be considered when setting system parameters in our proposed task scheduling strategy.

20.5   Performance Optimization

The economic analysis of cloud computing systems has recently been focusing increased attention on cloud providers. In this section, with an aim to providing an enhanced energy efficient strategy in a cloud environment and maintaining the SLA between the cloud providers and the cloud users, we establish a system cost function to improve the system performance.

Let f_1 be the impact factor of the average latency $E[Y_t]$ of tasks on the system cost. Let f_2 be the impact factor of the energy saving rate γ of the system on the system cost. The system cost function F is then given as follows:

$$F = f_1 \times E[Y_t] - f_2 \times \gamma \qquad (20.34)$$

where $E[Y_t]$ and γ are given in Eqs. (20.32) and (20.33), respectively.

Based on the analysis results in Sect. 20.4.1, we note that it is difficult to derive the closed-form solutions for the performance criteria. It is also an arduous task to address the strict monotonicity of the system cost function. The use of traditional optimization algorithms, such as the stochastic gradient descent method, the Lagrangian duality method, and the Gauss-Newton method, are inappropriate for obtaining the optimal system parameters. For the purpose of jointly optimizing the sleep parameter, the sleep-delay parameter and the wake-up threshold, we turn to an improved intelligent searching algorithm.

A Genetic algorithm is a method used to search for the globally optimal solution of objective function by simulating the natural evolutionary process. In a conventional Genetic algorithm, both the crossover probability and the mutation probability are set statically. We note that the fixed crossover probability and the mutation probability will make the Genetic algorithm premature and easy to become trapped in a local optimum. To improve the searching speed, and overcome premature and local convergence, we develop an improved Genetic algorithm by dynamically adjusting the crossover probability and the mutation probability. Furthermore, in order to make the initialization more diverse, we use chaotic equations to initialize the individuals in the population. The main steps for the improved Genetic algorithm are given as follows:

Step 1: Initialize the sleep parameter with the lower bound $\delta_l = 0.1$ and the upper bound $\delta_u = 0.9$, the sleep-delay parameter with the lower bound $\beta_l = 0.1$ and the upper bound $\beta_u = 10$.
Initialize the minimum crossover probability $P_{cl} = 0.001$ and the maximum crossover probability $P_{cm} = 0.1$, the minimum mutation probability $P_{ml} = 0.4$ and the maximum mutation probability $P_{mm} = 0.95$. Set the initial number of evolution generation as $gen = 1$, the maximum evolution generation as $gen_{max} = 50$. Set the initial wake-up threshold as $N = 1$, the maximum wake-up threshold as $N_{max} = 50$.

Step 2: Set the population size as $M = 100$, and initialize each individual as $(\delta, \beta)_i^N$,
$(i = 1, 2, 3, \ldots, M)$ in population S by using chaotic equations.
$(\delta, \beta)_1^N = \text{rand}(2, 1)$
for $i = 2 : M$
　　$(\delta, \beta)_i^N = r \times (\delta, \beta)_{i-1}^N \times \left(1 - (\delta, \beta)_{i-1}^N\right)$
endfor

% rand(2, 1) represents a 2×1 dimensional matrix, the value of the
elements are random between 0 and 1. r is the chaotic factor, $r = 3.85$.

Step 3: For each individual $(\delta, \beta)_i^N (i = 1, 2, 3, \ldots, M)$, calculate the fitness
$F\left((\delta, \beta)_i^N\right)$, the selection probability $P((\delta, \beta)_i^N)$ and the cumulative
probability $C\left((\delta, \beta)_i^N\right)$.

$$F\left((\delta, \beta)_i^N\right) = f_1 E[Y_t] - f_2 \gamma$$

$$P\left((\delta, \beta)_i^N\right) = \frac{F\left((\delta, \beta)_i^N\right)}{\sum\limits_{i=1}^{M} F\left((\delta, \beta)_i^N\right)}$$

$$C\left((\delta, \beta)_i^N\right) = \sum\limits_{j=1}^{i} P\left((\delta, \beta)_i^N\right)$$

% $E[Y_t]$ and γ are the average latency of tasks and the energy saving
rate of the system when the combination of the sleep parameter and the
sleep-delay parameter is $(\delta, \beta)_i^N$, respectively.

Step 4: Calculate the crossover probability P_c and the mutation probability P_m.

$$P_c = P_{cl} - \frac{(P_{cm} - P_{cl}) \times gen}{gen_{max}}$$

$$P_m = P_{ml} + \frac{(P_{mm} - P_{ml}) \times gen}{gen_{max}}$$

Step 5: Perform the genetic operation to update S.

    for $j = 1 : M$

        $slen = \text{selection} \left(S, C((\delta, \beta)_j^N)\right)$

        % Select two individuals with the maximum cumulative probability for
        crossing and mutating.

        $S = \text{crossover}(S, slen, P_c)$

        % Cross the selected individuals.

        $S = \text{mutation}(S, slen, P_m)$

        % Mutate the selected individuals.

    endfor

Step 6: Check the number of evolution generation.

    if $gen < gen_{max}$

        $gen = gen + 1$

        go to **Step 3**

    endif

Step 7: Select the optimal individual among the population S.

$$(\delta, \beta)^N = \underset{i \in \{1,2,3,\cdots,M\}}{\text{argmin}} \left\{ F\left((\delta, \beta)_i^N\right) \right\}$$

$$F((\delta, \beta)^N) = f_1 \times E[Y_t] - f_2 \times \gamma$$

% $(\delta, \beta)^N$ denotes the optimal parameter combination when the wake-up
threshold is N.

Step 8: Check the wake-up threshold N.

    if $N < N_{max}$

Table 20.2 Optimum combination of parameters in proposed strategy

Service rates μ	Optimum combinations (N^*, δ^*, β^*)	Minimum costs $F(N^*, \delta^*, \beta^*)$
0.07	(6, 0.1468, 7.9699)	4.2246
0.1	(6, 0.1269, 9.8577)	2.7236
0.13	(5, 0.1481, 8.5751)	1.7222
0.16	(4, 0.2306, 6.9526)	0.9801
0.19	(4, 0.1357, 6.1439)	0.4045

$$N = N + 1$$
go to **Step 2**
endif

Step 9: Choose the minimum cost in $F((\delta, \beta)^z)$, $z \in [1, N_{max}]$, record the corresponding wake-up threshold N^*, the sleep parameter δ^* and the sleep-delay parameter β^*, constitute the optimal parameter combination (N^*, δ^*, β^*).

Step 10: Output the optimal parameter combination (N^*, δ^*, β^*) and the minimum system cost $F(N^*, \delta^*, \beta^*)$.

By applying the system parameters used in Sect. 20.4.2 into the improved Genetic algorithm and setting the impact factors $f_1 = 0.3$ and $f_2 = 0.4$ as an example, we present numerical results for the optimal combination (N^*, δ^*, β^*) with the different service rates μ in Table 20.2.

20.6 Conclusion

By introducing the sleep-delay parameter and the wake-up threshold, in this chapter, we firstly proposed a sleep mode-based task scheduling strategy. The proposed strategy is representative of real-world cloud scenarios. We modeled the proposed task scheduling strategy as a vacation queueing system. Based on the model analysis in the steady state, we derived the average latency of tasks and the energy saving rate of the system. Taking into account numerous and different cloud services, we extended our presented model to investigate the strategy performance over a more complicated public cloud scenario. Numerical results with analysis and simulation showed that a larger sleep parameter, a smaller wake-up threshold and a smaller sleep-delay parameter may lead to a lower average latency of tasks, while a smaller sleep parameter, a larger wake-up threshold and a larger sleep-delay parameter may result in a higher energy saving rate of the system. Correspondingly, we constructed a system cost function to trade off different performance measures and developed an improved Genetic algorithm to search the optimal parameter combination. Numerical results for performance optimization indicated that an appropriate parameter setting can achieve the minimum cost.

Chapter 21
Energy-Efficient Virtual Machine Allocation Strategy

In this chapter, we propose a Virtual Machine (VM) allocation strategy with a sleep-delay and establish a corresponding mathematical model to achieve greener cloud computing for the open-source cloud platform. Taking into account the number of tasks and the state of the Physical Machine (PM), we construct a two-dimensional Markov chain, and derive performance measures of the system in terms of the average latency of tasks and the energy saving rate of the system. Moreover, we present numerical results to show the effectiveness of the proposed scheme. Furthermore, we study the Nash equilibrium behavior and the socially optimal behavior of tasks and develop an improved Genetic algorithm to obtain the socially optimal arrival rate of tasks. Finally, we propose a pricing policy by imposing an appropriate admission fee for tasks to socially optimize the system performance.

21.1 Introduction

As a commercial infrastructure paradigm, cloud computing has revolutionized the IT industry [Hani17, Sugu17]. However, the energy consumption of cloud computing in WCNs shows a rising trend, while the resources themselves are highly underutilized [Andr14, Krei17]. This presents a bottleneck that restricts the improvement of cloud computing and reveals the great importance of greening the networks.

Consolidation of VMs is an effective technique to minimize the excess energy consumption resulting from the diversity of workload. Many scholars have targeted solving the consolidation problem to improve resource utilization and reduce energy consumption over the cloud environment. In [Fard17], the authors presented a dynamic VM consolidation technique, in which the detections of server overload and server underload were supported. By calculating the deviation between the utilization and the threshold of the overload server, the VMs were consolidated until the number of VMs reached an upper threshold. In [Khos17b], the authors proposed a dynamic and adaptive energy-efficient VM consolidation mechanism by

© Springer Nature Singapore Pte Ltd. 2021
S. Jin, W. Yue, *Resource Management and Performance Analysis of Wireless Communication Networks*, https://doi.org/10.1007/978-981-15-7756-7_21

developing a two-phase VM allocation algorithm for the placement of new VMs and the consolidation of selected VMs. By using the consolidation methodologies mentioned above, energy conservation is achieved to a certain degree. However, the situation of all the VMs staying awake even though no tasks are to be processed remains.

In fact, VMs in a cloud computing system are usually underutilized to guarantee the Quality of Experience (QoE) of users. Extensive studies have consequently been conducted on how to reduce energy consumption during lower workload periods. In [Chen15b], the authors presented an energy saving task scheduling algorithm for a heterogeneous cloud computing system based on a vacation queueing model. In this model, the idle VMs were on vacation, and the vacation was similar to being in a sleep mode. In [Guo16], the authors conducted a theoretical study into the impact of a Base Station (BS) sleeping on both energy-efficiency and user-perceived delay, and presented three typical wake-up schemes in terms of Single Sleep (SS), multiple sleep and N-limited schemes. In [Yang16], the authors proposed a two-stage BS sleeping scheme in cellular networks. The sleeping mode was divided into a light sleeping stage and a deep sleeping stage according to whether there were tasks in the coverage of the BS. With the two-stage sleeping scheme, the BS frequently switched between the on state, the doze state and the shut-down state.

As discussed above, putting idle VMs into sleep mode can reduce the energy consumption. However, additional energy will be consumed and the user performance will be degraded due to the frequent state switches of PMs using a conventional sleep mode. Inspired by these observations, the trade-off between providing higher QoE to users and reducing energy consumption should be addressed.

In this chapter, we propose an energy saving VM allocation strategy with a synchronous multiple-sleep and a sleep-delay. By building a vacation queueing model, we evaluate and optimize the system performance of the proposed strategy.

The main contributions of this chapter can be listed as follows:

(1) We propose an energy saving VM allocation strategy with the constraint of response performance to aim a green cloud computing system. When the system is empty, all the VMs hosted on a PM, and the PM itself, keep awake for a period, rather than instantly switching into the sleep state, so the newly arriving tasks can receive timely service. In this way, the QoE of users can be guaranteed, while the additional energy consumption can be effectively reduced.

(2) We present a method to model the proposed strategy and evaluate the system performance mathematically. We establish a multi-server queueing model to capture the stochastic behavior of tasks in the Cloud Data Center (CDC) with the proposed strategy. By constructing a two-dimensional Markov chain, we evaluate the system performance in terms of the average latency of tasks and the energy saving rate of the system.

(3) We give a pricing policy charging for tasks to optimize the social benefit. Based on the reward for a processed task and the cost of a task waiting in the system buffer, we investigate the Nash equilibrium behavior. Considering also the saved income derived by a cloud service provider due to the energy conservation, we build a revenue function to investigate the socially optimal behavior of tasks. In

order to maximize the social benefit, we develop an improved Genetic algorithm to obtain the socially optimal arrival rate of tasks and impose an appropriate admission fee on tasks.

The chapter is organized as follows. In Sect. 21.2, we describe the energy-efficient VM allocation strategy with the constraint of response performance proposed in this chapter. Then, we present the system model in detail. In Sect. 21.3, we present a performance analysis of the system model, through the analysis of the transition rate matrix and the steady-state distribution. In Sect. 21.4, we obtain performance measures and present numerical results to evaluate the system performance. In Sect. 21.5, we firstly investigate the Nash equilibrium and the socially optimal behaviors of tasks in the energy-efficient VM allocation strategy proposed in this chapter. Then, we propose an appropriate pricing policy to maximize the value of the social benefit function for imposing an appropriate admission fee for tasks. Finally, we draw our conclusions in Sect. 21.6.

21.2 Energy-Efficient Virtual Machine Allocation Strategy and System Model

In this section, we propose an energy-efficient VM allocation strategy with the constraint of response performance within a cloud environment. Then, we present a method to model and evaluate the proposed strategy by constructing a two-dimensional Markov chain.

21.2.1 Energy-Efficient Virtual Machine Allocation Strategy

In a cloud computing system, many PMs constitute the real cloud servers. In order to process multiple tasks simultaneously and maintain an acceptable SLA, several identical VMs will be hosted on one PM. A VM is a software that works like a PM. In an open-source cloud platform, all the VMs hosted on a PM and the PM itself are always awake, even when there are no tasks to be processed. Thus, large amounts of power are wasted.

Sleep modes are intended to minimize the system power usage. In the sleep state, one or more operational components are powered down; only the event monitor stays active with a very low energy consumption. Obviously, by using the sleep mode, the energy consumption can be reduced. However, the system response performance will be degraded. With the aim of trading off the energy consumption against the response performance, we propose an energy-efficient VM allocation strategy with sleep-delay within a cloud environment.

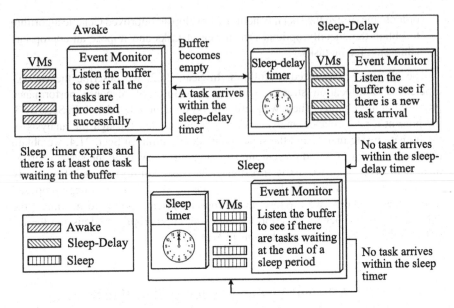

Fig. 21.1 State transition of proposed VM allocation strategy

In the proposed energy-efficient VM allocation strategy, the PM will switch among the awake state, the sleep-delay state and the sleep state. The state transition of the VM allocation strategy proposed in this chapter is illustrated in Fig. 21.1.

(1) *Awake State*: During the awake state of a PM, there is at least one VM busy with task processing. The tasks are processed continuously following a First-Come First-Served (FCFS) discipline. The event monitor deployed in the PM is mainly used for listening to the system to see if all the tasks are processed completely. Once all the tasks in the system have been processed completely, namely, the system becomes empty, under the control of the VM scheduler, all the VMs will remain awake for a period within the sleep-delay timer and be ready for providing service at any time. The PM will change into the sleep-delay state.

(2) *Sleep-Delay State*: Once the PM enters the sleep-delay state, a sleep-delay timer with a random duration will be activated to control the maximum time length of a sleep-delay period. During the sleep-delay state, the event monitor in the PM will listen to the system to see whether there are any new tasks arriving at the system.

If there is a task arrival before the sleep-delay timer expires, under the control of the VM scheduler, all the VMs will provide service immediately for the newly incoming task. The PM will switch to the awake state. In this way, the response performance of tasks will be guaranteed. In this case, the time period of the sleep-delay begins when the sleep-delay timer is activated and finishes the instant a new task arrives at the system.

If there are no task arrivals within the sleep-delay timer, under the control of the VM scheduler, all the VMs will go to sleep at the instant the sleep-delay timer expires. The PM will switch to the sleep state. In this case, the time period of the sleep-delay begins when the sleep-delay timer is activated and concludes when the sleep-delay timer expires.

(3) *Sleep State*: Once the PM enters the sleep state, a sleep timer with a random duration will be activated to control the time length of a sleep period. Tasks arriving within the sleep timer will queue in the buffer. At the end of a sleep period, the event monitor in the PM mainly listens to the buffer to see if there are tasks queueing in the system. If there are no tasks waiting in the system buffer, another sleep timer will be activated. Under the control of the VM scheduler, all the VMs will begin another sleep period. The PM will remain in the sleep state. Otherwise, all the VMs will wake up to serve all the tasks in the system one by one, and the PM will return to the awake state.

In the sleep state, all the VMs in the PM no longer consume memory and CPU. Thus, the energy consumption in the sleep state is lower than that in any other states.

21.2.2 System Model

In order to investigate the influence of arrival behaviors on the system performance under different sleep parameters and sleep-delay parameters, we establish a mathematical model based on the proposed energy-efficient VM allocation strategy.

We model the system with the proposed energy-efficient VM allocation strategy having a sleep-delay shown in Fig. 21.1 as a multi-server queueing system with a synchronous multi-vacation and a vacation-delay. The system buffer is supposed to have an infinite capacity.

We assume that the inter-arrival time of tasks follows an exponential distribution with mean $1/\lambda$, where $\lambda > 0$, called the arrival rate of tasks. We assume that the service time of a task follows an exponential distribution with mean $1/\mu$ seconds, where $\mu > 0$. We call μ the service rate. In addition, we assume the timer lengths of the sleep-delay period and the sleep period follow exponential distributions with means $1/\beta$ seconds and $1/\delta$ seconds, respectively, where $\beta > 0$ and $\delta > 0$. We call β the sleep parameter and δ the sleep-delay parameter.

In this chapter, we focus on all the identical VMs hosted on one PM. We suppose the number of VMs in the system is n and the system capacity is infinite. Let $X(t) = i$ ($i = 0, 1, 2, \ldots$) be the number of tasks in the system at the instant t. $X(t)$ is also called the system level. We let $Y(t) = j$ ($j = 0, 1, 2$) be the PM state at the instant t. $Y(t)$ is called the PM state. $j = 0$ means the PM is in the sleep state, $j = 1$ means the PM is in the awake state, and $j = 2$ means the PM is in the sleep-delay state.

Based on the assumptions above, $\{(X(t), Y(t)), \ t \geq 0\}$ constitutes a two-dimensional Markov chain. The state space of the Markov chain is given as follows:

$$\Omega = \{(0,0), (0,2)\} \cup \{(i,j) : i \geq 1, j = 0, 1, 2\}. \tag{21.1}$$

Let $\pi_{i,j}$ be the probability that the system level is i and the PM state is j in the steady state. $\pi_{i,j}$ is defined as follows:

$$\pi_{i,j} = \lim_{t \to \infty} \Pr\{X(t) = i, \ Y(t) = j\}, \quad (i,j) \in \Omega. \tag{21.2}$$

We define π_i as the probability vector of the system level being equal to i in the steady state. π_i can be partitioned as follows:

$$\pi_i = (\pi_{i,0}, \pi_{i,1}, \pi_{i,2}), \quad i = 0, 1, 2, \ldots. \tag{21.3}$$

The steady-state distribution Π of the two-dimensional Markov chain is composed of π_i $(i \geq 0)$. Π is then given as follows:

$$\Pi = (\pi_0, \pi_1, \pi_2, \ldots). \tag{21.4}$$

21.3 Performance Analysis

In this section, we present a performance analysis of the system model, through the analysis of the transition rate matrix and the steady-state distribution.

21.3.1 Transition Rate Matrix

One of the most important steps in analyzing the steady-state distribution of the Markov chain is to construct the transition rate matrix.

Let Q be the one-step state transition rate matrix of the two-dimensional Markov chain $\{(X(t), Y(t)), \ t \geq 0\}$, and $Q_{x,y}$ be the sub-matrix of Q for the system level changing from x $(x = 0, 1, 2, \ldots)$ to y $(y = 0, 1, 2, \ldots)$. Each sub-matrix $Q_{x,y}$ in the one-step state transition rate matrix Q will be dealt with in detail as follows.

(1) *System Level Increases*: $x = 0$ means that there are no tasks to be processed or being processed in the system. The PM can only be in the sleep state ($j = 0$) or the sleep-delay state ($j = 2$). For the case that the initial PM state is $j = 0$, if a task arrives at the system within the sleep timer, the system level increases by one, but the PM state remains fixed, and the transition rate will be λ. For the case that the initial PM state is $j = 2$, if a task arrives at the system within the

sleep-delay timer, the system level increases by one, and the PM returns to the awake state. The transition rate will also be λ.

Thus, the sub-matrix $Q_{0,1}$ can be written as follows:

$$Q_{0,1} = \begin{pmatrix} \lambda & 0 & 0 \\ 0 & 0 & 0 \\ 0 & \lambda & 0 \end{pmatrix}. \tag{21.5}$$

x ($x \geq 1$) means that there is at least one task in the system. The PM is in the sleep state ($j = 0$) or the awake state ($j = 1$). For these two cases, if a task arrives at the system, the system level increases by one, but the PM state remains fixed, and the transition rate will be λ.

Thus, the sub-matrix $Q_{x,x+1}$ can be written as follows:

$$Q_{x,x+1} = \begin{pmatrix} \lambda & 0 & 0 \\ 0 & \lambda & 0 \\ 0 & 0 & 0 \end{pmatrix}, \quad x \geq 1. \tag{21.6}$$

(2) *System Level Remains Fixed*: $x = 0$ means that the PM can only be in the sleep state ($j = 0$) or the sleep-delay state ($j = 2$). For the case that the initial PM state is $j = 0$, if the sleep timer expires and there are no tasks waiting at the system buffer, the PM will enter another sleep period, both the system level and the PM state will remain unchanged, and the transition rate will be $-\lambda$. For the case that the initial PM state is $j = 2$, if the sleep-delay timer expires and there are no tasks waiting at the system, the system level remains fixed, but the PM changes to the sleep state, and the transition rate is β. If there are no tasks arriving at the system within the sleep-delay timer, both the system level and the PM state remain fixed, and the transition rate will be $-(\lambda + \beta)$.

Thus, the sub-matrix $Q_{0,0}$ can be written as follows:

$$Q_{0,0} = \begin{pmatrix} -\lambda & 0 & 0 \\ 0 & 0 & 0 \\ \beta & 0 & -(\lambda + \beta) \end{pmatrix}. \tag{21.7}$$

x ($1 \leq x < n$) means that in the system, the number of tasks is less than the number of VMs. The PM can only be in the sleep state ($j = 0$) or the awake state ($j = 1$). For the case that the initial PM state is $j = 0$, if there are no tasks arriving at the system buffer within the sleep timer, both the system level and the PM state remain fixed, and the transition rate will be $-(\lambda + \delta)$. If the sleep timer expires and there is at least one task waiting in the system buffer, the system level remains fixed, but the PM changes to the awake state, and the transition rate will be δ. For the case that the initial PM state is $j = 1$, if neither an arrival nor a departure occurs, both the system level and the PM state remain unchanged, and the transition rate will be $-(\lambda + x\mu)$.

Thus, the sub-matrix $\boldsymbol{Q}_{x,x}$ can be written as follows:

$$\boldsymbol{Q}_{x,x} = \begin{pmatrix} -(\lambda+\delta) & \delta & 0 \\ 0 & -(\lambda+x\mu) & 0 \\ 0 & 0 & 0 \end{pmatrix}, \quad 1 \le x < n. \tag{21.8}$$

x $(x \ge n)$ means that in the system, the number of tasks is greater than the number of VMs. The PM can only be in the sleep state $(j = 0)$ or the awake state $(j = 1)$. For the case that the initial PM state is $j = 0$, the corresponding transition rates are the same as that given in Eq. (21.8). For the case that the initial PM state is $j = 1$, if neither an arrival nor a departure occurs, both the system level and the PM state remain fixed, and the transition rate will be $-(\lambda + n\mu)$.

Thus, the sub-matrix $\boldsymbol{Q}_{x,x}$ can be written as follows:

$$\boldsymbol{Q}_{x,x} = \begin{pmatrix} -(\lambda+\delta) & \delta & 0 \\ 0 & -(\lambda+n\mu) & 0 \\ 0 & 0 & 0 \end{pmatrix}, \quad x \ge n. \tag{21.9}$$

(3) *System Level Decreases*: In the case where the system level is decreased, the initial PM state can only be in the awake state $(j = 1)$.

$x = 1$ means that there is one task being processed in the system. If the only task in the system is completely processed, the system level decreases by one, and the PM switches to the sleep-delay state, and the transition rate will be μ.

Thus, the sub-matrix $\boldsymbol{Q}_{1,0}$ can be written as follows:

$$\boldsymbol{Q}_{1,0} = \begin{pmatrix} 0 & 0 & 0 \\ 0 & 0 & \mu \\ 0 & 0 & 0 \end{pmatrix}. \tag{21.10}$$

x $(1 < x \le n)$ means that all the tasks are being processed and the system buffer is empty. If one of the tasks is completely processed, the system level decreases by one, but the PM state remains fixed, and the transition rate will be $x\mu$.

Thus, the sub-matrix $\boldsymbol{Q}_{x,x-1}$ can be written as follows:

$$\boldsymbol{Q}_{x,x-1} = \begin{pmatrix} 0 & 0 & 0 \\ 0 & x\mu & 0 \\ 0 & 0 & 0 \end{pmatrix}, \quad 1 < x \le n. \tag{21.11}$$

x $(x > n)$ means that there are n tasks being processed and $(x - n)$ tasks waiting in the system buffer. If one of the tasks is completely processed, the first task queueing in the system buffer will occupy the just evacuated VM to receive service. The system level decreases by one, but the PM state remains fixed, and the transition rate will be $n\mu$.

Thus, the sub-matrix $Q_{x,x-1}$ can be written as follows:

$$Q_{x,x-1} = \begin{pmatrix} 0 & 0 & 0 \\ 0 & n\mu & 0 \\ 0 & 0 & 0 \end{pmatrix}, \quad x > n. \qquad (21.12)$$

Now, all the sub-matrices in Q have been addressed. For convenience of description, we denote $Q_{x,x}$ as A_x, $Q_{x,x-1}$ as B_x and $Q_{x,x+1}$ as C_x, where $0 \le x < n$. We note that starting from the system level n, all the sub-matrices of Q are repeated forever. Therefore, we denote the repetitive A_x, B_x and C_x as A, B and C, respectively. Using the lexicographical sequence for the system level and the system state, the transition rate matrix for the Markov chain is given as follows:

$$Q = \begin{pmatrix} A_0 & C_0 & & & & \\ B_1 & A_1 & C_1 & & & \\ & \ddots & \ddots & \ddots & & \\ & & B_{n-1} & A_{n-1} & C_{n-1} & \\ & & & B & A & C \\ & & & & \ddots & \ddots & \ddots \end{pmatrix}. \qquad (21.13)$$

From the matrix structure of Q given in Eq. (21.13), we find that the system state transitions occur only between adjacent levels. Hence, the two-dimensional Markov chain $\{(X(t), Y(t)), t \ge 0\}$ can be seen as a Quasi Birth-Death (QBD) process.

21.3.2 Steady-State Distribution

To analyze this QBD process, we need to solve for the minimal non-negative solution of the following matrix quadratic equation:

$$R^2 B + R A + C = 0. \qquad (21.14)$$

The solution R is called a rate matrix. The rate matrix R plays a crucial role in the analysis of the mathematical model. Under the condition that $\rho = \lambda(n\mu)^{-1} < 1$, Eq. (21.14) has a minimal non-negative solution R.

Since the coefficient matrices of Eq. (21.14) are all upper triangular, and all the elements of the third row and the third column are zero, the rate matrix R can be written in the following form:

$$R = \begin{pmatrix} r_{11} & r_{12} & 0 \\ 0 & r_{22} & 0 \\ 0 & 0 & 0 \end{pmatrix}. \tag{21.15}$$

Substituting Eq. (21.15) into Eq. (21.14) yields the following set of equations:

$$\begin{cases} n\mu(r_{11}r_{12} + r_{12}r_{22}) + \delta r_{11} - (\lambda + n\mu)r_{12} = 0 \\ n\mu r_{22}^2 - (\lambda + n\mu)r_{22} + \lambda = 0 \\ -(\lambda + \delta)r_{11} + \lambda = 0. \end{cases} \tag{21.16}$$

By solving Eq. (21.16), we can obtain $r_{11} = \lambda/(\lambda + \delta)$, $r_{12} = \rho$ and $r_{22} = \rho$. For this case, $Sp(R) = \max\{\lambda/(\lambda + \delta), \rho\} < 1$. Thus, the rate matrix R is derived as follows:

$$R = \begin{pmatrix} \dfrac{\lambda}{\lambda + \delta} & \rho & 0 \\ 0 & \rho & 0 \\ 0 & 0 & 0 \end{pmatrix}. \tag{21.17}$$

To use a matrix-geometric solution method, the transition rate matrix Q of the QBD process is re-partitioned as follows:

$$Q = \begin{pmatrix} H_{0,0} & H_{0,1} & & \\ H_{1,0} & A & C & \\ & B & A & C \\ & & \ddots & \ddots & \ddots \end{pmatrix} \tag{21.18}$$

where

$$H_{0,0} = \begin{pmatrix} A_0 & C_0 & & & \\ B_1 & A_1 & C_1 & & \\ & \ddots & \ddots & \ddots & \\ & & B_{n-2} & A_{n-2} & C_{n-2} \\ & & & B_{n-1} & A_{n-1} \end{pmatrix}_{3n \times 3n},$$

$$H_{1,0} = \begin{pmatrix} \mathbf{0} & B \end{pmatrix}_{3 \times 3n}, \quad H_{0,1} = \begin{pmatrix} \mathbf{0} \\ C \end{pmatrix}_{3n \times 3}.$$

Using the rate matrix R obtained from Eq. (21.17), we give a square matrix as follows:

$$B[R] = \begin{pmatrix} H_{00} & H_{01} \\ H_{10} & A + RB \end{pmatrix}.$$
(21.19)

Then, π_0 and π_1 satisfy the following set of linear equations:

$$\begin{cases} (\pi_0, \pi_1) B[R] = 0 \\ \pi_0 e + \pi_1 (I - R)^{-1} e = 1 \end{cases}$$
(21.20)

where e is a column vector with 3 elements and all elements of the vector are equal to 1.

Based on Eq. (21.20), we further construct an augmented matrix as follows:

$$(\pi_0, \pi_1) \left(B[R] \Big| \begin{matrix} e \\ (I - R)^{-1} e \end{matrix} \right) = \underbrace{(0, 0, 0, \ldots, 0, 1)}_{3(n+1)}.$$
(21.21)

By using the Gauss-Seidel method, we can obtain π_0 and π_1.

From the structure of the one-step state transition rate matrix Q given in Eq. (21.18), we know that π_i $(i = 2, 3, 4, \ldots)$ satisfies the matrix-geometric solution form as follows:

$$\pi_i = \pi_1 R^{i-1}, \quad i \geq 2.$$
(21.22)

By substituting π_1 obtained in Eq. (21.21) into Eq. (21.22), we can obtain π_i $(i = 2, 3, 4, \ldots)$. Then, the steady-state distribution $\Pi = (\pi_0, \pi_1, \pi_2, \ldots)$ of the system can be presented numerically.

21.4   Performance Measures and Numerical Results

In this section, we first derive performance measures of the system in terms of the average latency of tasks and the energy saving rate of the system, respectively. Then, we present numerical results to evaluate the performance of the system using the energy-efficient VM allocation strategy proposed in this chapter.

21.4.1   Performance Measures

We define the latency Y_t of a task as the duration from the instant a task arrives at the system to the instant when that task is completely processed. In other words, the latency of a task is the sum of the waiting time in the system buffer and the service time on the VM. The average latency of tasks is an important factor in measuring the QoE of cloud users.

Based on the steady-state distribution given in Sect. 21.3.2, we obtain the average latency $E[Y_t]$ of tasks as follows:

$$E[Y_t] = \frac{1}{\lambda} \sum_{i=0}^{\infty} i(\pi_{i,0} + \pi_{i,1} + \pi_{i,2}). \qquad (21.23)$$

We define the energy saving rate γ of the system as the energy conservation per second for our proposed energy-efficient saving VM allocation strategy. During the awake state and the sleep-delay state, energy will be consumed normally, while during the sleep state, energy can be saved. However, additional energy will be consumed when the PM switches from the sleep state to the awake state, and the VMs listen to the system buffer at the boundary of each sleep period.

Let g_1 be the energy consumption per second in the awake state and the sleep-delay state, and g_2 be the energy consumption per second in the sleep state. Let g_3 be the energy consumption for each switching from the sleep state to the awake state, and g_4 be the energy consumption for each listening. Without loss of generality, we suppose $g_1 > g_2$ in this chapter. Then, we give the energy saving rate γ of the system as follows:

$$\gamma = (g_1 - g_2) \sum_{i=0}^{\infty} \pi_{i,0} - g_3 \sum_{i=1}^{\infty} \pi_{i,0} \times \delta - g_4 \sum_{i=0}^{\infty} \pi_{i,0}\delta. \qquad (21.24)$$

21.4.2 Numerical Results

In order to investigate the impact of the system parameters on the performance of the system with the proposed energy-efficient VM allocation strategy, we present numerical results with analysis and simulations. All the experiments are performed on a computer configured with Intel (R) Core (TM) i7-4790 CPU @ 3.6 GHz, 6 GB RAM and 500G disk. Under the Matlab platform, the analysis results are calculated based on Eqs. (21.23) and (21.24). All the simulation results are obtained using MyEclipse 2014 with an average of 20 different runs. We create a JOB class with attributes in terms of UNARRIVE, WAIT, RUN and FINISH to record the task state. We also create a SERVER class with attributes in terms of BUSY, IDLE, SLEEP-DELAY and SLEEP to record the VM state. Good agreements between the analysis results and the simulation results are observed. With the help of real implementations, several values of parameters are assumed in the numerical results.

As an example in numerical results, the number of VMs is fixed as $n = 20$, the mean service time is assumed to be $\mu = 0.2 \text{ s}^{-1}$, the arrival rate is supposed to be a variable between 0.1 and 3.9 tasks per second. In addition, when investigating the change trend for the energy saving rate γ of the system, the presumptions are that $g_1 = 20$ mW, $g_2 = 4$ mW, $g_3 = 2$ mW and $g_4 = 12$ mW. However, all the values assumed above can be easily replaced by those from other real systems. They are merely utilized to demonstrate how the mathematical model works. In all of the following figures, the analytical and simulation results are shown by lines and markers, respectively. As seen in Figs. 21.2 and 21.3, the analytical and simulation results are in strong agreement.

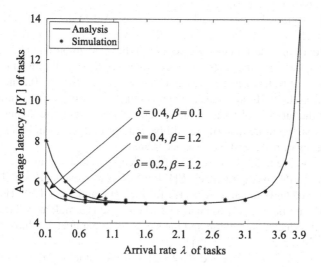

Fig. 21.2 Average latency of tasks versus arrival rate of tasks

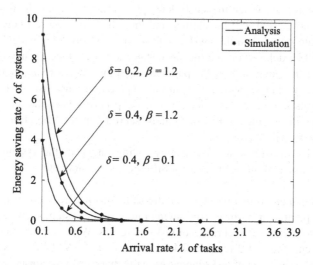

Fig. 21.3 Energy saving rate of system versus arrival rate of tasks

Figure 21.2 illustrates how the average latency $E[Y_t]$ of tasks changes in relation to the arrival rate λ of tasks for the different sleep parameters δ and sleep-delay parameters β.

As can be seen from Fig. 21.2, for the same sleep parameter δ and the same sleep-delay parameter β, as the arrival rate λ of tasks increases, the average latency $E[Y_t]$ of tasks firstly decreases slightly, then stabilizes at a certain value, and finally presents as an uptrend. When the arrival rate of tasks is smaller (such as $\lambda < 1.3$ for $\delta = 0.2$ and $\beta = 1.2$), as the arrival rate of tasks increases, the VMs are less likely to be asleep, so the service of a task is less likely to be delayed in the sleep state, and the average latency of tasks will decrease. When the arrival rate of tasks is moderate (such as $1.3 \leq \lambda \leq 2.5$ for $\delta = 0.2$ and $\beta = 1.2$), it is more likely that all the VMs are awake and no tasks are waiting in the system buffer. The latency of a task is only its service time, so the average latency of tasks will remain fixed. When the arrival rate of tasks further increases (such as $\lambda > 2.5$ for $\delta = 0.2$ and $\beta = 1.2$), even though all the VMs are awake, the processing ability of system is not sufficient to cater for the number of tasks in the system. For this case, the higher the arrival rate of tasks is, the more tasks there will be waiting in the system buffer. Thus, the average latency of tasks will increase accordingly.

We also observe that the influences of the sleep parameter δ and the sleep-delay parameter β on the average latency $E[Y_t]$ of tasks are reasonably small for a high arrival rate λ of tasks (such as $\lambda > 1.3$ for $\delta = 0.2$ and $\beta = 1.2$). The reason is that the VMs are more likely to be awake when there is a high arrival rate of tasks. Contrarily, the sleep parameter δ and the sleep-delay parameter β have remarkable influence on the average latency $E[Y_t]$ of tasks for a smaller arrival rate λ of tasks (such as $\lambda \leq 1.3$ for $\delta = 0.2$ and $\beta = 1.2$). For the same sleep parameter and the same arrival rate of tasks, the average latency of tasks increases as the sleep-delay parameter increases. When the sleep-delay parameter increases, the time length of the sleep-delay period will be shorter, and then the PM is more likely to enter to the sleep state rather than the awake state from the sleep-delay state. The process of a task will be delayed during the sleep state, so the average latency of tasks will increase. For the same sleep-delay parameter and the same arrival rate of tasks, the average latency of tasks decreases as the sleep parameter increases. When the sleep parameter gets larger, the time length of a sleep period gets shorter, namely, the tasks arriving during the sleep period can be served quicker. Therefore, the average latency of tasks will decrease.

Figure 21.3 illustrates how the energy saving rate γ of the system changes in relation to the arrival rate λ of tasks for different sleep parameters δ and sleep-delay parameters β.

As can be seen from Fig. 21.3, for the same sleep parameter δ and the same sleep-delay parameter β, the energy saving rate γ of the system shows a downtrend, decreasing to 0 as the arrival rate λ of tasks increases. When the arrival rate of tasks is lower (such as $\lambda \leq 1.3$ for $\delta = 0.4$ and $\beta = 1.2$), the VMs are less likely to be asleep as the arrival rate of tasks increases. Note that energy is conserved during the sleep state, so the energy saving rate of the system will decrease. When the arrival rate of tasks is higher (such as $\lambda > 1.3$ for $\delta = 0.4$ and $\beta = 1.2$), the energy saving rate of the system stabilizes at 0. For a higher arrival rate of tasks, the VMs are more likely to be awake. There is no energy conservation during the awake state, so the energy saving degree of the system reduces to 0.

We also observe that the sleep parameter δ and the sleep-delay parameter β have little or no impacts on the energy saving rate γ of the system for a higher arrival rate λ of tasks (such as $\lambda > 1.3$ for $\delta = 0.4$ and $\beta = 1.2$). As the arrival rate of tasks increases, all the VMs are more likely to be awake all the time, so no energy will be conserved. On the other hand, the sleep parameter δ and the sleep-delay parameter β have remarkable impacts on the energy saving rate γ of the system for a lower arrival rate λ of tasks (such as $\lambda \leq 1.3$ for $\delta = 0.4$ and $\beta = 1.2$). For the same sleep parameter and the same arrival rate of tasks, the energy saving rate of the system increases as the sleep-delay parameter increases. When the sleep-delay parameter increases, the PM will easily switch into the sleep state from the sleep-delay state, so the energy saving rate of the system will increase. For the same sleep-delay parameter and the same arrival rate of tasks, the energy saving rate of the system decreases as the sleep parameter increases. When the sleep parameter becomes greater, the time length of a sleep period gets shorter, which leads to a lower energy saving rate of the system.

Figures 21.2 and 21.3 confirm that a moderate arrival rate of tasks leads to the lowest average latency of tasks, while a lower arrival rate of tasks leads to the highest energy saving degree of the system. Determining how to regulate the arrival rate of tasks by trading off against the average latency of tasks and the energy saving rate of the system is an important issue to be addressed.

21.5   Analysis of Admission Fee

In this section, we first investigate the Nash equilibrium behavior and socially optimal behavior of tasks in the energy-efficient VM allocation strategy proposed in this chapter. Then, we propose a pricing policy for tasks to optimize the system performance socially. This issue can be addressed by imposing an appropriate admission fee for tasks.

21.5.1   Behaviors of Nash Equilibrium and Social Optimization

In the energy-efficient VM allocation strategy proposed in this chapter, all the tasks make decisions independently to access the system and get service quickly for the purpose of gaining a certain benefit. A task that joins the queue may cause future tasks to spend more time in the system. That is to say, the selfish behavior of all the tasks will lead to a higher arrival rate of tasks. However, a higher arrival rate of tasks will increase the average latency of tasks, and a higher latency of tasks will reduce the individual benefit to the point of creating a negative value. Without loss of generality, the decision to join or balk the system should be made from a social point of view.

In order to investigate the Nash equilibrium behavior and motivate the cloud users to accept the socially optimal behavior, we present some hypotheses as follows.

(1) The number of all the tasks, including the tasks queueing in the system buffer and the tasks being processed in the VMs, is unobservable for newly arriving tasks. But each task has to make a decision to join or balk the system. The tasks are risk neutral, namely, a decision to join the system is irrevocable, and reneging is not allowed.

(2) The reward of a task for completed service is R_{g1}. The cost of a task staying in the system is C_g per second. The income of the Cloud Service Provider (CSP) from energy conservation is R_{g2} per milliwatt. The benefits for the whole system, including all the tasks and the CSP, can be added together.

(3) In order to assure a positive benefit for an arriving task entering an empty system, we assume that

$$R_{g1} > \frac{C_g}{\mu} + \frac{C_g}{\delta}. \tag{21.25}$$

The individual benefit of a task is the difference between the reward for completed service and the cost for staying in the system. The individual benefit function $G_{ind}(\lambda)$ is then given as follows:

$$G_{ind}(\lambda) = R_{g1} - C_g E[Y_t]. \tag{21.26}$$

The social benefit of the system is the aggregation of the individual benefits of all the tasks and the income of the CSP. Thus, the social benefit function $G_{soc}(\lambda)$ can be calculated as follows:

$$G_{soc}(\lambda) = \lambda \times (R_{g1} - C_g E[Y_t]) + R_{g2} \times \gamma. \tag{21.27}$$

By applying system parameters given in Sect. 21.4.2, and setting $\delta = 0.3$, $\beta = 1.2$, $R_{g1} = 10$, $C_g = 1$ and $R_{g2} = 1$ as an example, we present numerical results to explore the changing trends of the individual benefit function $G_{ind}(\lambda)$ and the social benefit function $G_{soc}(\lambda)$ in relation to the arrival rate λ of tasks for different service rates μ of tasks in Figs. 21.4 and 21.5, respectively.

As can be seen from Fig. 21.4, for all the service rate μ of tasks, there is a unique arrival rate λ of tasks (marked with "■") subject to $G_{ind}(\lambda) = 0$. We call the unique value the Nash equilibrium arrival rate λ_e of tasks. When $\lambda < \lambda_e$, the newly arriving tasks is serviced quickly, so the cost of a task staying in the system is lower than its reward for a completed service. For this case, the individual benefit is positive. When $\lambda > \lambda_e$, the number of tasks waiting in the system buffer increases unbelievably, so

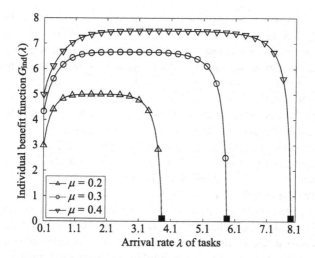

Fig. 21.4 Individual benefit function versus arrival rate of tasks

the cost of a task staying in the system is higher than its reward for a completed service. For this case, the individual benefit is negative.

We also find that in all the curves of the social benefit function $G_{soc}(\lambda)$ in Fig. 21.5, there is a unique arrival rate λ of tasks (marked with "□") subject to the maximum social benefit $G_{soc}(\lambda^*)$ for all the service rates μ of tasks. We call this unique value the socially optimal arrival rate λ^* of tasks. When the arrival rate of tasks is smaller (such as $\lambda < 0.4$ for $\mu = 0.2$), as the arrival rate of tasks increases, the individual benefit increases, whereas the energy saving rate of the

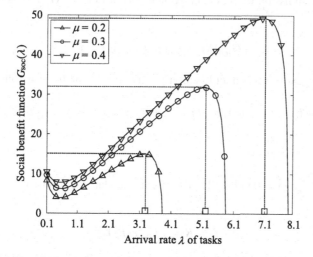

Fig. 21.5 Social benefit function versus arrival rate of tasks

system decreases. The uptrend of the individual benefit is slower than the downtrend of the energy saving rate of the system, so the social benefit function displays a downtrend. When the arrival rate of tasks becomes higher (such as $0.4 \leq \lambda \leq 2.5$ for $\mu = 0.2$), as the arrival rate of tasks increases, the individual benefit function tends to be fixed after a slow rise, whereas the energy saving rate of the system decreases continuously, so the social benefit function generally shows an uptrend. When the arrival rate of tasks increases to an even higher level (such as $\lambda > 2.5$ for $\mu = 0.2$), as the arrival rate of tasks further increases, the individual benefit decreases whereas the energy saving rate of the system stabilizes at 0, so the social benefit function displays a downtrend. Therefore, there is a peak value $G_{soc}(\lambda^*)$ for the social benefit function with the socially optimal arrival rate λ^* of tasks for all the service rates μ of tasks.

Comparing the numerical results of the individual benefit function $G_{ind}(\lambda)$ in Fig. 21.4 and the social benefit $G_{soc}(\lambda)$ in Fig. 21.5, we find that the Nash equilibrium arrival rate λ_e of tasks is greater than the socially optimal arrival rate λ^* of tasks. That is to say, the Nash equilibrium behavior cannot lead to social optimization. In order to motivate tasks to accept the socially optimal strategy, a reasonable fee should be imposed on tasks to restrain their enthusiasm on receiving the cloud service.

21.5.2 Pricing Policy

In order to align the Nash equilibrium arrival rate λ_e of tasks with the socially optimal arrival rate λ^* of tasks, we present a pricing policy whereby tasks are charged an appropriate admission fee f.

With the admission fee f, the individual benefit function $G'_{ind}(\lambda)$ is modified as follows:

$$G'_{ind}(\lambda) = R_{g1} - C_g E[Y_t] - f. \tag{21.28}$$

Accordingly, the social benefit function $G'_{soc}(\lambda)$ can be modified as follows:

$$G'_{soc}(\lambda) = \lambda(R_{g1} - C_g \times E[Y_t] - f) + \lambda f + R_{g2}\gamma = \lambda \times (R_{g1} - C_g E[Y_t]) + R_{g2} \times \gamma.$$

$$\tag{21.29}$$

Setting $G'_{ind}(\lambda) = 0$ in Eq. (21.28), we can calculate the admission fee f of tasks as follows:

$$f = R_{g1} - C_g \times E[Y_t]. \tag{21.30}$$

It is arduous work to address the strict monotonicity of the social benefit function $G_{\text{soc}}(\lambda)$ of the system based on Eqs. (21.27) and (21.29). This implies that with the traditional optimization method, it is difficult to obtain the socially optimal arrival rate λ^* of tasks and the maximum social benefit $G_{\text{soc}}(\lambda^*)$. For this, we turn to an intelligent optimization algorithm to search for the socially optimal arrival rate λ^* of tasks and the maximum social benefit $G_{\text{soc}}(\lambda^*)$.

A Genetic algorithm is a method used to search for a globally optimal solution of objective function by simulating the natural evolutionary process. In a conventional Genetic algorithm, both the crossover probability and the mutation probability are fixed. We note that when these probabilities are higher, the Genetic algorithm will become a random algorithm. Contrarily, when these probabilities are lower, the Genetic algorithm will converge quite slowly. For this reason, we develop an improved Genetic algorithm by dynamically adjusting the crossover probability and the mutation probability. The main steps for the improved Genetic algorithm are given as follows:

Step 1: Initialize the search space with the upper boundary λ^u and the lower boundary λ^l, the population size N, the minimum and maximum crossover probability P_{cl}, P_{cm}, the minimum and maximum mutation probability P_{ml}, P_{mm}. Set the initial number of evolution generations as $gen = 1$, and the maximum evolution generation as $gen_{\max} = 50$.

Step 2: Initialize the population $S = \{\lambda_1, \lambda_2, \lambda_3, \ldots, \lambda_N\}$, the constraint condition $[\lambda^l, \lambda^u]$.

Step 3: For each individual λ_i, $i \in \{1, 2, 3, \ldots, N\}$, calculate the fitness $G_{\text{soc}}(\lambda_i)$, the selection probability P_i and the cumulative probability F_i.

$$G_{\text{soc}}(\lambda_i) = \lambda_i(R_{g1} - C_g E[Y_t])$$

$$P_i = \frac{G_{\text{soc}}(\lambda_i)}{\displaystyle\sum_{i=1}^{N} G_{\text{soc}}(\lambda_i)}$$

$$F_i = \sum_{j=1}^{i} P_j$$

Step 4: Calculate the crossover probability P_c and the mutation probability P_m.

$$P_c = P_{\text{cl}} - \frac{(P_{\text{cm}} - P_{\text{cl}}) \times gen}{gen_{\max}}$$

$$P_m = P_{\text{ml}} + \frac{(P_{\text{mm}} - P_{\text{ml}}) \times gen}{gen_{\max}}$$

Step 5: Perform the genetic operation to update the population S.

    for $j = 1 : N$

        $slen = \text{selection}(S, F_i)$

        % Select two individuals to cross and mutate.

        $S = \text{crossover}(S, slen, P_c)$

        % Cross the selected individuals.

Table 21.1 Numerical results for admission fee

Service rates μ	Optimum arrival rates λ^* of tasks	Admission fees f
0.2	3.2606	4.6058
0.3	5.1785	6.1496
0.4	7.0318	7.0082

       $S = \text{mutation}(S, slen, P_m)$
       % Mutate the selected individuals.
    endfor
Step 6: Check the number of evolution generations.
    if $gen < gen_{\max}$
       $gen = gen + 1$
       go to **Step 3**
    endif
Step 7: Select the optimum individual among the population S.
$$\lambda^* = \underset{i \in \{1,2,3,\dots,N\}}{\text{argmax}} \ \{G_{\text{soc}}(\lambda_i)\}$$
$$G_{\text{soc}}(\lambda_i^*) = \lambda^*(R_{g1} - C_g E[Y_t](\lambda^*))$$
Step 8: Output λ^* and $G_{\text{soc}}(\lambda_i^*)$.

Applying the parameters used in Sect. 21.5.1 into the improved Genetic algorithm, we can obtain the socially optimal arrival rate λ^* of tasks with different service rates μ of tasks. Substituting the socially optimal arrival rate λ^* of tasks into Eq. (21.23), we calculate the average latency $E[Y_t]$ of tasks. Furthermore, we can obtain the admission fee f of tasks with Eq. (21.30). The numerical results of the socially optimal arrival rate λ^* of tasks and the admission fee f of tasks are shown in Table 21.1. All the values of λ^* and f are exactly to four decimal places.

From Table 21.1, we observe that the larger the service rate μ of tasks is, the higher the socially optimal arrival rate λ^* of tasks is, and the higher the admission fee f is. As the service rate μ of tasks increases, the time for a task's service to be completed becomes shorter, and the waiting cost is lowered. This rouses the enthusiasm of tasks to join the system, so the admission fee f should be set higher.

21.6   Conclusion

In this chapter, we focused on how to achieve greener cloud computing under the constraints of a SLA. For this purpose, we proposed an energy-efficient VM allocation strategy with the constraint of response performance. We presented a method to model and evaluate the proposed strategy by constructing a two-dimensional Markov chain. The proposed model quantified the effects of changes to different sets of parameters, such as the sleep parameter and the sleep-delay parameter. These effects were evaluated using two important performance measures in terms of the average latency of tasks and the energy saving rate of the system. With numerical results, the impacts of the system parameters on the system performance were revealed and the effectiveness of the proposed strategy was validated. Moreover, after we investigated the Nash equilibrium and the socially optimal behaviors of tasks, we proposed an appropriate pricing policy to maximize the value of the social benefit function. This issue can be addressed by imposing an appropriate admission fee for tasks.

References

[3GPP] Project Co-ordination Group: 3GPP TS 36.321, Technical Specification Group Radio Access Network, Evolved Universal Terrestrial Radio Access (E-UTRA): Medium Access Control (MAC) Protocol Specification (2008)

[Abdu14] Abdulhamid, S., Latiff, M., Bashir, M.: On-demand grid provisioning using cloud infrastructures and related virtualization tools: A survey and taxonomy. Int. J. Adv. Stud. Comput. Sci. Eng. 3(1), 49–59 (2014)

[Abdu16] Abdullahi, M., Ngadi, M., Abdulhamid, S.: Symbiotic organism search optimization based task scheduling in cloud computing environment. Futur. Gener. Comput. Syst. 56, 640–650 (2016)

[Abed17] Abedi, M., Mokari, N., Javan, M., Yanikomeroglu, H.: Secure communication in OFDMA-based cognitive radio networks: An incentivized secondary network coexistence approach. IEEE Trans. Veh. Technol. 66(2), 1171–1185 (2017)

[Abet10] Abeta, S.: Toward LTE commercial launch and future plan for LTE enhancements (LTE-advanced). In: Proceedings of the IEEE International Conference on Communication Systems, pp. 146–150 (2010)

[Ahn18] Ahn, S., Lee, J., Park, S., Newaz, S., Choi, J.: Competitive partial computation offloading for maximizing energy efficiency in mobile cloud computing. IEEE Access 6, 899–912 (2018)

[Akyi06] Akyildiz, I., Lee, W., Vuran, M.: Next generation/dynamic spectrum access/cognitive radio wireless networks: A survey. Computer Networks 50(13), 2127–2159 (2006)

[Alfa10] Alfa, A.: Queueing Theory for Telecommunications: Discrete Time Modeling of a Single Node System. Springer (2010)

[Altr14] Altrad, O., Muhaidat, S., Al-Dweik, A., Shami, A., Yoo, P.: Opportunistic spectrum access in cognitive radio networks under imperfect spectrum sensing. IEEE Trans. Veh. Technol. 63(2), 920–925 (2014)

[Anan14] Anand, S., Sengupta, S., Hong, K.: Exploiting channel fragmentation and aggregation/bonding to create security vulnerabilities. IEEE Trans. Veh. Technol. 63(8), 3867–3874 (2014)

[Andr14] Andrews, J., Buzzi, S., Choi, W., Hanly, S., Lozano, A., Soong, A., Zhang, J.: What will 5G be?. IEEE J. Sel. Areas Commun. 32(16), 1065–1082 (2014)

[Aria17] Arianyan, E., Taheri, H., Khoshdel, V.: Novel fuzzy multi objective DVFS-aware consolidation heuristics for energy and SLA efficient resource management in cloud data centers. J. Netw. Comput. Appl. 78, 43–61 (2017)

[Atma13] Atmaca, S.: Improving TDMA channel utilization in random access cognitive radio networks by exploiting slotted CSMA. Wirel. Pers. Commun. **71**(4), 2417–2430 (2013)

[Bae10] Bae, Y., Alfa, A., Choi, B.: Performance analysis of modified IEEE 802.11-based cognitive radio networks. IEEE Commun. Lett. **14**(10), 975–977 (2010)

[Baek11a] Baek, S., Choi, B.: Performance of an efficient sleep mode operation for IEEE 802.16m. J. Ind. Manag. Optim. **7**(3), 623–639 (2011)

[Baek11b] Baek, S., Choi, B.: Analysis of discontinuous reception with both downlink and uplink packet arrivals in 3GPP LTE. J. Ind. Manag. Optim. **8**(4), 841–860 (2011)

[Bala14] Balapuwaduge, I., Lei, J., Pla, V., Li, Y.: Channel assembling with priority-based queues in cognitive radio networks: Strategies and performance evaluation. IEEE Trans. Wirel. Commun. **13**(2), 630–645 (2014)

[Behe15] Behera, S., Seth, D.: Efficient resource allocation in cognitive radio network under imperfect spectrum sensing and unsecured environment. In: Proceedings of the IEEE International Conference on Electrical, Electronics, Signals, Communication and Optimization, pp. 1–5 (2015). https://doi.org/10.1109/EESCO.2015.7253850

[Bhow14] Bhowmick, A., Das, M., Biswas, J., Roy, S., Kundu, S.: Throughput optimization with cooperative spectrum sensing in cognitive radio network. In: Proceedings of the 4th IEEE International Advance Computing Conference, pp. 329–332 (2014)

[Boch14] Bochechka, G., Tikhvinskiy, V.: Spectrum occupation and perspectives millimeter band utilization for 5G networks. In: Proceedings of the ITU Kaleidoscope Academic Conference: Living in a Converged World-Impossible Without Standards?, pp. 69–72 (2014)

[Bont09] Bontu, C., Illidgee, E.: DRX mechanism for power saving in LTE. IEEE Commun. Mag. **47**(6), 48–55 (2009)

[Budd19] Buddala, R., Mahapatra, S.: An integrated approach for scheduling flexible job-shop using teaching-learning-based optimization method. Int. J. Ind. Eng. **15**, 181–192 (2019)

[Buyy10] Buyya, R., Beloglazov, A., Abawajy, J.: Energy-efficient management of data center resources for cloud computing: A vision, architectural elements, and open challenges. Eprint Arxiv. **12**, 6–17 (2010)

[Cao16] Cao, H., Xu, J., Ke, D., Jin, C., Deng, S., Tang, C., Cui, M., Liu, J.: Economic dispatch of micro-grid based on improved particle-swarm optimization algorithm. In: Proceedings of the North American Power Symposium, pp. 1–6 (2016). https://doi.org/10.1109/NAPS.2016.7747875

[Chak15] Chakraborty, T., Misra, I.: Design and analysis of channel reservation scheme in cognitive radio networks. Comput. Electr. Eng. **42**, 148–167 (2015)

[Chen08] Chen, D., Zhang, Q., Jia, W.: Aggregation aware spectrum assignment in cognitive ad-hoc networks. In: Proceedings of the 3rd International Conference on Cognitive Radio Oriented Wireless Networks and Communications, pp. 1–6 (2008). https://doi.org/10.1109/CROWNCOM.2008.4562548

[Chen09a] Chen, T., Chen, J., Chen, Y.: Maximizing unavailability interval for energy saving in IEEE 802.16e wireless MANs. IEEE Trans. Mobile Comput. **8**(4), 465–487 (2009)

[Chen09b] Chen, T., Chen, J.: Extended maximizing unavailability interval (eMUI): Maximizing energy saving in IEEE 802.16e for mixing type I and type II PSCs. IEEE Commun. Lett. **13**(2), 151–153 (2009)

[Chen10] Chen, C., Hsu, C., Feng, K.: Performance analysis and comparison of sleep mode operation for IEEE 802.16m advanced broadband wireless networks. In: Proceedings of the 21st Annual IEEE International Symposium on Personal, Indoor and Mobile Radio Communications, pp. 1425–1430 (2010)

[Chen11a] Chen, Y., Zhang, S., Xu, S., Li, G.: Fundamental trade-offs on green wireless networks. IEEE Commun. Mag. **49**(6), 30–37 (2011)

[Chen11b] Chen, T., Yang, Y., Zhang, H., Kim, H., Horneman, K.: Network energy saving technologies for green wireless access networks. IEEE Wirel. Commun. **18**(5), 30–38 (2011)

[Chen14] Chen, Y., Wang, N., Shih, Y. Lin, J.: Improving low-energy adaptive clustering hierarchy architectures with sleep mode for wireless sensor networks. Wirel. Pers. Commun. **75**(1), 349–368 (2014)

[Chen15a] Chen, H., Zhang, Q., Zhao, F.: Energy-efficient base station sleep scheduling in relay-assisted cellular networks. KSII Trans. Internet Inf. Syst. **9**(3), 1074–1086 (2015)

[Chen15b] Cheng, C., Li, J., Wang, Y.: An energy-saving task scheduling strategy based on vacation queueing theory in cloud computing. Tsinghua Sci. Technol. **20**(1), 28–39 (2015)

[Chen16] Chen, Y., Chang, M., Liang, W., Lee, C.: Performance and energy efficient dynamic voltage and frequency scaling scheme for multicore embedded system. In: Proceedings of the IEEE International Conference on Consumer Electronics, pp. 58–59 (2016)

[Choi15] Choi, Y., Lee, J., Back, J., Park, S., Chung, Y.: Energy efficient operation of cellular network using on/off base stations. Int. J. Distrib. Sens. Netw. **2015**, 7 p. (2015). https://doi.org/10.1155/2015/108210

[Chou16] Chou, C., Wong, D., Bhuyan, L.: DynSleep: Fine-grained power management for a latency-critical data center application. In: Proceedings of the ACM/IEEE International Symposium on Low Power Electronics and Design, pp. 212–217 (2016)

[Dabb15a] Dabbagh, M., Hamdaoui, B., Guizani, M.: Toward energy-efficient cloud computing: Prediction, consolidation and overcommitment. IEEE Network **29**(2), 56–61 (2015)

[Dabb15b] Dabbagh, M., Hamdaoui, B., Guizani, M., Rayes, A.: Energy-efficient resource allocation and provisioning framework for cloud data centers. IEEE Trans. Netw. Serv. Manag. **12**(3), 377–391 (2015)

[Davi15] David, C., Ricardo, A., Raquel, L.: Business-driven long-term capacity planning for SaaS applications. IEEE Trans. Cloud Comput. **3**(3), 290–303 (2015)

[Dhan15] Dhanwate, R., Bhagat, V.: Improving energy efficiency on android using cloud based services. Int. J. Adv. Res. Comput. Sci. Manag. Stud. **3**, 75–79 (2015)

[Dini13] Dini, P., Miozzo, M., Bui, N., Baldo, N.: A model to analyze the energy savings of base station sleep mode in LTE HetNets. In: Proceedings of the IEEE International Conference on Green Computing, Communications and Electrical Engineering, pp. 1375–1380 (2013)

[Do12] Do, C., Tran, N., Hong, C.: Throughput maximization for the secondary user over multi-channel cognitive radio networks. In: Proceedings of the International Conference on Information Networking, pp. 65–69 (2012)

[Do18] Do, N., Van, D., Melikov, A.: Equilibrium customer behavior in the M/M/1 retrial queue with working vacations and a constant retrial rate. Oper. Res. (2018). https://doi.org/10.1007/s12351-017-0369-7

[Dosh86] Doshi, B.: Queueing systems with vacations: A survey. Queueing Systems **1**, 29–66 (1986)

[Duan15] Duan, L., Zhan, D., Hohnerlein, J.: Optimizing cloud data center energy efficiency via dynamic prediction of CPU idle intervals. In: Proceedings of the 8th International Conference on Cloud Computing, pp. 985–988 (2015)

[Duan18] Duan, K., Fong, S., Siu, S., Song, W., Guan, S.: Adaptive incremental genetic algorithm for task scheduling in cloud environments. Symmetry **10**(5), 168–180 (2018)

[Fara15] Farahnakian, F., Ashraf, A., Pahikkala, T.: Using ant colony system to consolidate VMs for green cloud computing. IEEE Trans. Serv. Comput. **8**(2), 187–198 (2015)

[Fard17] Fard, S., Ahmadi, M., Adabi, S.: Erratum to: A dynamic VM consolidation technique for QoS and energy consumption in cloud environment. J. Supercomput. **73**(10), 4369–4372 (2017)

[Fati19] Fatima, A., Javaid, N., Butt, A., Sultana, T., Hussain, W., Bilal, M., Aqeel, M., Hashmi, R., Akbar, M., Ilahi, M.: An enhanced multi-objective gray wolf optimization for virtual machine placement in cloud data centers. Electronics **8**, 218–249 (2019)

[Fede13] Coase, R: Federal communications commission. J. Law Econ. **56**(4), 879–915 (2013)

[Fowl11] Fowler, S.: Study on power saving based on radio frame in LTE wireless communication system using DRX. In: Proceedings of the IEEE GLOBECOM Workshops, pp. 1062–1066 (2011)

[Fowl12] Fowler, S., Bhamber, R., Mellouk, A.: Analysis of adjustable and fixed DRX mechanism for power saving in LTE/LTE-advanced. In: Proceedings of the IEEE International Conference on Communications, pp. 1964–1969 (2012)

[Gao11] Gao, S., Tian, H., Zhu, J., Chen, L.: A more power-efficient adaptive discontinuous reception mechanism in LTE. In: Proceedings of the IEEE 38th Vehicular Technology Conference (VTC Fall), pp. 1–5 (2011). https://doi.org/10.1109/VETECF.2011.6092961

[Gao12] Gao, P., Curtis, A., Wong, B., Keshav, S.: It's not easy being green. ACM SIGCOMM Comput. Commun. Rev. **42**(2), 211–222 (2012)

[Gary00] Gary, W., Wang, P., Scott, M.: A vacation queueing model with service breakdowns. Appl. Math. Modell. **24**(5), 391–400 (2000)

[Ge15] Ge, S., Jin, S., Yue, W.: Throughput analysis for the opportunistic channel access mechanism in CRNs with imperfect sensing results. In: Proceedings of the 10th Queueing Theory and Network Applications, pp. 55–62 (2015)

[Ghor16] Ghorbani, M., Rad, M.: ML and Bayes estimation in a two-phase tandem queue with a second optional service and random feedback. Commun. Stat. Theory Methods **45**(9), 2576–2591 (2016)

[Ghos14] Ghosh, G., Chatterjee, S., Das, P.: Cognitive radio and dynamic spectrum access: A study. Int. J. Next Gener. Netw. **6**(1), 43–60 (2014)

[Gree97] Greenbaum, A.: Iterative methods for solving linear systems. Soc. Ind. Appl. Math. **2**(1), 123–181 (1997)

[Gui13] Gui, Z., Xia, J., Zhou, N., Huang, Q.: How to Choose Cloud Services: Toward a Cloud Computing Cost Model. CRC Press (2013)

[Guo11] Guo, Z., Song, M., Wang, Q.: Policy-based market-oriented cloud service management architecture. In: Proceedings of the International Conference on Information and Management Engineering, pp. 284–291 (2011)

[Guo16] Guo, X., Niu, Z., Zhou, S., Kumar, P.: Delay-constrained energy-optimal base station sleeping control. IEEE J. Sel. Areas Commun. **34**(2), 1073–1085 (2016)

[Hadd17] Haddar, I., Raouyane, B., Bellafkih, M.: Generating a service broker framework for service selection and SLA-based provisioning within network environments. In: Proceedings of the 9th International Conference on Ubiquitous and Future Networks, pp. 630–635 (2017)

[Hame16] Hameed, A., Khoshkbarforoushha, A., Ranjan, R., Jayaraman, P., Kolodziej, J., Balaji, P., Zeadally, S., Malluhi, Q., Tziritas, N., Vishnu, A., Khan, S., Zomaya, A.: A survey and taxonomy on energy efficient resource allocation techniques for cloud computing systems. Computing **98**, 751–774 (2016)

[Han06] Han, K., Choi, S.: Performance analysis of sleep mode operation in IEEE 802.16e mobile broadband wireless access systems. In: Proceedings of the IEEE 63rd Vehicular Technology Conference (VTC Spring), pp. 1–5 (2006). https://doi.org/10.1109/VETECS.2006.1683013

[Hani17] Hanini, M., Kafhali, S.: Cloud computing performance evaluation under dynamic resource utilization and traffic control. In: Proceedings of the 2nd International Conference on Big Data, Cloud and Applications, pp. 1–6 (2017). https://doi.org/10.1145/3090354.3090359

[Hasa17] Hasan, S., Kouki, Y., Ledoux, T., Pazat, J.: Exploiting renewable sources: When green SLA becomes a possible reality in cloud computing. IEEE Trans. Cloud Comput. **5**(2), 249–262 (2017)

[Hass03] Hassin, R., Haviv, M.: To Queue or Not to Queue: Equilibrium Behavior in Queueing Systems. Springer (2003)

[Hint16] Hintemann, R., Clausen, J.: Green cloud? The current and future development of energy consumption by data centers, networks and end-user devices. In: Proceedings of the 4th International Conference on ICT for Sustainability, pp. 109–115 (2016)

[Hong09] Hong, X., Wang, C., Chen, H.: Secondary spectrum access networks. IEEE Veh. Technol. Mag. 2(4), 36–43 (2009)

[Hoss15] Hosseinimotlagh, S., Khunjush, F., Samadzadeh, R.: SEATS: Smart energy-aware task scheduling in real-time cloud computing. J. Supercomput. 71(1), 45–66 (2015)

[Hu05] Hu, Q., Yue, W.: Two new optimal models for controlling discrete event systems. J. Ind. Manag. Optim. 1(1), 65–80 (2005)

[Hu06] Hu, Q., Yue, W.: Optimal control for resource allocation in discrete event systems. J. Ind. Manag. Optim. 2(1), 63–80 (2006)

[Hu08] Hu, F., Jin, Y.: Research on the selection of the optimal relaxation factor selection method for SOR method. J. Southwest Normal Univ. 33(5), 43–36 (2008)

[Hu13a] Hu, H., Zhang, H., Xu, Y., Li, N.: Minimum transmission delay via spectrum sensing in cognitive radio networks. In: Proceedings of the IEEE Wireless Communications and Networking Conference, pp. 4101–4106 (2013)

[Hu13b] Hu, J., Deng, J., Wu, J.: A green private cloud architecture with global collaboration. Telecommunication Systems 52(2), 1269–1279 (2013)

[Hu14] Hu, H., Zhang, H., Xu, Y.: Efficient spectrum sensing with minimum transmission delay in cognitive radio networks. Mobile Netw. Appl. 19(4), 487–501 (2014)

[Huan14] Huang, Q.: Development of a SaaS application probe to the physical properties of the Earth's interior: An attempt at moving HPC to the cloud. Comput. Geosci. 70, 147–153 (2014)

[Huan15a] Huang, S., Jiau, M., Lin, C.: A genetic-algorithm-based approach to solve carpool service problems in cloud computing. IEEE Trans. Intell. Transp. Syst. 16(1), 352–364 (2015)

[Huan15b] Huang, X., Han, T., Ansari, N.: On green-energy-powered cognitive radio networks. IEEE Commun. Surv. Tutorials 17(2), 827–842 (2015)

[Huo09] Huo, Z., Yue, W., Tian, N., Jin, S.: Performance evaluation for the sleep mode in the IEEE 802.16e based on a queueing model with close-down time and multiple vacations. J. Ind. Manag. Optim. 5(3), 511–524 (2009)

[Huo11] Huo, Z., Liu, X.: Performance analysis of sleep-mode with uplink/downlink traffics in IEEE 802.16m. In: Proceedings of the International Conference on Web Information Systems and Mining, pp. 426–433 (2011)

[Huss15] Hussein, S., Alkabani, Y., Mohamed, H.: Green cloud computing: Datacenters power management policies and algorithms. In: Proceedings of the International Conference on Computer Engineering and Systems, pp. 421–426 (2015)

[Hwan07] Hwang, E., Choi, B.: A survey on power saving schemes in the IEEE 802.16. In: Proceedings of the 2nd International Conference on Digital Telecommunications, pp. 273–287 (2007)

[Hwan09a] Hwang, E., Kim, K., Son, J., Choi, D.: The power saving mechanism with binary exponential traffic indications in the IEEE 802.16e/m. Queueing Systems 62, 197–227 (2009)

[Hwan09b] Hwang, E., Kim, K., Choi, B.: Delay distribution and loss probability of bandwidth requests under truncated binary exponential backoff mechanism in IEEE 802.16e over gilbert-elliot error channel. J. Ind. Manag. Optim. 5(3), 525–540 (2009)

[Hwan09c] Hwang, E., Kim, K., Son, J., Choi, B.: The power saving mechanism with periodic traffic indications in the IEEE 802.16e/m. IEEE Trans. Veh. Technol. 59(1), 319–334 (2009)

[IEEE06a] IEEE Computer Society and the IEEE Microwave Theory and Techniques Society: IEEE Std 802.16e-2006, Part 16: Air Interface for Fixed and Mobile: Amendment for Physical and Medium Access Control Layers for Combined Fixed and Mobile Operation in Licensed Bands and Corrigendum 1 (2006)

[IEEE06b] IEEE Computer Society and the IEEE Microwave Theory and Techniques Society: IEEE Std 802.16e-2005, Part 16: Air Interface for Fixed and Mobile Broadband Wireless Access Systems: Amendment 2 for Physical and Medium Access Control Layers for Combined Fixed and Mobile Operation in Licensed Bands and Corrigendum 1 (2006)

[IEEE09] IEEE Computer Society and the IEEE Microwave Theory and Techniques Society: IEEE Std 802.16-2009, Part 16: Air Interface for Fixed and Mobile Broadband Wireless Access Systems-Amendment 3 for Advanced Air Interface (2009)

[IEEE11] IEEE Computer Society and the IEEE Microwave Theory and Techniques Society: IEEE Std 802.16m-2011, Part 16: Air Interface for Broadband Wireless Access Systems-Amendment 3 for Advanced Air Interface (2011)

[Jain17] Jain, A. Jain, M.: Multi-server machine repair problem with unreliable server and two types of spares under asynchronous vacation policy. Int. J. Math. Oper. Res. **10**(3), 286–315 (2017)

[Jha12] Jha, S., Koc, A., Vannithamby, R.: Optimization of discontinuous reception (DRX) for mobile Internet applications over LTE. In: Proceedings of the IEEE Vehicular Technology Conference (VTC Fall), pp. 1–5 (2012). https://doi.org/10.1109/VTCFall.2012.6399035

[Ji17] Ji, X., Ye, H., Zhou, J.: An improved teaching-learning-based optimization algorithm and its application to a combinatorial optimization problem in foundry industry. Appl. Soft Comput. **57**, 504–516 (2017)

[Jian12] Jiang, Y., Jiang, J., Zhang, Y.: A novel fuzzy multiobjective model using adaptive genetic algorithm based on cloud theory for service restoration of shipboard power systems. IEEE Trans. Power Syst. **27**(2), 612–620 (2012)

[Jin07] Jin, S., Yue, W.: Performance analysis for a system of connection oriented internet service with a release delay. In: Proceedings of the 7th international conference on Computational Science, pp. 3083–3094 (2007)

[Jin09] Jin, S., Han, L., Yue, W.: Performance evaluation for the power saving class of type III with sleep-delay in IEEE 802.16e. In: Proceedings of the 4th International Conference on Queueing Theory and Network Applications, pp. 1–6 (2009). https://doi.org/10.1145/1626553.1626554

[Jin10] Jin, S., Yue, W.: Performance analysis and evaluation for power saving class type III in IEEE 802.16e network. J. Ind. Manag. Optim. **6**(3), 691–708 (2010)

[Jin11a] Jin, S., Yue, W.: Performance analysis for power saving class type III of IEEE 802.16 in WiMAX. Computer Networks **55**(16), 3734–3743 (2011)

[Jin11b] Jin, S., Chen, X., Qiao, D., Choi, S.: Adaptive sleep mode management in IEEE 802.16m wireless metropolitan area networks. Computer Networks **55**(16), 3774–3783 (2011)

[Jin12a] Jin, S., Yue, W.: Modeling and analysis of a sleep mode in IEEE 802.16 with switching procedure and correlated traffic. Pac. J. Optim. **8**(3), 577–594 (2012)

[Jin12b] Jin, S., Yue, W., Li, H.: Performance analysis and evaluation of an enhanced power saving class type III in IEEE 802.16 with self-similar traffic. J. Global Optim. **52**(2), 353–367 (2012)

[Jin12c] Jin, S., Zhang, X., Yue, W.: Performance evaluation of the sleep mode in IEEE 802.16m with multiple heterogeneous vacation queueing model. ICIC Exp. Lett. B **3**(4), 717–724 (2012)

[Jin13] Jin, S., Yue, W.: System modeling and performance analysis of the power saving class type II in BWA networks. J. Global Optim. **56**(4), 1375–1391 (2013)

[Jin14] Jin, S., Yue, W., Zhang, X.: Performance analysis of the sleep mode in WiMAX 2 networks with multimedia application. J. Appl. Math. **2014**, 10 p. (2014). https://doi.org/10.1155/2014/976179

[Jin15a] Jin, S., Ma, X., Yue, W.: Energy saving strategy in cognitive networks based on software defined radio. In: Proceedings of the 21st IEEE International Workshop on Local and Metropolitan Area Networks, pp. 1–3 (2015). https://doi.org/10.1109/LANMAN.2015.7114728

[Jin15b] Jin, S., Yue, W., Meng, C., Zsolt, S.: A novel active DRX mechanism in LTE technology and its performance evaluation. J. Ind. Manag. Optim. **11**(3), 849–866 (2015)

[Jin16a] Jin, S., Ma, X., Yue, W.: Energy-saving strategy for green cognitive radio networks with an LTE-advanced structure. J. Commun. Netw. **18**(4), 610–618 (2016)

[Jin16b] Jin, S., Yue, W., Li, G.: Stochastic model and system optimization for a channel aggregation strategy in cognitive radio networks. Pac. J. Optim. **12**(4), 775–793 (2016)

[Jin16c] Jin, S., Yue, W., Zhang, J., Zhang, D.: Performance optimization of the mini-slotted spectrum allocation strategy with imperfect sensing. Int. J. Oper. Res. **13**(1), 1–12 (2016)

[Jin16d] Jin, X., Zhang, F., Vasilakos, A., Liu, Z.: Green data centers: A survey, perspectives, and future directions (2016). arXiv: 1608.00687 [cs.DC]

[Jin17a] Jin, S., Hao, S., Yue, W.: Energy-efficient strategy with a speed switch and a multiple-sleep mode in cloud data centers. In: Proceedings of the 12th International Conference on Queueing Theory and Network Applications, pp. 143–154 (2017)

[Jin17b] Jin, S., Hao, S., Wang, B.: Virtual machine scheduling strategy based on dual-speed and work vacation mode and its parameter optimization. J. Commun. **38**(12), 10–20 (2017) (in Chinese)

[Jin17c] Jin, S., Yue, W., Ge, S.: Equilibrium analysis of an opportunistic spectrum access mechanism with imperfect sensing results. J. Ind. Manag. Optim. **13**(3), 1255–1271 (2017)

[Jin18] Jin, S., Wang, X., Yue, W.: A task scheduling strategy with a sleep-delay timer and a waking-up threshold in cloud computing. In: Proceedings of the 13th International Conference on Queueing Theory and Network Applications, pp. 115–123 (2018)

[Jin19a] Jin, S., Hao, S., Qie, X., Yue, W.: A virtual machine scheduling strategy with a speed switch and a multi-sleep mode in cloud data centers. J. Syst. Sci. Syst. Eng. **28**(2), 194–210 (2019)

[Jin19b] Jin, S., Qie, X., Zhao, W., Yue, W. Takahashi. Y.: A clustered virtual machine allocation strategy based on a sleep-mode with wake-up threshold in a cloud environment. Ann. Oper. Res. (2019). https://doi.org/10.1007/s10479-019-03339-3

[Jin19c] Jin, S., Wu, H., Yue, W., Takahashi, Y.: Performance evaluation and Nash equilibrium of a cloud architecture with a sleeping mechanism and an enrollment service. J. Ind. Manag. Optim. (2019). https://doi.org/10.3934/jimo.2019060

[Jin19d] Jin, S., Wu, H., Yue, W.: Pricing policy for a cloud registration service with a novel cloud architecture. Cluster Computing **22**(1), 271–283 (2019)

[Josh13] Joshi, D., Popescu, D., Dobre, O.: Dynamic spectral shaping in LTE-advanced cognitive radio systems. In: Proceedings of the 7th IEEE Radio and Wireless Symposium, pp. 19–21 (2013)

[Juso13] Jusoh, M., Jamlos, M., Kamarudin, M., Sabapathy, T.: A reconfigurable WiMAX antenna for directional and broadside application. Int. J. Antennas Propag. **2013**, 8 p. (2013). https://doi.org/10.1155/2013/405943

[Kahv13] Kahvand, M., Soleimani, M., Dabiranzohouri, M.: Channel selection in cognitive radio networks: A new dynamic approach. In: Proceedings of the 11th IEEE Malaysia International Conference on Communications, pp. 407–411 (2013)

[Kall12] Kalle, R., Nandan, A., Das, D.: La VoLTE: Novel cross layer optimized mechanism of video transmission over LTE for DRX. In: Proceedings of the IEEE 75th Vehicular Technology Conference (VTC Spring), pp. 1–5 (2012). https://doi.org/10.1109/VETECS.2012.6240102

[Kemp15] Kempa, W.: Time-dependent analysis of transmission process in a wireless sensor network with energy efficient mechanism based on threshold waking up. In: Proceedings of the 16th International Workshop on Signal Processing Advances in Wireless Communications, pp. 26–30 (2015)

[Khai10] Khaitan, S., McCalley, J., Raju, M.: Numerical methods for on-line power system load flow analysis. Energy Systems 1(3), 273–289 (2010)

[Khoj18] Khojandi, A., Shylo, O., Zokaeinikoo, M.: Automatic EEG classification: a path to smart and connected sleep interventions. Ann. Oper. Res. (2018). https://doi.org/10.1007/s10479-018-2823-1

[Khos17a] Khosravi, A., Andrew, L., Buyya, R.: Dynamic VM placement method for minimizing energy and carbon cost in geographically distributed cloud data centers. IEEE Trans. Sustain. Comput. 2(2), 183–196 (2017)

[Khos17b] Khoshkholghi, M., Derahman, M., Abdullah, A., Subramaniam, S., Othman, M.: Energy-efficient algorithms for dynamic virtual machine consolidation in cloud data centers. IEEE Access 5, 10709–10722 (2017)

[Kim15] Kim, J., Hwang, G.: Cross-layer modeling and optimization of multi-channel cognitive radio networks under imperfect channel sensing. J. Ind. Manag. Optim. 11(3), 763–777 (2015)

[Knop09] Knopov, P., Pardalos, P.: Simulation and Optimization Methods in Risk and Reliability Theory. Nova Science Publishers (2009)

[Ko14] Ko, H., Lee, J., Kim, C.: The optimal spectrum sensing time for maximizing throughput of 802.11-based MAC protocol for cognitive radio networks under unsaturated traffic conditions. Wirel. Pers. Commun. 77(2), 1397–1414 (2014)

[Koc13] Koc, A., Jha, S., Vannithamby, R., Torlak, M.: Optimizing DRX configuration to improve battery power saving and latency of active mobile applications over LTE-A network. In: Proceedings of the Wireless Communications and Networking Conference, pp. 568–573 (2013)

[Kong06] Kong, L., Tsang, D.: Performance study of power saving classes of types I and II in IEEE 802.16e. In: Proceedings of the 31st IEEE Conference on Local Computer Networks, pp. 20–27 (2006)

[Krei17] Krein, P.: Data center challenges and their power electronics. CPSS Trans. Power Electron. Appl. 2(1), 39–46 (2017)

[Kuo11] Kuo, F., Ting, K., Tseng, C.: Energy-efficient DRX scheduling for QoS traffic in LTE networks. In: Proceedings of the IEEE 9th International Symposium on Parallel and Distributed Processing with Applications, pp. 213–218 (2011)

[Lawa16] Lawanyashri, M., Balusamy, B., Subha, S.: Threshold-based workload control for an under-utilized virtual machine in cloud computing. Int. J. Intell. Eng. Syst. 9(4), 234–241 (2016)

[Lee06] Lee, B., Lee, H., Rhee, S., Kwon, J., Ahn, J.: Cumulative-TIM method for the sleep mode in IEEE 802.16e wireless. In: Proceedings of the International Conference on Information Networking, pp. 502–511 (2006)

[Lee08] Lee, J., Cho, D.: An optimal power-saving class II for VoIP traffic and its performance evaluations in IEEE 802.16e. Computer Communication 31(14), 3204–3208 (2008)

[Lee11] Lee, K., Mun, Y.: Enhanced power saving mechanism for type I and type II power saving classes in IEEE 802.16e. IEICE Trans. Commun. E94-B(9), 2642–2645 (2011)

[Lei10] Lei, J., Plat, V., Li, F.: Analysis on channel bonding/aggregation for multi-channel cognitive radio networks. In: Proceedings of the European Wireless Conference, pp. 468–474 (2010)

[Li99] Li, H., Yang, T.: Steady-state queue size distribution of discrete-time PH/Geo/1 retrial queues. Math. Comput. Modell. 30(3–4), 51–63 (1999)

[Li07a] Li, J., Tian, N., Liu, W.: Discrete-time GI/Geo/1 queue with multiple working vacations. Queueing Systems **56**, 53–63 (2007)

[Li07b] Li, J., Tian, N.: The discrete-time GI/Geo/1 queue with working vacations and vacation interruption. Appl. Math. Comput. **185**(1), 1–10 (2007)

[Li09] Li, C., Chen, W., Zhang, X., Yang, D.: Analysis and simulation for spectrum aggregation in LTE-advanced system. In: Proceedings of the IEEE 70th Vehicular Technology Conference (VTC Fall), pp. 1–6 (2009). https://doi.org/10.1109/VETECF.2009.5378825

[Li10] Li, R., Zhang, C., Ning, X.: Improved algorithm of DRX mechanism in long term evolution system. J. Comput. Appl. **30**(12), 3187–3190 (2010)

[Li13] Li, Y., Ma, Y., Wang, Y., Zhao, W.: Base station sleeping with dynamical clustering strategy of CoMP in LTE-advanced. In: Proceedings of the 2019 IEEE International Conference on Green Computing and Communications, pp. 157–162 (2013)

[Li14a] Li, R., Zhao, Z., Zhou, X., Zhang, H.: Energy savings scheme in radio access networks via compressive sensing-based traffic load prediction. Trans. Emerg. Telecommun. Technol. **25**(4), 468–478 (2014)

[Li14b] Li, X., Garraghan, P., Jiang, X., Wu, Z., Xu J.: Holistic virtual machine scheduling in cloud datacenters towards minimizing total energy. IEEE Trans. Parallel Distrib. Syst. **29**(6), 1317–1331 (2014)

[Li15a] Li, F., Liu, F., L., Zhu, J.: Reputation-based secure spectrum situation fusion in distributed cognitive radio networks. J. China Univ. Posts Telecommun. **22**(3), 110–117 (2015)

[Li15b] Li, S., Xiao, S., Zhang, M., Zhang, X.: Power saving and improving the throughput of spectrum sharing in wideband cognitive radio networks. J. Commun. Netw. **17**(4), 394–405 (2015)

[Li15c] Li, S., Zhou, Y., Jiao, L., Yan, X., Wang, X., Michael R.: Towards operational cost minimization in hybrid clouds for dynamic resource provisioning with delay-aware optimization. IEEE Trans. Serv. Comput. **8**(3), 398–409 (2015)

[Li16] Li, K.: Improving multicore server performance and reducing energy consumption by workload dependent dynamic power management. IEEE Trans. Cloud Comput. **4**(2), 122–137 (2016)

[Li17] Li, L., Weng, W., Fujimura, S.: An improved teaching-learning-based optimization algorithm to solve job shop scheduling problems. In: Proceedings of the 3rd International Conference on Computer and Information Sciences, pp. 797–801 (2017)

[Lian08] Liang, Y., Zeng, Y., Peh, E., Hoang, A.: Sensing-throughput tradeoff for cognitive radio networks. IEEE Trans. Wirel. Commun. **7**(4), 1326–1337 (2008)

[Lian11] Liang, Y., Chen, K., Li, G., Mahonen, P.: Cognitive radio networking and communications: An overview. IEEE Trans. Veh. Technol. **60**(7), 3386–3407 (2011)

[Liao15a] Liao, Y., Song, L., Han, Z.: Full duplex cognitive radio: A new design paradigm for enhancing spectrum usage. IEEE Commun. Mag. **53**(5), 138–145 (2015)

[Liao15b] Liao, D., Li, K., Sun, G., Anand, V., Gong, Y., Tan, Z.: Energy and performance management in large data centers: A queueing theory perspective. In: Proceedings of the International Conference on Computing, Networking and Communications, pp. 287–291 (2015)

[Lin14] Lin, F., Zhou, X., Huang, D., Song, W., Han, D.: Service scheduling in cloud computing based on queueing game model. KSII Trans. Internet Inf. Syst. **8**(5), 1554–1566 (2014)

[Liu12] Liu, Y., Zhu, H., Lu, K. Wang, X.: Self-adaptive management of the sleep depths of idle nodes in large scale systems to balance between energy consumption and response times. In: Proceedings of the 4th IEEE International Conference on Cloud Computing Technology and Science, pp. 633–639 (2012)

[Liu13] Liu, F., Xu, Y., Guo, X., Zhang, W., Zhang D., Li, C.: A spectrum handoff strategy based on channel reservation for cognitive radio network. In: Proceedings of the 3rd International Conference on Intelligent System Design and Engineering Applications, pp.179–182 (2013)

[Liu16] Liu, Z., Wang, P., Xia, Y., Yang, H., Guan, X.: Chance-constraint optimization of power control in cognitive radio networks. Peer-to-Peer Networking and Applications. 9(1), 1–9 (2016)

[Liu18] Liu, J., Jin, S., Yue W.: A novel adaptive spectrum reservation strategy in CRNs and its performance optimization. Optimization Letters. 12(6), 1215–1235 (2018)

[Liu19] Liu, J., Jin, S., Yue, W.: Performance evaluation and system optimization of green cognitive radio networks with a multiple-sleep mode. Ann. Oper. Res. 277(2), 371–391 (2019)

[Liu20] Liu, J., Jin, S.: An imperfect sensing-based channel reservation strategy in CRNs and its performance evaluation. J. Ind. Manag. Optim. 16(3), 1149–1169 (2020)

[Lu09]   Lu, Y., He, H., Wang, J., Li, S.: Opportunistic spectrum access with channel bonding. In: Proceedings of the 4th International Conference on Communications and Networking in China, pp. 1–5 (2009). https://doi.org/10.1109/CHINACOM.2009.5339750

[Luo11] Luo, J., Liu, C., Shao, Z.: The analysis of next generation mobile network discontinuous reception mechanism on mobile Internet business performance. Telecommunications Science. 27(7), 29–34 (2011) (in Chinese)

[Luo17] Luo, J., Zhang, S., Yin, L., Guo, Y.: Dynamic flow scheduling for power optimization of data center networks. In: Proceedings of the IEEE International Conference on Cloud and Big Data Computing, pp. 57–62 (2017)

[Ma07]   Ma, Z., Tian, N., Chen, L., Liu, M.: A Geom/G/1 gate service system with multiple adaptive vacation. Information and Management Sciences. 18(3), 209–221 (2007)

[Ma09]   Ma, Z., Yue, W., Tian, N.: Performance and cost analysis of a Geom/G/1 (G, SV) system. Optimization and Engineering. 10, 239–251 (2009)

[Ma11]   Ma, Z., Yue, W., Su, X.: Performance analysis of a Geom/Geom/1 queueing system with variable input probability. J. Ind. Manag. Optim. 7(3), 641–653 (2011)

[Ma17]   Ma, Z., Wang, P., Yue, W.: Performance analysis and optimization of a pseudo-fault Geo/Geo/1 repairable queueing system with N-policy, setup time and multiple working vacations. J. Ind. Manag. Optim. 13(3), 1493–1507 (2017)

[Mada00] Madan, K.: An M/G/1 queue with second optional service. Queueing Systems. 34, 37–46 (2000)

[Madn16] Madni, S., Latiff, M., Coulibaly, Y., Abdulhamid, S.: Resource scheduling for infrastructure as a service (IaaS) in cloud computing: Challenges and opportunities. J. Netw. Comput. Appl. 68, 173–200 (2016)

[Madn17a] Madni, S., Latiff, M., Coulibaly, Y.: Recent advancements in resource allocation techniques for cloud computing environment: A systematic review. Cluster Computing. 20(3), 2489–2533 (2017)

[Madn17b] Madni, S., Abd, L., Abdullahi, M., Abdulhamid, S., Usman, M.: Performance comparison of heuristic algorithms for task scheduling in IaaS cloud computing environment. PLOS ONE. 12(5), 1–26 (2017)

[Mare18] Marek, R., Hoon, K.: Cognitive systems and operations research in big data and cloud computing. Ann. Oper. Res. 265(2), 183–186 (2018)

[Mari12] Marinho, J., Monteiro, E.: Cognitive radio: Survey on communication protocols, spectrum decision issues, and future research directions. Wireless Networks. 18(2), 147–164 (2012)

[Mcba16] Mcbay, C., Parr, G., Mcclean, G.: Energy efficient in data center servers using optimal scheduling to ensure QoS. In: Proceedings of the 7th International Conference on Cloud Computing, GRIDs, and Virtualization, pp. 56–60 (2016)

[Miho10] Mihor, Y., Kasser, K., Tsankov, B.: Analysis and performance evaluation of the DRX mechanism for power saving in LTE. In: Proceedings of the 26th Convention of Electrical and Electronics Engineers in Israel, pp. 520–524 (2010)

[Min07] Min, M., Pardalos, P.: Total energy optimal multicasting in wireless ad hoc networks. Journal of Combinatorial Optimization. 13(4), 365–378 (2007)

[Mond19] Mondal, S., Das, G., Wong, E.: An analytical cost-optimal cloudlet placement framework over fiber-wireless networks with quasi-convex latency constraint. Electronics. 8, 18 p. (2019). https://doi.org/10.3390/electronics8040404

[Mont01] Montemanni, R., Smith, D., Allen, S.: Lower bounds for fixed spectrum frequency assignment. Ann. Oper. Res. 107(1–4), 237–250 (2001)

[Muth13] Muthumeenakshi, K., Radha, S.: Distributed cognitive radio spectrum access with imperfect sensing using CTMC. Int. J. Distrib. Sens. Netw. 2013, 11 p. (2013). https://doi.org/10.1155/2013/257801

[Naka17] Nakamura, L., Azevedo, L., Batista, B., Meneguette, R., Toledo, C., Estrella, J.: An analysis of optimization algorithms designed to fully comply with SLA in cloud computing. IEEE Latin America Transactions. 15(8), 1497–1505 (2017)

[Neut81a] Neuts, M.: Matrix-Geometric Solutions in Stochastic Models. Johns Hopkins University Press (1981)

[Neut81b] Neuts, M.: Matrix-Geometric Solutions in Stochastic Models: An Algorithmic Approach. Courier Dover Publications (1981)

[Nga11] Nga, D., Lim, H.: Power-saving mechanism with delay bound for mobile WiMAX systems. IET Communications. 5(13), 1854–1859 (2011)

[Nguy14] Nguyen, T., Pham, A., Nguyen, V.: Medium access control design for cognitive radio networks: A survey. IEICE Transactions on Communications. E97-B(2), 359–374 (2014)

[Niu07] Niu, Z., Zhu, Y., Benetis, V.: A Phase-type based Markov chain model for IEEE 802.16e sleep mode and its performance analysis. In: Proceedings of the International Teletraffic Congress, pp. 791–802 (2007)

[Oh10] Oh, E., Krishnamachari, B.: Energy savings through dynamic base station switching in cellular wireless access networks. In: Proceedings of the Global Telecommunications Conference, pp. 1–5 (2010). https://doi.org/10.1109/GLOCOM.2010.5683654

[Olok17] Olokunde, T., Misra, S., Adewumi, A.: Quality model for evaluating platform as a service in cloud computing. In: Proceedings of the Information and Software Technologies, pp. 280–291 (2017)

[Ozaw13] Ozawa, T.: Asymptotics for the stationary distribution in a discrete-time two-dimensional quasi-birth-and-death process. Queueing Systems. 74, 109–149 (2013)

[Park19] Park, S., Hwang, G., Choi, J.: Optimal throughput analysis of multiple channel access in cognitive radio networks. Ann. Oper. Res. 277(2), 345–370 (2019)

[Peng13] Peng, Q., Peng, Y., Wu, W., Rao H., Liu, G.: Dynamic spectrum access scheme of variable service rate and optimal buffer-based in cognitive radio. Communications and Network. 5, 232–237 (2013)

[Peng14] Peng, J., Hong, P., Xue, K.: Stochastic analysis of optimal base station energy saving in cellular networks with sleep mode. IEEE Commun. Lett. 18(4), 612–615 (2014)

[Pham14] Pham, C., Tran, N., Do, C., Hong, C.: Spectrum handoff model based on hidden Markov model in cognitive radio networks. In: Proceedings of the 28th International Conference on Information Networking, pp. 406–411 (2014)

[Qava14] Qavami, H., Jamali, S., Akbari, M., Javadi, B.: Dynamic resource provisioning in cloud computing: A heuristic Markovian approach. In: Proceedings of the International Conference on Cloud Computing, pp. 102–111 (2014)

[Qiao12] Qiao, J., Liu, J., Wang, W., Zhang, Y.: Spectrum-driven sleep scheduling algorithm based on reliable theory in cognitive radio sensor networks. Journal of China Universities of Posts and Telecommunications. 19(11), 47–51 (2012)

[Qiu15] Qiu, M., Ming, Z., Li, J., Gai, K., Zong, Z.: Phase-change memory optimization for green cloud with genetic algorithm. IEEE Transactions on Computers. 64(2), 3528–3540 (2015)

[Qu14] Qu, Y., Wang, M., Hu, J.: A new energy-efficient scheduling algorithm based on particle swarm optimization for cognitive radio networks. In: Proceedings of the International Conference on Signal Processing, Communications and Computing, pp. 467–472 (2014)

[Rao13] Rao, R., Kalyankar, V.: Parameter optimization of modern machining processes using teaching-learning-based optimization algorithm. Engineering Applications of Artificial Intelligence. **26**(1), 524–531 (2013)

[Rao16] Rao, R.: Jaya: A simple and new optimization algorithm for solving constrained and unconstrained optimization problems. Int. J. Industrial Engineering Computations. **7**(1), 19–34 (2016)

[Rehm16] Rehman, A., Yang, L., Hanzo, L.: Performance of cognitive hybrid automatic repeat request: Go-Back-N. In: Proceedings of the IEEE 83rd Vehicular Technology Conference (VTC Spring), pp. 1–5 (2016). https://doi.org/10.1109/VTCSpring.2016. 7504417

[Rese06] Resende, M., Pardalos, P.: Handbook of Optimization in Telecommunications. Springer (2006)

[Rhee97] Rhee, K., Pearce, C.: On some basic properties of the inhomogeneous quasi-birth-and-death process. Communications of the Korean Mathematical Society. **12**(1), 177–191 (1997)

[Ross95] Ross, S.: Stochastic Processes. John Wiley & Sons (1995)

[Rubi87] Rubino, G., Marie, R.: M/M/1 multiclass FIFO queues and generalizations. 1 page. http://www.researchgate.net/publication/281598000 (1987)

[Sala17a] Salameh, O., De, K., Turck, K., Bruneel, H. Blondia, C., Wittevrongel, S.: Analysis of secondary user performance in cognitive radio networks with reactive spectrum handoff. Telecommunication Systems. **65**, 539–550 (2017)

[Sala17b] Salami, Z., Ahmadian-Attari, M., Jannati, H., Aref, M.R.: A location privacy-preserving method for spectrum sharing in database-driven cognitive radio networks. Wirel. Pers. Commun. **95**(4), 3687–3711 (2017)

[Sale15] Saleem, Y., Salim, F., Rehmani, M.: Routing and channel selection from cognitive radio network's perspective: A survey. Computers and Electrical Engineering. **42**, 117–134 (2015)

[Sali12] Salimian, L., Safi, F., Nadimi, M.: An adaptive fuzzy threshold-based approach for energy and performance efficient consolidation of virtual machines. Computing. **98**, 641–660 (2012)

[Sama16] Samarakoon, S., Bennis, M., Saad, W., Latva-Aho, M.: Opportunistic sleep mode strategies in wireless small cell networks. In: Proceedings of the 14th IEEE International Conference on Communications, pp. 707–712 (2016)

[Saff10] Saffer, Z., Telek, M.: Analysis of BMAP vacation queue and its application to IEEE 802.16e sleep mode. J. Ind. Manag. Optim. **6**(3), 661–690 (2010)

[Sarj11] Sarji, I., Ghali, C., Chehab, A., Kayssi, A.: CloudESE: Energy efficiency model for cloud computing environments. In: Proceedings of International Conference on Energy Aware Computing, pp. 1–6 (2011). https://doi.org/10.1109/ICEAC.2011. 6136680

[Shen17] Shen, Y., Bao, Z., Qin, X., Shen, J.: Adaptive task scheduling strategy in cloud: When energy consumption meets performance guarantee. World Wide Web-Internet and Web Information systems. **20**, 155–173 (2017)

[Shi17] Shi, P., Wang, H., Yue, X., Yang, S., Fu, X., Peng, Y.: Corporation architecture for multiple cloud service providers in joint cloud computing. In: Proceedings of the International Conference on Distributed Computing Systems Workshops, pp. 294–298 (2017)

[Sing11] Singh, C. Jain, M., Kumar, B.: Queueing model with state-dependent bulk arrival and second optional service. Int. J. Mathematics in Operational Research. **3**(3), 322–340 (2011)

[Sing16a] Singh, S., Chana, I.: Resource provisioning and scheduling in clouds: QoS perspective. J. Supercomput. **72**(3), 926–960 (2016)

[Sing16b] Singh, D., Devgan, M.: Task scheduling with multilayer hybrid energy efficient approach in green cloud computing. Int. J. for Scientific Research and Development. **4**, 814–818 (2016)

[Slei00] Sleijpen, G., Vorst, H.: A Jacobi-davidson iteration method for linear eigenvalue problems. SIAM Review. **42**(2), 267–293 (2000)

[Sole13] Soleimani, M., Kahvand, M., Sarikhani, R.: Handoff reduction based on prediction approach in cognitive radio networks. In: Proceedings of the International Conference on Communication Technology, pp. 319–323 (2013)

[Son17] Son, J., Dastjerdi, A., Calheiros, R., Buyya, R.: SLA-aware and energy-efficient dynamic overbooking in SDN-based cloud data centers. IEEE Trans. Sustain. Comput. **2**(2), 76–89 (2017)

[Spag15] Spagnuolo, A., Petraglia, A., Vetromile, C., Formosi, R., Lubritto, C.: Monitoring and optimization of energy consumption of base transceiver stations. Energy. **81**, 286–293 (2015)

[Šten10] Štencel, V., Muller, A., Frank, P.: LTE advanced: A further evolutionary step for next generation mobile networks. In: Proceedings of the 20th International Conference Radioelektronika, pp. 15–19 (2010)

[Sugu17] Sugumaran, R., Armstrong, M.: Cloud Computing. The International Encyclopedia of Geography (2017)

[Sult16] Sultana, A., Fernando, X., Zhao, L.: An overview of medium access control strategies for opportunistic spectrum access in cognitive radio networks. Peer-to-Peer Netw. Appl. **10**, 1113–1141 (2017)

[Syed14] Syed, A., Yau, K.: Spectrum leasing in cognitive radio networks: A survey. Int. J. Distrib. Sens. Netw. **2014**, 22 p. (2014). https://doi.org/10.1155/2014/329235

[Taka93] Takagi, H.: Queueing Analysis: A Foundation of Performance Evaluation, Volume 3: Discrete-Time Systems. North-Holland (1993)

[Tama13] Tamal, C., Saha, M.: An analytical framework for channel reservation scheme in cognitive radio network. In: Proceedings of the International Conference on Advances in Computing, pp. 127–132 (2013)

[Tan13] Tan, S., Zeidler, J., Rao, B.: Opportunistic spectrum access for cognitive radio networks with multiple secondary users. IEEE Trans. Wirel. Commun. **12**(12), 6214–6227 (2013)

[Tang06] Tang, P., Chew, Y., Ong, L., Haldar, M.: Performance of secondary radios in spectrum sharing with prioritized primary access. In: Proceedings of the IEEE Military Communications Conference, pp. 2127–2159 (2006)

[Tang13] Tang, W., Zhou, J., Yu, J.: Modeling and analysis of opportunistic spectrum sharing with considering access strategy. In: Proceedings of the International Conference on Communications Circuits and Systems, pp. 7–11 (2013)

[Teng13] Teng, Y., Xu, H.: An energy efficiency heuristic algorithm for joint optimization in cognitive radio networks. In: Proceedings of the IEEE International Conference on Communications Workshops, pp. 469–473 (2013)

[Tian99] Tian, N., Li, Q., Gao, J.: Conditional stochastic decompositions in the $M/M/c$ queue with server vacation. Stochastic Models **14**(2), 367–377 (1999)

[Tian01] Tian, N., Gao, Z., Zhang, Z.: The equilibrium theory for queueing system $M/M/c$ with asynchronous vacations. Acta Math. Appl. Sin. **24**(2), 185–194 (2001)

[Tian06] Tian, N., Zhang, G.: Vacation Queueing Models: Theory and Applications. Springer (2006)

[Tian18] Tian, D.: Particle swarm optimization with chaos-based initialization for numerical optimization. Intell. Autom. Soft Comput. **24**(2), 331–342 (2018)

[Ting11] Ting, K., Wang, H., Tseng, C., Kuo, F.: Energy-efficient DRX scheduling for QoS traffic in LTE network. In: Proceedings of the IEEE 9th International Symposium on Parallel and Distributed Processing with Applications, pp. 213–218 (2011)

[Ting13] Ting, K., Kuo, F., Hwang, B., Wang, H., Lai, F.: An accurate power analysis model based on MAC layer for the DCF of 802.11n. J. Chin. Inst. Eng. **36**(1), 17–26 (2013)

[Tran13] Tran, N., Do, C., Moon, S., Hong, C.: Pricing mechanisms and equilibrium behaviors of noncooperative users in cognitive radio networks. In: Proceedings of the IEEE Global Communications Conference, pp. 913–918 (2013)

[Tsen11] Tseng, Y., Chen, J., Yang, Y.: Managing power saving classes in IEEE 802.16 wireless MANs: A fold-and-demultiplex method. IEEE Trans. Mobile Comput. **10**(9), 1237–1247 (2011)

[Turc07] Turck, K., Vuyst, S., Fiems, D., Wittevrongel, S.: An analytic model of IEEE 802.16e sleep mode operation with correlated traffic. In: Proceedings of the Next Generation Teletraffic and Wired/Wireless Advanced Networking, pp. 109–120 (2007)

[Usui94] Usui, M., Niki, H., Kohno, T.: Adaptive Gauss-Seidel method for linear systems. Int. J. Comput. Math. **51**(1-2), 119–125 (1994)

[Vara15] Varade, P., Ravinder, Y.: Optimal spectrum allocation in cognitive radio using genetic algorithm. In: Proceedings of the 11th India Conference, pp. 1–5 (2015). https://doi.org/10.1109/INDICON.2014.7030480

[Wang11a] Wang, B., Liu, K.: Advances in cognitive radio networks: A survey. IEEE J. Sel. Top. Sign. Process. **5**(31), 5–23 (2011)

[Wang11b] Wang, Y., Xie, Q., Ammari, A., Pedram, M.: Deriving a near-optimal power management policy using model-free reinforcement learning and Bayesian classification. In: Proceedings of the 48th Design Automation Conference, pp. 41–46 (2011)

[Wang12] Wang, C., Wang, L.: Analysis of reactive spectrum handoff in cognitive radio networks. IEEE J. Sel. Areas Commun. **30**(10), 2016–2028 (2012)

[Wang13a] Wang, J., Huang, A., Wang, W.: Admission control in cognitive radio networks with finite queue and user impatience. IEEE Wireless Commun. Lett. **2**(2), 175–178 (2013)

[Wang13b] Wang, S., Zhou, Z., Ge, M., Wang, C.: Resource allocation for heterogeneous cognitive radio networks with imperfect spectrum sensing. IEEE J. Sel. Areas Commun. **31**(3), 464–475 (2013)

[Wang14a] Wang, J., Lin, M., Hong, X.: QoS-guaranteed capacity of centralized cognitive radio networks with interference averaging techniques. KSII Trans. Internet Inf. Syst. **8**(1), 18–34 (2014)

[Wang14b] Wang, Y., Li, J., Huang, L., Jing, Y., Georgakopoulos, A., Demestichas, P.: 5G mobile: Spectrum broadening to higher-frequency bands to support high data rates. IEEE Veh. Technol. Mag. **9**(3), 39–46 (2014)

[Wang15a] Wang, Y., Li, C., Wen, T., Wei, X.: Dynamic channel reservation for cognitive radio networks. In: Proceedings of the IEEE International Conference on Computational Intelligence Communication Technology, pp. 339–343 (2015)

[Wang15b] Wang, L., Sheng, M. Zhang, Y.: Robust energy efficiency maximization in cognitive radio networks: The worst-case optimization approach. IEEE Trans. Commun. **63**(1), 51–65 (2015)

[Wang19] Wang, X., Zhu, J., Jin, S., Yue, W., Takahashi, Y.: Performance evaluation and social optimization of an energy-saving virtual machine allocation scheme within a cloud environment. J. Oper. Res. Soc. China (2019). https://doi.org/10.1007/s40305-019-00272-x

[Wei16] Wei, C., Cai, L., Wang, J.: A discrete-time Geom/G/1 retrial queue with balking customers and second optional service. OPSEARCH **53**, 344–357 (2016)

[Wiga09] Wigard, J., Kolding, T., Dalsgaard, L., Coletti, C.: On the user performance of LTE UE power saving schemes with discontinuous reception in LTE. In: Proceedings of the IEEE International Conference on Communications Workshops, pp. 1–5 (2009). https://doi.org/10.1109/ICCW.2009.5208042

[Whit78] Whitmore, G., Findly, M.: Stochastic Dominance. Lexington Books (1978)

[Wu10] Wu, W., Gao, X., Pardalos, P., Du, D.: Wireless networking, dominating and packing. Optimization Letters **4**(3), 347–358 (2010)

[Wu14] Wu, X., Xu, J., Chen, M., Wang, J.: Optimal energy-efficient sensing in cooperative cognitive radio networks. EURASIP J. Wirel. Commun. Netw. (2014). https://doi.org/10.1186/1687-1499-2014-173

[Wu15] Wu, G., Dong, L., Qin, Z., Xu, Z.: Dynamic programming-based pico base station sleep mode control in heterogeneous networks. Int. J. Commun. Syst. (2015). https://doi.org/10.1002/dac.2967

[Xia15] Xia, Y., Zhou, M., Luo, X., Pang, S., Zhu, Q.: A stochastic approach to analysis of energy-aware DVS-enabled cloud datacenters. IEEE Trans. Syst. Man Cybern. Syst. **45**(1), 73–83 (2015)

[Xiao05] Xiao, Y.: Energy saving mechanism in the IEEE 802.16e wireless MAN. IEEE Commun. Lett. **9**(7), 595–597 (2005)

[Xiao13] Xiao, Y., Zhang, S., Cao, J.: Energy-preserving spectrum access strategy in cognitive radio net-works. In: Proceedings of the IEEE Wireless Communications and Networking Conference, pp. 738–743 (2013)

[Xie12] Xie, R., Yu, F., Ji, H.: Dynamic resource allocation for heterogeneous services in cognitive radio networks with imperfect channel sensing. IEEE Trans. Veh. Technol. **61**(2), 770–780 (2012)

[Qie19] Qie, X., Jin, S., Yue, W.: An Energy-efficient strategy for virtual machine allocation over cloud data centers. J. Netw. Syst. Manag. **27**(4), 860–882 (2019)

[Xu17] Xu, C., Zheng, M., Liang, W., Yu, H., Liang, Y.C.: End-to-end throughput maximization for underlay multi-hop cognitive radio networks with RF energy harvesting. IEEE Trans. Wirel. Commun. **16**(6), 3561–3572 (2017)

[Xue11] Xue, J., Watada, J.: Short-term power load forecasting method by radial-basis-function neural network with support vector machine model. ICIC Express Lett. **5**(5), 1523–1528 (2011)

[Yang12] Yang, C., Sheng, M., Li, J.: Energy-aware joint power and rate control in overlay cognitive radio networks: A Nash bargaining perspective. In: Proceedings of the International Conference on Intelligent Networking and Collaborative Systems, pp. 520–524 (2012)

[Yang13a] Yang, M., Li, Y., Jin, D., Yuan, J., Su, L., Zeng, L.: Opportunistic spectrum sharing based resource allocation for wireless virtualization. In: Proceedings of the International Conference on Innovative Mobile and Internet Services in Ubiquitous Computing, pp. 51–58 (2013)

[Yang13b] Yang, Y., Chen, L., Wang, W.: A novel energy saving scheme based on base stations dynamic configuration in green cellular networks. In: Proceedings of the IEEE 78th Vehicular Technology Conference (VTC Fall), pp. 1–5 (2013). https://doi.org/10.1109/VTCFall.2013.6692051

[Yang16] Yang, J., Zhang, X., Wang, W.: Two-stage base station sleeping scheme for green cellular networks. J. Commun. Netw. **18**(4), 600–609 (2016)

[Yazd14] Yazdani, S., Nezamabadi-Pour, H., Kamyab, S.: A gravitational search algorithm for multimodal optimization. Swarm Evol. Comput. **14**, 1–14 (2014)

[Ye10] Ye, K., Huang, D., Jiang, X., Chen, H., Wu, S.: Virtual machine based energy-efficient data center architecture for cloud computing: A performance perspective, In: Proceedings of the IEEE/ACM International Conference on Green Computing and Communications, pp. 171–178 (2010)

[Yin12] Yin, F.: An application aware discontinuous reception mechanism in LTE-advanced with carrier aggregation consideration. Ann. Telecommun. **67**(3, 4), 147–159 (2012)

[You16] You, C., Huang, K., Chae, H.: Energy efficient mobile cloud computing powered by wireless energy transfer. IEEE J. Sel. Areas Commun. **2**, 1757–1771 (2016)

[Yu12] Yu, Y., Feng, K.: Traffic-based DRX cycles adjustment scheme for 3GPP LTE systems. In: Proceedings of the IEEE 75th Vehicular Technology Conference (VTC Spring), pp. 1–5 (2012). https://doi.org/10.1109/VETECS.2012.6240136

[Yu16]   Yu, K., Wang, X., Wang, Z.: An improved teaching-learning-based optimization algorithm for numerical and engineering optimization problems. J. Intell. Manuf. **27**(4), 831–843 (2016)

[Yue02]   Yue, W., Matsumoto, Y.: Performance Analysis of Multi-Channel and Multi-Traffic on Wireless Communication Networks. Kluwer Academic Publishers (2002)

[Zaka17] Zakarya, M., Gillam, L.: Energy efficient computing, clusters, grids and clouds: A taxonomy and survey. Sustain. Comput. Inform. Syst. **14**, 13–33 (2017)

[Zapp13] Zappaterra, L., Gomes, J., Arora, A., Choi, H.: Resource discovery algorithms for channel aggregation in cognitive radio networks. In: Proceedings of the IEEE Wireless Communications and Networking Conference, pp. 309–314 (2013)

[Zhan06] Zhang, Y., Fujise, M.: Energy management in the IEEE 802.16e MAC. IEEE Commun. Lett. **10**(4), 311–313 (2006)

[Zhan13a] Zhang, H., Cai, J., Li, X.: Energy-efficient base station control with dynamic clustering in cellular network. In: Proceedings of the International Conference on Communications and Networking, pp. 384–388 (2013)

[Zhan13b] Zhang, Y., Jin, S., Yue, W.: An enhanced energy saving strategy for an active DRX in LTE wireless networks. Int. J. Innov. Comput. Inf. Control **9**(11), 4373–4387 (2013)

[Zhan14a] Zhang, G., Huang, A., Shan, H.: Design and analysis of distributed hopping-based channel access in multi-channel cognitive radio systems with delay constraints. IEEE J. Sel. Areas Commun. **32**(11), 2026–2038 (2014)

[Zhan14b] Zhang, D., Jin, S., Yue, W.: Performance analysis of the spectrum allocation strategy with mini-slot structure. Intell. Inform. Ser. Mem. Konan Univ. **7**(2), 119–130 (2014)

[Zhan16] Zhang, Y., Zhao, Y., Fu, X., J, Xu.: A feature extraction method of the particle swarm optimization algorithm based on adaptive inertia weight and chaos optimization for Brillouin scattering spectra. Optics Communications **376**(1), 56–66 (2016)

[Zhan18] Zhang, Y., Yao, J., Guan, H.: Intelligent cloud resource management with deep reinforcement learning. IEEE Cloud Comput. **4**(6), 60–69 (2018)

[Zhao07] Zhao, Q., Sadler, B.: A survey of dynamic spectrum access. IEEE Signal Process. Mag. **24**(3), 79–89 (2007)

[Zhao13] Zhao, Y., Jin, S., Yue, W.: Performance optimization of a dynamic channel bonding strategy in cognitive radio networks. Pac. J. Optim. **9**(4), 679–696 (2013)

[Zhao15a] Zhao. Y., Jin. S., Yue, W.: An adjustable channel bonding strategy in centralized cognitive radio networks and its performance optimization. Qual. Technol. Quant. Manag. **12**(3), 291–310 (2015)

[Zhao19] Zhao, W., Wang, X., Jin, S., Yue, W., Takahashi, Y.: An energy efficient task scheduling strategy in a cloud computing system and its performance evaluation using a two-dimensional continuous time Markov chain model. Electronics **8**, 16 p. (2019). https://doi.org/10.3390/electronics8070775

[Zhou18] Zhou, Z., Abawajy, J., Li, F.: Fine-grained energy consumption model of servers based on task characteristics in cloud data center. IEEE Access **6**, 27080–27090 (2018)

Index

A

Active DRX mechanism, 122
    energy saving rate, 130, 154
    sleep mode, 130
    system cost function, 136
Admission fee, 23
    energy saving strategy, 283
    opportunistic channel access mechanism, 225
    virtual machine allocation strategy, 437
Average latency, 10
    average latency of anonymous users, 393
    average latency of SU packets, 175, 198, 220, 240, 261, 281, 299
    average latency of tasks, 327, 350, 372, 416, 433
Average response time, 10
    average response time of data packets, 45, 64, 81, 95, 108, 130, 154
Awake mode, 2
    power saving class type I, 38

B

Base station (BS), 2, 88
    power saving class type I, 37
    power saving class type II, 54
    power saving class type III, 71
Bat algorithm, 30, 401
Blocking rate, 11, 154, 174, 198, 220
Broadband wireless access (BWA), 1
    awake mode, 2, 38
    base station, 2, 35, 52, 71, 88
    energy saving rate, 11, 43, 64, 80, 108, 130, 154, 350

IEEE 802.16e, 24, 35, 52, 87
IEEE 802.16m, 52, 88, 100
LTE, 2, 119, 141
mobile station, 2, 51, 69, 97
power saving class type I, 2
power saving class type II, 2
power saving class type III, 2
sleep mode, 3, 35, 51, 69, 88, 98, 141
vacation-delay, 24, 70, 122, 146
Busy cycle, 43, 57, 78, 93, 100, 128

C

Channel aggregation strategy, 168
    admission fee, 185
    average latency, 175
    Nash equilibrium behavior, 181
    pricing policy, 184
    socially optimal behavior, 183
    system cost function, 176
Channel reservation strategy, 249
    algorithm for solving QBD process, 259
    Gauss-Seidel method, 257
    imperfect sensing, 249
Chaotic mapping mechanism, 334, 354, 376
Cloud architecture, 384
    free service, 384
    registration fee, 397
    registration service, 384
    second optional service, 385
    sleep mode, 385
    transition rate matrix, 386
Cloud computing, 1, 315, 381, 405, 425
    anonymous users, 385

© Springer Nature Singapore Pte Ltd. 2021
S. Jin, W. Yue, *Resource Management and Performance Analysis of Wireless
Communication Networks*, https://doi.org/10.1007/978-981-15-7756-7

cloud computing system, 1, 30, 361, 384, 408, 424
  free cloud service, 385
  registration fee, 397
  registration service, 385
Cloud computing system, 4, 30, 361, 384, 408, 424
  base-line module, 317
  cloud architecture, 384
  cloud data center, 317, 340, 408, 424
  clustered virtual machine, 30, 361
  reserve module, 318
  speed switch, 317, 338
  task schedule, 408
  virtual machine allocation, 337, 361, 424
Cloud data center (CDC), 315, 340, 361, 408
Clustered virtual machine allocation strategy, 361
  average latency, 372
  energy saving rate, 372
  Gauss-Seidel method, 371
  sleep mode, 361
  system cost function, 376
Code division multiple access (CDMA), 6
Cognitive radio networks (CRNs), 1
  admission fee, 185
  centralized channel allocation strategy, 4
  channel aggregation, 167
  distributed channel allocation strategy, 4
  in-band sensing, 3
  out-of-band sensing, 3
  spectrum holes, 3
  spectrum sensing, 3
Continuous-time Markov chain (CTMC), 16, 166, 252, 295, 320

D

Discrete-time Markovian arrival process (D-MAP), 25, 102
Dynamic power management (DPM), 316
Dynamic spectrum allocation, 6
  dynamic exclusive use model, 7
  hierarchical access model, 7
  open sharing model, 7
Dynamic voltage and frequency scaling (DVFS), 316, 338, 406

E

Early arrival system (EAS), 169, 191, 214, 234, 275
Embedded Markov chain, 40, 56
  average response time, 45, 81, 95, 107, 130

  busy cycle, 44, 59, 78, 93, 107, 128
  embedded Markov points, 56, 74, 91, 102, 124
  $Geom^X/G/1$, 75
  $Geom/G/1$, 24, 36, 53, 77, 92, 107
  number of data packets, 40, 91, 107, 127
  queue length, 75, 127
  standard deviation, 108
  waiting time, 62, 77, 92, 128
Energy consumption, 30, 130, 281, 300, 317, 350, 372, 394, 434
Energy-efficient task scheduling strategy, 408
  average latency, 416
  energy saving rate, 417
  Gauss-Seidel method, 416
  sleep mode, 408
  system cost function, 420
Energy-efficient virtual machine allocation strategy, 425
  average latency, 433
  energy saving rate, 434
  Gauss-Seidel method, 433
  Nash equilibrium behavior, 438
  pricing policy, 440
  sleep mode, 425
  socially optimal behavior, 438
Energy saving rate, 11, 44, 63, 80, 94, 108, 130, 154, 299, 350, 372, 394, 417, 434
Energy saving strategy, 273, 294
  average latency, 281, 306
  energy saving rate, 300
  Gauss-Seidel method, 280
  gravitation-gravitational search algorithm, 272
  multiple-sleep mode, 294
  pricing policy for SU packets, 287
  single-sleep mode, 273
  sleep mode, 274
  system cost function, 308
Enhanced Active DRX mechanism, 143
  downlink transmission, 143
  logical channels, 142
  on-duration timer, 144
  sleep-delay, 145
  sleep interval, 145
  transition probability sub-matrices for case I, 148
  transition probability sub-matrices for case II, 151
Enhanced energy saving strategy, 121
  busy cycle, 128
  busy period, 125
  queue length of data packets, 127
  sleep-delay, 129

sleep mode, 130
waiting time of data packets, 128
Evolved Node B (eNodeB), 120

F

5th generation (5G), 3, 209
Firefly algorithm, 317
First-come first-served (FCFS), 15, 39, 56, 90,
      101, 123, 190, 231, 250, 274, 409,
      426
Frequency division multiplexing (FDM), 6

G

Gauss-Seidel method, 20
    Gauss-Seidel iteration formula, 21
    linear equation, 21
Genetic algorithm, 30, 406, 441
Geom/G/1, 24, 36, 53, 77, 92, 107
Green cloud computing, 5
    asynchronous vacation, 340, 363, 384
    average latency, 10
    dual-threshold, 318
    energy efficiency, 52, 293, 317, 361, 407
    energy saving level, 11, 281, 326
    energy saving rate, 11, 108
    multiple-sleep mode, 28, 292
    Nash equilibrium behavior, 23, 181, 283,
        397, 437
    sleep mode, 11, 44, 54
    sleep-delay, 30, 70, 121, 407, 425
    socially optimal behavior, 28, 183, 283,
        397, 425
    synchronous vacation, 320, 409, 427
    virtual machine allocation, 425
    virtual machine scheduling, 317
    wake-up thresholds, 29, 361, 407

H

Handover rate, 11, 63, 80, 129

I

Infrastructure as a service (IaaS), 4
Interruption rate, 11, 239, 299

J

Jacobi iterative method, 19
Jaya algorithm, 28, 293

L

Laplace-Stieltjes transform (LST), 12
Late arrival system (LAS), 39, 56, 90, 101, 146
L'Hôspital rule, 42, 76
Little's law, 10, 45, 107, 128, 154, 198, 239,
      261, 281, 299, 327, 372, 393, 416,
      433
Long term evolution (LTE), 2, 120, 141
LTE-advanced (LTE-A), 120, 166, 271

M

Matrix-geometric solution method, 15, 325
    matrix equation, 299
    transition probability matrix, 215, 235, 276
    transition rate matrix, 298, 349
Medium access control (MAC), 120, 230
Mini-slotted spectrum allocation strategy, 231
    average latency, 231
    imperfect sensing, 233
M/M/1, 14
Mobile station (MS), 2, 51, 71, 98
Multiple-class customers, 13
Multiple-server vacation, 12
Multi-thresholds automated configuration drx
    (M-ADRX), 142

N

Nash equilibrium, 22, 181, 225, 283, 397, 437
Non-real-time variable rate (NRT-VR), 2, 35
Number of customers in the system, 10
    number of data packets, 107
    number of SU packets, 262
    number of tasks, 416
Number of customers in the system buffer, 10
    number of anonymous users, 393
    number of data packets, 127
    number of tasks, 327

O

Opportunistic spectrum access mechanism,
    211
    activity of PU packets, 211
    activity of SU packets, 212
    average latency, 212
    imperfect sensing, 212
    mistake detections and false alarms, 212
    pricing policy for SU packets, 227
Optimization, 21
    individual benefit, 22, 23, 182, 225, 284,
        398, 437

social benefit, 22, 183, 225, 284, 398, 438
social optimization, 22, 181, 225, 283, 397,
    437
system cost function, 45, 67, 136, 176, 205,
    354, 376, 420
system profit function, 143, 243, 332

P

Particle swarm optimization (PSO), 29, 338
Performance measures, 10, 154, 299, 393, 433
Platform as a Service (PaaS), 4
Power saving class, 2
    power saving class type I, 24, 35
    power saving class type II, 24, 51
    power saving class type III, 24, 69
Power saving class type I, 36
    average busy cycle, 44
    average length of the busy period, 44
    average number of sleep cycles, 44
    awake mode, 38
    close-down, 36
    downlink traffic, 37
    IEEE 802.16e, 37
Power saving class type II, 51
    awake state, 53
    busy period in awake state, 55
    busy periods in listening state, 55
    listening state, 53
Power saving class type III, 69
    batch arrival, 70
    busy cycle, 78
    Pareto distribution, 72
    self-similar traffic, 70
    sleep-delay, 70
Priority queue, 169, 191, 214, 233, 252
    matrix-geometric solution, 298
    one-step transition rate matrix, 296
    preemptive priority, 295
    priority queue with multiple vacations, 295
    priority queue with single vacation, 275
Primary user (PU), 3, 166, 187, 210, 229, 247,
    272, 291
Private cloud, 5, 382
Probability generating function (PGF), 12, 39,
    56, 73, 90, 103, 124
Public cloud, 5, 422

Q

Quality of experience (QoE), 424
Quality of service (QoS), 6, 64, 81, 88, 188,
    249, 315, 383

Quasi birth-death (QBD), 17, 196, 219, 237,
    257, 279, 324, 348, 369, 391, 414,
    431
Queueing model, 9
    arrival process, 36, 90, 102
    birth-death process, 16
    Independent and identically distributed
        (i.i.d.), 9, 169, 252, 295, 343
    Probability generating function (PGF), 12,
        39, 56, 73, 90, 103, 124
    queue length, 10, 75, 225
    service process, 9
    sojourn time, 10

R

Real-time variable rate (RT-VR), 2, 53
Registration fee, 401
    average latency, 393
    cloud architecture, 384
    energy saving rate, 394
    Nash equilibrium behavior, 397
    sleep mode, 384
    socially optimal behavior, 398
Resource management, 6
    BWA network, 35, 53, 69
    cloud computing, 315, 339, 361, 384, 408,
        425
    CRN, 166, 189, 211, 231, 249, 273, 293
    dynamic spectrum allocation, 166
    QoS, 38, 64, 97, 108, 190, 249, 293, 318,
        383
    static spectrum allocation, 165
    virtual machine, 317, 338, 360, 382, 406,
        423
    virtual machine migration, 8

S

Secondary user (SU), 3, 166, 187, 210, 229,
    247, 272, 291
Service level agreement (SLA), 338
Signal-to-noise ratio (SNR), 191, 215, 233, 249
Single-server vacation, 12
Single-threshold automated configuration drx
    (S-ADRX), 142
Sleep mode, 3, 88, 98, 141
    power saving class type I, 35
    power saving class type II, 51
    power saving class type III, 70
Sleep mode for power saving class type I, 37
    average response time, 45
    energy saving rate, 44
    system cost function, 45

Sleep mode for power saving class type II, 51
  average response time, 64
  busy cycle, 59
  energy saving rate, 64
  handover rate, 63
  optimal sleep window length, 66
  system cost function, 67
Sleep mode for power saving class type III, 70
  busy cycle, 78
  data packets and batches, 74
  energy saving rate, 80
  handover rate, 80
  queue length, 75
  waiting time, 77
Social benefit function, 23
  pricing policy, 185, 227, 287, 440
  pricing policy for registration service, 400
  social benefit of the system, 183, 225, 284,
      398, 438
  socially optimal arrival rate, 183, 225, 285,
      398, 439
Software as a service (SaaS), 4
Spectrum allocation, 6
  dynamic spectrum allocation, 7, 166
  static spectrum allocation, 6, 165
Spectrum reservation strategy, 189
  average latency, 190
  perfect-sensing, 187
  retrial feedback, 187
  system cost function, 205
Speed switch and multiple-sleep mode, 317
  average latency, 327
  Gauss-Seidel method, 325
  optimization of sleep parameter, 332
  sleep mode, 317
  virtual machine scheduling strategy, 317
Static spectrum allocation, 6
  CDMA, 6
Static spectrum allocation FDM, 6
Static spectrum allocation TDM, 6
Steady-state distribution, 18
  batch size, 73, 81
  number of anonymous users, 385
  number of busy periods, 59
  number of data packets, 40, 103, 125
  number of PU packets, 252
  number of SU packets, 252, 275, 296
  number of tasks, 320, 343, 410, 427
  transition probability matrix, 215, 235, 276
  transition rate matrix, 391, 411, 428
  transmission time, 56, 73, 90, 103, 124
Successive over relaxation (SOR), 248
Switching rate, 11, 239, 262
System cost function, 22

average latency, 176, 200, 308, 354, 376,
    420
average response time, 110, 136
blocking rate, 176, 205
close-down time, 45
energy saving rate, 110, 136, 308, 354, 376,
    420
handover rate, 136
sleep cycle, 110
sleep window, 67
trade-off, 66
System energy, 11
  energy saving degree, 281
  energy saving level, 326
  energy saving rate, 43, 63, 80, 108, 130,
      154, 299, 350, 372, 394
System profit function, 22, 155, 243, 332

T
Teaching-learning-based optimization (TLBO),
    27, 188, 248, 360
Teaching-learning-based optimization-
    successive over relaxation
    (TLBO-SOR), 248
  population-based intelligent algorithm, 258
  QBD process, 248
  relaxation factor, 258
  sparse linear system, 258
  successive over relaxation, 248
Third generation partnership project (3GPP),
    3, 119
Throughput, 11, 199, 220, 239, 261, 299
Time division multiplexing (TDM), 6

U
Underlying Markov chain (UMC), 102
Unsolicited grant service (UGS), 2, 52
User equipment (UE), 3, 120, 141
Utilization, 11, 81, 175, 261, 301

V
Vacation queue, 11, 72, 90, 101, 123, 146, 275,
    343, 363, 385, 409, 427
Very important person (VIP), 381
Virtualization, 6, 382
  hardware virtualization, 7
  programming language abstraction, 8
Virtual machine (VM), 6, 316, 338, 360, 382,
    406, 423
  virtual machine allocation, 340, 361, 425
  virtual machine schedule, 317

Virtual machine allocation strategy, 340
   average latency, 350
   Gauss-Seidel method, 349
   sleep mode, 340
   system cost function, 354
Virtual machine migration, 8
   live migration, 382
   migrating to another PM, 8
   migrating to other storage on the same PM,
     8
   non-live migration, 8
   Physical machine (PM), 338, 360, 384, 408,
     425
Virtual machine scheduling strategy, 317
   average latency, 327
   energy saving level, 326
   multi-sleep mode, 317
   speed switch, 318
   transition rate matrix, 321

W
Waiting time, 10, 62, 77, 92, 393
   waiting time in awake state, 61
   waiting time in listening state, 60

Wireless communication networks (WCNs), 1
   energy consumption, 2, 130, 281, 300, 317,
     372, 394, 434
   energy efficiency, 120, 292, 317, 339, 361,
     384
   optimal design, 53, 99
   resource management, 6
   spectrum efficiency, 3, 273
   wireless spectrum, 7, 165, 187, 209, 271,
     291
Worldwide interoperability for microwave
     access (WiMAX), 2, 87, 97
   average response time, 95, 108
   Bernoulli arrival, 87, 99
   correlation parameter, 109
   embedded Markov points, 91, 108
   energy saving rate, 96, 108
   listening window, 89
   Markovian arrival process (MAP), 98
   sleep cycle, 89, 99
   sleep mode, 87, 98
   standard deviation, 108
   system cost function, 110
   WiMAX 2, 87, 98

Printed in the United States
by Baker & Taylor Publisher Services

Printed in the United States
by Baker & Taylor Publisher Services